Hydrometeorology: Advances in Weather and Climate Sciences

Hydrometeorology: Advances in Weather and Climate Sciences

Editor: Peter Spencer

R CALLISTO
REFERENCE

www.callistoreference.com

Callisto Reference,
118-35 Queens Blvd., Suite 400,
Forest Hills, NY 11375, USA

Visit us on the World Wide Web at:
www.callistoreference.com

ISBN: 978-1-64116-566-2 (Hardback)

Cataloging-in-Publication Data

Hydrometeorology : advances in weather and climate sciences / edited by Peter Spencer.
   p. cm.
Includes bibliographical references and index.
ISBN 978-1-64116-566-2
1. Hydrometeorology. 2. Weather. 3. Climatology. 4. Hydrology. 5. Meteorology. I. Spencer, Peter.
GB2803.2 .H93 2022
551.57--dc23

Table of Contents

Preface

Hydrometeorology is an interdisciplinary branch of meteorology and hydrology. It studies the transfer of water and energy between the land surface and the lower atmosphere. An important aspect of hydrometeorology is the prediction and mitigation of the effects of high precipitation events. Nowcasting, numerical weather prediction and statistical techniques are the three primary ways to model meteorological phenomena in weather forecasting. Nowcasting predicts events a few hours out. It utilizes observations and live radar data to combine them with numerical weather prediction models. Numerical weather prediction is the primary technique used to forecast weather. It uses mathematical models to account for the atmosphere, ocean, and many other variables for producing forecasts. Statistical techniques use regressions and other statistical methods to create long-term projections. This book provides comprehensive insights into the field of hydrometeorology. The various studies that are constantly contributing towards advancing technologies and evolution of this field are examined in detail. This book is appropriate for students seeking detailed information in this area as well as for experts.

This book unites the global concepts and researches in an organized manner for a comprehensive understanding of the subject. It is a ripe text for all researchers, students, scientists or anyone else who is interested in acquiring a better knowledge of this dynamic field.

I extend my sincere thanks to the contributors for such eloquent research chapters. Finally, I thank my family for being a source of support and help.

Editor

Hydrometeorological effects of historical land-conversion in an ecosystem-atmosphere model of Northern South America

R. G. Knox[1,*], M. Longo[2,**], A. L. S. Swann[3], K. Zhang[2,***], N. M. Levine[2,****], P. R. Moorcroft[2], and R. L. Bras[4]

[1]Massachusetts Institute of Technology, Cambridge, Massachusetts, USA

[2]Harvard University, Cambridge, Massachusetts, USA

[3]University of Washington, Seattle, Washington, USA

[4]Georgia Institute of Technology, Atlanta, Georgia, USA

[*]now at: Lawrence Berkeley National Laboratory, Berkeley, California, USA

[**]now at: EMBRAPA Satellite Monitoring, Campinas, São Paulo, Brazil

[***]now at: Cooperative Institute for Mesoscale Meteorological Studies, University of Oklahoma, Oklahoma, USA

[****]now at: University of Southern California, Los Angeles, California, USA

Correspondence to: R. G. Knox (rgknox@lbl.gov)

Abstract. This work investigates how the integrated land use of northern South America has affected the present day regional patterns of hydrology. A model of the terrestrial ecosystems (ecosystem demography model 2: ED2) is combined with an atmospheric model (Brazilian Regional Atmospheric Modeling System: BRAMS). Two realizations of the structure and composition of terrestrial vegetation are used as the sole differences in boundary conditions that drive two simulations. One realization captures the present day vegetation condition that includes deforestation and land conversion, the other is an estimate of the potential structure and composition of the region's vegetation without human influence. Model output is assessed for differences in resulting hydrometeorology.

The simulations suggest that the history of land conversion in northern South America is not associated with a significant precipitation bias in the northern part of the continent, but has shown evidence of a negative bias in mean regional evapotranspiration and a positive bias in mean regional runoff. Also, negative anomalies in evaporation rates showed pattern similarity with areas where deforestation has occurred.

In the central eastern Amazon there was an area where deforestation and abandonment had lead to an overall reduction of above-ground biomass, but this was accompanied by a shift in forest composition towards early successional functional types and grid-average-patterned increases in annual transpiration.

Anomalies in annual precipitation showed mixed evidence of consistent patterning. Two focus areas were identified where more consistent precipitation anomalies formed, one in the Brazilian state of Pará where a dipole pattern formed, and one in the Bolivian Gran Chaco, where a negative anomaly was identified. These locations were scrutinized to understand the basis of their anomalous hydrometeorologic response. In both cases, deforestation led to increased total surface albedo, driving decreases in net radiation, boundary layer moist static energy and ultimately decreased convective precipitation. In the case of the Gran Chaco, decreased precipitation was also a result of decreased advective moisture transport, indicating that differences in local hydrometeorology may manifest via teleconnections with the greater region.

1 Introduction

It has been held that massive and widespread Amazonian deforestation would lead to regional reductions in precipitation, evaporation, and moisture convergence, with slight increases in surface temperature (Henderson-Sellers et al., 1993; Nobre et al., 1991; Lean and Warrilow, 1989; Dickinson and Henderson-Sellers, 1988). The Amazon Basin and its forest ecosystems are also an important component of the global circulation of energy (Gedney and Valdes, 2000), where changes (complete deforestation in either Amazonia or all tropical broadleaf forests) are thought to teleconnect beyond the continent (Avissar and Werth, 2005; Snyder, 2010). The literature documenting Amazonian land conversion and the surrounding areas is significant, the reader is referred to a small selection of non-exhaustive references for some background (Cardille and Foley, 2003; Skole and Tucker, 1993; INPE, 2003; Geist and Lambin, 2002; Laurance et al., 2001; Nepstad et al., 2001; Soares-Filho et al., 2006). The work presented here is motivated by a need to better understand how the history of land conversion has influenced the hydrology of the region. As will be outlined further, the mechanistic relationships between land conversion and hydrometeorological response is complex and has benefited from study with newer generations of land–atmosphere models with increased granularity and increased complexity in representing physical process.

There are several direct hydrologic mechanisms that connect changes in tropical forest structure (i.e., deforested versus intact canopies) to the regional climate system. Leaves, stems, and bare earth have variable light-scattering properties, such that intact forest canopies composed of dark vegetation typically have lower shortwave radiation albedo than areas with exposed soil (Chapin et al., 2002). This directly impacts the surface energy balance via net radiation. Forest canopies have a complex relationship with the surface moisture balance and mediate the transport of water in numerous ways. Model studies have shown that the representation of canopy interception can substantially impact the partitioning of evapotranspiration and surface runoff (Pitman et al., 1990; Wang et al., 2007; Crockford and Richardson, 2000). Some studies have found that forest canopies increase the interception of precipitation (Asdak et al., 1998; Dietz et al., 2006), and further that crown structure influences turbulent transport and evaporation of wet leaves (Dietz et al., 2006). Yet some have indicated that canopy interception increased in degraded forests (Chappell et al., 2001). Pastures and converted agricultural systems are generally associated with soil degradation such as decreased infiltration rates, nutrient loss and increased surface runoff, subject to variability and factors such as the soil texture and the existence of perennial under-story vegetation (Benegas et al., 2014). Practices such as grazing and agriculture promote soil compaction and decreased infiltration (Martinez and Zinck, 2004; Lal, 1996), and intense fires used for clearing lands may reduce soil or-

ganic matter that may favor infiltration (Kennard and Gholz, 2001). Forests with deep rooted trees draw from deeper soil moisture pools, which have different periodicity in available water and therefore alter the timing of latent heat flux via transpiration compared to grasslands (Kleidon and Heimann, 2000; Nepstad et al., 1994). Canopy structure also influences the turbulent exchange of heat, moisture and momentum with the atmosphere (Raupach et al., 1996).

The higher surface temperatures associated with widespread deforestation, as reported with the first generations of general circulation models and beyond, (Henderson-Sellers et al., 1993; Nobre et al., 1991; Lean and Warrilow, 1989; Dickinson and Henderson-Sellers, 1988) are thought to be the result of losses in evaporative cooling associated with cleared vegetation. The decrease in evaporative cooling is also thought to drive reduced precipitation, and subsequently reduces the heat released to the atmosphere through condensation (Eltahir and Bras, 1993). This has the potential to outcompete the effect of increased surface albedo of deforested lands (Eltahir, 1996), which suppresses net radiation thereby promoting surface cooling and divergence (Eltahir and Bras, 1993; Lean and Warrilow, 1989). Positive surface temperature anomalies induce convergent circulations coincident with a decrease in surface pressure. Decreased precipitation heating anomalies reduce the tendency towards convergence.

In the southwestern Amazonian dry season, statistical connections have been made between pastures and higher incidents of shallow cumulus clouds, compared to intact forests where shallow clouds are less frequent yet deep precipitating convective events are more frequent (Wang et al., 2009). The higher rate of deep precipitating convection over forests was associated with larger values of convective available potential energy (CAPE) (Williams and Renno, 1993), which in this case was driven by increased humidity and moisture flux from intact forest canopies. The increased frequency of shallow convection was attributed to more vigorous mesoscale circulations associated with deforestation induced land-surface heterogeneity (Wang et al., 2000; Souza et al., 2000).

Regional land–atmosphere simulations that can parameterize convective clouds indicate that structured land-conversion scenarios elicit shifts in mean basin precipitation, albeit less so than traditional coarse scale general circulation model studies (Silva et al., 2008). Coherent land surface patterns may strengthen convergence zones on the surface, creating vertical wind triggers to thunderstorms. For instance, Avissar and Werth (2005) determined that coherent land surface patterns transfer heat, moisture and wave energy to the higher latitudes through thunderstorm activity. Moreover, mesoscale simulations are found to capture key cloud feedback processes which fundamentally alter the atmospheric response to land-surface heterogeneities (Medvigy et al., 2011). The mesoscale simulation's ability to represent land-use scenarios at finer resolutions can impact spatial patterning of rainfall. For example, western propagating squall lines from the Atlantic are thought to dissipate over regions of wide-spread deforestation (Silva et al., 2008; d'Almeida et al., 2007). Evidence has also shown that convection can be driven by localized convergent air circulations triggered by land-surface heterogeneities, and that the likelihood and quality of resulting events are both dependent on the scale of heterogeneity and the position relative to disturbed and intact landscapes (Pielke, 2001; Dalu et al., 1996; Baldi et al., 2008; Anthes, 1984; Knox et al., 2011; Wang et al., 2009).

Regional scale-coupled land–atmosphere models can capture feed-backs resulting from land conversion at the scale of tens of kilometers and lower, particularly through improved resolution and the parameterization of atmospheric physics (such as convection and radiation scattering). At the land-surface, there is variability in canopy structure at the gap (size of a single large tree crown) scale and below. Processes that occur at these scales may be important to predicting ecosystem response and land–atmosphere exchange, as discussed above. Physics-based land surface models have non-linear representations of hydrologic and thermodynamic processes; therefore, using average canopy structure (such as in the "big-leaf" approach) to represent processes uniformly may provide different results compared to explicitly capturing these processes with sub-canopy and gap-scale structure.

This research uses the Brazilian Regional Atmospheric Modeling System (BRAMS, a variant of the Regional Atmospheric Modeling System (RAMS), Cotton et al., 2003) coupled with the ecosystem demography model 2 (ED2 or EDM2, Moorcroft et al., 2001; Medvigy et al., 2009), to explore the sensitivity of hydrologic climate of northern South America in response to present day land conversion. This modeling system can explicitly represent the processes of energy and mass transfer in the canopy and soil system, with sub-grid variability along ecosystem age-structured and vegetation size-structured axes. An experiment is conducted by comparing simulations that singularly differ in their representation of regional vegetation cover, one which captures the present day vegetation condition that includes deforestation and land-conversion, the other being an estimate of the potential structure and composition of the region's vegetation without human influence. Section 2 of this manuscript will detail experiment design of the coupled model experiment. The model system and experiment design is verified by comparison of model output with observations, see Appendices B–F.

In Sect. 3 we evaluate the hydrometeorological response to the changes in land-use history in a regional context. In Sect. 4 we evaluate the processes underlying the observed changes in hydrometeorology in two focus areas. A discussion and conclusion of the results follows.

2 Experiment design

The main task of this experiment is to conduct two regional simulations of the South American biosphere and atmosphere. The defining difference between the two simulations is how the land-surface model (ED2) represents the structure (the distribution of plant sizes) and composition (the distribution of plant types) of the region's terrestrial ecosystems, as a consequence of two different disturbance regimes. In one simulation, the vegetation reflects a structure and composition that has no effects of human land use, henceforth referred to as a *potential vegetation* (PV) condition. In the other simulation, the model will incorporate an estimate of modern (e.g., 2008) human land use, henceforth referred to as an *actual vegetation* (AV) condition. The procedure is broken down into steps and elaborated upon.

2.1 Description of the vegetation model – ED2

The ecosystem demography model 2 predicts the changes in the terrestrial vegetation structure, as modulated by the physically based conservation of water, carbon and enthalpy. Its central design philosophy assumes that the stochastic representation of plant communities integrated over a large sample can be portrayed deterministically as land fractions and plant groups, with explicit size (of the plants) and age (time since a patch of land housing the plants has experienced major disturbance) structure. By discretely representing the distribution of plant sizes and types, it can estimate vertical canopy structure, which directly impacts radiation scattering (throughfall interception), and in-canopy transport of scalars. By discretely representing variable disturbance history, the model can also explicitly simulate energy balance over a wide array of canopy types (closed canopies, recovering forests, grasslands, etc.) that exist within the footprint of driving meteorological data. In this experiment, the ED2 model resolves five different relevant tropical plant functional types (PFT): C4 grasses, early successional tropical evergreens, mid-successional tropical evergreens, late successional tropical evergreens and tropical C2 grasses. In the ED2 system, PFTs are used as sets of attributes that can be applied to numerous explicitly resolved plant groups of different size and in different parts of the disturbance strata.

2.2 Generation of surface boundary conditions

The creation of the initial vegetation conditions used a "spin-up" process. The spin-up process is an off-line dynamic ED2 simulation, where the driving atmospheric information comes from a pre-compiled forcing data set. The vegetation is initialized with an equal assortment of newly recruited (saplings) plant types. The off-line model is integrated over several centuries by sampling from the climate data set as the vegetation reaches an equilibrium. We identify equilibrium when the total biomass of each plant func-

Table 1. Simulation constraints describing the spin-up process creating the initial boundary conditions.

Specification	Value
Climate data	modified DS314*
Soil data	Quesada et al. (2011) + IGBP-DIS
Plant types	late succession tropical evergreens mid succession tropical evergreens early succession tropical evergreens subtropical grasses C4 grasses
Simulation period	508 years
Spatial resolution	gridded 1°
Bounding domain	30° S–15° N, 85° W–30° W
Tree allometry (DBH, height) (crown properties)	Chave et al. (2001), Baker et al. (2004b) Poorter et al. (2006), Dietze et al. (2008)
Turbulent transport	Beljaars and Holtslag (1991) atmospheric boundary Massman (1997) within canopy
Photosynthesis & leaf conductance	Collatz et al. (1991) Collatz et al. (1992) Leuning (1995)
Canopy radiation scattering	Zhao and Qualls (2005, 2006)
Soil hydrology	Walko et al. (2000), Tremback and Kessler (1985) Medvigy et al. (2009)

* Modified DS314 data is derived from Sheffield et al. (2006), precipitation down-scaling and radiation interpolation is applied, see footnote for data availability.

tional type in a grid-cell does not change more than 0.5 % over a period of 40 years. If equilibrium within this threshold was not achieved, the spin-up was allowed to continue to 508 years before stopping. For reference, ED2 simulations in old-growth central Amazonian forests take roughly 250 years to reach equilibrium biomass. The ED2 vegetation structure and composition at the end of the multi-century simulation was saved as the *potential vegetation* (PV) initial condition.

A summary of the simulation conditions in the spin-up is covered in Table 1. The model soil textures were derived from a combination of databases. Within the Amazon Basin, soil data were retrieved from Quesada et al. (2011); outside the basin soil data were retrieved from a combination of RADAMBRASIL and IGBP-DIS (Scholes et al., 1995; Rossato, 2001). The climate data used to drive the spin-up process was derived from the UCAR DS314 product

(Sheffield et al., 2006)[1]. The DS314 is based on the National Center for Environmental Prediction's Reanalysis Product (NCEP) and maintains the same global and temporal coverage period but has bias corrections and increased resolution based on the assimilation of composite data sets. The DS314 surface precipitation record was further processed such that grid cell average precipitation was downscaled to reflect the point-scale statistical qualities of local rain gauges. This technique used methods of Lammering and Dwyer (2000), and is explained in more detail in Knox (2012). The native NCEP reanalysis and European Center for Medium Range Weather Forecasting (ECMWF) 40 year Reanalysis (ERA-40) were also tested as driver data sets. The downscaled DS314 was ultimately chosen due to better agreement of estimated equilibrium biomass with observations (not shown).

The *actual vegetation* (AV) was created by continuing the simulation that produced the *potential vegetation*, and assumed that the starting year was 1900. This simulation was continued for another 108 years (until 2008) while incorporating human driven land-use change. Throughout the 508 year *potential vegetation* spin-up, as well as the 108 year continuation with human land use, the atmospheric carbon dioxide concentration was held constant at 378 ppm, which

---

[1] Original data sets used in the DS314 are from the Research Data Archive (RDA) which is maintained by the Computational and Information Systems Laboratory (CISL) at the National Center for Atmospheric Research (NCAR). The original Sheffield/DS314 data are available from the RDA (http://dss.ucar.edu) in data set number ds314.0.

approximates concentrations at the turn of the millennium (present day).

The model applies human land use by reading an externally compiled data set of land-use transition matrices (Albani et al., 2006) that defines the area fractions of which various land-cover types will change to another type over the course of the year. Two external data sets are used to create the land-use transition matrices, the global land-use data set (GLU) (Hurtt et al., 2006) and the SIMAMAZONIA-1 data set (Soares-Filho et al., 2006). The GLU data set incorporates the SAGE-HYDE 3.3.1 data set and provides land-use transitions in its native format globally, on a $1°$ grid from the years 1700 to 1999[2]. The SIMAMAZONIA 1 product provides a more intensive assessment of forest cover and deforestation focused in the Amazon Basin, starting in the year 2000. The data is formatted as yearly 1 km forest cover grids (forest, non-forest, and natural grasslands). The fraction of forest and non-forest cells that fall within each ED2 model simulation grid-cell are counted for each year. This enables the calculation of a rate of change equivalent to the transition matrix format of the GLU data set. The transitions from the GLU data set from 1990 to 1999 were linearly scaled to have continuity with the SIMAMAZONIA data set that is introduced in 2000. Land use reported in the GLU data prior to 1900 were lumped into a single combined transition and applied at the year 1900. A map of the fraction of the land surface containing human land use is provided (see Fig. 1).

Regional maps of above-ground biomass for the *potential vegetation* (PV) scenario and the differences between the two scenarios (*actual vegetation–potential vegetation*, or AV–PV) are provided in Fig. 2. The majority of above-ground biomass in the *potential* simulation is concentrated in the Amazon Basin and the Atlantic Forest of southern Brazil. Late successional broadleaf evergreens comprise most of the above-ground biomass in these regions. Early successional broadleaf evergreens are a prevalent but secondary contributor to biomass in the Amazon Basin. The early succession's contribute the majority of biomass in Cerrado (savanna like ecosystem, mixed open canopy forests with grasses) ecotones found roughly in central Brazil on the southern border of the Amazon rain-forest. This is consistent with their competition and resource niche which emphasizes fast growth and colonization of disturbed areas (such as fire and drought prone Cerrado). The model-estimated equilibrium above-ground biomass (AGB) and basal area (BA) that represent the initial condition are compared with a collection of census measurements in Baker et al. (2004a, b) (see Appendix B).

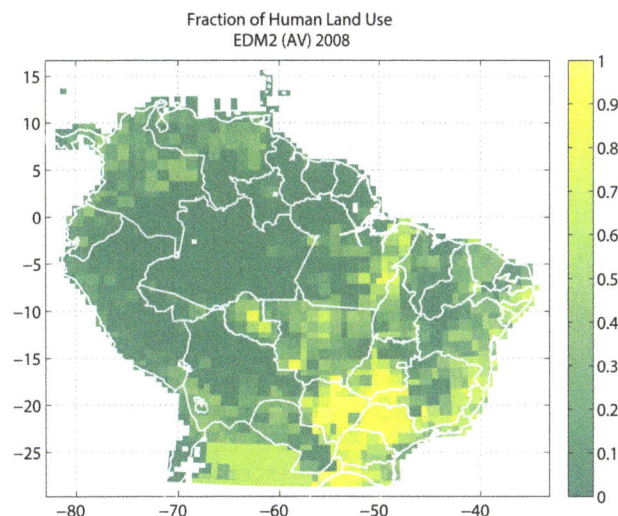

Figure 1. Fraction of the land surface with human land use. Output is taken from the ED2 the *actual vegetation* (AV) simulation, which was driven with global land-use (GLU) and SIMAMAZONIA-1-land-use transition data.

2.3　Land–atmosphere coupled simulations

The two coupled land–atmosphere simulations were conducted over 4 years, from January 2002 through December 2005. These 4 years were chosen because of the availability of lateral boundary conditions and validation data sets. With the exception of differing vegetation structure at the lower boundary, the lateral boundary conditions, model parameters, initialization of the atmospheric state, and timing are all identical between the two. The lateral boundary conditions (air temperature, specific humidity, geopotential, meridional wind speed, zonal wind speed) are taken from the European Centre for Medium Range Weather Forecasting's Interim Reanalysis (ERA-Interim) product (Dee et al., 2011). The data is interpolated from the ERA-Interim's model native Reduced Gaussian Grid (N128, which has an equatorial horizontal resolution of $0.75°$).

The *actual* and *potential* boundary conditions utilized a dynamic model process to generate structure and composition. However, when applied to the coupled simulations, the land-surface dynamics including the processes of mortality, recruitment and growth are turned off and only phenology is left to vary in time. The motivation for this decision is to create a more simple comparison and efficient simulation. Further, the length of the coupled simulations are not long enough to generate large changes in above-ground biomass. As an example, Lewis (2006) and Baker et al. (2004a) estimate that in recent decades, the Amazon has sequestered approximately $0.6 \pm 0.2 \, \mathrm{Mg \, C \, ha^{-1} \, yr^{-1}}$. Over a course of 4 years, this is less than $3 \, \mathrm{Mg \, C \, ha^{-1}}$, which is on the order of 1–2 % of total forest biomass.

---

[2]The use of the SAGE-HYDE 3.3.1 global land-use data set acknowledges the University of New Hampshire, EOS-WEBSTER Earth Science Information Partner (ESIP) as the data distributor for this data set.

Table 2. Run time parameters and specifications in the ED2-BRAMS coupled simulations.

Specification	Value
Simulation period	January 2002–December 2005
Grid projection	polar stereographic
Grid dimensions	98° (E–W), 86° (N–S), 56° (vertical)
Horizontal grid resolution	64 km
Vertical grid resolution	110 m (lowest) stretching to 1500 m at 7 %
Atmospheric time step	30 s
Atmospheric acoustic time step	10 s
Land–surface model time step	120 s
Method of calculating Updraft base	level of maximum sum of mean and variance of vertical velocity
Number of prototype cloud scales	2
Mean radius of cloud 1	20 000 m
Minimum depth of cloud 1	4000 m
Mean radius of cloud 2	800 m
Minimum depth of cloud 2	80 m
Cumulus convective scheme	Grell and Dévényi (2002)
Cumulus convective trigger	pressure differential between updraft base and LFC* < 100 hpa
Cumulus dynamic control	Kain and Fritsch (1990), Kain (2004)
Condensate to precipitation conversion efficiency	3 %
Cloud # concentrations and distribution parameters	Medvigy et al. (2010)
Turbulent closure	Nakanishi and Niino (2006)
Shortwave radiation scattering	Harrington and Olsson (2001)
Longwave radiation scattering	Chen and Cotton (1983)
Advection	monotonic, Walcek and Aleksic (1998) & Freitas et al. (2012)
Cumulus feedback on radiation?	yes

* LFC = level of free convection.

A group of modeling parameters associated with convective parameterization and the radiation scattering of convective clouds were tuned using a manual binary search procedure. The parameters of mean cloud radius, mean cloud depth, cumulus convective trigger mechanism, dynamic control method and the condensate to precipitation conversion efficiency were calibrated against the Tropical Rainfall Measurement Mission 3B43 product and the surface radiation from the Global Energy and Water Cycle Experiment–Surface Radiation Budget (SRB) product version 2.5. Fitness metrics include monthly mean spatial bias, mean squared error and the variance ratios (i.e., the spatial variance of mean model output over the spatial variance of the observations). Manual binary search calibration was chosen because of the complexity of the parameter space, the need for human supervision and sanity checks, and the non-trivial computation requirements for each simulation. Fifty-four iterations were performed, utilizing a reduced domain in the first group of iterations to facilitate a more rapid calibration. A table of the finalized coupled model runtime conditions is provided

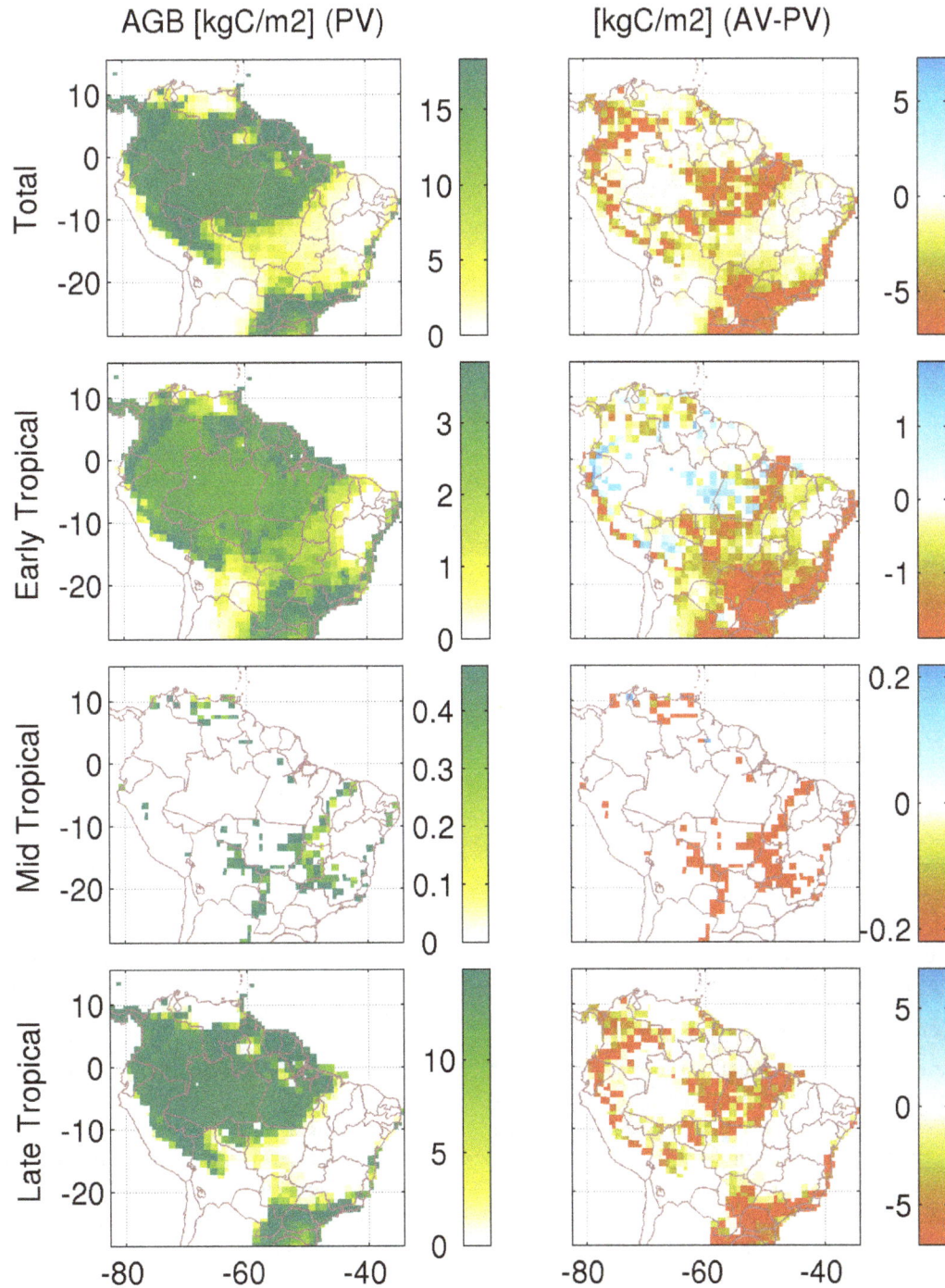

Figure 2. Regional maps of total above-ground biomass (AGB) (kg m^{-2}) from the ED2 initial condition. The left column indicates results are from the *potential vegetation* condition, the right column is the relative differences between the *actual* and *potential* scenarios, (AGB$_{AV}$− AGB$_{PV}$). Each row represents the partitioning of the above-ground biomass into respect plant functional types. "Early Tropical", "Mid Tropical" and "Late Tropical" refer to broadleaf tropical evergreen plant functional types.

in Table 2. Model output was then compared with observations of atmospheric thermodynamic variables, mean regional surface fluxes (precipitation, radiation, latent heat flux and sensible heat flux) and mean cloud cover profiles; these comparisons are provided in Appendices C–F, respectively.

3   Regional analysis of the actual and potential scenarios

The following analysis of results will repeatedly refer to anomalies, here defined as the subtracted differences of the

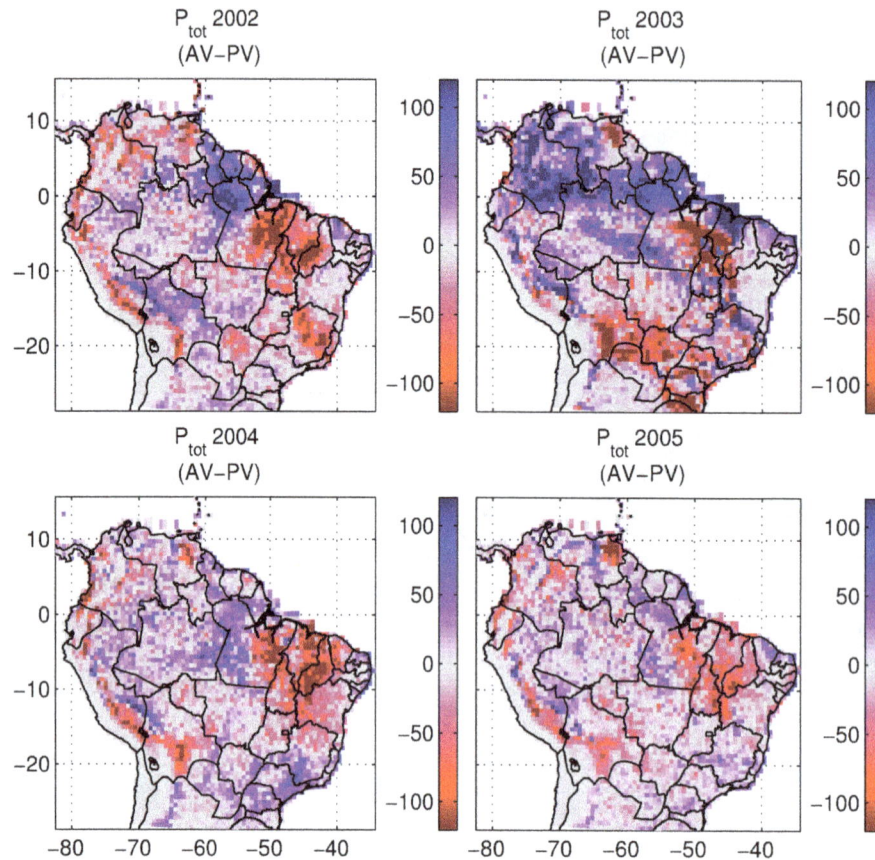

Figure 3. Differences in total annual surface precipitation (mm), 2002–2005. The *potential vegetation* (PV) condition is subtracted from the *actual vegetation* (AV) condition, (AV–PV).

potential vegetation scenario model output from the *actual* vegetation scenario model output (or alternatively, AV–PV).

3.1   Emergence of patterning and continental biases

The annual precipitation accumulations for each simulation were mapped in space and the anomalies between the two were compared for consistent patterning (see Fig. 3). Each anomaly appears to feature a dipole structure with a positive lobe (more *actual* scenario precipitation) in the north and west of the Amazon delta, and a negative lobe in the south and east of the Amazon delta region. Pattern differences also appear on the Peruvian–Bolivian border, although whether or not it can be considered a dipole is left to the reader. In each year, the precipitation anomaly shows increases on the foothills of the Andes Mountains in southern Peru and the northern tip of Bolivia. There is also a negative anomaly in precipitation in southern and central Bolivia. However, in each year there is also noise among the pattern. For instance in 2002, 2003 and 2005, there are locations in southern Bolivia that show increases in the precipitation anomaly adjacent to the area of decrease.

The patterning in downwelling shortwave surface radiation showed opposite behavior to that of precipitation (not shown), the response is strongly influenced by increased

cloud optical depth where convective precipitation has increased, and vice-versa. The atmospheric model did not incorporate dynamics of aerosols or atmospheric gases other than multi-phase water; therefore, variability in multi-phase water explains the variability and differences seen in optical depth. Maximum mean annual differences in surface irradiance peak at about $10 \, \mathrm{W \, m^{-2}}$, and are strongest over the dipole associated with the precipitation anomaly, as well as over the eastern Brazilian dry lands (41° W).

The mean annual continental bias in accumulated precipitation, evapotranspiration and total runoff is presented in Fig. 4. There is little evidence of an overall continental bias in accumulated precipitation. However, the human land-use scenario generated a negative continental evapotranspiration anomaly and a positive runoff anomaly in each of the 4 years. Consistent patterning in evapotranspiration and transpiration anomaly were also evident (see Fig. 5). Generally speaking, a negative anomaly in transpiration and total evapotranspiration is evident over the "arc of Amazonian deforestation" (starting at 48° W 2° S going clockwise to 62° W 10° S, also see forest biomass differences in Fig. 2). The spatial correlation between the biomass and evapotranspiration anomalies ($R^2 = 0.4$), suggests that the variability in evapotranspiration cannot be explained purely by first order effects from

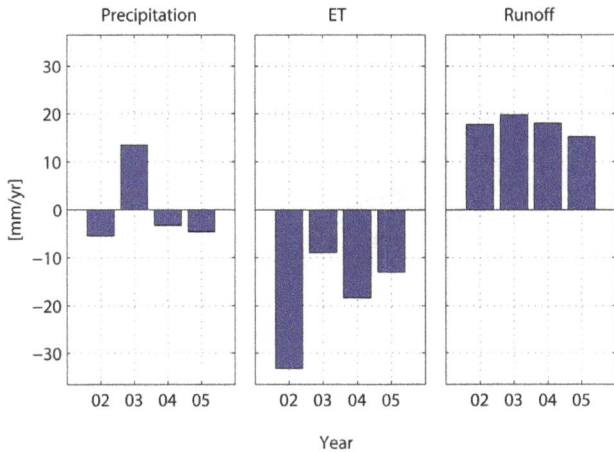

Figure 4. Difference in mean continental precipitation, evaporation and total runoff, between the *actual vegetation* case and *potential vegetation* case (AV–PV).

changes to forest structure. Second order effects and complex system feedbacks account for a portion of the variability. These effects include differences in precipitation, and potentially the effects of differences in surface heating and turbulent transport of scalars (heat and water).

3.2 Connecting hydrologic anomalies and ecosystem response

The availability of root-zone soil moisture, photosynthetically active radiation (PAR) and nutrients are examples of resource limitations that can potentially mediate the response of vegetation to changes in climate. Light and water are critical limiters of plant growth, disturbance (particularly through fire), and mortality (which can be functionally related to growth). However, the significance of these limiters in how they may drive ecosystem response is dependent on various factors other than the mean, such as the consistency of change (inter-annual variance), when the changes occur (seasonality) and how large the differences are relative to the total. A standard score "ζ" is one way to evaluate consistency, calculated as the inter-annual mean difference (denoted by brackets "$<>$") divided by the first standard deviation of the normalized difference η of variable x for year t (mean annual precipitation or downwelling shortwave radiation).

$$\eta_{(t)} = \frac{x_{AV,(t)} - x_{PV,(t)}}{0.5\left(x_{AV,(t)} + x_{PV,(t)}\right)} \quad (1)$$

$$\zeta = \frac{\langle\eta\rangle}{\sigma_\eta} \quad (2)$$

The spatial maps of the standard scores for precipitation and shortwave radiation anomalies are provided in the upper panels of Fig. 6. For reference, a standard difference of 1 suggests that the normalized difference is equal to its inter-annual standard deviation. The maps indicate that the dif-

ferences in precipitation and radiation from the two scenarios are relatively consistent at the two locations previously identified (Pará Brazil and northern Bolivia). They also indicate that the negative precipitation anomaly, and the positive radiation anomaly over the regions of intense deforestation (i.e., the arc of deforestation) are consistent.

The susceptibility of ecosystems to anomalous precipitation forcing may be derived from an ED2 model mechanic called the "moisture stress index" (MSI). This metric is simply the fraction of time that ED2 vegetation cohorts (plants) are actively keeping their stomata closed due to water limitations. For an ecosystem with N plant groups (also known as cohorts) indexed i, the mean land-surface moisture stress index is calculated by the leaf area index (LAI) weighting of the open-fraction f_o' of stomata for each plant group in the community. Brackets "$<>$" denote an averaging in space and time. The stomatal open fraction f_o', is based on the ratio of the plant's "demand" for root zone soil moisture, and the "supply" of water the roots are capable of extracting at that time. The demand requirement is driven by the maximum transpiration the plant would generate given the existing light, carbon and vapor pressure conditions with unlimited soil moisture.

$$msi = 1 - \langle\frac{\sum_{i=1}^{N} LAI_{(i)} f_{o(i)}'}{\sum_{i=1}^{N} LAI_{(i)}}\rangle \quad (3)$$

$$f_{o(i)}' = \frac{1}{1 + \frac{Demand}{Supply}} \quad (4)$$

Vegetation communities that have experienced high moisture stress indices in the past are more likely to respond structurally to changes in precipitation, because subsequent changes in soil moisture availability will have immediate impacts on photosynthesis and the assimilation of carbon. The lower left panel of Fig. 6 shows the mean moisture stress index for the *actual* vegetation scenario. The lower right panel of Fig. 6 shows a map of above-ground biomass as a reference to the extents of the modeled Amazon tropical forests. Moisture stress is low in areas where there is copious precipitation (the supply term). Note that in the interior of the Amazon Basin, moisture stress has little to no influence on stomatal regulation (and subsequently photosynthesis). The open canopy dry forests in southern Brazil an Bolivia, as well as the Cerrado, have higher moisture stress.

4 Hydrometeorological focus areas – Pará Brazil and the Gran Chaco

Two locations that coincide with the pattern differences in precipitation are highlighted in Fig. 6. Each location shows decreases in normalized precipitation and increases in down-

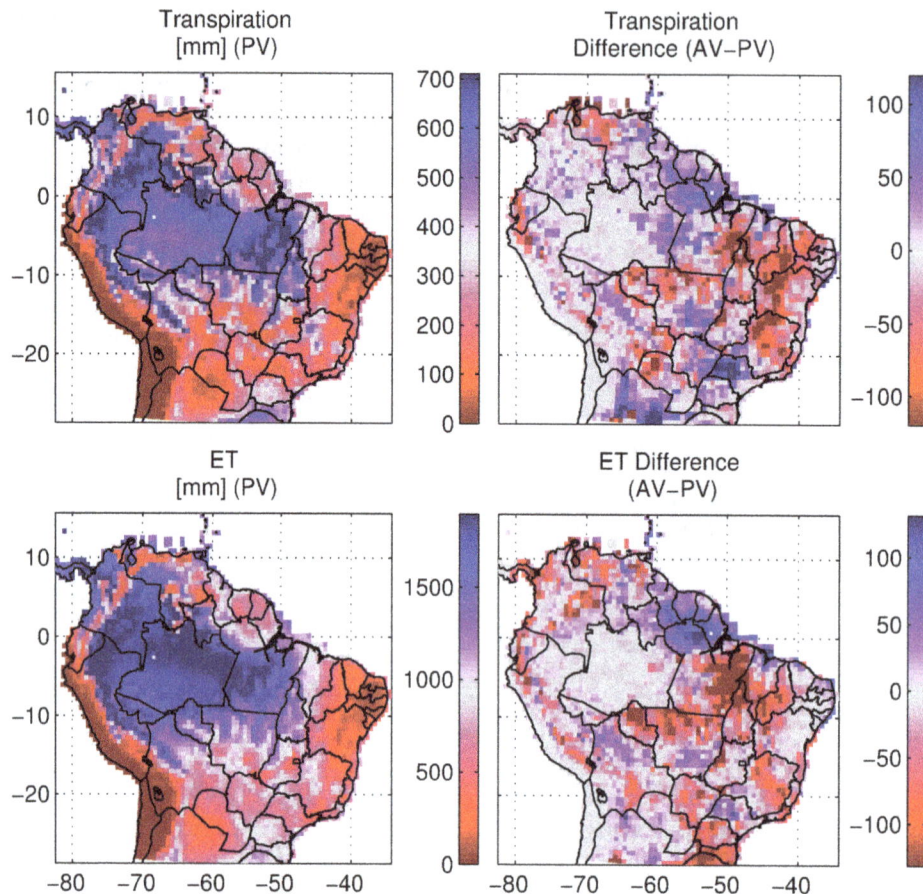

Figure 5. Left panels: mean annual transpiration and evapotranspiration in the *potential vegetation* (PV) scenario, from 2002 to 2005 (mm). Right panels: difference in mean annual transpiration and evapotranspiration between the *actual vegetation* case and *potential vegetation* case (AV–PV).

welling shortwave radiation associated with the *actual vegetation* scenario. The vegetation of these locations also show a degree of seasonal moisture stress according to the MSI metric presented in section 3.2. One site is centered on 4.5° S 50.5° W in the Brazilian state of Pará. The other site is centered on 19.5° S 63.5° W in the northwestern part of the Bolivian Gran Chaco where it meets the Andes mountains (sometimes referred to as the Montane Gran Chaco). These two locations are chosen as areas of focused evaluation of hydrology and hydrometeorology. For simplicity, these will be referred to as the *Pará* and Gran Chaco focus areas.

A representation of the vegetation demographics at the centroids of the two focus areas, as estimated by the ED2 model, are provided in Fig. 7. The natural landscapes at the Pará focus areas are dominated by tropical evergreen forests, and are close to (but not within) the ecotone transition between tropical forests and Cerrado. The offline model spin-up of the *actual* vegetation scenario imposed pastures on approximately one-third of the land-cover. Roughly 10 % of the landscape contains old-growth forests that have gone 200 years since the last disturbance. The focus area in the

Gran Chaco is located in a region influenced by the outlet of the South American Low Level Jet. The continental precipitation recycling ratio in this area is relatively high compared to the rest of the continent (Eltahir and Bras, 1994). This exact location in the Gran Chaco is a dry forest ecosystem that borders adjacent ecotones of tropical rainforests to the north, montane ecosystems to the west and grasslands to the south. The ED2 model estimated a *potential* vegetation demographic that is fairly consistent with the depiction of dry forests, a sparse cover of short trees with grasses in the understory. The *actual vegetation* simulation of the Gran Chaco, driven by the GLU data set (Hurtt et al., 2006), forced 25 % of the natural landscape to pasture (grasses), with an accompanying 20 % of abandoned and degraded lands. Human land use, as represented in the ED2 model at this specific location, led to a collapse of the estimated tree cover, which includes natural landscapes. This specific site is undoubtedly a more aggressive representation of the differences between the *actual* and *potential* scenario ecosystems in this region. As a whole, of course, human land-conversion has not lead to a collapse of the Gran Chaco's dry-forest ecosystems.

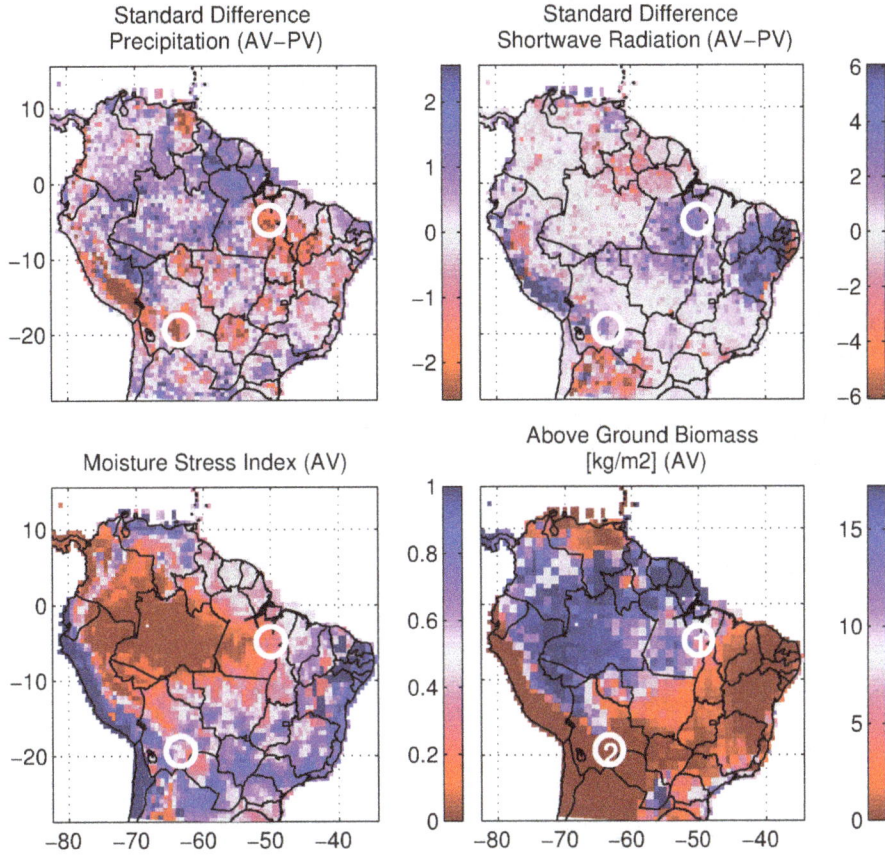

Figure 6. Combined assessment of the regional significance in differences between precipitation and radiation, and the susceptibility of the ecosystems. Upper panels show standard scores for consistency of differences *actual* and *potential* (AV–PV) surface precipitation and surface downwelling shortwave radiation. The lower left panel shows the moisture stress index for the *actual* (AV) scenario, see Eq. (3). For reference, *actual* (AV) scenario above-ground biomass is provided in the bottom right panel.

4.1 Canopy water and energy balance – Pará

Simulated annual precipitation at the Pará focus area was typically around $1500 \, \text{mm} \, \text{yr}^{-1}$, the surface energy flux was dominated by leaf evaporation and transpiration. Transpiration dominated vapor flux in the dry season (May–November). Runoff in the form of drainage through the lower soil column occurred mostly during the wet season. The time series water and energy balance at the land-surface is summarized in Fig. 8. Accumulated water fluxes from the *potential* vegetation scenario are shown in the upper left panel a, anomalous accumulations are shown in the upper right panel b. The *actual* vegetation scenario experienced roughly 10 % less surface precipitation at the Pará focus area. However, the site experienced a small net increase $(30 \, \text{mm} \, \text{yr}^{-1})$ in precipitation throughfall, due to a proportionally stronger decrease in leaf interception surfaces. There is also increased drainage in the *actual* vegetation scenario, which appears to be symptomatic of both increased throughfall and the decrease in the root-zone soil-moisture sink from transpiration.

The *actual* vegetation scenario receives more total shortwave and longwave radiation $(R_{\text{SD}} + R_{\text{LD}})$, which is directly attributable to the decrease in mean convective cloud albedo associated with the decrease in convective rainfall at the site. Although the site receives more total incoming radiation in the *actual* vegetation simulation, the surface albedo increases with the conversion of forests to pasture. This results in more reflected radiation and a decrease in combined sensible and latent heat flux $(H + L)$, see the bottom right panel d of Fig. 8.

4.2 Canopy water and energy balance – Gran Chaco

The annual precipitation at Gran Chaco in the *potential* simulation ranged from 500 to nearly 1000 mm. Annual precipitation was roughly 15 % lower in the *actual* vegetation simulation. A summary of the hydrologic response and the anomalies are shown in Fig. 9. Like the Pará site, land-conversion drove a decrease in leaf area, and therefore a significant decrease in leaf interception of precipitation in the *actual* (AV) simulation. However, in this case, the relative decrease in interception surfaces due to deforestation

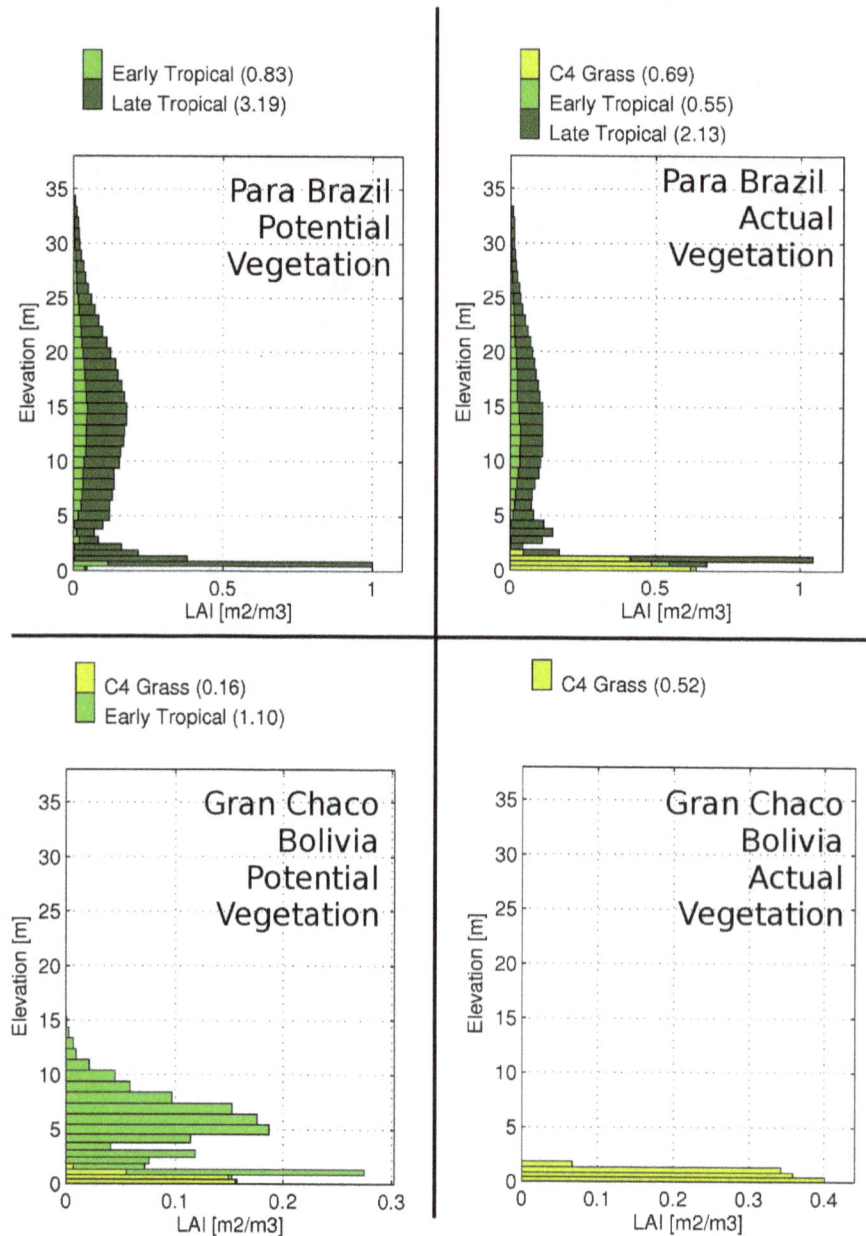

Figure 7. Mean vertical leaf area index profiles ($m^2\,m^{-3}$) estimated by the ecosystem demography model 2 at the two focus areas. Vertically integrated leaf area index ($m^2\,m^{-2}$) per each represented functional type of plant is shown in the key above each plot.

was less affecting than the decrease in total precipitation due to land–atmosphere feedbacks. Therefore, the *actual* vegetation simulation experienced a decrease in total precipitation throughfall. Soil evaporation accounted for half of the water losses, while leaf evaporation and transpiration equally combined to represent the other half. The relatively low precipitation rates promoted almost no detectable runoff. Transpiration decreased by 20 % in the *actual* simulation, which is a direct consequence of decreased stomatal density and precipitation throughfall.

Notwithstanding the decreased precipitation throughfall in the *actual* simulation, surface evaporation increased. Despite decreased precipitation throughfall, upper soil-column moisture from rain events has a longer residence time in the root zone, as shown in Fig. 10. This is an effect of decreased transpiration, and thus moisture in the grass root zone lasts comparatively longer into the dry season. Note that the relative reduction in precipitation throughfall nearly balanced the reduction in transpiration. There is little evidence to suggest that increased soil evaporation rates would maintain indef-

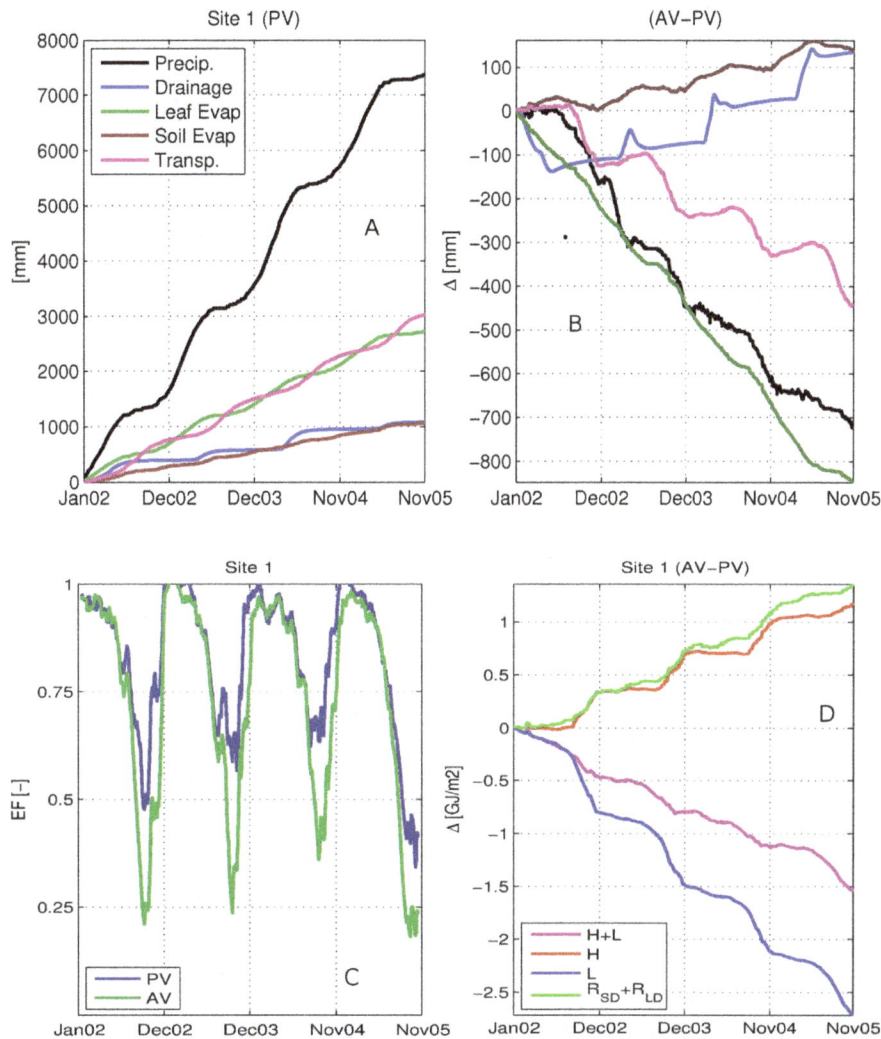

Figure 8. Time series analysis of the surface water and energy balance at the Pará focus site, 2002–2005. (a) Accumulated water flux for the *potential* scenario. (b) Accumulated differential water flux between the *actual* (AV) and *potential* (PV) scenarios. (c) Mean evaporative fraction, latent heat flux (L) divided by the sum of latent and sensible (H) heat flux ($L/(L + H)$). (d) Accumulated differential energy flux in gigajoules per square meter. R_{SD} is downwelling shortwave radiation incident on the surface, R_{LD} is downwelling longwave radiation incident on the surface.

initely in the *actual* scenario, which could alternatively be associated with transient changes in soil column storage.

Like the Pará site, land-conversion at Gran Chaco also drove an increase in total surface albedo, a direct effect due to the loss of dark foliage. More incident radiation is reflected, which reduces net radiation. Unlike the Pará focus area, the albedo effect is stronger than the increase in incident radiation, which leads to decreased sensible heat flux (see Fig. 9) and slightly cooler annual surface temperatures (not shown).

4.3 Land–atmosphere coupling – Pará

A box is constructed around this site for the month of September 2003 that contains the extents of a continuous space with negative precipitation anomaly (see Fig. 11). Ta-

ble 3 shows a selection of spatiotemporal mean indicators from the bounded domain. To summarize the differences in surface fluxes, the results are consistent with the single site time-series analysis, where the *actual* vegetation scenario experienced a decrease in net radiation ($-10 \, \text{W} \, m^{-2}$, despite increased incident shortwave radiation) and an increased mean surface albedo.

The decreased precipitation and surface energy flux of the *actual* scenario are accompanied by a boundary layer with lower equivalent potential temperature (see Fig. 12). The decrease in boundary layer equivalent potential temperature has a strong physical connection to explaining the decrease in precipitation, particularly since the vast majority of the precipitation was generated through the convective parameterization (data not shown). This was verified by recording a log of failures in deep convection (precipitating convective clouds) generated by the convective parameterization. All of the bias in these convective failures occurring in the *actual* simulation was attributed to the generation of convection that resulted in clouds that were too thin to be classified as precipitating deep convection clouds. This is indicative of the how much convective available potential energy (CAPE) can be released through convective buoyancy, which is controlled by the moist static energy in the surface parcels as well as the moist static energy of the mean atmosphere over the depth of the troposphere. Alternatively, there was no positive bias in the logs associated with the inability to trigger parcel buoyancy.

It is questioned if the driving force behind the reduced equivalent potential temperature profiles of the *actual* scenario is solely the result of local surface fluxes or caused by changes in the regional energy circulation. Both scenarios net a negative moisture convergence (divergent) budget for the month (total integrated water mass flux through the box boundaries, normalized by the box area). The *potential vegetation* scenario loses more water ($-51.32 \, \text{kg m}^{-2}$) through its lateral boundaries than the AV scenario ($-37.14 \, \text{kg m}^{-2}$), see Table 3. This can be visualized by flux vectors as well (see Fig. 13). In the *potential vegetation* case shown in the left panel, the flow vectors run east-to-west and up the gradient, which means the advecting air mass is gaining moisture and is consistent with the net water divergence described in Table 3.

4.4 Land–atmosphere coupling – Gran Chaco

Similar to the case study in Pará, a bounding box was constructed around the Gran Chaco site in April 2003 that captures a spatially continuous negative anomaly in precipitation. The boundaries are shown in black against the mean total and mean anomaly in monthly precipitation and evapotranspiration (see Fig. 14). Mean statistics are shown in Table 4. The *actual* scenario experienced less than half as much precipitation ($41 \, \text{kg m}^{-2}$ compared to $85 \, \text{kg m}^{-2}$). The evaporation anomaly between the two scenarios was not as strong ($111 \, \text{kg m}^{-2}$ in the *potential* scenario compared to $83 \, \text{kg m}^{-2}$ for the *actual* scenario).

The *actual vegetation* scenario experienced more downwelling shortwave radiation yet less net surface radiation, which was influenced by an increased surface albedo. Like the Pará case, the *actual* scenario experienced a lower equivalent potential temperature over the boundary layer (see Fig. 15). The convective parameterization logs accounted that the *actual* scenario experienced a great deal more failed convective events associated with an inability to generate deep clouds (the same reason as the Pará case). Note that the model's cloud depth parameterization is controlled by convective available potential energy. However, in this case, about 25 % of the bias in failures was also explained by an inability to trigger convection. This is interesting because this

case was slightly different from the Pará in that the *actual* scenario did not experience higher levels of turbulent kinetic energy over the boundary layer.

An analysis of moisture convergence and advective flux was used to better understand the local versus regional controls that drive convective precipitation. Both scenarios showed negative moisture convergence, typical during the onset of the dry season in this region, refer to Table 4. The *potential* scenario showed less moisture divergence ($-37 \, \text{kg m}^{-2}$) than the *actual* scenario ($-52 \, \text{kg m}^{-2}$). The moisture advected into the Gran Chaco site comes via northerly winds from the moist Amazonian air mass, see the left panel of Fig. 16. Moisture transport from the north decreases in the *actual* scenario, see the right panel of Fig. 16.

5 Discussion

5.1 Secondary forests and evapotranspiration patterning

The maps of evapotranspiration and transpiration anomaly showed pattern similarity with the differences in above-ground biomass, as compared to a lack of pattern similarity between precipitation and above-ground biomass anomalies. The stronger correlation between the evapotranspiration and forest biomass anomalies can be rationalized by understanding how the ED2 model resolves canopy hydrologic process. In ED2, closed canopy tropical broadleaf evergreen forests have higher leaf area indices than grasses, with possible exceptions during drought deciduous leaf drop. Deforestation of closed canopy forests will therefore decrease total leaf area. Rainfall that is intercepted in ED2 has two outcomes: it can either re-evaporate or drip to the land surface (one shortcoming of this assumption is that epiphytes may store water directly from leaf interception). Throughfall precipitation has multiple outcomes: it can evaporate from the surface, become stored in the soil and vegetation indefinitely, or leave via transpiration or leave via runoff. The evaporation rates between the leaf and soil surfaces with canopy air space are regulated by two factors, the aerodynamic resistance and the effective vapor pressure deficit between the respective surfaces with the air space. In the ED2 model formulation, which scales in-canopy wind speeds following Massman (1997), the aerodynamic resistance in the forest canopy will attenuate from top to bottom as wind speeds monotonically decrease. Moreover, water that becomes bound in the soil matrix will have a decreased vapor pressure deficit with the adjacent air (compared to leaf water which is not bound, and assumed saturated) due to the effects of pore spaces at the soil surface (Lee and Pielke, 1992). Therefore in ED2, precipitation that is intercepted in the canopy has both an extra opportunity (simply considering the order of process) to evaporate back to the atmosphere, but also has a tendency towards both de-

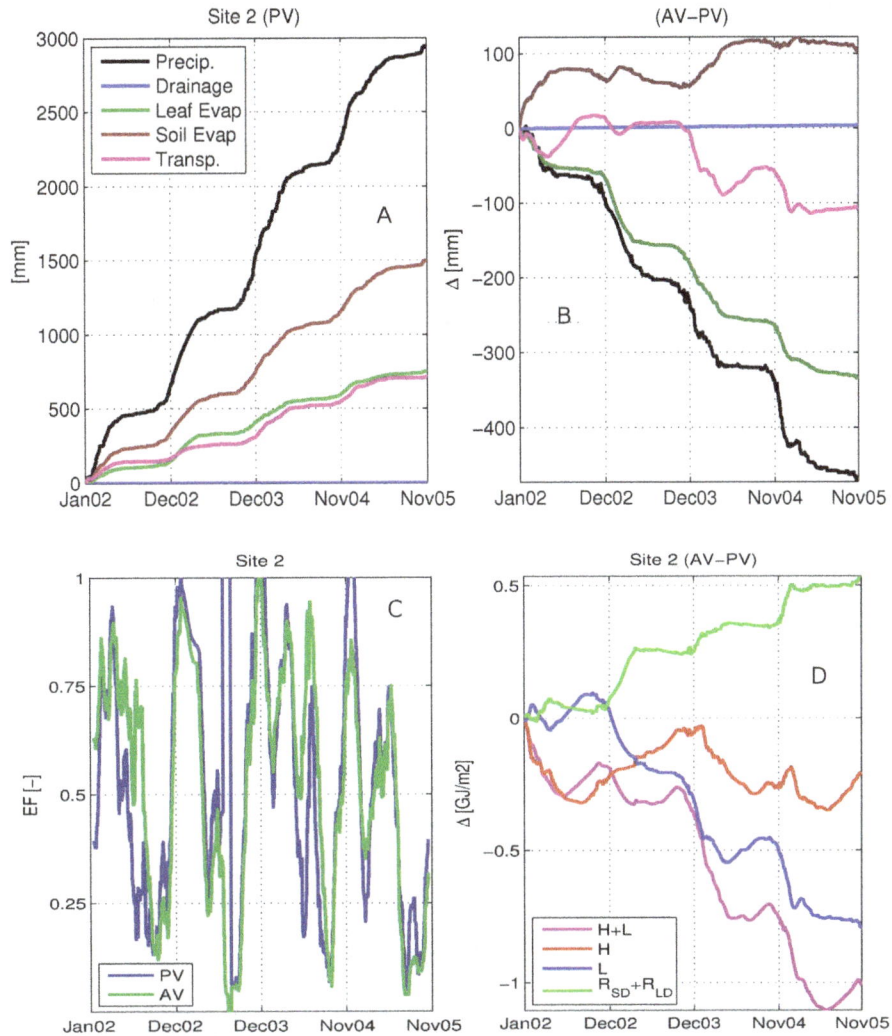

Figure 9. Time series analysis of the surface water and energy balance at the Gran Chaco site, 2002–2005. (**a**) Accumulated water flux for the *potential* scenario. (**b**) Accumulated differential water flux between *actual* (AV) and *potential* (PV) scenarios. (**c**) Mean evaporative fraction, latent heat flux (L) divided by the sum of latent and sensible (H) heat flux ($L/(L + H)$). (**d**) Accumulated differential energy flux in GigaJoules per square meter. R_{SD} is downwelling shortwave radiation incident on the surface, R_{LD} is downwelling longwave radiation incident on the surface.

creased aerodynamic resistance and increased vapor pressure deficit that drives evaporation rate.

However this explanation of process only considers the immediate structural effects of deforestation, which is not static but also has phases of recovery when left unmanaged. During the recovery cycle of tropical forests following *natural* disturbance, successful new growth in the canopy is typically dominated by pioneer species. Pioneer species have lower wood density, higher maximum photosynthetic capacity and quicker vertical growth than late successional species (Laurance et al., 2004; Poorter et al., 2006; Chave et al., 2006). The canopy leaf area may flush to previous levels within a century, yet it may take several centuries for total forest biomass to rebound. This type of behavior was observed in the model spin-up, where newly disturbed patches

of land in the central Amazon reached maximum leaf area over a span of a few decades, compared to the length of time (more than a century) it took for biomass to stabilize.

The results presented here support that secondary forests undergoing recovery from deforestation can drive detectable pattern increases in total transpired water across the region. It is rationalized that at these locations, photosynthetic capacity is scaled by leaf area, as well as a distribution of members skewed towards rapid growth and high photosynthetic capacity. If there is sufficient available soil water there would be an expected increase in total transpiration. In the *actual* model scenario containing deforestation effects, the model estimated an increase in early successional tropical evergreens (pioneers) in the recovering forests of northern Pará and eastern Amazonas (centered on 5° S 58° W) (see Fig. 2

Table 3. Hydrologic monthly means within the bounded area above the Pará case study, September 2003. Total change in column precipitable water for the month per square meter ΔM_{pw}, evapotranspiration ET, precipitation P and resolved moisture convergence Mc, 55 m air temperature T, mixing ration (55 m) r, equivalent potential temperature θ_e, surface albedo to shortwave radiation α, downwelling shortwave radiation R_{SD}, downwelling longwave radiation R_{LD}, upwelling longwave radiation R_{LU}, net surface radiation R_{net}, sensible heat flux SHF and latent heat flux LHF.

Case	ΔM_{pw}	ET	P	Mc	T	r
Units	$kg\,m^{-2}$	$kg\,m^{-2}$	$kg\,m^{-2}$	$kg\,m^{-2}$	$°C$	$g\,kg^{-1}$
AV	−3.457	63.1	29.8	−37.14	32.83	12.18
PV	−3.515	94.7	47.3	−51.32	32.35	12.93

Case	θ_e	α	R_{SD}	R_{LD}	R_{LU}	R_{net}	SHF	LHF
Units	K	–	$W\,m^{-2}$	$W\,m^{-2}$	$W\,m^{-2}$	$W\,m^{-2}$	$W\,m^{-2}$	$W\,m^{-2}$
AV	336.8	0.262	300.2	443.3	513.2	180.9	139.25	70.97
PV	338.2	0.257	285.6	443.9	498.0	187.8	114.50	106.45

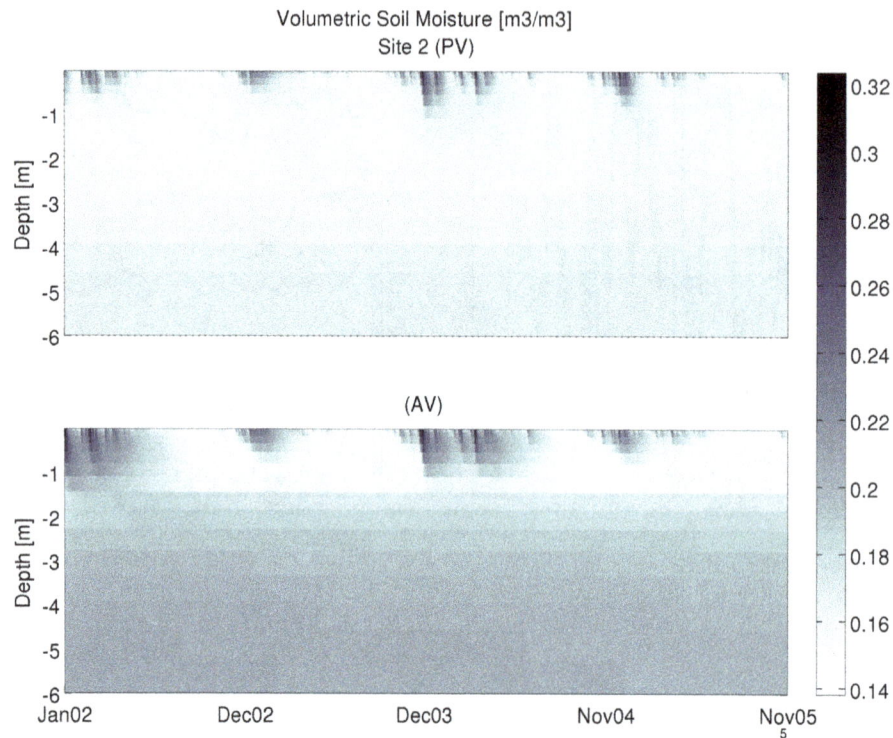

Figure 10. Time series profile of volumetric soil water at the Gran Chaco focus area. Both scenarios, *potential* (PV) and *actual* (AV), are shown separately.

second row right panel). There is an increase in regional transpiration here (Fig. 5) that has a strong pattern match with the increase in early successional biomass, moreover there is little evidence of influence from pattern precipitation here (see Fig. 3).

5.2 Regional surface water balance and runoff generation

There was a clear and consistent bias in the total annual evapotranspiration (negative) and runoff (positive) estimated

in the *actual* scenario when integrated over the entire domain of northern South America. There is some rudimentary explanation of the first order biases in evapotranspiration explained above. The regional runoff bias appears to have several potential explanations and remaining questions. It is clear that in the *actual* scenario, decreased canopy interception promotes a first order effect of increased canopy throughfall. However, we saw in comparing the Pará and Gran Chaco case studies that changes in canopy interception can be offset by changes in incident precipitation. Therefore increased canopy throughfall from deforestation is not ubiq-

Table 4. Hydrologic monthly means within the bounded area above the Gran Chaco case study, April 2003. Total change in column precipitable water for the month per square meter ΔM_{pw}, evapotranspiration ET, precipitation P and resolved moisture convergence Mc, 55 m air temperature T, mixing ration (55 m) r, equivalent potential temperature θ_{e}, surface albedo to shortwave radiation α, downwelling shortwave radiation R_{SD}, downwelling longwave radiation R_{LD}, upwelling longwave radiation R_{LU}, net surface radiation R_{net}, sensible heat flux SHF and latent heat flux LHF.

Case	ΔM_{pw}	ET	P	Mc	T	r
Units	kg m^{-2}	kg m^{-2}	kg m^{-2}	kg m^{-2}	$^\circ\text{C}$	g kg^{-1}
AV	-11.42	82.95	41.89	-52.49	25.98	12.73
PV	-11.02	111.89	85.91	-36.99	27.36	15.15

Case	θ_{e}	α	R_{SD}	R_{LD}	R_{LU}	R_{net}	SHF	LHF
Units	K	–	W m^{-2}	W m^{-2}	W m^{-2}	W m^{-2}	W m^{-2}	W m^{-2}
AV	334.4	0.330	252.6	400.2	466.9	111.74	38.54	91.0
PV	342.0	0.297	218.7	424.9	462.6	130.2	28.15	122.5

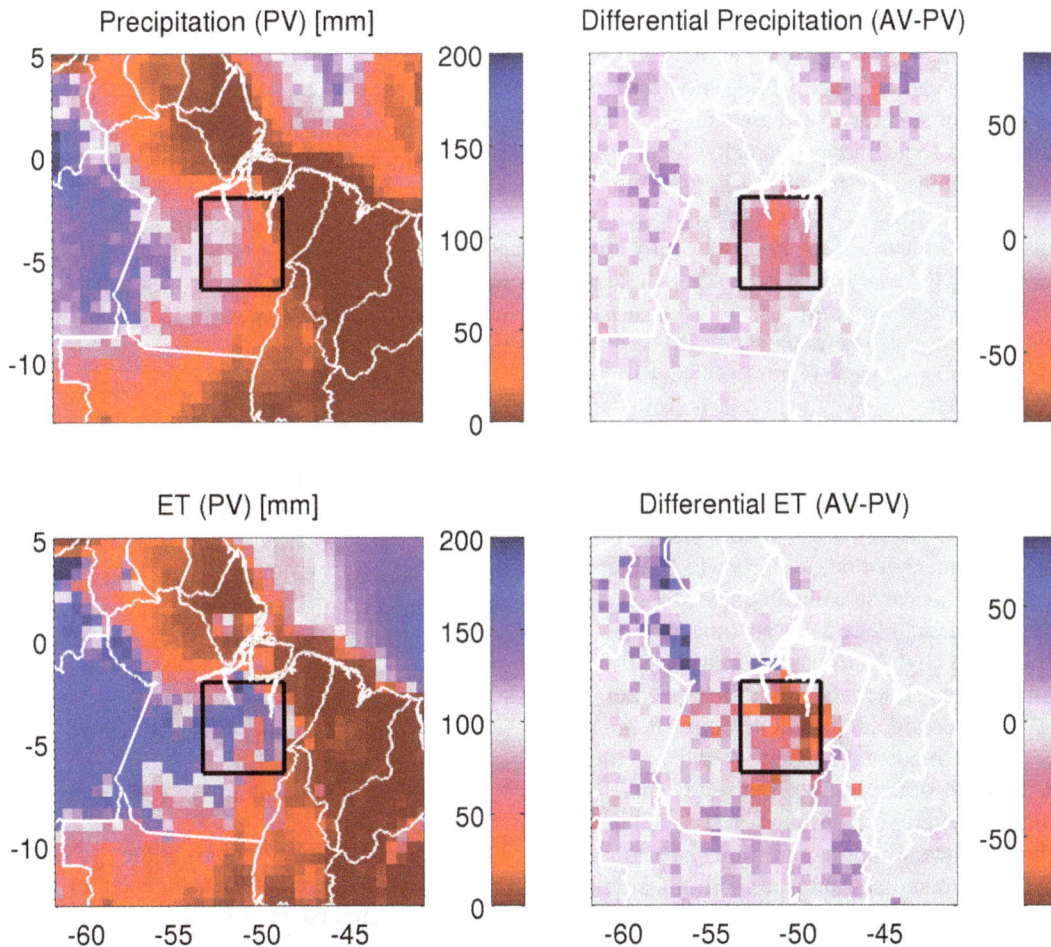

Figure 11. Monthly integrated surface water fluxes over the Pará focus region, September 2003. Upper left panel: map of integrated monthly precipitation for the *potential vegetation* simulation (PV). Upper right panel: map of the integrated difference in monthly precipitation, *actual vegetation* case minus the *potential vegetation* case (AV–PV). Lower left panel: map of integrated monthly evapotranspiration for the *potential vegetation* simulation. Lower right panel: map of the integrated difference in monthly evapotranspiration, *actual vegetation* case minus the *potential vegetation* case.

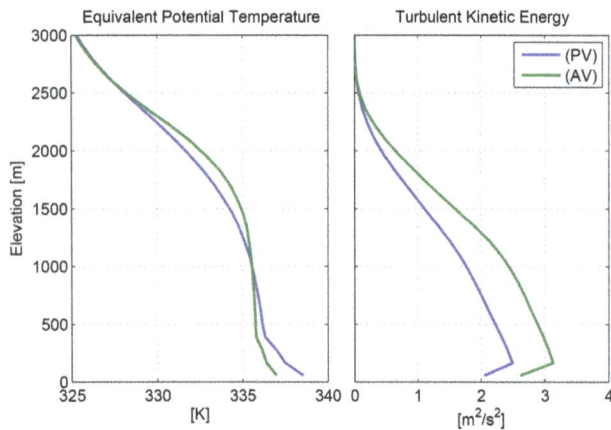

Figure 12. Mean profiles of Equivalent Potential Temperature and Turbulent Kinetic Energy at 15Z within the bounded domain at Pará, September 2003.

uitously associated with regional increases in runoff. In a regional water balance analysis, d'Almeida et al. (2007) also observed that wide-spread regional deforestation promoted decreased evapotranspiration and increased runoff. Similar to this study, they also found that precipitation feedback response to deforestation had the potential to impact the water cycling on par with direct effects of surface hydrologic parameters (although in their results, bi-directionally weakening or strengthening the water cycle depending on heterogeneity and land-cover fractionation).

In the simulations presented here, the increased continental runoff from the *actual* scenario is driven by higher mean annual soil moisture. The regional mean soil moisture depth simulated in the *actual* scenario oscillated around a inter-seasonal mean of 1.40 m (over an 8 m medium), averaging 5 cm greater than the *potential* case. As increased runoff has a negative feedback on increased soil water, and there was no consistent bias in precipitation, it is most likely that the positive shift in mean annual moisture in the *actual* scenario is driven by the decreases in regional evapotranspiration.

This experiment highlighted the use of relatively sophisticated vegetation biophysical processes, which incorporated variable vegetation structure, composition, rooting depth and uptake. However, the modeling framework did not incorporate lateral transport of any surface moisture. Therefore, these results must be interpreted with the understanding that lateral re-infiltration, lateral vadose-zone flow, interflow and water table dynamics could not influence soil moisture dynamics. In light of this, this experiment provides a gauge on the strength of the control that evaporation response to deforestation can have on regional water balance and runoff generation. There has been some evidence that soil hydrologic properties can be affected by land conversion in the tropics, Zimmermann et al. (2006) found that both infiltrability and upper root-zone saturated hydraulic conductivity was highest in intact rainforest compared to pasture and tree planta-

tions. Decreased infiltration in pastures has been related to increased runoff generation as well (Muñoz-Villers and McDonnell, 2013). But there seems less certainty in the literature in quantifying the evaporative response from canopy and soil to regional Amazonian deforestation, degradation and recovery.

5.3   Intersection of seasonal hydrology and represented plant functional types on canopy process

The two case studies showed that the structure of the vegetation canopy can influence the seasonal cycle of moisture storage and land–atmosphere moisture flux. At the Gran Chaco site, transpiration was greater in the *potential vegetation* scenario during the wet season when the deeper roots and higher stomatal density of the open canopy forest could access available soil moisture. Alternatively, total evapotranspiration was greater in the *actual vegetation* scenario at the onset of the dry-season, mostly due to the fact that the grasslands had more available water stored in the upper root zone (recall Fig. 10).

The natural vegetation at the Gran Chaco site is represented "in model" with early successional broadleaf evergreen plant functional types, with accompanying C4 grasses. While the demographic size structure, composition, and the openness of the canopy shows some similarity with dry-forest structure, it must be realized that the wider range of water conservation strategies observed in nature could influence how the differences in surface-to-atmosphere energy fluxes play out at this site.

These findings suggest that the next generations of earth system models may benefit from improvements in representing plant diversity. The seasonal flux of surface-to-atmosphere water vapor is regulated by plants, and can potentially impact the hydrometeorological dynamic of the region. Total evapotranspiration during the transition from the late wet season to early dry season (April) at the Gran Chaco site was larger in the potential vegetation scenario. As shown in the hydrometeorological analysis in Sect. 4.4, this was a time in the seasonal cycle that exhibited relatively strong differences in the instability profiles in the atmosphere, albeit from competing local and regionally driven mechanisms.

5.4   Drivers of anomalous convective precipitation

The negative precipitation bias at the two focus areas in the *actual* scenario were both accompanied by reductions in net radiation, decreased annual latent heat flux (evapotranspiration) and increased albedo. They also experienced boundary layers with lower mean equivalent potential temperature, which was then related to fewer cumulus events that lacked sufficient convective available potential energy to generate deep clouds (diagnosed through convective logs). It is believed that differences in convective available potential energy underlies the convective precipitation anomaly.

Total Mixing Ratio Flux Vectors
and Gradient in Precipitable Water (PV)

Total Mixing Ratio Flux (AV–PV)
(Magnified 12x)

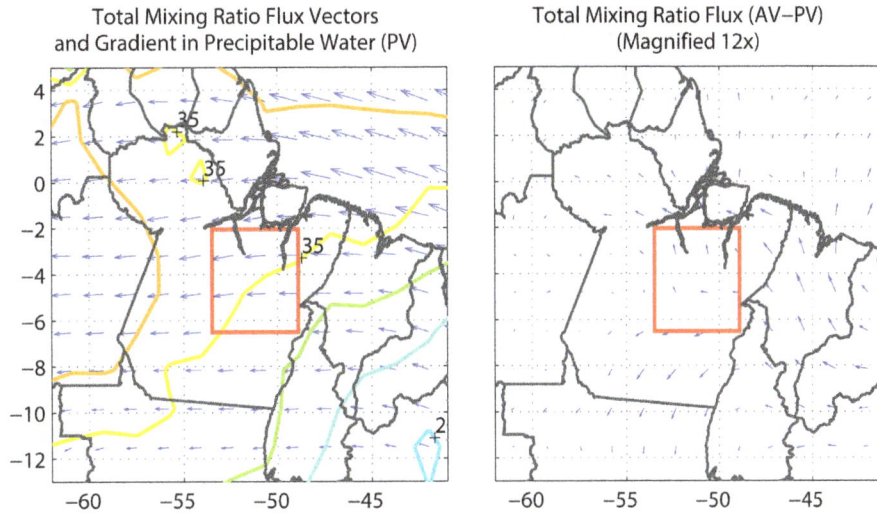

Figure 13. Left panel: map of vertically integrated total water advective flux vectors (quivers) and vertically integrated precipitable water (contours) for the *potential vegetation* case (PV), region near the Pará site, September 2003. Quivers are scaled and convey only directionality and relative magnitude. Contours of low precipitable water are shown by cool colors (blues) and high precipitable water with warm colors (reds). Right panel: the differential in vertically integrated advection of total precipitable water, *actual vegetation* minus *potential vegetation* (AV–PV). Quivers are scaled to 12 times relative to the left panel. In both panels the sub-domain bounding the Pará focus region is shown with a red box.

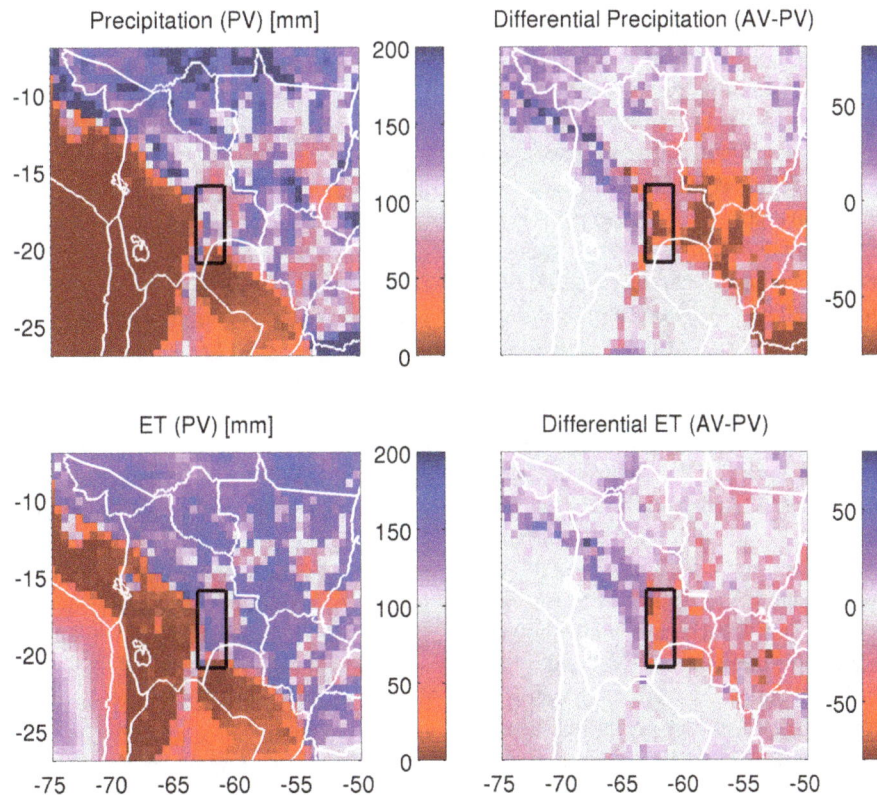

Precipitation (PV) [mm]

Differential Precipitation (AV-PV)

ET (PV) [mm]

Differential ET (AV-PV)

Figure 14. Monthly integrated surface water fluxes over the Gran Chaco focus region, April 2003. Upper left panel: map of integrated monthly precipitation for the *potential vegetation* simulation (PV). Upper right panel: map of the integrated difference in monthly precipitation, *Actual* vegetation case minus the *potential vegetation* case (AV–PV). Lower left panel: map of integrated monthly evapotranspiration for the *potential vegetation* simulation. Lower right panel: map of the integrated difference in monthly evapotranspiration, *actual vegetation* case minus the *potential vegetation* case.

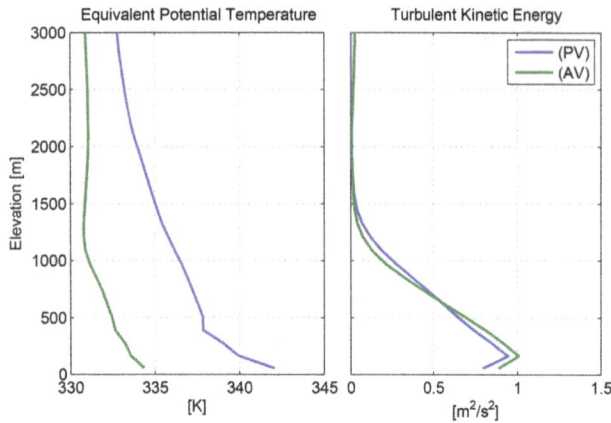

Figure 15. Mean profiles of equivalent potential temperature and turbulent kinetic energy at 15Z within the bounded domain at Gran Chaco, April 2003.

Positive anomalies in sensible heat flux for the *actual* scenario are concurrent with increased turbulent kinetic energy, which is thought to promote the circulations and boundary layer development that lifts air parcels to trigger convection (Wang et al., 2009; Fisch et al., 2004). The case study at Pará did show increased boundary layer turbulent kinetic energy and sensible heat flux, and it is possible that increased boundary layer turbulence may have helped recoup some losses in precipitation. This is deductive reasoning; however, following that the Gran Chaco case study did not have increased turbulent kinetic energy associated with the *actual* scenario and it did experience more failed convective events associated with the inability to overcome convective inhibition (where there was no such trend at Pará).

The two case studies offer evidence that the negative precipitation anomaly, can be mediated by by primarily local effects and also by a combination of local effects and regional circulation effects. The Pará case study suggests that the negative precipitation anomaly is driven primarily by changes in the local surface energy flux. The dry season prevailing winds at Pará flowed up the moisture gradient (i.e., gaining moisture, not losing moisture). And according to the flux vectors in Fig. 13, despite the decreased precipitation experienced in the *actual* scenario, the prevailing winds fluxed more moisture into the domain when compared to the *potential* case. In contrast, the Gran Chaco case study showed evidence that the precipitation anomaly was responding to a change in the regional circulation as well as changes in the local surface energy fluxes. Here, the prevailing winds came out of the north and flowed down the moisture gradient (i.e., losing moisture). While the domain was a net source of moisture (divergent) the net flux into the domain via advection was positive. The *actual* scenario experienced a relative decrease in advected moisture flux from the prevailing winds (see Fig. 16).

5.5 Uncertainty in model estimates

Estimation uncertainty in coupled regional models exist in many sources, including the initial condition, the boundary conditions, the scale limitations of the resolved processes, the mechanics of the model processes and the parameters that govern the processes. In limited area models, there is also variability in the lateral boundary conditions. For the current research, this variability can affect the differences detected between the two scenarios, potentially impacting results. Ideally, this variability space can be explored in depth, perhaps using ensembles over multiple decades, including differing starting years and perturbations. The range of the variability space that can be sampled is limited by computational expense of the simulations. The 4–7 year simulations presented here took approximately 2 months each using 96 parallel computational cores with high-speed interconnects, the computational time being an result of the highly memory intensive ecosystem model and atmospheric time stepping that is relatively short (30 s) compared to general circulation models. However, we maintain that the the the pattern differences in precipitation and evaporation between the two scenarios showed consistency enough to merit commentary.

The simulations used a number of different external data sets, each of which contained information at different spatial scales. This includes the soils information (variable scales), the human disturbance transitions (1°), the scale at which the forest structure was "spun-up" from climate driver data (1°) versus the scale at which that data was re-sampled (using nearest neighbor) in the coupled simulation (64 km). The uncertainty associated with the dynamics of the coupled simulations are subject to the scale of the information provided by these external data sets, as well as any biases that might be inherent in those data sets. This may be particularly true in the case of the lateral boundary condition data.

It is also necessary to acknowledge that there is uncertainty inherent in model process. For instance, in the absence of explicitly resolving buoyant updrafts (which is not possible in mesoscale simulations), the successful triggering of convection was based on the negative energy between the level of updraft and the level of free convection. There are alternative methods for estimating where updrafts start and how a parcel may or may not overcome inhibition to reach free buoyancy. We chose a straight forward parameterization that compares the negative buoyancy to a threshold. There are other trigger mechanisms that can be used, such as estimating the statistical distribution of vertical kinetic energy present in eddy motions at the level of updraft, and using that to estimating the likelihood that eddies will overcome negative buoyancy through force balance computations. This is simply an example of how the simulations were undoubtedly influenced by these choices, and this is something that must be considered when interpreting the output from any complex numerical model.

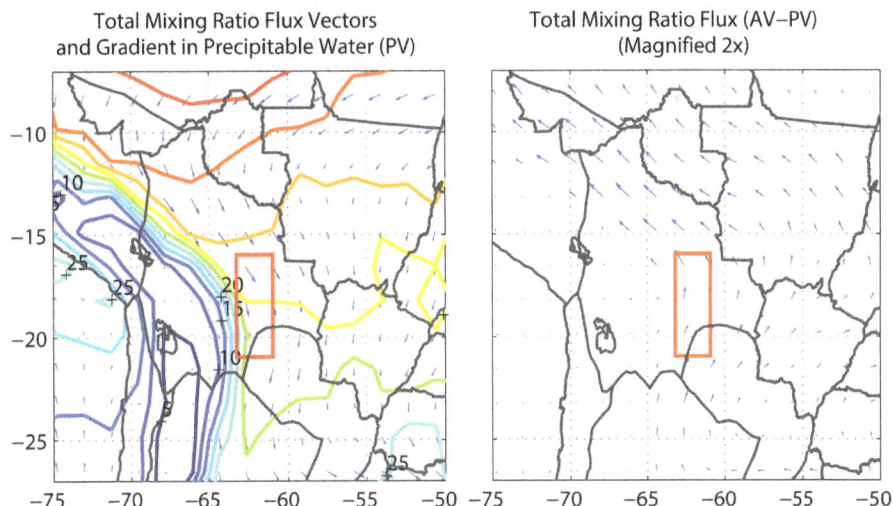

Figure 16. Left panel: map of vertically integrated total water advective flux vectors (quivers) and vertically integrated precipitable water (contours) for the *potential vegetation* case (PV), region near the Gran Chaco site, April 2003. Quivers are scaled and convey only directionality and relative magnitude. Contours of low precipitable water are shown by cool colors (blues) and high precipitable water with warm colors (reds). Right panel: the differential in vertically integrated advection of total precipitable water, *actual vegetation* minus *potential vegetation* (AV–PV). Quivers are scaled to 2 times relative to the left panel. In both panels the sub-domain bounding the Gran Chaco focus region is shown with a red box.

In the acknowledging uncertainty inherent in the simulations presented here, we have also tried to verify if these simulations can show agreement with observations. The *actual* scenario's simulations were compared to a group of different observations with the intention of verifying the modeling system's ability to represent key processes. The model's regional demographic of vegetation biomass was compared to field inventory data, atmospheric thermodynamic profiles were compared to soundings, mean all-sky profiles of cloud water were compared to satellite estimates and the seasonality of precipitation, net radiation, latent heat flux and sensible heat flux were compared with multi-data composite data products (see Appendices B–F). The comparisons with observations (which harbor their own uncertainty) suggest the model system is adequate to make a meaningful comparison between the two scenarios, yet not without room for improvement. In particular, this modeling study (and regional coupled simulations in general) would stand to benefit from improvements in how large-scale precipitation (weak spatiotemporal variability compared to point scale measurements) and canopy interception is handled.

A small selection of published research has conducted similar simulations to those reported here, with results that offer limited comparison. Perhaps most similar, Bagley et al. (2014) compared different land-cover scenarios with some similarities to our *potential* and *actual vegetation*, finding a generally weaker patterning of precipitation anomaly. However, evidence of the strongest pattern differences also occurred south of the Amazon delta, which may be viewed as correlated with the depression in the precipitation dipole presented in this work. The simulations conducted by Bagley et al. (2014) also used a different modeling system (The Weather Research and Forecasting System WRF) and land-surface model (Noah LSM), suggesting that model intercomparison would be a useful endeavor in rectifying the regional hydrometeorological effects of South American land-cover change.

Other studies have used similar simulation approaches to the work described here, but targeted land–atmosphere effects under future conditions. Global simulations conducted by Medvigy et al. (2011) found that future business as usual (aggressive) deforestation would generate dipole precipitation differential with respect to conservative deforestation scenarios. The length scales of the patterning in the dipoles (100s of kilometers) showed similarity with results presented here. However, the locations of differences they presented were different, as of course were the driving land-use scenarios as well. Walker et al. (2009) also evaluated the effects of future deforestation scenarios on the regional hydrometeorology, finding that massive deforestation outside of protected areas will lead to basin-wide changes in rainfall, both positive and negative. In summary, there are some commonalities between the various regional simulations, albeit from differing land-use scenarios; impacts on hydrometeorological climate are expressed most strongly as changes in dry-season precipitation, and that patterning exists to varying degrees at large length scales (> 100 km) (Walker et al., 2009; Bagley et al., 2014; Medvigy et al., 2011).

6   Conclusions

The simulations presented here produced negative anomalies in evapotranspiration rates that showed pattern similarity with areas where deforestation had driven noticeable differences in aboveground biomass. The results showed mixed support that historical land conversion has had influence on the patterning of South American precipitation. Over 4 years of simulation most of the region showed little consistency in annual anomalies. However, patterns in the precipitation anomaly emerged at specific locations, where mean annual precipitation showed moderate yet consistent differences as evaluated through a standard score. One pattern showed a dipole structure that occurred near eastern Pará Brazil. The other pattern, showed a negative anomaly in the Gran Chaco region of central and southern Bolivia. In this area, their was positive precipitation anomaly at the Peruvian–Bolivian border, yet the general patterning may not-necessarily be considered a dipole.

The simulations suggest land conversion in South America has not had a large impact on mean precipitation in the region as a whole. Mean regional precipitation was lower in the *actual* scenario in only 3 of the 4 simulation years. In contrast, differences in mean continental runoff (increased with human land conversion) and total evapotranspiration (decreased with human land conversion) were both consistent from year to year and showed greater differences compared to precipitation.

It was also observed that a large-scale shift in forest composition to early successional tropical evergreens following deforestation and abandonment in central eastern Amazon produced a noticeable increase in grid-average transpiration. This increase was striking not only because the pattern difference matched the pattern shift in composition, but particularly because the overall biomass in this area had decreased compared to the natural (*potential*) condition, suggesting that the physiological differences of the vegetation had influenced regional fluxes and not necessarily the quantity.

Beyond characterizing mean and pattern differences in the regional water balance, it was identified that changes (be they moderate) in precipitation may have occurred in locations where terrestrial vegetation actively regulates photosynthesis due to water limitations and that the changes in precipitation can be attributed to combinations of changes in local surface fluxes and changes in the regional atmospheric circulation. An assessment of the regional vegetation's response to moisture stress has indicated that both these locations show some susceptibility to changes in root-zone soil moisture (plant stomata were actively regulated to conserve water over a broad range depending on exact location, varying from 20–80 % of the time); however, the ecosystems of Gran Chaco are generally dryer and show greater susceptibility.

In both focus cases, deforestation led to increases in total surface albedo, driving decreases in net radiation, boundary layer moist static energy and ultimately convective precipitation. However, the differences in precipitation at Pará Brazil are more strongly connected with these localized differences in land-surface energy flux. The hydrometeorological analysis near Gran Chaco suggests that human land-conversion has had some impact on the strength of the South American moisture circulation in the southwestern portion of the Amazon, which claims partial responsibility along with differences in surface fluxes for an estimated decrease in annual precipitation in the Gran Chaco.

Appendix A: Land–atmosphere model coupling

The atmospheric model (BRAMS) and the land surface model (ED2) are loosely and asynchronously coupled. This assumes that the two models pass each others' boundary conditions at a frequency that captures the natural variability of the flux, yet the fluxes between models are not dynamically changing as prognostic variables within the numerical integrator sub-stepping. The atmosphere (BRAMS) provides information at the beginning of the ED2 forward step, while ED2 assumes this information is constant over the duration of its forward step and makes a time average of the fluxes during the step to return.

Mesoscale atmospheric models at spatial resolutions of tens of kilometers typically perform integrations on the order of tens of seconds to maintain numerical stability and convergence. The simulations in this work used an atmospheric time step of 30 s (for non-acoustic dynamics, acoustics were 10 s). Ideally, the land surface and atmospheric models operate at time steps that approach the infinitesimally small. The ED2 has a non-trivial computational overhead due to the large number of vegetation cohorts experiencing energy balancing within each grid cell. Ultimately, the ED2.1 model used a 120 s time step while coupled to BRAMS.

Atmospheric variables such as air temperature, humidity and wind-speed are provided to ED2 at a reference height of 70 m. This is required because ED2 internally calculates turbulent transport of heat, moisture and momentum through the canopy and into the inertial sub-layer of the lower atmosphere. The turbulent transport of scalars through the sub-layer above the canopy were calculated following Beljaars and Holtslag (1991). These calculations relied on gradients that were based on the enthalpy, density and specific humidity of air at reference temperature and within the canopy's interstitial air-space. A list of the variables required to drive ED2 is provided in Table A1.

The ED2 model passes boundary fluxes at the grid scale to the atmospheric surface layer as an area-weighted average across patches. There are three groups of information the land surface must provide the atmosphere model: (1) the topography which governs the geometry of the atmosphere's coordinate system and drag, (2) a lower boundary condition for turbulent closure, i.e., the vertical velocity perturbations of momentum, heat, moisture and carbon, and (3) a surface albedo for the atmospheric radiative transfer calculations. These variables are listed in Table A2. Because the ED2 model prognoses spatial variables at the patch sale (such as canopy temperature, humidity, etc), spatial averaging is required for flux variables. For any generic variable S, at a grid cell with M patches of area A_{patch}, the area averaged quantity is straight forward.

$$S_{\text{grid}} = \sum_{j}^{M} S_{\text{patch},j} A_{\text{patch},j} \tag{A1}$$

$$\sum_{j}^{M} A_{\text{patch},j} = 1 \tag{A2}$$

More detailed explanation of how turbulent fluxes are calculated in the ED2 model can be found in Medvigy et al. (2009) and Knox (2012).

Table A1. Atmospheric boundary conditions provided by BRAMS, that drive the ED2 model.

Symbol	Units	Description
u_x	[m s^{-1}]	Zonal wind speed
u_y	[m s^{-1}]	Meridional wind speed
T_a	[K]	Air temperature
q_a	[kg kg^{-1}]	Air specific humidity
$\dot{m}_{\text{pcp}}$	[kg s^{-1}]	Precipitation mass rate
z_{ref}	[m]	Height of the reference point
R_{ld}	[w m^{-2}]	Downward longwave radiation
R_{vb}	[w m^{-2}]	Downward shortwave radiation, visible beam
R_{vd}	[w m^{-2}]	Downward shortwave radiation, visible diffuse
R_{nb}	[w m^{-2}]	Downward shortwave radiation, near infrared beam
R_{nd}	[w m^{-2}]	Downward shortwave radiation, near infrared diffuse

Table A2. ED2 flux variables providing the lower boundary condition for the BRAMS atmospheric model.

Symbol	Units	Description
$\overline{(u'w')}$	[m^2 s^{-2}]	Average vertical flux of horizontal wind velocity perturbations
$\overline{(w'w')}$	[m^2 s^{-2}]	Average vertical flux of vertical wind velocity perturbations
$\overline{(t'w')}$	[mK s^{-2}]	Average vertical flux of temperature perturbations
$\overline{(q'w')}$	[kg m^{-2} s^{-2}]	Average vertical flux of moisture perturbations
$\overline{(c'w')}$	[μmol m^{-2} s^{-2}]	Average vertical flux of carbon perturbations
χ_s	[$-$]	Average total shortwave albedo
χ_l	[$-$]	Average total longwave albedo
R_{lu}	[w m^{-2}]	Average upwelling longwave radiation

Appendix B: Regional above-ground biomass

The model estimated live above-ground biomass (AGB) and basal area (BA) that represent the initial condition is compared with a collection of census measurements in Baker et al. (2004a, b) (see Fig. B1). A map is provided showing the locations of the plot experiments (see Fig. B2). The coordinates from the measurement stations were matched with ED2 nearest neighbor grid cells. Consistent with observations, only ED2 plants greater than 10 cm in primary forests were included in the comparison. The published measurements were taken at different times over the previous decades, the lag between these sites and the time of the simulation initial condition (January 2008) varies and is reported. Tree diameters in ED2 is diagnosed allometrically from structural carbon, similar to allometric equations in Chave et al. (2001) and Baker et al. (2004b), with differences in parameterization to reflect functional groups.

Please note that the forest inventory data makes no assumption that aboveground biomass is in equilibrium. This comparison is only intended to evaluate the present day static representation of forest structure. As applied to the coupled simulation, the were treated as static. As stated earlier, the length of the simulation did not merit the need to incorporate dynamics.

The majority of sites show fair agreement with model estimated above-ground biomass and basal area. The exception to this is the cluster of sites located in eastern Bolivia at Huanchaca Dos (HCC), Chore 1 (CHO), Los Fierros Bosque (LFB) and Cerro Pelao (CRP). There are several potential reasons for this discrepancy attributed to the model such as: variability in (1) climate forcing data (most notably precipitation and vapor pressure deficit as these are water limited growth conditions), (2) edaphic conditions and (3) plant functional parameters. These plots are close to the southern Amazonian transition between tropical rain-forests and Cerrado type open canopy forests where gradients in vegetation types are large and uncertainty is expected to be greater. Large spatial gradients in biomass are also reflected in the differences among the cluster of plots (HCC, CHO, LFB and CRP), (124.8 Mg ha^{-1} above-ground biomass at CHO and 260.0 Mg ha^{-1} AGB at HCC). The sharp gradient in forest biomass suggests that the omission of sub-pixel variability in the modified DS314 climate forcing could explain a portion of this difference, along with any persistent biases.

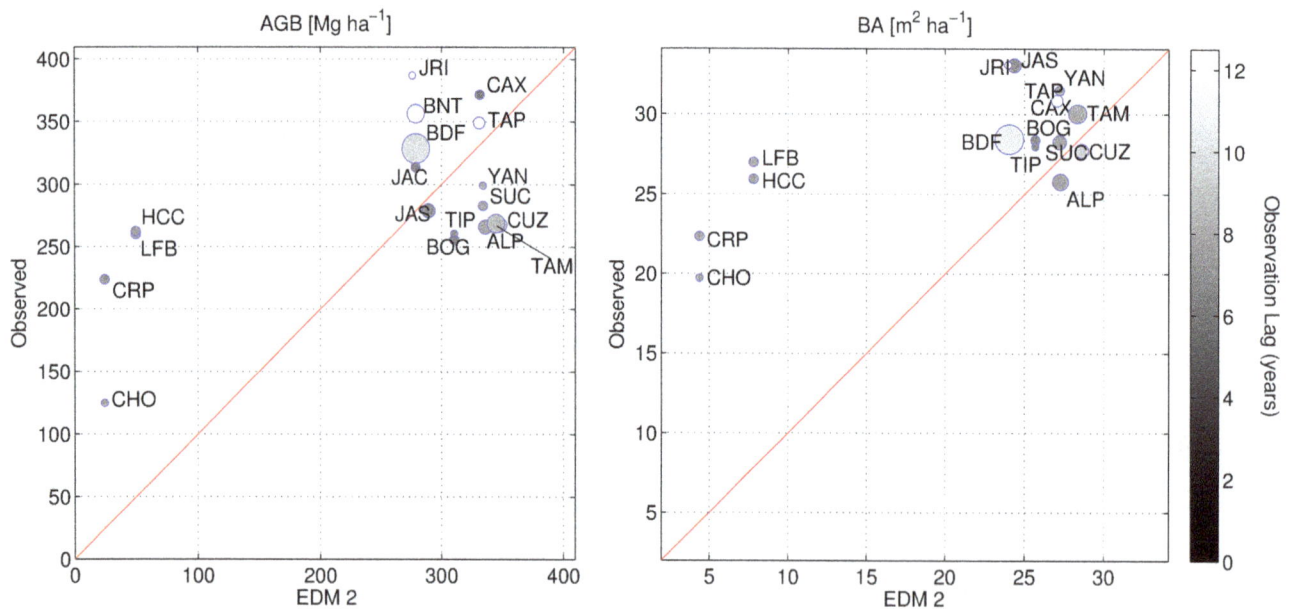

Figure B1. Comparison of model estimated mean above-ground biomass (AGB) and basal area (BA) with measurements presented in Baker et al. (2004a, b). Circle size shows relative approximation of the number of census sites used in the field measurements reported in Baker – maximum = 11 separate plots at BDF (code for the Biological Dynamics Forest Fragments Project). Darker circles indicate that measurements were taken more recently and therefore have less time lag in the comparison with the ED2 initial condition (January 2008). In accord with methods of Baker et al. (2004a, b), model estimates were filtered to include only primary forests and ignored vegetation less than 10 cm diameter. Coarse woody debris was excluded from comparison, only live stems were accounted for.

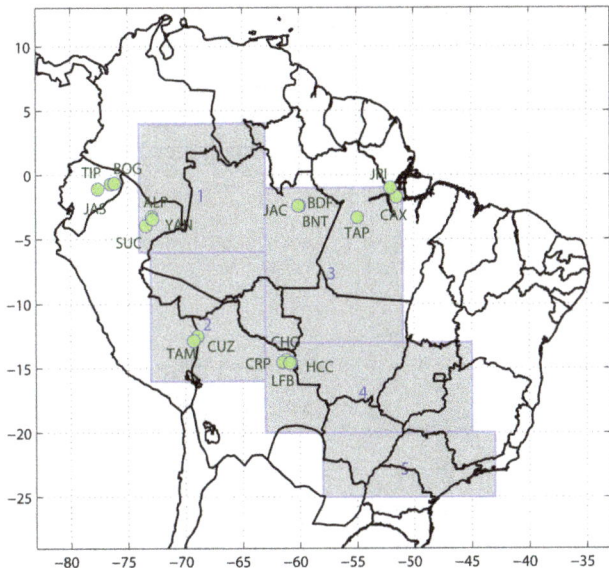

Figure B2. Locations of zones and sites of analysis. Zones are numbered 1 through 5, and reflect geographic areas where model and observation spatial means are compared for validation (see Appendix D and E). Forest census plot sites from Baker et al. (2004a, b) are referenced with green markers and their station code. Station codes designate the following site names: Allpahuayo (ALP), BDFFP (BDF), Bionte (BNT), Bogi (BOG), Caxiuana (CAX), Chore (CHO), Cerro Pelao (CRP), Cuzco Amazonico (CUZ), Huanchaca Dos (HCC), Jacaranda (JAC), Jatun Sacha (JAS), Jari (JRI), Los Fierros Bosque (LFB), Sucusari (SUC), Tambopata (TAM), Tapajos (TAP), Tiputini (TIP) and Yanamono (YAM).

Appendix C: Thermodynamic mean profiles

The objectives of this experiment require that modeling system output match mean observations to such a degree that there is trust in the model's ability to represent physical processes. It is believed that the relative differences between the two simulations have validity if model processes reproduce mean observed tendencies in the land surface and atmosphere.

Mean monthly profiles of model estimated air temperature, specific humidity and moist static energy are compared with mean radiosonde data over Manaus Brazil (see Figs. C1 and C2). Comparisons are made at 00:00 UTC (20:00 LT) and 12:00 UTC (20:00 LT). The model estimates a consistently warmer atmosphere, in the range of about two degrees both morning and evening. Model estimated specific humidity in the lower troposphere is lower than the radiosondes

(see Fig. C1). Moist static energy is slightly underestimated by the model in the lower troposphere and then overestimated in the mid to upper troposphere. This may suggest that the model is convecting relatively large quantities of warm, moist air at the surface and entraining it to the upper atmosphere.

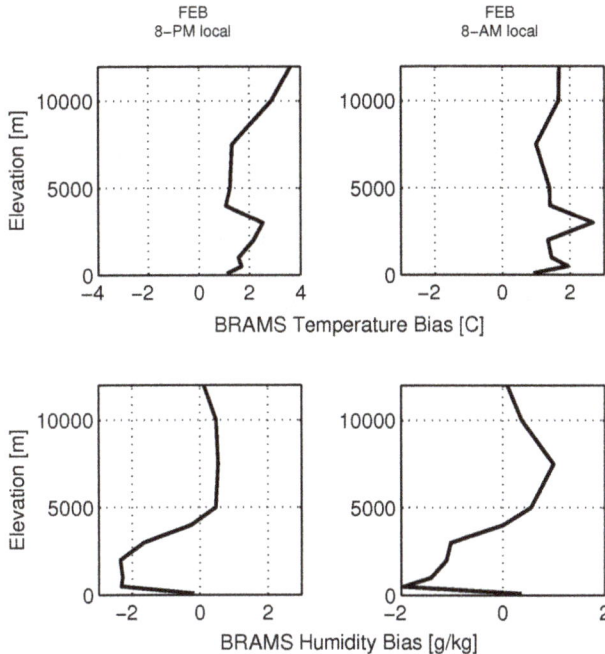

Figure C1. Comparison of model estimates with radiosonde data, differences in mean air temperature and specific humidity. Manaus, February 2003.

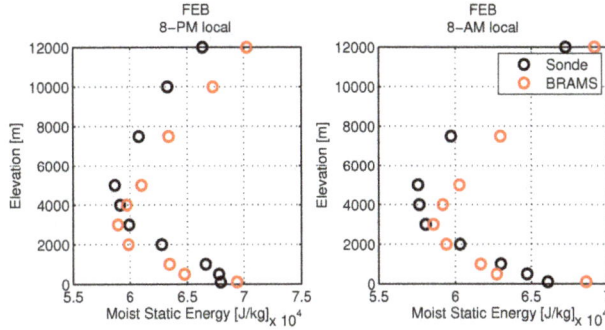

Figure C2. Comparison of model estimated mean moist static energy with rawinsonde measurements. Manaus, February 2003.

Appendix D: Inter-seasonal precipitation and surface radiation

Monthly precipitation and downwelling shortwave radiation in the model is evaluated as spatial means within five separate zones of analysis. The boundaries of the zones of analysis are shown in Fig. B2. Monthly mean model estimates are again compared to precipitation estimates from the Tropical Rainfall Measurement Mission (TRMM) 3B43 product and surface radiation from the Global Energy and Water Exchanges Project (GEWEX) Surface Radiation Budget (SRB) product version 2.5.[3] There is generally acceptable agreement between the model and TRMM estimated precipitation. The seasonal variability in both data sets is greater than their differences (see Fig. D1). The largest differences are reflected in the strength of of the wet-season peak precipitation in Zone 3 (central eastern Amazon) and the severity of the dry season precipitation in Zone 5 (southern Brazil). The timing of peak and minimum rainfall show generally good

agreement, particularly in Zones 2–5. The lower estimate of mean precipitation in southern Brazil is consistent with decreased cloud albedo and increased downwelling shortwave radiation at the surface (see Fig. D2). Surface shortwave radiation is modestly over-estimated compared with the SRB estimates for most other cases.

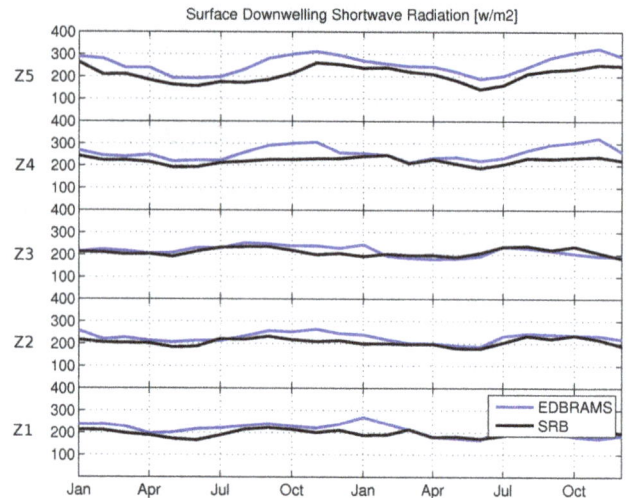

Figure D2. Mean monthly surface radiation from ED2-BRAMS and the Surface Radiation Budget (SRB) product version 2.5, years 2002–2003. Spatial means are taken within zones according to Fig. B2.

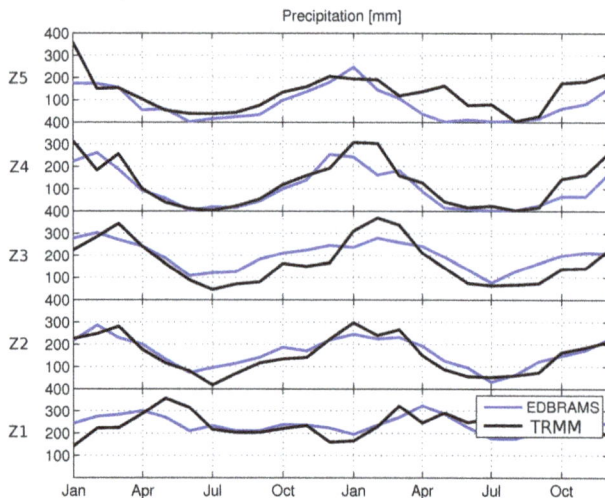

Figure D1. Mean monthly precipitation from ED2-BRAMS and the Tropical Rainfall Measurement Mission (TRMM) 3B43 product, years 2002–2003. Spatial means are taken within zones according to Fig. B2.

---

[3]These data were obtained from the NASA Langley Research Center Atmospheric Sciences Data Center NASA/GEWEX SRB Project.

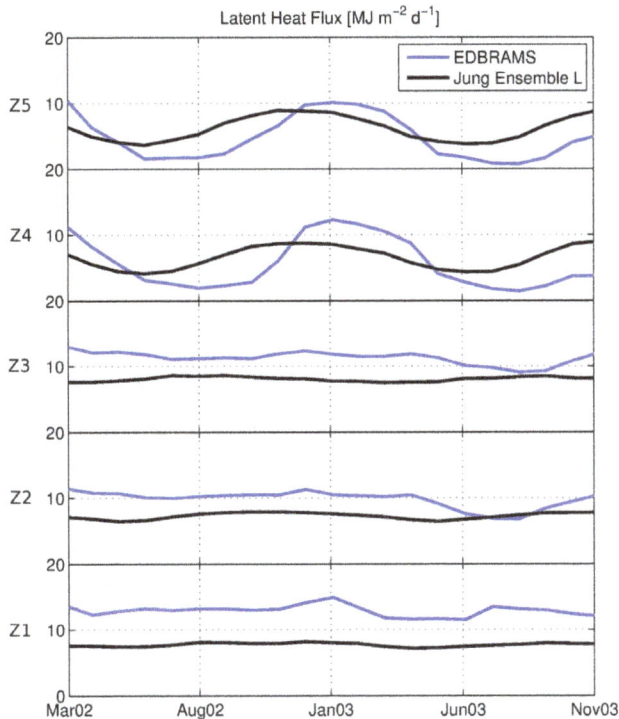

Figure E1. Mean monthly surface-to-atmosphere latent heat flux from ED2-BRAMS and the synthesis product from Jung et al. (2011, 2009). Spatial means are calculated over the zones shown in Fig. B2.

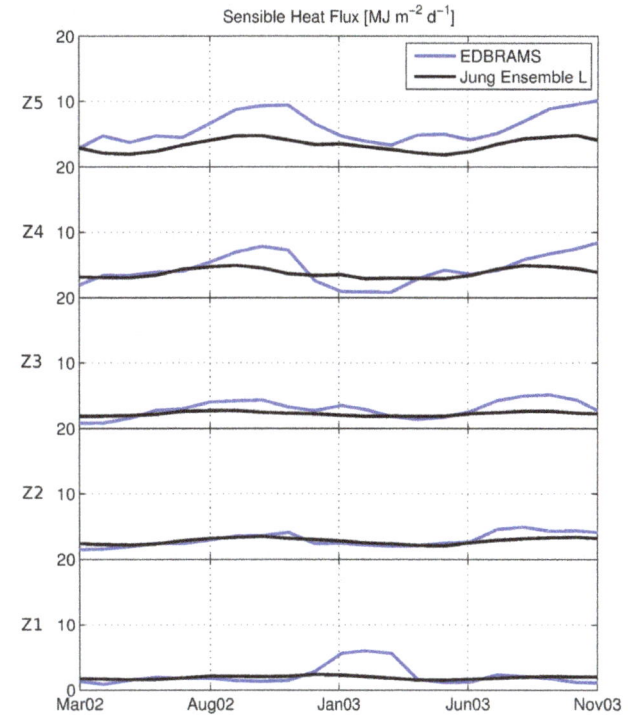

Figure E2. Mean monthly surface-to-atmosphere sensible heat flux from ED2-BRAMS and the synthesis product from Jung et al. (2011, 2009). Spatial means are calculated over the zones shown in Fig. B2.

Appendix E: Inter-seasonal latent and sensible heat flux

Similar to the comparison in Appendix D, monthly mean model estimated latent (see Fig. E1) and sensible heat flux (see Fig. E2) are compared with means from a benchmark. In this case, we compare output with the products of Jung et al. (2011). These products are based off of surface observations, which are upscaled using gridded explanatory variables from various sources including the Climate Research Unit (CRU), Global Precipitation Climatology Centre (GPCC) as well as the ERA-Interim product used in this study. The inter-seasonal means also incorporate spatial means within the domains shown in Fig. B2.

The purpose of this comparison is to give some benchmark of the ecosystem model's ability to partition energy flux at the land surface. The patterned flux of energy is dependant on the atmospheric model and its lateral boundary conditions as well. Compared to the synthesis product, the model system estimated a stronger inter-seasonal variability of latent and sensible heat flux over southern Brazil and the South American Convergence zone. Both model and observations showed relatively lower seasonal variability over the Amazon. In these regions, the model estimated small biases in sensible heat flux.

The greatest discrepancies between benchmark and model output was a latent heat flux bias (higher in the EDBRAMS

model) among the three zones covering the Amazon Basin. There are several possible explanations for this, outside of uncertainty inherent in the benchmark product. Latent and sensible heat flux contribute a portion of the total energy flux balance through the land surface, which also includes contributions from change in storage, diffusive ground heat flux, net radiation and the enthalpy contained in the mass flux of precipitation and runoff (enthalpy flux from density and pressure changes can be assumed near zero). Latent heat flux also contributes to a portion of the water mass balance at the land surface, which also includes precipitation mass flux, change in storage and total runoff. During previous experiments with the ED2 model used in offline simulations, we found that the surface water balance was sensitive to the scale of the precipitation input. When driving the land-surface model with precipitation resolved at coarse scales (such as native NCEP and ECMWF products), the leaf evaporation rates were disproportionately high. It was found that low but continuous precipitation rates from these products promoted a slow wetting of canopy leaves, and as a result the canopy leaves overflowed to the point of generating throughfall with less frequency and magnitude. Point scale precipitation rates from rain gauges in the Amazon showed a much larger variability in precipitation intensity. After using downscaling routines based on Lammering and Dwyer (2000) to preserve the monthly volume of grid-cell precipitation and creating point-scale precipitation intensity, these products (specifically the DS314 from UCAR) elicited a shift in canopy throughfall rates thereby decreasing latent heat flux and increasing surface runoff. Precipitation scale and how it affects the distribution of intensity, storm duration and the time between storms is a challenge in couple model simulation, that cannot be overcome using the same techniques as offline simulations. The spatial and temporal resolution of the simulations used in this work (40 km with 15 min time step between convective precipitation calls) are smaller than the reanalysis models (larger than 1°), yet they cannot generate point-scale precipitation. There are various approaches to ameliorating precipitation scale effects, such as using multi-atmosphere multi-land (MAML) sub-grid methods and by employing creative ways at the land-surface to generate throughfall volumes that match observations even when driven with precipitation rates that cannot match those that are observed. Regional and mesoscale couple simulations such as the work presented here, could benefit greatly from advances in this area.

Appendix F: All-sky cloud water content profiles

Cloud profile validation data sets were constructed from CloudSat cloud water content (2B-CWC-RO) and cloud classification (2B-CLDCLASS-LIDAR) data sets.[4] Overpasses during February 2007–2011 that intersected the geographic subset between 3° N–12° S and 70–55° W were collected and interpolated to a constant vertical datum above the surface. Overpasses typically occurred near 17:00 UTC.

Making a rigorous comparison of the model estimated cloud water and observation is challenging. Consider that the simulation time frame does not overlap with the CloudSat mission time frame, so these comparisons are treated as proxies to climatology and not weather validation. CloudSat measurements are known to have signal loss, attenuation and clutter during moderate to intense rainfall; events such as these could not be filtered from the comparison. It must also be assumed that the cloud classification algorithm is not without error. Nonetheless, the purpose of the comparison was to get a sense of whether the simulations estimated reasonable mean ranges of water contents and cloud fractions, and also if the phase transitions (liquid to ice) were occurring at reasonable elevations.

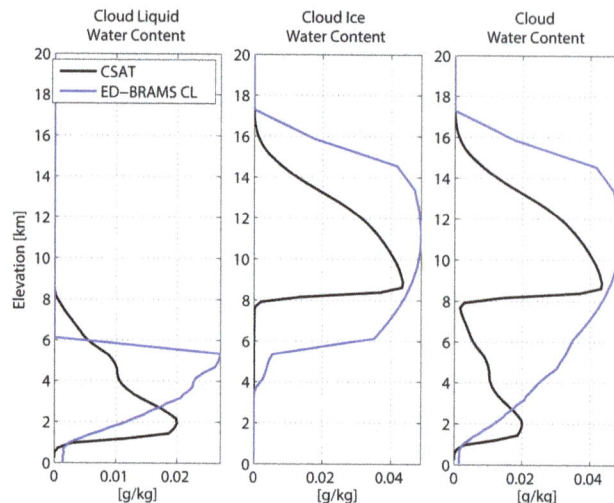

Figure F1. CloudSat climatological water content profiles and model estimated water content profiles for February 2003, 17:00 UTC, 3° N–12° S and 70–55° W.

The all-sky cloud water content profiles for both cumulus and non-cumulus clouds are provided in Fig. F1. The peaks in model estimated mean cloud water content showed reasonable agreement across liquid and ice cloud types. The model estimated generally more water content in both phases, skewed towards higher altitudes and showed a unimodal shape in the vertical distribution. It is possible that CloudSat relative underestimation could be explained by the omission of precipitating clouds.

Acknowledgements. This work was made possible through both the National Science Foundation Grant ATM-0449793 and National Aeronautics and Space Administration Grant NNG06GD63G. The authors would like to thank D. Entekhabi and E. A. B. Eltahir for their generous discussion and council. Further, the peer reviewers provided us with thorough, challenging and fair recommendations. We believe their efforts have contributed to significant improvements to this manuscript and we thank them.

Edited by: P. Regnier

References

Albani, M., Medvigy, D., Hurtt, G. C., and Moorcroft, P. R.: The contributions of land-use change, CO_2 fertilization, and climate variability to the Eastern US carbon sink, Global Change Biol., 12, 2370–2390, 2006.

Anthes, R. A.: Enhancement of convective precipitation by mesoscale variations in vegetative covering in semiarid regions, J. Clim. Appl. Meteorol., 23, 541–554, 1984.

---

[4]CloudSat data sets were provided on-line by CloudSat Data Processing Center, courtesy of NASA, Colorado State University and their partners.

Asdak, C., Jarvis, P. G., van Gardingen, P., and Fraser, A.: Rainfall interception loss in unlogged and logged forest areas of Central Kalimantan, Indonesia, J. Hydrol., 206, 237–244, 1998.

Avissar, R. and Werth, D.: Global hydroclimatological teleconnections resulting from tropical deforestation, J. Hydrometeorol., 6, 134–146, 2005.

Bagley, J., Desai, A., Harding, K., Synder, P., and Foley, J.: Drought and deforestation: Has land cover change influenced recent precipitation extremes in the Amazon?, J. Climate, 27, 345–361, 2014.

Baker, T., Phillips, O., Malhi, Y., Almeida, S., Arroyo, L., Fiore, A. D., Erwin, T., amd T.J. Killeen, N. H., Laurance, S., Laurance, W., Lewis, S., Monteagudo, A., Neill, D., Vargas, P., Pitman, N., Silva, N., and Vasquez-Martinez, R.: Increasing biomass in Amazonian forest plots, Philos. T. Roy. Soc. Lond. B, 359, 353–365, 2004a.

Baker, T., Phillips, O., Malhi, Y., Almeida, S., Arroyo, L., Fiore, A. D., Erwin, T., Killeen, S., Laurance, S., Laurance, W., Lewis, S., Lloyd, J., Monteagudo, A., Neill, D., Patino, S., Pitman, N., Silva, N., and Martinez, R. V.: Variation in wood density determines spatial patterns in Amazonian forest biomass, Global Change Biol., 10, 545–562, 2004b.

Baldi, M., Dalu, G. A., and Pielke, R. A.: Vertical velocities and available potential energy generated by landscape variability – theory, J. Appl. Meteorol. Clim., 47, 397–410, 2008.

Beljaars, A. C. M. and Holtslag, A. A. M.: Flux Parameterization over Land Surfaces for Atmospheric Models, J. Appl. Meteorol., 30, 327–341, 1991.

Benegas, L., Ilstedt, U., Roupsard, O., Jones, J., and Malmer, A.: Effects of trees on infiltrability and preferential flow in two contrasting agroecosystems in Central America, Agr. Ecosyst. Environ., 183, 185–196, 2014.

Cardille, J. and Foley, J.: Agricultural Land-use Change in Brazilian Amazonia Between 1980 and 1995: Evidence from Integrated Satellite and Census Data, Remote Sens. Environ., 87, 551–562, 2003.

Chapin, F., Matson, P., and Mooney, H.: Principles of Terrestrial Ecosystem Ecology, Springer-Verlag, New York, 2002.

Chappell, N., Bidin, K., and Tych, W.: Modelling rainfall and canopy controls on net-precipitation beneath selectively-logged tropical forest, Plant Ecology, 153, 215–229, 2001.

Chave, J., Riera, B., and Dubois, M.: Estimation of biomass in a neotropical forest of French Guiana: spatial and temporal variability, J. Trop. Ecol., 17, 79–96, 2001.

Chave, J., Muller-Landau, H. C., Baker, T. R., Easdale, T. A., Ter Steege, H., and Webb, C. O.: Regional and phylogenetic variation of wood density across 2456 neotropical tree species, Ecol. Appl., 16, 2356–2367, 2006.

Chen, C. and Cotton, W.: A One-Dimensional Simulations of the Stratocumulus-Capped Mixed Layer, Bound.-Lay. Meteorol., 25, 289–321, 1983.

Collatz, G. J., Ball, J., Grivet, C., and Berry, J. A.: Physiological and environmental regulation of stomatal conductance, photosynthesis and transpiration: a model that includes a laminar boundary layer, Agr. Forest Meteorol., 54, 107–136, 1991.

Collatz, G. J., Ribas-Carbo, M., and Berry, J.: Coupled Photosynthesis-Stomatal Conductance Model for Leaves of C_4 Plants, Aust. J. Plant Physiol., 19, 519–538, 1992.

Cotton, W., Pielke, R., Walko, R., Liston, G., Tremback, C., Jiang, H., McAnelly, R., Harrington, J., Nicholls, M., Carrio, G., and McFadden, J.: RAMS 2001: Current status and future directions, Meteorol. Atmos. Phys., 82, 5–29, 2003.

Crockford, R. and Richardson, D.: Partitioning of rainfall into throughfall, stemflow and interception: effect of forest type, ground cover and climate, Hydrol. Process., 14, 2903–2920, 2000.

d'Almeida, C., Vörösmarty, C. J., Hurtt, G. C., Marengo, J. A., Dingman, S. L., and Keim, B. D.: The effects of deforestation on the hydrological cycle in Amazonia: a review on scale and resolution, Int. J. Climatol., 27, 633–647, 2007.

Dalu, G. A., Pielke, R. A., Baldi, M., and Zeng, X.: Heat and momentum fluxes induced by thermal inhomogeneities, J. Atmos. Sci., 53, 3286–3302, 1996.

Dee, D. P., Uppala, S. M., Simmons, A. J., Berrisford, P., Poli, P., Kobayashi, S., Andrae, U., Balmaseda, M. A., Balsamo, G., Bauer, P., Bechtold, P., Beljaars, A. C. M., van de Berg, L., Bidlot, J., Bormann, N., Delsol, C., Dragani, R., Fuentes, M., Geer, A. J., Haimberger, L., Healy, S. B., Hersbach, H., Hólm, E. V., Isaksen, L., Kållberg, P., Köhler, M., Matricardi, M., McNally, A. P., Monge-Sanz, B. M., Morcrette, J.-J., Park, B.-K., Peubey, C., de Rosnay, P., Tavolato, C., Thépaut, J.-N., and Vitart, F.: The ERA-Interim reanlysis: configuration and performance of the data assimilation system, Q. J. Roy. Meteorol. Soc., 137, 553–597, 2011.

Dickinson, R. and Henderson-Sellers, A.: Modeling Tropical Deforestation: A Study of GCM Land-Surface Parameterization, Q. J. Roy. Meteorol. Soc., 114, 439–462, 1988.

Dietz, J., Hoelscher, D., Leuschner, C., and Hendrayanto: Rainfall partitioning in relation to forest structure in differently managed montane forest stands in Central Sulawesi, Indonesia, Forest Ecol. Manage., 237, 170–178, 2006.

Dietze, M., Wolosin, M., and Clark, J.: Capturing diversity and inerspecific variability in allometries: A hierarchical approach, Forest Ecol. Manage., 256, 1939–1948, 2008.

Eltahir, E.: Role of Vegetation in Sustaining Large-Scale Atmospheric Circulations in the Tropics, J. Geophys. Res.-Atmos., 101, 4255–4268, 1996.

Eltahir, E. and Bras, R.: On the Response of the Tropical Atmosphere to Large-Scale Deforestation, Q. J. Roy. Meteorol. Soc., 119, 779–793, 1993.

Eltahir, E. A. B. and Bras, R.: Precipitation recyling in the Amazon basin, Q. J. Roy. Meteorol. Soc., 120, 861–880, 1994.

Fisch, G., Tota, J., Machado, L., Dias, M., Lyra, R., Nobre, C., Dolman, A., and Gash, J.: The convective boundary layer over pasture and forest in Amazonia, Theor. Appl. Climatol., 78, 47–59, 2004.

Freitas, S. R., Rodrigues, L. F., Longo, K. M., and Panetta, J.: Impact of a monotonic advection scheme with low numerical diffusion on transport modeling of emissions from biomass burning, J. Adv. Model. Earth Syst., 4, Q1, doi:10.1029/2011MS000084, 2012.

Gedney, N. and Valdes, P.: The effect of Amazonian deforestation on the northern hemisphere circulation and climate, Geophys. Res. Lett., 27, 3053–3056, 2000.

Geist, H. and Lambin, E.: Proximate Causes of Underlying Driving Forces of Tropical Deforestation, Bioscience, 52, 143–150, 2002.

Grell, G. A. and Dévényi, D.: A generalized approach to parameterizing convection combining ensemble and data assimilation techniques, Geophys. Res. Lett., 29, 38-1–38-4, 2002.

Harrington, J. and Olsson, P.: A Method for the Parameterization of Cloud Optical Properties in Bulk and Bin Microphysical Models. Implications for Arctic Cloud Boundary Layers, Atmos. Res., 57, 51–80, 2001.

Henderson-Sellers, A., Dickinson, R., Durbridge, T., Kennedy, P., McGufie, K., and Pitman, A.: Tropical Deforestation: Modeling Local to Regional Scale Climate Change, J. Geophys. Res., 98, 7289–7315, 1993.

Hurtt, G., Frolking, S., Fearon III, M. B. M., Shevialokova, E., Malyshew, S., Pacala, S., and Houghton, R.: The underpinnings of land-use history: three centuries of global gridded land-use transitions, wood harvest activity and resulting secondary lands, Global Change Biol., 12, 1–22, 2006.

INPE: Monitoring of the Amazon forest by satellite 2001–2002, Instituto Nacional de Pesquisas Espaciais, Technical Paper, Sao Jose Dos Campos, Brazil, 2003.

Jung, M., Reichstein, M., and Bondeau, A.: Towards global empirical upscaling of FLUXNET eddy covariance observations: validation of a model tree ensemble approach using a biosphere model, Biogeosciences, 6, 2001–2013, doi:10.5194/bg-6-2001-2009, 2009.

Jung, M., Reighstein, M., Margolis, H., Cescatti, A., Richardson, A., Arain, M. A., Arneth, A., Bernhofer, C., Bonal, D., Chen, J., Gianelle, D., Gobron, N., Kiely, G., Kutsch, W., Lasslop, G., Law, B., Lindroth, A., Merbold, L., Montagnani, L., Moors, E. J., Papale, D., Sottoconrola, M., Vaccari, F., and Williams, C.: Global Patterns of Land-Atmosphere Fluxes of Carbon Dioxide, Latent Heat, and Sensible Heat Derived from Eddy Covariance, Satellite and Meteorological Observations, J. Geophys. Res., 116, 1–16, 2011.

Kain, J.: The Kain-Fritsch Convective Parameterization: An Update, J. Appl. Meteorol., 43, 170–181, 2004.

Kain, J. and Fritsch, J.: A One-Dimensional Entraining-Detraining Plume Model and Its Application in Convective Parameterization, J. Atmos. Sci., 47, 2784–2802, 1990.

Kennard, D. and Gholz, H.: Effects of high- and low-intensity fires on soil properties and plant growth in a Bolivian dry forest, Plant Soil, 234, 119–129, 2001.

Kleidon, A. and Heimann, M.: Assessing the Role of Deep Rooted Vegetation in the Climate System with Model Simulations: Mechanism, Comparison to Observations and Implications for Amazonian Deforestation, Clim. Dynam., 16, 183–199, 2000.

Knox, R.: Land Conversion in Amazonia and Northern South America; Influences on Regional Hydrology and Ecosystem Response, PhD thesis, Massachusetts Institute of Technology, Cambridge, Massachusetts, USA, 2012.

Knox, R., Bisht, G., Wang, J., and Bras, R.: Precipitation Variability over the Forest-to-Nonforest Transition in Southwestern Amazonia, J. Climate, 24, 2368–2377, 2011.

Lal, R.: Deforestation and land-use effects on soil degradation and rehabilitation in western Nigeria, 1. Soil physical and hydrological properties, Land Degred. Develop., 7, 19–45, 1996.

Lammering, B. and Dwyer, I.: Improvement of Water Balance in Land Surface Schemes by Random Cascade Disaggregation of Rainfall, Int. J. Climatol., 20, 681–695, 2000.

Laurance, W., Cochrane, M., Bergen, S., Fearnside, P., Delamonica, P., Barber, C., D'Angelo, S., and Fernandes, T.: The future of the Brazilian Amazon, Science, 291, 438–439, doi:10.1126/science.291.5503.438 2001.

Laurance, W., Nascimento, H., Laurance, S., Condit, R., D'Angelo, S., and Andrade, A.: Inferred longevity of Amazonian rainfor-

est trees based on a long-term demographic study, Forest Ecol. Manage., 190, 131–143, 2004.

Lean, J. and Warrilow, D.: Simulation of the Regional Climatic Impact of Amazon Deforestation, Nature, 342, 411–413, 1989.

Lee, T. and Pielke, R.: Estimating the Soil Surface Specific-Humidity, J. Appl. Meteorol., 31, 480–484, 1992.

Leuning, R.: A critical appraisal of a combined stomatal-photosynthesis model for C3 plants, Plant Cell Environ., 18, 339–355, 1995.

Lewis, S.: Tropical forests and the changing earth system, Philos. T. Roy. Soc. B, 361, 195–210, 2006.

Martinez, L. and Zinck, J.: Temporal variation of soil compaction and deterioration of soil quality in pasture areas of Colombian Amazonia, Soil Till. Res., 75, 3–17, 2004.

Massman, W.: An Analytical One-Dimensional Model of Momentum Transfer by Vegetation of Arbitrary Structure, Bound.-Lay. Meteorol., 83, 407–421, 1997.

Medvigy, D., Wofsy, S., Munger, J., Hollinger, D., and Moorcroft, P.: Mechanistic scaling of ecosystem function and dynamics in space and time: Ecosystem Demography model version 2, J. Geophys. Res., 114, 1–21, 2009.

Medvigy, D., Walko, R., Otte, M., and Avissar, R.: The Ocean-Land-Atmosphere-Model: Optimization and Evaluation of Simulated Radiative Fluxes and Precipitation, Mon. Weather Rev., 138, 1923–1939, 2010.

Medvigy, D., Walko, R., and Avissar, R.: Effects of Deforestation on Spatiotemporal Distributions of Precipitation in South America, J. Climate, 24, 2147–2163, 2011.

Moorcroft, P., Hurtt, G., and Pacala, S.: A Method for Scaling Vegetation Dynamics: The Ecosystem Demography Model, Ecol. Monogr., 71, 557–586, 2001.

Muñoz-Villers, L. E. and McDonnell, J. J.: Land use change effects on runoff generation in a humid tropical montane cloud forest region, Hydrol. Earth Syst. Sci., 17, 3543–3560, doi:10.5194/hess-17-3543-2013, 2013.

Nakanishi, M. and Niino, H.: An improved Mellor-Yamada level-3 model with condensation physics: its numerical stability and application to a regional prediction of advection fog, Bound.-Lay. Meteorol., 119, 397–407, 2006.

Nepstad, D., de Carvalho, C., Davidson, E., Jipp, P., Lefebvre, P., Negreiros, H., dal Silva, E., Stone, T., Trubore, S., and Vieira, S.: The Role of Deep Roots in the Hydrological and Carbon Cycles of Amazonian Forests and Pastures, Nature, 372, 666–669, 1994.

Nepstad, D., Carvalho, G., Barros, A., Alencar, A., Capobianco, J., amd P. Moutinho, J. B., Lefebvre, P., Silva, U. L., and Prins, E.: Road paving, Fire Regime Feedbacks and the Future of Amazon Forests, Forest Ecol. Manage., 154, 395–407, 2001.

Nobre, C., Sellers, P., and Shukla, J.: Amazonian Deforestation and Regional Climate Change, J. Climate, 4, 957–988, 1991.

Pielke, R.: Influence of the spatial distribution of vegetation and soils on the prediction of cumulus convective rainfall, Rev. Geophys., 39, 151–171, 2001.

Pitman, A., Henderson-Sellers, A., and Yang, Z.: Sensitivity of Regional Climates to Localized Precipitation in Global-Models, Nature, 346, 734–737, 1990.

Poorter, L., Bongers, L., and Bongers, F.: Architecture of 54 moist-forest tree species: traits, trade-offs and functional groups, Ecology, 87, 1289–1301, 2006.

Quesada, C. A., Lloyd, J., Anderson, L. O., Fyllas, N. M., Schwarz, M., and Czimczik, C. I.: Soils of Amazonia with particular reference to the RAINFOR sites, Biogeosciences, 8, 1415–1440, doi:10.5194/bg-8-1415-2011, 2011.

Raupach, M., Finnigan, J., and Brunet, Y.: Coherent eddies and turbulence in vegetation canopies: The mixing-layer analogy, Bound.-Lay. Meteorol., 78, 351–382, 1996.

Rossato, L.: Estimativa da capacidade de armazenamento de água no solo do Brasil, Msc. thesis, Instituto Nacional de Pesquisas Espaciais (INPE), São José dos Campos, Brazil, 2001.

Scholes, R., Skole, D., and (eds.), J. I.: A Global Database of Soil Properties: Proposal for Implementation, Report of the Global Soils Task Group, Tech. Rep. IGBP-DIS Working Paper 10a, International Geosphere-Biosphere Programme – Data and Information System (IGBP-DIS), University of Paris, Paris, France, 1995.

Sheffield, J., Goteti, G., and Wood, E.: Development of a 50-Year High-Resolution Global Dataset of Meteorological Forcings for Land Surface Modeling, J. Climate, 19, 3088–3111, 2006.

Silva, R. R. D., Werth, D., and Avissar, R.: Regional impacts of future land-cover changes on the Amazon basin wet-season climate, J. Climate, 21, 1153–1170, 2008.

Skole, D. and Tucker, C.: Tropical Deforestation and Habitat Fragmentation in the Amazon: Satellite Data from 1978 to 1988, Science, 260, 1905–1910, 1993.

Snyder, P. K.: The Influence of Tropical Deforestation on the Northern Hemisphere Climate by Atmospheric Teleconnections, Earth Interact., 14, 1–34, doi:10.1175/2010EI280.1, 2010.

Soares-Filho, B. S., Nepstad, D. C., Curran, L. M., Cerqueira, G. C., Garcia, R. A., Ramos, C. A., Voll, E., McDonald, A., Lefebvre, P., and Schlesinger, P.: Modelling conservation in the Amazon basin, Nature, 440, 520–523, 2006.

Souza, E. P., Renno, N. O., and Silva-Dias, M. A. F.: Convective circulations induced by surface heterogeneities, J. Atmos. Sci., 57, 2915–2922, 2000.

Tremback, C. and Kessler, R.: A surface temperature and moisture parameterization for use in mesoscale models, Preprints, Seventh Conf. on Numerical Weather Prediction, Montreal, PQ, Canada, Amer. Meteor. Soc., 355–358, 1985.

Walcek, C. and Aleksic, N.: A simple but accurate mass conservative, peak-preserving, mixing ratio bounded advection algorithm with Fortran code, Atmos. Environ., 32, 3863–3880, 1998.

Walker, R., Moore, N., Arima, E., Perz, S., Simmons, C., Caldas, M., Vergara, D., and Bohrer, C.: Protecting the Amazon with protected areas, P. Natl. Acad. Sci. USA, 106, 10582–10586, 2009.

Walko, R., Band, L., Baron, J., Kittel, T., Lammers, R., Lee, T., Ojima, D., Pielke, R., Taylor, C., Tague, C., Tremback, C., and Vidale, P.: Coupled atmosphere-biophysics-hydrology models for environmental modeling, J. Appl. Meteorol., 39, 931–944, 2000.

Wang, D., Wang, G., and Anagnostou, E. N.: Evaluation of canopy interception schemes in band surface models, J. Hydrol., 347, 308–318, 2007.

Wang, J., Bras, R., and Eltahir, E.: The Impact of Observed Deforestation on the Mesoscale Distribution of Rainfall and Clouds in Amazonia, J. Hydrometeorol., 1, 267–286, 2000.

Wang, J., Chagnon, F., Williams, E., Betts, A., Renno, N., Machado, L., Bisht, G., Knox, R., and Bras, R.: The impact of deforestation in the Amazon basin on cloud climatology, P. Natl. Acad. Sci., 106, 3670–3674, 2009.

Williams, E. and Renno, N.: An Analysis of the Conditional Stability of the Tropical Atmosphere, Mon. Weather Rev., 121, 21–36, 1993.

Zhao, W. and Qualls, R. J.: A multiple-layer canopy scattering model to simulate shortwave radiation distribution within a homogeneous plant canopy, Water Resour. Res., 41, W08409, doi:10.1029/2005WR004016, 2005.

Zhao, W. and Qualls, R. J.: Modeling of long-wave and net radiation energy distribution within a homogeneous plant canopy via multiple scattering processes, Water Resour. Res., 42, W08436, doi:10.1029/2005WR004581, 2006.

Zimmermann, B., Elsenbeer, H., and De Moraes, J.: The influence of land-use changes on soil hydraulic properties: Implications for runoff generation, Forest Ecol. Manage., 222, 29–38, 2006.

A conceptual prediction model for seasonal drought processes using atmospheric and oceanic standardized anomalies: application to regional drought processes in China

Zhenchen Liu[1], Guihua Lu[1], Hai He[1], Zhiyong Wu[1], and Jian He[2]

[1]Institute of Water Problem, College of Hydrology and Water Resources, Hohai University, Nanjing, China

[2]Hydrology and Water Resources Investigation Bureau of Jiangsu Province, Nanjing, China

Correspondence: Hai He (hehai_hhu@hhu.edu.cn)

Abstract. Reliable drought prediction is fundamental for water resource managers to develop and implement drought mitigation measures. Considering that drought development is closely related to the spatial–temporal evolution of large-scale circulation patterns, we developed a conceptual prediction model of seasonal drought processes based on atmospheric and oceanic standardized anomalies (SAs). Empirical orthogonal function (EOF) analysis is first applied to drought-related SAs at 200 and 500 hPa geopotential height (HGT) and sea surface temperature (SST). Subsequently, SA-based predictors are built based on the spatial pattern of the first EOF modes. This drought prediction model is essentially the synchronous statistical relationship between 90-day-accumulated atmospheric–oceanic SA-based predictors and SPI3 (3-month standardized precipitation index), calibrated using a simple stepwise regression method. Predictor computation is based on forecast atmospheric–oceanic products retrieved from the NCEP Climate Forecast System Version 2 (CFSv2), indicating the lead time of the model depends on that of CFSv2. The model can make seamless drought predictions for operational use after a year-to-year calibration. Model application to four recent severe regional drought processes in China indicates its good performance in predicting seasonal drought development, despite its weakness in predicting drought severity. Overall, the model can be a worthy reference for seasonal water resource management in China.

1 Introduction

Drought is an economically and ecologically disruptive natural hazard that profoundly impacts water resources, agriculture, ecosystems, and basic human welfare (Dai, 2011). In recent years, extreme drought events have had disastrous impacts worldwide. The 2011 eastern African drought led to famine and severe food crises in several countries, affecting over 9 million people (Funk, 2011). As part of the 2011–2014 California Drought, the drought in 2014 alone cost California USD 2.2 billion in damages and 17 000 agricultural jobs (Howitt et al., 2014). China has also suffered from extreme drought events, such as the 2009–2010 severe drought in southwestern China (Yang et al., 2012), 2011 spring drought in the Yangtze River basin (Lu et al., 2014), and 2014 summer drought in northern China (Wang and He, 2015). Because drought is a costly and disruptive natural hazard, reliable drought prediction is fundamental for water resource managers to develop and implement feasible drought mitigation measures. In the present study, drought prediction is restricted to relatively long-term drought, which is associated with season-scale precipitation deficits.

Drought is generally predicted using two types of methods: model-based dynamical forecasting and statistical prediction. Dynamical forecasting primarily relies on computed drought indicators, such as the standardized precipitation index (SPI; McKee and Kleist, 1993), based on forecast precipitation retrieved from seasonal climate forecast systems (Dutra et al., 2013, 2014; Mo and Lyon, 2015; Yoon et al., 2012). Although dynamically predicted precipitation is useful information for drought situations, especially for short-term forecasting 1 month ahead, it also contains high levels of uncertainty and limited skill with respect to long lead times (Wood et al., 2015; Yoon et al., 2012; Yuan et al., 2013). In contrast, statistical drought prediction is an additional source of prospective drought information (Behrangi et al., 2015; Hao et al., 2014). Different from the physical, complex processes in coupled atmosphere–ocean models used for dynamical prediction, statistical drought prediction models are relatively simple but also perform well. They consist of input variables, methodology, and prediction targets (Mishra and Singh, 2011).

Reasons for good and effective performance of statistical models include methodology improvements and drought-related climate indices used as input variables. To date, much attention has been paid to methodology improvements. Taking advantage of probabilistic and temporal-evolution features of input variables, statistical drought prediction models are primarily forced with probability or machine-learning methods, such as the ensemble streamflow prediction (ESP) method (AghaKouchak, 2014), Markov chain- and Bayesian network-based models (Aviles et al., 2015, 2016; Shin et al., 2016), neural network, and support vector models (Belayneh et al., 2014). In addition to method improvement, climate indices represent large-scale atmospheric or oceanic drivers of precipitation, partly responsible for effective model performance. These climate indices include typical atmospheric and oceanic circulation patterns, such as the North Atlantic Oscillation (NAO; Hurrell, 1995) and El Niño–Southern Oscillation (ENSO; Ropelewski and Halpert, 1987), which have been widely used for drought prediction in different seasons and regions (Behrangi et al., 2015; Bonaccorso et al., 2015; Chen et al., 2013; Mehr et al., 2014; Moreira et al., 2016).

Climate indices, such as the NAO index and NINO 3.4 index, are simple, explicit, and widely used. Therefore, they are the primary indices used for drought prediction. Additionally, based on the relationship between drought indices and potential atmospheric or oceanic circulation patterns, some researchers have also discovered large-scale circulation patterns closely related to regional droughts or have structured new drought predictors (Funk et al., 2014; Kingston et al., 2015). For instance, after discovering the two dominant modes of the eastern African boreal spring rainfall variability that are tied to SST fluctuations, Funk et al. (2014) further determined that the first- and second-mode SST correlation structures were related to two SST indices that could be used to predict eastern African spring droughts.

Similarly, potential atmospheric and oceanic circulation patterns, which are closely related to regional droughts, are also used to construct drought predictors in the present study. Considering that the development of drought processes is closely related to the spatial–temporal evolution of large-scale circulation patterns, we constructed predictors based on anomalous spatial patterns. Because precipitation-inducing circulation patterns usually occur in the troposphere, predictors can be built based on sea surface temperature (SST) and 200 and 500 hPa geopotential height (HGT), reflecting information from different levels of the troposphere. Subsequently, all these predictors from different drought processes and the 3-month SPI, updated daily (hereafter, SPI3), were used to calibrate a synchronous stepwise-regression relationship. The model can be forced with dynamically forecast SST and 200 and 500 hPa HGT conditions, indicating that the lead time depends on that of the climate forecast models. Based on predicted prospective 90-day SPI3 curves, we developed angle-based rules for the drought outlook, which can make the drought outlook easily accessible to water resource managers.

Overall, the objective of this study is to build a conceptual prediction model of seasonal drought processes. The essential and important steps are to (1) structure predictors on the basis of drought-related atmospheric and oceanic circulation patterns, (2) build the synchronous statistical predictor-SPI3 relationship forced with reanalysis and operationally forecast datasets, (3) simulate and predict four severe seasonal drought processes in China to investigate model performance, and (4) propose an objective angle-based method for drought outlook.

Considering the proposed conceptual model consists of several important parts, a brief but general introduction with sequential procedures is presented (Fig. 1), prior to specific descriptions in Sects. 3–8. In Sect. 3, historical extreme and severe drought processes are identified with SPI3. These drought processes usually go through one or several dry and wet spells, in which precipitation deficit characteristics and circulation patterns vary. Therefore, process-split rules, according to dry and wet spells, are designed to assign drought process segments to different dry and wet spells in Sect. 4. Gridded values in the fields of 200 and 500 hPa HGT and SST are transformed into gridded values of standardized anomalies (SAs) in Sect. 5. Maps of atmospheric–oceanic SAs during drought process segments within the same dry and wet spells are important inputs to the construction of the predictors. After empirical orthogonal function (EOF) analyses are conducted on these SA-based maps, the first leading EOF modes are used to generate predictors (Sect. 5). Further, synchronous statistical relationships between SA-based predictors and SPI3 are calibrated with the stepwise regression method in Sect. 6. The National Centers for Environmental Prediction/National Center for Atmospheric Research (NCEP/NCAR) reanalysis datasets and the NCEP Climate Forecast System Version 2 (CFSv2) operational forecast datasets are used to force the synchronous statistical relationship. Simulated and predicted 90-day prospective SPI3 time series are presented in Sect. 7. With the aid of angle-based rules for seasonal drought outlook, simulated and predicted SPI3 time series are transformed to five types of

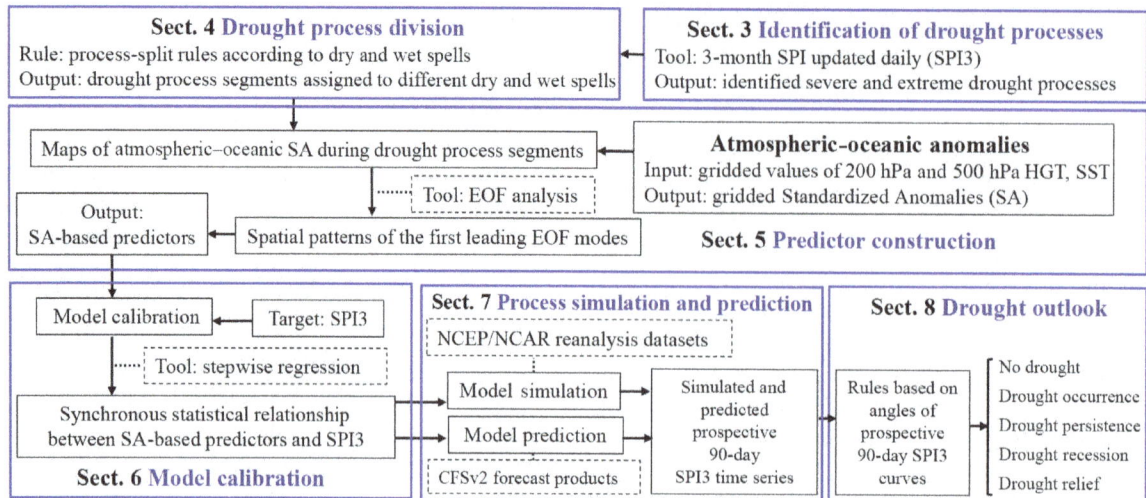

Figure 1. A brief introduction of the sequential procedures described in the sections of this study for drought prediction model construction.

drought outlooks (Sect. 8), which are easily accessible to water resource managers.

In particular, although drought process predictions in northern, eastern, and southwestern China are all the targets in the present study, only the historical drought processes in northern China are used to introduce the model construction and calibration in Sect. 3–6. Similar procedures were also applied to drought processes in eastern and southwestern China. However, for the sake of conciseness, these procedures, together with intermediate results, are not shown in this study.

2   Data

The precipitation data used were the second-version Dataset of Observed Daily Precipitation Amounts at each $0.5° \times 0.5°$ grid point in China for 1961–2014 (http://data.cma.cn/data/detail/dataCode/SURF_CLI_CHN_PRE_DAY_GRID_0.5.html), which was kindly provided by the Climate Data Centre (CDC) of the National Meteorological Information Centre, China Meteorological Administration (CMA). It was initially used to calculate area-averaged precipitation over northern China, eastern China, and southwestern China (Fig. 2), which are the three Chinese drought regions investigated in this study. They cover areas of approximately 0.69, 0.91, and 1.12 million km^2, respectively. Atmospheric anomalies were diagnosed with respect to the NCEP/NCAR reanalysis datasets, which has a resolution of $2.5° \times 2.5°$ at 17 pressure levels, extending from January 1948 to the present (Kalnay et al., 1996). The National Oceanic and Atmospheric Administration (NOAA) high-resolution SST dataset, with a spatial resolution of $0.25° \times 0.25°$ and extending from September 1981 to present (Reynolds et al., 2007), was used for SST anomaly analysis.

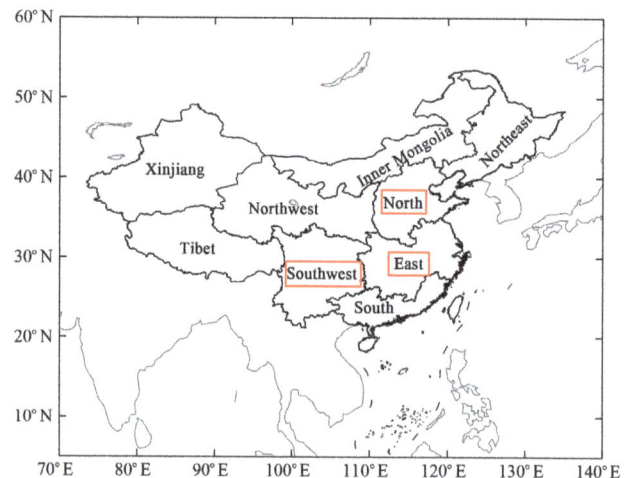

Figure 2. The geographical distribution of China's nine drought regions (black solid curves). The three regions labeled with red boxes are the focus in the present study.

The NCEP Climate Forecast System Version 2 (CFSv2; Saha et al., 2014) was used to verify operational performance of the proposed conceptual model. Since CFSv2 began on 1 April 2011, some drought processes that occurred before this date were forced with the CFS reforecast output. All the relevant reforecast and forecast datasets are accessible on the website (https://nomads.ncdc.noaa.gov/modeldata/). In particular, we focus on the prospective 90-day seasonal drought process prediction during four severe drought processes in this study. To achieve this, prospective 90-day forecast data subsets for 200 and 500 hPa HGT and SST are retrieved from CFSv2 and CFS products, which are used for the predictor calculation. Details can be found in Sect. S1 in the Supplement.

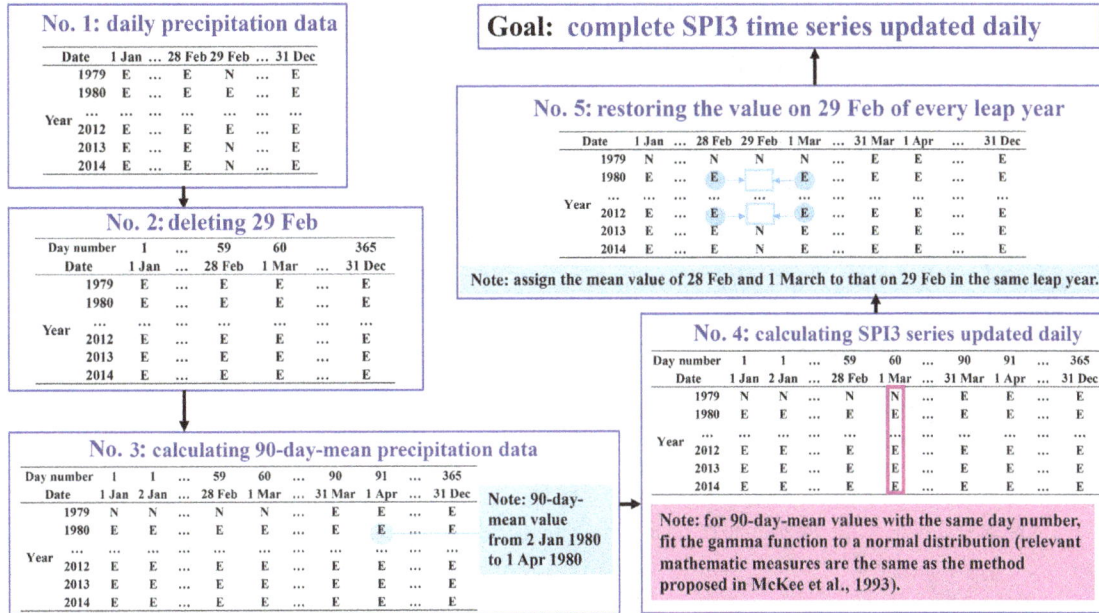

Figure 3. Illustration indicating the steps for calculating daily-updated SPI3. The letter "E" represents value existence, while the letter "N" represents no relevant data.

3 Identification of drought processes

3.1 Three-month SPI updated daily

SPI3 was used as the drought index for seasonal drought recognition and prediction in this study. The calculation period is 1979–2014. The daily area-averaged precipitation datasets were first computed over the three study regions. Traditionally, SPI3 values vary on a monthly timescale, i.e., each month a new value is determined from the precipitation totals of the previous 3 months (McKee and Kleist, 1993). In this study, we chose to update SPI3 daily, which was also recommended by the World Meteorological Organization (2012), i.e., every day a new value is determined from the precipitation totals of the previous 90 days. Specified illustration and details for calculating daily-updated SPI3 are shown in Fig. 3.

3.2 Drought process identification and grade classification

Similar to the rules for SPI grade division recommended by the World Meteorological Organization (2012), the rules in our study are shown in Table 1. Drought processes are identified when the daily SPI3 values are below −0.50 for more than 30 consecutive days.

Each daily SPI3 value for a recognized drought process was assigned to the corresponding SPI3 grade. Starting from the extremely dry grade to the slightly dry grade, the ratio between the duration of a particular SPI3 grade and the total days of the entire drought process is calculated. When the

Table 1. Rules for SPI3 grade classification.

Daily SPI3 value	Grade
0.50 and more	wet
−0.49 to 0.49	near normal
−0.99 to −0.50	slightly dry
−1.49 to −1.00	moderately dry
−1.99 to −1.50	severely dry
−2.00 and less	extremely dry

proportion increases beyond 35 %, the corresponding grade is assigned to the entire drought process. For example, as shown in Fig. 4, the proportion of the severely dry days is beyond 35 %. Accordingly, the 2001 summer drought in northern China corresponded to the severe grade.

Therefore, we identified severe and extreme drought processes for 1979–2008 in northern China. As shown in Table 2, persistent drought periods from 1997 to 2002 in northern China were found, in agreement with other associated studies (Rong et al., 2008; Wei et al., 2004). Relevant results of identified drought processes in eastern and southwestern China are not shown in the paper.

4 Drought process division according to dry and wet spells

Identified drought processes usually go through one or several dry and wet spells. Different dry and wet spells usually correspond to various precipitation deficit characteristics and

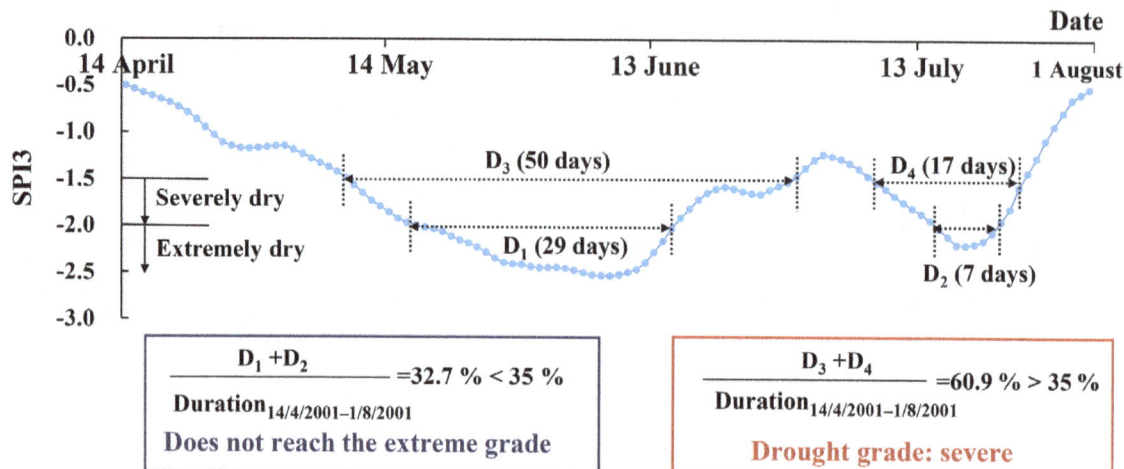

Figure 4. An example of grade classification for one complete drought process: the 2001 summer drought in northern China.

Table 2. Identified severe and extreme drought processes from 1979 to 2008 in northern China.

Extreme Drought	12 Jun 1997–28 Nov 1997
	2 Nov 1998–11 Apr 1999
Severe Drought	15 Jan 1984–14 May 1984
	9 Nov 1988–9 Jan 1989
	17 Jul 1999–1 Nov 1999
	23 Mar 2000–27 Jun 2000
	14 Apr 2001–1 Aug 2001
	3 Aug 2002–4 Dec 2002
	26 Dec 2005–2 Feb 2006

atmospheric–oceanic circulation patterns. Therefore, we divided drought processes into different segments according to dry and wet spells, in order to further analyze atmospheric–oceanic anomalies during drought segments within the same dry and wet spells. Additionally, SPI3 on the start date of an identified drought process indicates that SPI3 is initially less than −0.5 and a severe drought process indeed follows, which actually reflects drought-inducing precipitation information for the previous 90 days. Therefore, the start date of the drought process is advanced to the past 90th day, preceding the drought process division. This measure can contribute to introducing early drought-inducing information to predictor construction.

Using northern China as an example, the specified procedures for the division process are as follows. Similar to general seasonal classification, we divided the annual period into four dry and wet spells (Table 3) according to the temporal evolution of the daily precipitation rate in northern China (Fig. 5). It is evident that the wet spell (one-fourth of the annual duration) accounts for over 50 % of total precipitation, while the dry spell (one-third of the annual duration) accounts for about 6 %.

Table 3. Dates of dry and wet spells and their associated proportions of annual total precipitation in northern China. Both Wet–Dry and Dry–Wet represent corresponding transition spells.

Spell	Period	Precipitation Proportion (%)
Wet	21 June–10 September	56.4
Wet–dry	11 September–20 November	14.9
Dry	21 November–20 March	6.3
Dry–wet	21 March–20 June	22.4

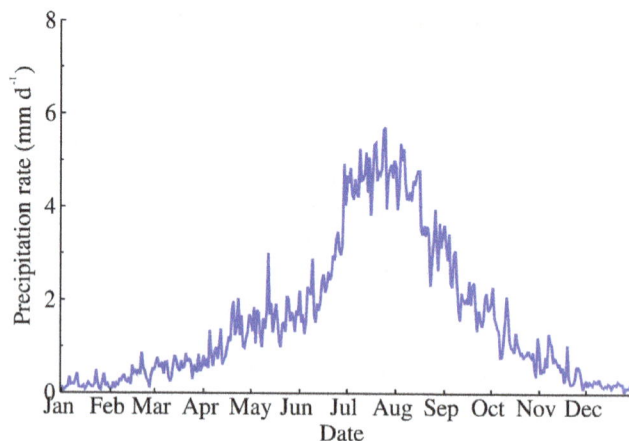

Figure 5. Temporal evolution of daily precipitation rate in northern China, averaged from 1961 to 2010.

Based on these dry and wet spells, process-split rules (Fig. 6) are constructed using the intersection proportion (IP) and critical proportion (P, set as 40 %). Herein, IP is the proportion of initial segments accounting for relevant dry and wet spells, and the initial segments (e.g., D_1, D_3, and D_4 in Fig. 6) refer to parts of one drought process split with dry and wet spells. As shown in Fig. 6, one complete process is first transformed into several initial segments according to and critical proportion (P, set as 40 %). Herein, IP is the pro-

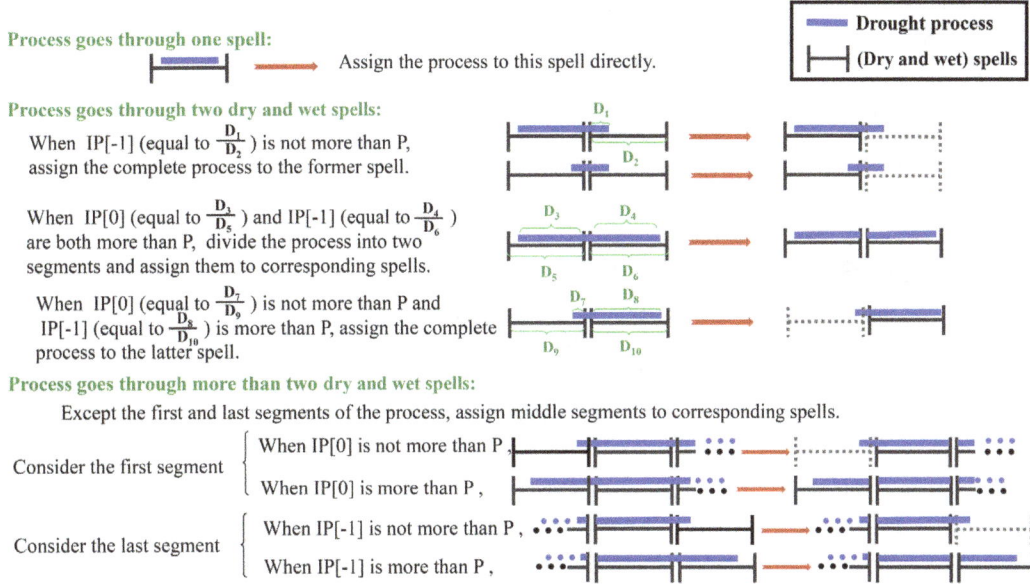

Figure 6. Process-split rules for one drought process according to dry and wet spells. IP represents intersection proportion, while P refers to critical proportion. The terms "IP[0]" and "IP[−1]" express the IP at the start and end segments, respectively.

Table 4. Drought process segments assigned to dry and wet spells during 1979–2008 in northern China.

Drought Grades	Dry spell	Dry–Wet spell	Wet spell	Wet–Dry spell
Extreme	21 Nov 1998–11 Apr 1999 –	14 Mar 1997–20 Jun 1997 –	21 Jun 1997–10 Sep 1997 4 Aug 1998–10 Sep 1998	11 Sep 1997–28 Nov 1997 11 Sep 1998–20 Nov 1998
Severe	21 Nov 1983–20 Mar 1984 21 Nov 1988–9 Jan 1989 24 Dec 1999–20 Mar 2000 14 Jan 2001–20 Mar 2001 21 Nov 2005–2 Feb 2006	21 Mar 1984–14 May 1984 18 Apr 1999–20 Jun 1999 21 Mar 2000–27 Jun 2000 21 Mar 2001–20 Jun 2001 5 May 2002–20 Jun 2002	21 Jun 1999–10 Sep 1999 21 Jun 2001–1 Aug 2001 21 Jun 2002–10 Sep 2002 – –	17 Oct 1983–20 Nov 1983 11 Aug 1988–20 Nov 1988 11 Sep 1999–1 Nov 1999 11 Sep 2002–4 Dec 2002 27 Sep 2005–20 Nov 2005

portion of initial segments accounting for relevant dry and wet spells, and the initial segments (e.g., D_1, D_3, and D_4 in Fig. 6) refer to parts of one drought process split with dry and wet spells. As shown in Fig. 6, one complete process is first transformed into several initial segments according to dry and wet spells. Second, "IP[0]" and "IP[−1]" are calculated, which express IP at the start and end segments, respectively. Third, based on a comparison of IP and P results, these initial segments can be assigned to different dry and wet spells.

Following the process-split rules shown in Fig. 6, we divided these drought processes according to dry and wet spells in northern China (Table 3). Detailed procedures of relevant IP calculations and comparisons can be found in Fig. S1 in the Supplement, while final assignments of initial drought segments are shown in Table 4. In addition, to highlight the importance of extreme droughts, severe and extreme drought segments are considered in turn.

5 Predictor construction

5.1 Atmospheric and oceanic standardized anomalies

To describe atmospheric and oceanic anomalies objectively, we chose the SA method. It was first used to effectively identify high-impact weather events (Grumm and Hart, 2001; Hart and Grumm, 2001). Subsequently, the SA method has also provided significant values for the analysis of extreme precipitation events (Duan et al., 2014; Jiang et al., 2016). In the present study, the SA of a meteorological variable was defined in Hart and Grumm (2001), described as

$$SA = \frac{X - \mu}{\sigma}, \tag{1}$$

where X represents daily grid-point atmospheric–oceanic circulation pattern variables, which are 200 and 500 hPa HGT and SST in this study. The terms μ and σ are the daily grid-point mean value and daily grid-point standard deviation, respectively. The climatological periods are 1979–2008

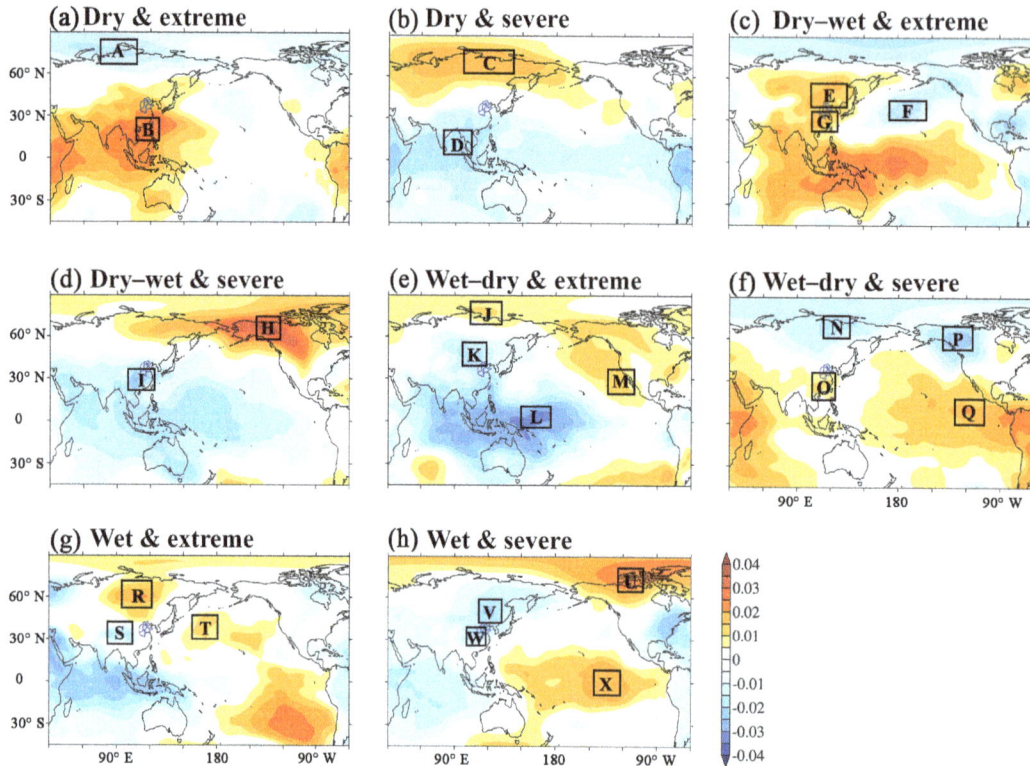

Figure 7. The first leading empirical orthogonal function (EOF) modes of standardized anomalies (SAs) for 500 hPa geopotential height fields (HGT) during all severe and extreme drought process segments during different dry and wet spells in northern China, which is the region described with blue curves. The black boxes outline the selected areas used to structure predictors for northern China, while capital letters refer to the selected area codes.

for 200 and 500 hPa HGT and 1982–2008 for SST, respectively. For example, with respect to one certain grid point, both the mean 1 January 500 hPa HGT value and associated standard deviation are computed on the 1 January 500 hPa HGT datasets observed during 1979–2008 at each grid point.

5.2 The first EOF leading modes of SA

Empirical orthogonal function analysis (Wilks, 2011) is introduced to decompose spatial–temporal datasets of drought-related atmospheric–oceanic SA into spatially stationary coefficients (leading modes) and time-varying coefficients (principal component). Considering that the first leading EOF modes reflect the largest fraction of drought-related atmospheric–oceanic spatial variability, we focus on them in this study. In addition, in order to highlight the importance of extreme droughts, EOF analysis is conducted on atmospheric–oceanic SAs during severe and extreme drought segments. With the same dry and wet spells and drought grade, SA-based maps during all drought process segments are used for EOF analysis. For example, SA-based maps of 500 hPa HGT during all three severe segments during wet spells in northern China (Table 4) are analyzed with the EOF method, and the first EOF lead mode is shown in Fig. 7h. Identical EOF analysis is conducted on atmospheric–oceanic

SA of 200 and 500 hPa HGT and SST during all four dry and wet spells in northern China. Relevant results for northern China are shown in Figs. 7, 8, and S2. In addition, the relevant results of EOF analysis for eastern and southwestern China are different, but for the sake of conciseness, they are not shown in the paper.

5.3 Pattern-based predictor construction

Positive and negative pattern areas in the first EOF leading modes are used to build predictors, which resemble the pattern-based definition of atmospheric teleconnection indices (Wallace and Gutzler, 1981). As shown in Fig. 7a, a large area of positive pattern (region B) occurs over southeastern China, while a negative pattern area (region A) appears to the north of Eurasia. Generally, the predictor is area-averaged over all gridded SA-based variables in selected areas, such as A and B, considering the reversed signs indicated with different colors. Results from the pattern-based predictor construction are shown in Table 5.

As shown in Fig. 7, the spatial pattern of different phases in the 500 hPa HGT fields were adequately considered, including low–high latitude differences (e.g., $P_{HGT500,0}$ in Table 5) and ocean–continent differences (e.g., $P_{HGT500,3}$ in Ta-

Table 5. Predictor-structured results based on the first leading empirical orthogonal function (EOF) modes for SAs of 200 hPa HGT, 500 hPa HGT, and SST fields during different dry and wet spells in northern China. Capital letters refer to the code for selected areas in Figs. 7, 8, and S2. In the term "$P_{XXX,Y}$", P, XXX, and Y refer to predictors, atmospheric or oceanic elements, and the code of new predictors, respectively.

Dry	Dry–Wet	Wet–Dry	Wet
$P_{SST,0} = A - B$	$P_{SST,5} = L + K - I$	$P_{SST,9} = Q$	$P_{SST,12} = T$
$P_{SST,1} = D - B$	$P_{SST,6} = J - I$	$P_{SST,10} = R$	$P_{SST,13} = U - V$
$P_{SST,2} = A - C$	$P_{SST,7} = M - P$	$P_{SST,11} = S$	$P_{SST,14} = W - X$
$P_{SST,3} = F - E$	$P_{SST,8} = N - O$	$P_{HGT500,5} = J - K$	$P_{HGT500,9} = R - S$
$P_{SST,4} = H - G$	$P_{HGT500,2} = E - F$	$P_{HGT500,6} = M - L$	$P_{HGT500,10} = T - S$
$P_{HGT500,0} = B - A$	$P_{HGT500,3} = G - F$	$P_{HGT500,7} = O - N$	$P_{HGT500,11} = U - V$
$P_{HGT500,1} = C - D$	$P_{HGT500,4} = H - I$	$P_{HGT500,8} = Q - P$	$P_{HGT500,12} = X - W$
$P_{HGT200,0} = A - B$	$P_{HGT200,2} = F - E$	$P_{HGT200,6} = K - L$	$P_{HGT500,13} = U - W$
$P_{HGT200,1} = C - D$	$P_{HGT200,3} = F - G$	$P_{HGT200,7} = K - M$	$P_{HGT200,10} = R - S$
–	$P_{HGT200,4} = H - I$	$P_{HGT200,8} = O - N$	$P_{HGT200,11} = X - T$
	$P_{HGT200,5} = H - J$	$P_{HGT200,9} = Q - P$	$P_{HGT200,12} = V - U$
	–	–	$P_{HGT200,13} = W - U$

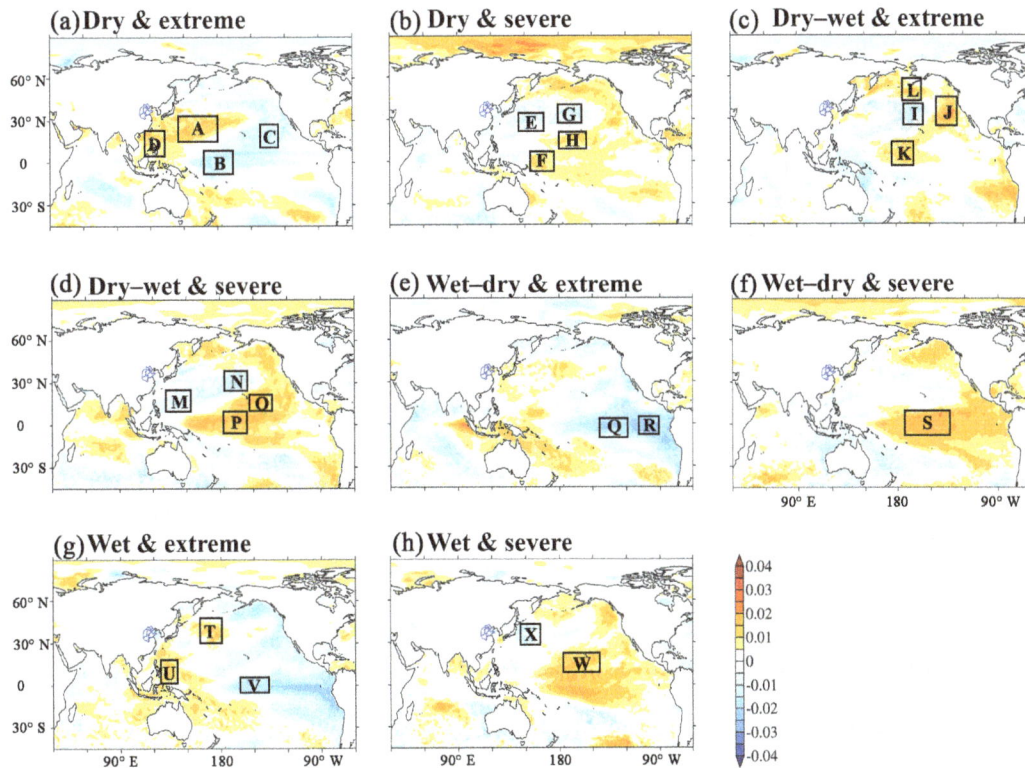

Figure 8. Same as Fig. 7, but for standardized anomalies (SAs) of SST fields associated with droughts in northern China.

ble 5). In addition, the spatial pattern of different phases surrounding the prediction-targeted region (e.g., regions R, S, and T in Fig. 7g) was intentionally used to construct predictors, such as $P_{HGT500,9}$ and $P_{HGT500,10}$ in Table 5. Because the first EOF modes of 200 hPa HGT (Fig. S2) were similar to those of 500 hPa HGT, the specified illustrations were not shown here but were considered in the analysis. Additionally, the positive and negative pattern areas in the Pacific SST SA

fields were also used, especially in the subtropical gyre zone (Fig. 8a–d) and El Niño region (Fig. 8e and f). Furthermore, some regions, such as the El Niño Regions R, Q, and S, were separately used for the construction of the predictors. In addition to the predictors constructed for northern China (Table 5), different predictor-structured results for eastern and southwestern China were also obtained but not shown in the paper.

Table 6. Statistical parameters of stepwise-regression equations used for prediction during different calibration periods in northern China.

Calibration period (1 Jan 1983–)	Simulation or prediction period	Numbers of selected/ initial predictors	Multiple correlation coefficient
31 Dec 2008	1 Jan–31 Dec 2009	38/43	0.76
31 Dec 2009	1 Jan–31 Dec 2010	37/43	0.76
31 Dec 2010	1 Jan–31 Dec 2011	39/43	0.75
31 Dec 2011	1 Jan–31 Dec 2012	39/43	0.76
31 Dec 2012	1 Jan–31 Dec 2013	38/43	0.76
31 Dec 2013	1 Jan–31 Dec 2014	39/43	0.75

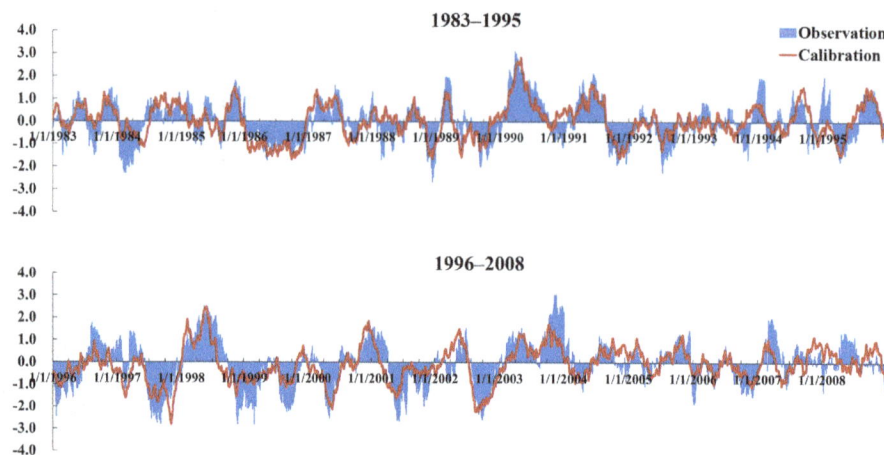

Figure 9. Temporal evolution of observed and calibrated SPI3 during the calibration period between 1 January 1983 and 31 December 2008 in northern China.

6 Model calibration

6.1 Synchronous statistical relationship

Stepwise regression (Afifi and Azen, 1972) is a method for fitting multiple linear regression models, in which a predictive variable is considered for addition to or subtraction from a set of explanatory variables according to statistically significant extent or loss. In this study, it is used to build the synchronous statistical relationship between all 90-day-accumulated SA-based predictors and the prediction target SPI3. SA-based predictors are calculated with the NCEP/NCAR reanalysis dataset (Kalnay et al., 1996). Essentially, the conceptual model, aimed at seasonal drought process prediction, is a synchronous stepwise relationship.

6.2 Rolling calibration year by year

To meet the practical requirements of operational service departments, model calibration is also running year by year (Table 6). For example, the seasonal drought prediction model, calibrated from 1 January 1983 to 31 December 2011, is used for initial daily prediction time in the entire 2012 year. For every initial drought prediction in the year 2013, the corresponding drought model is calibrated from 1 January 1983 to

31 December 2012. In addition, detailed information about selected predictors and relevant coefficients can be found in Table S1 in the Supplement.

The calibration period increases year by year, therefore, the number of samples used for calibration also increases year by year. Multiple correlation coefficients in six drought prediction models are no less than 0.75. Statistical parameters and their total numbers show slight changes across the six calibration experiments (Table 6). Furthermore, calibrated SPI3 curves are almost consistent with the observation data (Fig. 9), especially with respect to turning points and trends. Different parameter sets and results of model calibration for eastern and southwestern China are not shown in the paper.

7 Drought process simulation and prediction

7.1 Model forcing

Because the conceptual model is essentially a synchronous statistical relationship, the model itself has no lead time. Therefore, model simulation and prediction have to be further forced with reanalysis and forecast datasets. During the periods of model simulation, the synchronous statistical rela-

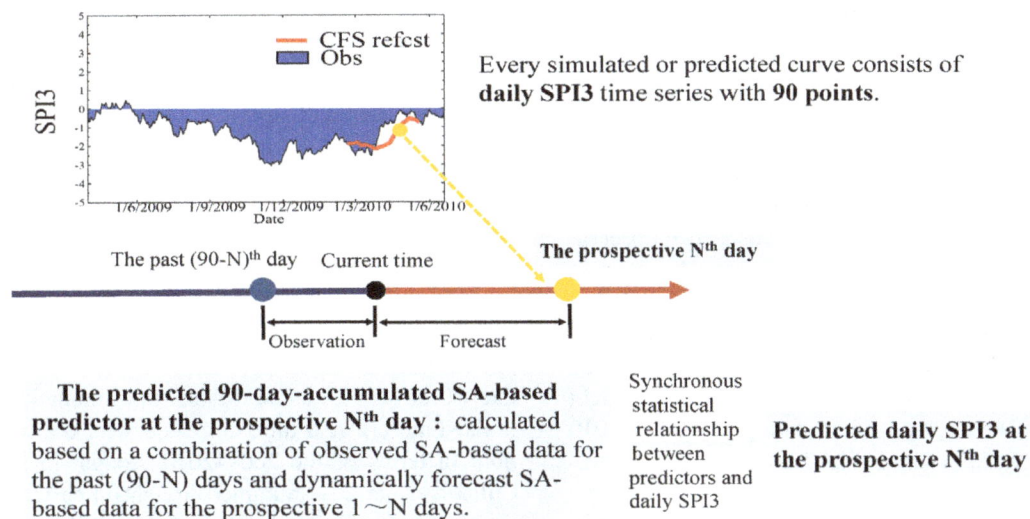

Figure 10. Illustration about how to calculate the prospective 90-day daily SPI3 time series. "Reforecast" is denoted by "refcst".

tionship is forced with the NCEP/NCAR reanalysis dataset. For model prediction, it is operationally forced with CFSv2 forecast datasets, together with the NCEP/NCAR reanalysis dataset. Therefore, the lead time for the conceptual model depends on that of the climate forecast models.

In particular, because we focus on the prospective 90-day drought process prediction (predicted daily SPI3 time series with 90 points), it is necessary to illustrate how to calculate every forecast point (daily SPI3). As shown in Fig. 10, predicted daily SPI3 at the prospective Nth day is originally based on a combination of observed and dynamically forecast SA-based data. The computation of these SA-based data follow Sect. 5.1, and the observed data are also retrieved from the NCEP/NCAR reanalysis data. In addition, when the lead time is longer, more dynamically forecast data are included and corresponding daily SPI3 value contains larger uncertainty.

7.2 Drought processes simulated with the NCEP/NCAR reanalysis datasets

To assess model performance of severe seasonal droughts, we take four recent drought processes in southwestern China, eastern China, and northern China as examples. First, southwestern China experienced two severe droughts (the black boxes in Fig. 11c). Although the simulated SPI3 does not reach its peak during the 2009–2010 drought, it indicates the state transformation from drought occurrence to persistence and eventually to relief. In terms of the 2011 summer drought in the southwestern China, the simulated SPI3 indicates that the state remains wet and gradually becomes wetter, indicating no valuable information consistent with observations. Nevertheless, during the phase of drought recession, the simulated development is quite similar to the observed development. This comparison indicates that the conceptual

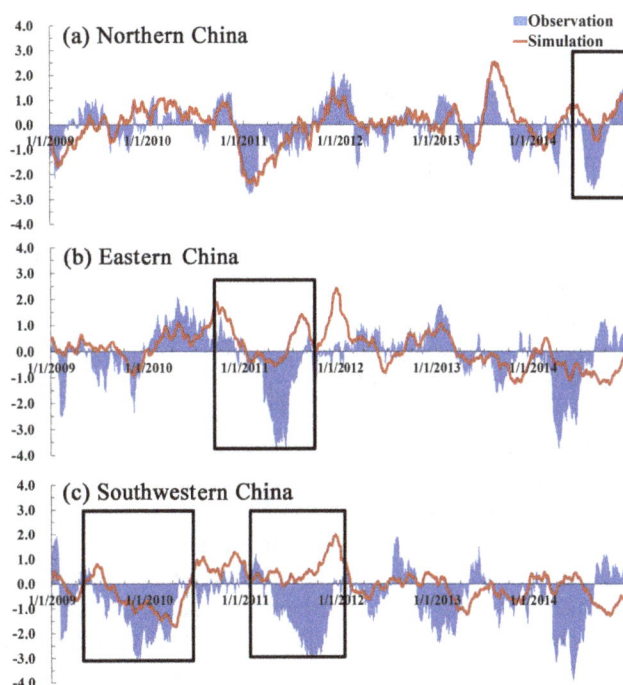

Figure 11. Temporal evolution of observed and simulated SPI3 processes during the period from 1 January 2009 to 31 December 2014. The black boxes in **(a)**–**(c)** indicate the 2014 summer and autumn drought in northern China, 2011 spring drought in eastern China, 2009–2010 drought in southwestern China, and 2011 summer drought in southwestern China. Red curves refer to simulated SPI3, while curves filled with light blue represent observed SPI3.

model performs well in development but is weak in severity. This distinct feature also appears in the simulation of the 2011 drought in eastern China (the black box in Fig. 11b) and 2014 drought in northern China (the black box in Fig. 11a).

7.3 Drought processes predicted with the CFSv2 forecast datasets

Compared with drought simulation, operationally predicted results may bring some uncertainties into the prospective drought processes. As shown in Fig. 12b, predicted curves perform worse than the simulated curves near the peak of the 2011 eastern China drought, as the prospective observation tendency is rising rather than decreasing. However, in the other three droughts, the predicted curves are good at indicating drought development to some different degrees, resembling the simulated results quite well. For example, the presented operationally reforecast curves indicate drought occurrence, persistence, and relief during the 2009–2010 drought in southwestern China (Fig. 12a).

8 Drought outlook

8.1 Angle-based rules

Compared with the predicted prospective SPI3 time series, the drought outlook is a convenient and valuable attachment product for water resource managers. To create the drought outlook, angle-based rules are developed to transform the predicted prospective 90-day SPI3 curves into different drought tendencies. Three essential technical points are as follows.

First, some variables must be defined to describe drought development. Similar to the slope of curves, angles of predicted 90-day SPI3 curves are used to describe the prospective drought situation. Generally, positive angles of SPI3 curves indicate wetter tendencies, while negative angles represent drier tendencies.

The second is two general classifications of drought outlook on the basis of the current drought situation. For no current drought (see sketch map I in Fig. 13), the prospective situation tends to be no drought or drought occurrence. In this case, a critical angle α_1 can be used to help distinguish between these two types of drought outlook. A calculated SPI3 curve angle α that is less than α_1 results in the prospective development of drought occurrence; otherwise, the non-drought situation persists. Similarly, for a current condition of being in drought (see sketch map II in Fig. 13), a comparison of critical angles α_2 (equal to zero) and α_3 defines the other three types of drought outlook, which are drought persistence (α less than α_2), drought recession (α more than α_2, but less than α_3), and drought relief (α more than α_3).

Third, it is necessary to explain the practical calculation for curve angles and how to conduct an angle-based drought outlook. Except the constant critical angle α_2 (equal to zero), both α_1 and α_3 represent angles between the horizontal line and arrow from the original point (initial prediction time) to the points on the time axis (see red dashed arrowed lines in Fig. 13a–e). Similarly, α represents angles between the hor-

izontal line and arrow from the original point to the points on the predicted SPI3 curve (see green solid arrowed lines in Fig. 13a–e). However, considering the predicted period of SPI3 time series is prospectively 90 days, curve angle α_i and critical angles α_{1i}, α_{2i} and α_{3i} ($i = 1, 2, \ldots, 90$) can be calculated. Finally, according to the angle-based rules shown in Table 7, a drought outlook can be performed.

8.2 Simulated and predicted results

Following the method in Sect. 8.1, drought outlook is conducted based on angle comparison of the simulated prospective 90-day SPI3 curve (Table 8). Simulations at every initial time are real-time corrected with the current situation. In terms of the 2009–2010 drought in southwestern China and the 2011 summer drought in eastern China, the simulated drought outlook performs well with respect to drought occurrence, persistence, and recession before 2 December 2009 and 1 May 2011. In addition, the simulation of the 2011 drought in southwestern China performs well in August 2011. The 2014 summer drought in northern China lasts for a relatively short time, resulting in an observed drought outlook that maintains a state of drought relief during the first month of the drought process. Even so, the simulation can also capture it. Additionally, these four drought outlooks remain weak in simulating the development of drought relief after 31 January 2010, 11 May 2011, 11 September 2011, and 21 July 2014, respectively. Weak performance in simulating severity leads to the development of drought recession rather than drought relief.

For predicted drought outlooks, operationally predicted results (Table 9) in southwestern China and eastern China are relatively similar to the simulated ones (Table 8). In comparison, predicted drought outlook during the first month of the 2014 drought in northern China performs worse than simulated results.

9 Discussion

Considering that the development of drought processes is closely related to the spatial–temporal evolution of atmospheric and oceanic anomalies, a conceptual prediction model of seasonal drought processes is proposed in our study. Despite its weakness in predicting drought severity, the model performs well in simulating and predicting drought development. Because the proposed model is a new attempt, several associated discussion issues are as follows.

First, process prediction and outlook of seasonal drought are the focus of our study. To date, a considerable number of studies have focused on predicting discrete drought classes (Aviles et al., 2016; Bonaccorso et al., 2015; Chen et al., 2013; Moreira et al., 2016) and the probability of drought occurrence within certain classes (AghaKouchak, 2014, 2015; Hao et al., 2014). Compared with these studies, prediction

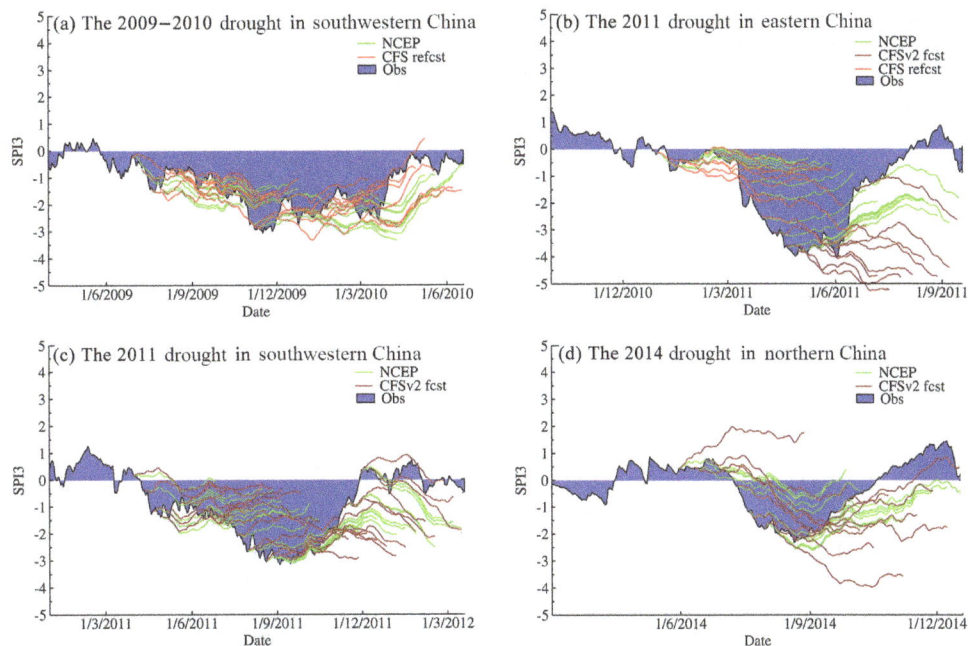

Figure 12. Simulation and prediction results of four recent severe drought processes in China. Every unfilled curve represents simulated or predicted prospective 90-day SPI3, with an interval of initial prediction time of about 10 days. The curves filled with blue refer to observed SPI3. Dark red and bright red curves refer to SPI3 predicted with CFSv2 and CFS products, respectively. Light green curves represent SPI3 simulated with the NCEP/NCAR reanalysis datasets. Every simulated or predicted curve consists of daily SPI3 time series with 90 points. "Reforecast" is denoted by "refcst".

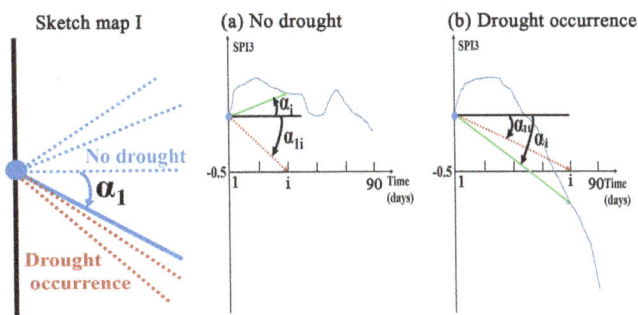

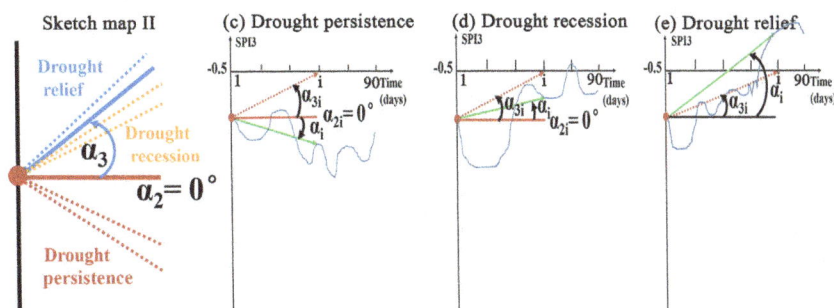

Figure 13. Rules of drought outlook based on angle comparison of prospective 90-day SPI3 curves. Sketch maps I and II show general drought outlook based on the current drought situation. Panels **(a)**–**(b)** and **(c)**–**(e)** express different situations of drought outlook associated with the rules regarding critical angles in Table 7.

Table 7. Specific rules for drought outlook based on angle comparison. R1 represents the ratio of days when α_i is less than the critical angle $\alpha_{1i}(\alpha_{3i})$ to the total 90 days. R2 represents the proportion of specific days in the period to the predicted prospective 46–90 days. In R2 calculation, these specific days meet the criteria that α_i is greater than critical angle α_{3i}.

Current SPI3	Current condition	R1	R2	Drought outlook
Greater than −0.5	no drought	less than 10 %	–	no drought
		greater than 10 %	–	drought occurrence
Less than −0.5	in drought	greater than 90 %	less than 90 %	drought persistence
		greater than 90 %	greater than 90 %	drought recession
		less than 90 %	–	drought relief

Table 8. Simulation assessment of recent severe drought processes in China forced with the NCEP/NCAR reanalysis datasets. The numbers 0–4 in the below table represent different drought states: no drought (0), drought occurrence (1), drought persistence (2), drought recession (3), and drought relief (4). As well as this, the abbreviation "Simul." and "Obs." represent the simulated and observed drought outlooks, respectively. The abbreviation "Assess." in the column refers to whether the simulation and observation agree or not.

Drought Processes	Initial Time	Simul.	Obs.	Assess.	Initial Time	Simul.	Obs.	Assess.	Initial Time	Simul.	Obs.	Assess.
	30 Jun 2009	1	2	–	28 Sep 2009	3	2	–	11 Jan 2010	2	3	–
	10 Jul 2009	2	2	yes	18 Oct 2009	3	2	–	21 Jan 2010	2	3	–
	20 Jul 2009	2	3	–	2 Nov 2009	3	3	yes	31 Jan 2010	3	4	–
The 2009–2010	30 Jul 2009	2	3	–	12 Nov 2009	3	3	yes	10 Feb 2010	3	4	–
drought in	9 Aug 2009	2	2	yes	22 Nov 2009	3	3	yes	20 Feb 2010	3	4	–
southwestern China	19 Aug 2009	2	2	yes	2 Dec 2009	3	3	yes	2 Mar 2010	3	4	–
	29 Aug 2009	2	2	yes	12 Dec 2009	2	3	–	12 Mar 2010	3	4	–
	8 Sep 2009	2	2	yes	22 Dec 2009	2	3	–	22 Mar 2010	3	4	–
	18 Sep 2009	2	2	yes	1 Jan 2010	2	3	–	–	–	–	–
	1 Jan 2011	1	1	yes	2 Mar 2011	1	1	yes	1 May 2011	3	3	yes
The 2011 summer	11 Jan 2011	1	1	yes	12 Mar 2011	3	2	–	11 May 2011	3	4	–
drought in	21 Jan 2011	1	1	yes	22 Mar 2011	3	2	–	21 May 2011	3	4	–
eastern China	31 Jan 2011	1	1	yes	1 Apr 2011	3	3	yes	1 Jun 2011	3	4	–
	10 Feb 2011	0	1	–	11 Apr 2011	3	3	yes	11 Jun 2011	3	4	–
	20 Feb 2011	1	1	yes	21 Apr 2011	3	3	yes	21 Jun 2011	3	4	–
	11 Apr 2011	1	1	yes	1 Jul 2011	3	2	–	21 Sep 2011	3	4	–
	21 Apr 2011	2	2	yes	11 Jul 2011	3	2	–	1 Oct 2011	3	4	–
The 2011 summer	1 May 2011	2	2	yes	21 Jul 2011	3	2	–	11 Oct 2011	3	4	–
drought in	11 May 2011	2	2	yes	1 Aug 2011	3	3	yes	21 Oct 2011	3	4	–
southwestern China	21 May 2011	4	2	–	11 Aug 2011	3	3	yes	1 Nov 2011	3	4	–
	1 Jun 2011	3	2	–	21 Aug 2011	3	3	yes	11 Nov 2011	3	4	–
	11 Jun 2011	3	2	–	1 Sep 2011	3	3	yes	21 Nov 2011	2	4	–
	21 Jun 2011	3	2	–	11 Sep 2011	3	4	–	–	–	–	–
The 2014 summer	1 Jun 2014	4	4	yes	11 Jul 2014	3	3	yes	21 Aug 2014	3	4	–
drought in	11 Jun 2014	4	4	yes	21 Jul 2014	3	4	–	1 Sep 2014	3	4	–
northern China	21 Jun 2014	4	4	yes	1 Aug 2014	3	4	–	11 Sep 2014	3	4	–
	1 Jul 2014	1	1	yes	11 Aug 2014	3	4	–	21 Sep 2014	4	4	yes

of regional drought processes is another valuable attempt, which is beneficial from the moving window of SPI3 extended from 1 month to 1 day. It performs relatively well in predicting the development of seasonal drought processes (Fig. 12). In addition, it can indicate drought occurrence, persistence, and relief relatively well (Tables 8 and 9), which is meaningful for seasonal water resource management.

Second, the proposed model is essentially one stepwise-regression equation. Despite its simplicity, it incorporates drought-related spatial and temporal information as integrally as possible. Because precipitation-related synoptic systems appear in the troposphere, SST, 500 hPa HGT, and 200 hPa HGT are chosen as representatives of the low, middle, and upper levels of the troposphere, respectively. Furthermore, all drought process segments assigned to different dry and wet spells are used for EOF analysis within the same dry and wet spells (shown in Sect. 5.2). Therefore, adequate drought-related spatial–temporal information has been included in these drought predictors.

Third, the reasons for acceptable performance of operationally predicted results need to be illustrated. Compared

Table 9. Same as Table 8 but for predicted results forced with the operational output from CFSv2. The abbreviation "Predi." represents the predicted drought outlook. The abbreviation "Assess." in the column refers to whether the prediction and observation agree or not.

Drought Processes	Initial Time	Predi.	Obs.	Assess.	Initial Time	Predi.	Obs.	Assess.	Initial Time	Predi.	Obs.	Assess.
The 2009–2010 drought in southwestern China	30 Jun 2009	1	2	–	28 Sep 2009	3	2	–	11 Jan 2010	3	3	yes
	10 Jul 2009	2	2	yes	18 Oct 2009	2	2	yes	21 Jan 2010	3	3	yes
	20 Jul 2009	3	3	yes	2 Nov 2009	3	3	yes	31 Jan 2010	3	4	–
	30 Jul 2009	3	3	yes	12 Nov 2009	3	3	yes	10 Feb 2010	4	4	yes
	9 Aug 2009	2	2	yes	22 Nov 2009	3	3	yes	20 Feb 2010	3	4	–
	19 Aug 2009	2	2	yes	2 Dec 2009	3	3	yes	2 Mar 2010	3	4	–
	29 Aug 2009	2	2	yes	12 Dec 2009	3	3	yes	12 Mar 2010	3	4	–
	8 Sep 2009	3	2	–	22 Dec 2009	3	3	yes	22 Mar 2010	3	4	–
	18 Sep 2009	2	2	yes	1 Jan 2010	3	3	yes	–	–	–	–
The 2011 summer drought in eastern China	1 Jan 2011	1	1	yes	2 Mar 2011	1	1	yes	1 May 2011	2	3	–
	11 Jan 2011	1	1	yes	12 Mar 2011	2	2	yes	11 May 2011	2	4	–
	21 Jan 2011	1	1	yes	22 Mar 2011	2	2	yes	21 May 2011	2	4	–
	31 Jan 2011	1	1	yes	1 Apr 2011	2	3	–	1 Jun 2011	2	4	–
	10 Feb 2011	1	1	yes	11 Apr 2011	2	3	–	11 Jun 2011	3	4	–
	20 Feb 2011	1	1	yes	21 Apr 2011	2	3	–	21 Jun 2011	3	4	–
The 2011 summer drought in southwestern China	11 Apr 2011	0	1	–	1 Jul 2011	4	2	–	21 Sep 2011	3	4	–
	21 Apr 2011	3	2	–	11 Jul 2011	3	2	–	1 Oct 2011	3	4	–
	1 May 2011	3	2	–	21 Jul 2011	3	2	–	11 Oct 2011	3	4	–
	11 May 2011	3	2	–	1 Aug 2011	3	3	yes	21 Oct 2011	3	4	–
	21 May 2011	4	2	–	11 Aug 2011	3	3	yes	1 Nov 2011	3	4	–
	1 Jun 2011	4	2	-	21 Aug 2011	3	3	yes	11 Nov 2011	4	4	yes
	11 Jun 2011	4	2	–	1 Sep 2011	3	3	yes	21 Nov 2011	2	4	–
	21 Jun 2011	3	2	–	11 Sep 2011	3	4	–	–	–	–	–
The 2014 summer drought in northern China	1 Jun 2014	0	4	–	11 Jul 2014	1	3	–	21 Aug 2014	3	4	–
	11 Jun 2014	1	4	–	21 Jul 2014	2	4	–	1 Sep 2014	4	4	yes
	21 Jun 2014	1	4	–	1 Aug 2014	3	4	–	11 Sep 2014	3	4	–
	1 Jul 2014	1	1	yes	11 Aug 2014	2	4	–	21 Sep 2014	4	4	yes

with those forced with the NCEP/NCAR reanalysis datasets (green curves in Fig. 12), the predicted developments of drought processes forced with CFSv2 or CFS datasets (red curves in Fig. 12) are relatively similar, especially with respect to the former segment of every predicted prospective 90-day SPI3 curve. Essentially, the 90-day-accumulated SA-based predictors strengthen the good performance of operational use. This indicates that observed information from atmospheric and oceanic anomalies are involved to different degrees (Fig. 10). With the incorporation of observed data, its operational application provides relatively accurate and valuable information. However, it is also worthwhile to investigate how changing the length of the predicted period can make predicted drought processes relatively accurate and acceptable, such as the prospective 1–30-day or the prospective 1–60-day periods. The relevant comparison results with different predicted periods are shown in Fig. 14. It appears that the 2009–2010 drought in southwestern China and 2014 drought in northern China can be predicted and simulated well even for the prospective 1–75-day period. In contrast, the prospective 1–45-day period may be a feasible and acceptable lead time for simulation and prediction of the 2011 droughts in southwestern China and eastern China, after which the simulated and predicted developments clearly change.

Fourth, the weak performance in predicting the severity of drought, including drought peak and drought relief, is an important issue. Similar to the concluding remarks regarding a probabilistic drought prediction model, the weak performance in predicting the severity of the drought peak is due to the typical problem of an inherent averaging effect depressing the extremes (Behrangi et al., 2015). With the help of real-time correction from operational application, the prediction of drought peaks can be improved. In addition, the prediction of drought relief should also be considered. As listed in both Tables 8 and 9, the simulated and predicted results for drought relief are unsatisfying. This weak performance may be associated with precipitation-causing weather patterns during drought relief. They are unsteady and change dramatically compared with those features during drought persistence. Because the period of drought relief is a relatively short phase of the drought process, the relevant information may not be involved in the first EOF modes (Sect. 5.2). Generally, three measures for potential improvement are as follows. (1) More secondary EOF modes, including precipitation-causing circulation patterns during drought relief, can be incorporated when building initial predictors. (2) The rapid change index (Otkin et al., 2015) could be introduced to describe temporal changes during drought relief on subseasonal timescales. (3) The empirical factor can

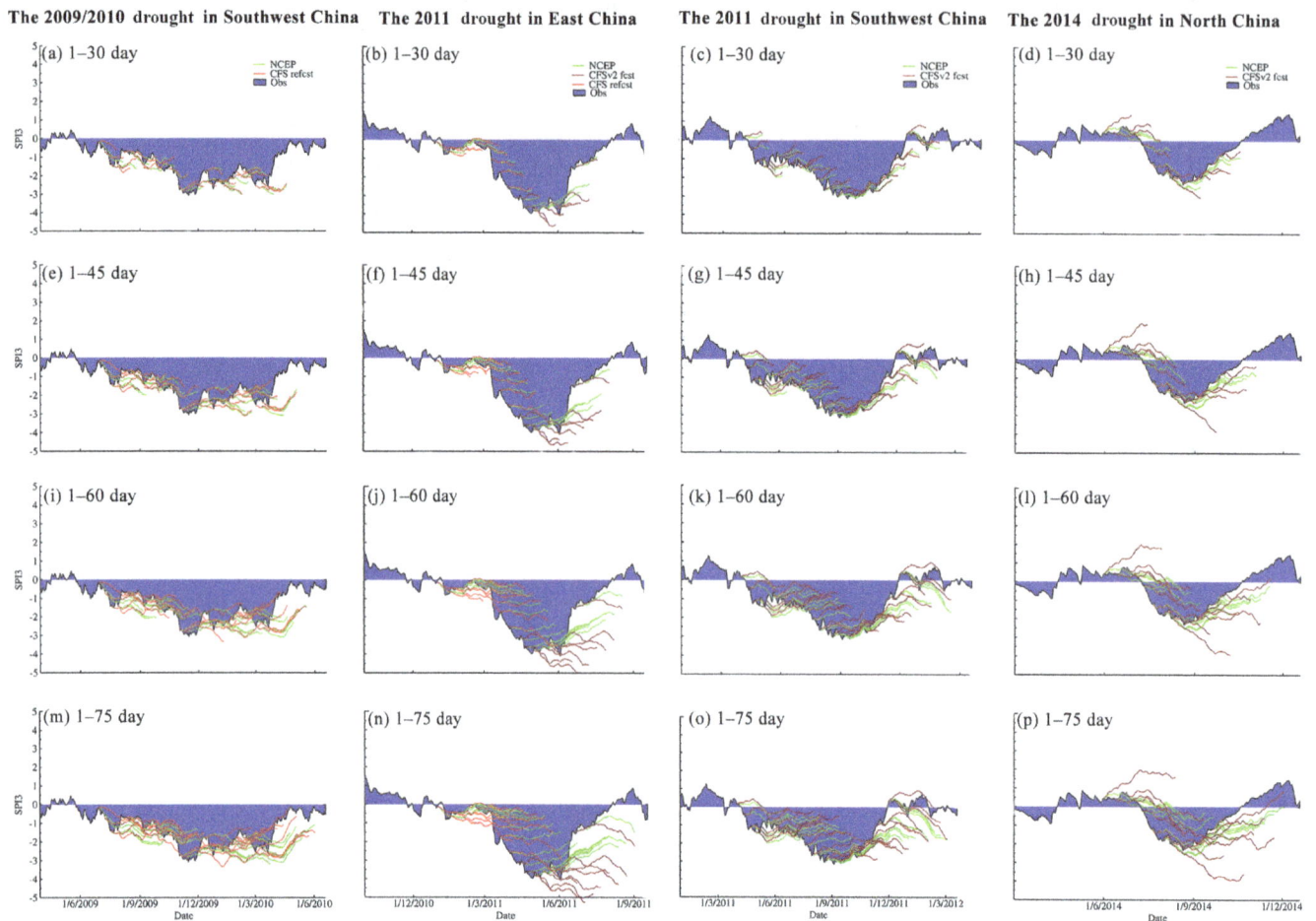

Figure 14. Same as Fig. 12 but for different predicted periods, which are namely the prospective **(a)**–**(d)** 1–30-day, **(e)**–**(h)** 1–45-day, **(i)**–**(l)** 1–60-day, and **(m)**–**(p)** 1–75-day periods. "Reforecast" is denoted by "refcst" and "forecast" by "fcst".

be introduced to improve drought-relief prediction. The predicted SPI3 during the phase of drought relief could be multiplied by empirical factors to strengthen drought relief development.

Fifth, it is necessary to explain the method of predictor construction. The predictor-structured method in our study is similar to the definition of teleconnection indices (Wallace and Gutzler, 1981). It is more goal-directed, because these structured predictors are directly related to synchronous atmospheric–oceanic anomalous circulation patterns during different drought segments within the same dry and wet spells. However, to design geographical ranges of anomalous areas and combine them is subjective, which leads to considerable uncertainties. Accordingly, an objective anomaly-recognized method with explicit critical values needs to be developed. This will contribute to auto-run feasibility of this conceptual prediction model without artificial interaction.

The sixth issue to illustrate is synchronous SST anomalies used in EOF analysis and model construction. Traditionally, SST anomalies a few months ahead influence the subsequent regional droughts. However, it is also feasible and common that synchronous SST anomalies are used in the investigation of regional drought events in southwestern China (Feng et al., 2014), the Yangtze River basin (Lu et al., 2014), and northern China (Wang and He, 2015), which may shape synchronous drought-related circulation patterns. In addition, this is convenient for operational application, while forecast SST and 200 and 500 hPa HGT can be retrieved together from CFSv2 products simultaneously.

Finally, the timescale of the drought index needs to be explained. SPI3 is the index used for drought identification and prediction in the study, which provides a seasonal estimation of precipitation and tends to be a good indication of soil moisture conditions as the growing season begins in some primary agricultural regions worldwide (WMO, 2012). However, to meet operational requirements of seasonal hydrological forecasting, the indices such as 6-month up to 24-month SPI can be also used for hydrological drought analyses and applications (WMO, 2012). With the increasing timescale of SPI, its lead time with given accuracy requirements might be longer, together with the smoother temporal evolution. Accordingly, atmospheric and oceanic anomalies used for model calibration need to be changed from 90-day accumulated to 6- or 24-month accumulated.

10   Conclusions

Drought prediction is fundamental for seasonal water management. In this study, we constructed a conceptual prediction model of seasonal drought processes based on synchronous standardized anomalies of 200 and 500 hPa geopotential height and sea surface temperature; we considered that drought development is closely related to the spatial–temporal evolution of large-scale atmospheric–oceanic circulation patterns. We used northern China as an example to introduce the method and used four recent severe regional drought processes in China for model application. This model can be used for seamless drought prediction and drought outlook, forced with seasonal climate forecast models. The main process is as follows. (1) A 3-month SPI updated daily (SPI3) was used to capture severe and extreme drought processes. (2) Empirical orthogonal function analysis was applied to SA of 200 and 500 hPa HGT and SST during drought process segments within the same dry and wet spells. Subsequently, spatial patterns of the first EOF modes were used to structure SA-based predictors. (3) The synchronous stepwise-regression relationship between SPI3 and all 90-day-accumulated SA-based predictors were calibrated using the NCEP/NCAR reanalysis datasets. (4) To achieve a prospective 90-day drought outlook, we further developed an objective method based on angles of the predicted prospective 90-day SPI3 curves. (5) Finally, simulation and prediction of seasonal drought processes, together with drought outlook, were forced with the NCEP/NCAR reanalysis datasets and the NCEP Climate Forecast System Version 2 (CFSv2) operationally forecast datasets, respectively. Model application during four recent severe drought processes in China revealed that the model is good at development prediction but weak in severity prediction. These results indicate that the proposed conceptual drought prediction model is another potentially valuable addition to current research on drought prediction.

Competing interests. The authors declare that they have no conflict of interest.

Special issue statement. This article is part of the special issue "Sub-seasonal to seasonal hydrological forecasting". It does not belong to a conference.

Acknowledgements. This work is supported by the Special Public Sector Research Program of Ministry of Water Resources (grant nos. 201301040 and 201501041), Fundamental Research Funds for the Central Universities (grant no. 2015B20414), Program for New Century Excellent Talents in University (grant no. NCET-12-0842), National Natural Science Foundation of China (grant no. 51579065), and Natural Science Foundation of Jiangsu Province of China (grant no. BK20131368). In addition, we are grateful for the Handling Editor (Maria-Helena Ramos) and the two anonymous referees. Their comments and suggestions help improve the clarity of the paper and made us think about the research work more deeply.

Edited by: Maria-Helena Ramos

References

Afifi, A. A. and Azen, S. P.: Statistical analysis: a computer oriented approach, Academic press, 1972.

AghaKouchak, A.: A baseline probabilistic drought forecasting framework using standardized soil moisture index: application to the 2012 United States drought, Hydrol. Earth Syst. Sci., 18, 2485–2492, https://doi.org/10.5194/hess-18-2485-2014, 2014.

AghaKouchak, A.: A multivariate approach for persistence-based drought prediction: Application to the 2010-2011 East Africa drought, J. Hydrol., 526, 127–135, https://doi.org/10.1016/j.jhydrol.2014.09.063, 2015.

Aviles, A., Celleri, R., Paredes, J., and Solera, A.: Evaluation of Markov Chain Based Drought Forecasts in an Andean Regulated River Basin Using the Skill Scores RPS and GMSS, Water Resour. Manag., 29, 1949–1963, 10.1007/s11269-015-0921-2, 2015.

Aviles, A., Celleri, R., Solera, A., and Paredes, J.: Probabilistic Forecasting of Drought Events Using Markov Chain- and Bayesian Network-Based Models: A Case Study of an Andean Regulated River Basin, Water, 8, 16 pp., https://doi.org/10.3390/w8020037, 2016.

Behrangi, A., Hai, N., and Granger, S.: Probabilistic Seasonal Prediction of Meteorological Drought Using the Bootstrap and Multivariate Information, J. Appl. Meteorol. Clim., 54, 1510–1522, https://doi.org/10.1175/jamc-d-14-0162.1, 2015.

Belayneh, A., Adamowski, J., Khalil, B., and Ozga-Zielinski, B.: Long-term SPI drought forecasting in the Awash River Basin in Ethiopia using wavelet neural network and wavelet support vector regression models, J. Hydrol., 508, 418–429, 10.1016/j.jhydrol.2013.10.052, 2014.

Bonaccorso, B., Cancelliere, A., and Rossi, G.: Probabilistic forecasting of drought class transitions in Sicily (Italy) using Standardized Precipitation Index and North Atlantic Oscillation Index, J. Hydrol., 526, 136–150, 10.1016/j.jhydrol.2015.01.070, 2015.

Chen, S. T., Yang, T. C., Kuo, C. M., Kuo, C. H., and Yu, P. S.: Probabilistic Drought Forecasting in Southern Taiwan Using El Nino-Southern Oscillation Index, Terr. Atmos. Ocean. Sci., 24,

911–924, 2013.

Dai, A. G.: Drought under global warming: a review, Clim. Change, 2, 45–65, 2011.

Duan, W. L., He, B., Takara, K., Luo, P. P., Nover, D., Yamashiki, Y., and Huang, W. R.: Anomalous atmospheric events leading to Kyushu's flash floods, July 11–14, 2012, Nat. Hazards, 73, 1255–1267, 2014.

Dutra, E., Di Giuseppe, F., Wetterhall, F., and Pappenberger, F.: Seasonal forecasts of droughts in African basins using the Standardized Precipitation Index, Hydrol. Earth Syst. Sci., 17, 2359–2373, https://doi.org/10.5194/hess-17-2359-2013, 2013.

Dutra, E., Pozzi, W., Wetterhall, F., Di Giuseppe, F., Magnusson, L., Naumann, G., Barbosa, P., Vogt, J., and Pappenberger, F.: Global meteorological drought – Part 2: Seasonal forecasts, Hydrol. Earth Syst. Sci., 18, 2669–2678, https://doi.org/10.5194/hess-18-2669-2014, 2014.

Feng, L., Li, T., and Yu, W.: Cause of severe droughts in Southwest China during 1951–2010, Clim. Dynam., 43, 2033–2042, https://doi.org/10.1007/s00382-013-2026-z, 2014.

Funk, C.: We thought trouble was coming, Nature, 476, 7–7, https://doi.org/10.1038/476007a, 2011.

Funk, C., Hoell, A., Shukla, S., Bladé, I., Liebmann, B., Roberts, J. B., Robertson, F. R., and Husak, G.: Predicting East African spring droughts using Pacific and Indian Ocean sea surface temperature indices, Hydrol. Earth Syst. Sci., 18, 4965–4978, https://doi.org/10.5194/hess-18-4965-2014, 2014.

Grumm, R. H. and Hart, R.: Standardized anomalies applied to significant cold season weather events: Preliminary findings, Weather Forecast., 16, 736–754, https://doi.org/10.1175/1520-0434(2001)016<0736:saatsc>2.0.co;2, 2001.

Hart, R. E. and Grumm, R. H.: Using normalized climatological anomalies to rank synoptic-scale events objectively, Mon. Weather Rev., 129, 2426–2442, https://doi.org/10.1175/1520-0493(2001)129<2426:uncatr>2.0.co;2, 2001.

Hurrell, J. W.: Decadal trends in the north Atlantic oscillation: regional temperatures and precipitation, Science (New York, NY), 269, 676–679, https://doi.org/10.1126/science.269.5224.676, 1995.

Jiang, N., Qian, W. H., Du, J., Grumm, R. H., and Fu, J. L.: A comprehensive approach from the raw and normalized anomalies to the analysis and prediction of the Beijing extreme rainfall on July 21, 2012, Nat. Hazards, 84, 1551–1567, https://doi.org/10.1007/s11069-016-2500-0, 2016.

Kalnay, E., Kanamitsu, M., Kistler, R., Collins, W., Deaven, D., Gandin, L., Iredell, M., Saha, S., White, G., Woollen, J., Zhu, Y., Chelliah, M., Ebisuzaki, W., Higgins, W., Janowiak, J., Mo, K. C., Ropelewski, C., Wang, J., Leetmaa, A., Reynolds, R., Jenne, R., and Joseph, D.: The NCEP/NCAR 40-year reanalysis project, B. Am. Meteorol. Soc., 77, 437–471, https://doi.org/10.1175/1520-0477(1996)077<0437:tnyrp>2.0.co;2, 1996.

Kingston, D. G., Stagge, J. H., Tallaksen, L. M., and Hannah, D. M.: European-Scale Drought: Understanding Connections between Atmospheric Circulation and Meteorological Drought Indices, J. Climate, 28, 505–516, https://doi.org/10.1175/jcli-d-14-00001.1, 2015.

Lu, E., Liu, S. Y., Luo, Y. L., Zhao, W., Li, H., Chen, H. X., Zeng, Y. T., Liu, P., Wang, X. M., Higgins, R. W., and Halpert, M. S.: The atmospheric anomalies associated with the drought over

the Yangtze River basin during spring 2011, J. Geophys. Res.-Atmos., 119, 5881–5894, 2014.

McKee, T. B., Doesken, N. J., and Kleist, J.: The relationship of drought frequency and duration to time scales, 8th Conference on Applied Climatology, Anaheim, California, 17–22 January, 1993.

Mehr, A. D., Kahya, E., and Ozger, M.: A gene-wavelet model for long lead time drought forecasting, J. Hydrol., 517, 691–699, https://doi.org/10.1016/j.jhydrol.2014.06.012, 2014.

Mishra, A. K. and Singh, V. P.: Drought modeling – A review, J. Hydrol., 403, 157–175, 2011.

Mo, K. C. and Lyon, B.: Global Meteorological Drought Prediction Using the North American Multi-Model Ensemble, J. Hydrometeorol., 16, 1409–1424, 2015.

Moreira, E. E., Pires, C. L., and Pereira, L. S.: SPI Drought Class Predictions Driven by the North Atlantic Oscillation Index Using Log-Linear Modeling, Water, 8, 18 pp., https://doi.org/10.3390/w8020043, 2016.

Otkin, J. A., Anderson, M. C., Hain, C., and Svoboda, M.: Using Temporal Changes in Drought Indices to Generate Probabilistic Drought Intensification Forecasts, J. Hydrometeorol., 16, 88–105, https://doi.org/10.1175/jhm-d-14-0064.1, 2015.

Reynolds, R. W., Smith, T. M., Liu, C., Chelton, D. B., Casey, K. S., and Schlax, M. G.: Daily high-resolution-blended analyses for sea surface temperature, J. Climate, 20, 5473–5496, https://doi.org/10.1175/2007jcli1824.1, 2007.

Rong, Y., Duan, L., and Xu, M.: Analysis on Climatic Diagnosis of Persistent Drought in North China during the Period from 1997 to 2002, Arid Zone Res., 25, 842–850, 2008.

Ropelewski, C. F. and Halpert, M. S.: Global and Regional Scale Precipitation Patterns Associated with the El Niño/Southern Oscillation, Mon. Weather Rev., 115, 1606–1626, https://doi.org/10.1175/1520-0493(1987)115<1606:GARSPP>2.0.CO;2, 1987.

Saha, S., Moorthi, S., Wu, X. R., Wang, J., Nadiga, S., Tripp, P., Behringer, D., Hou, Y. T., Chuang, H. Y., Iredell, M., Ek, M., Meng, J., Yang, R. Q., Mendez, M. P., Van Den Dool, H., Zhang, Q., Wang, W. Q., Chen, M. Y., and Becker, E.: The NCEP Climate Forecast System Version 2, J. Climate, 27, 2185–2208, 2014.

Shin, J. Y., Ajmal, M., Yoo, J., and Kim, T.-W.: A Bayesian Network-Based Probabilistic Framework for Drought Forecasting and Outlook, Adv. Meteorol., 2016, 9472605, https://doi.org/10.1155/2016/9472605, 2016.

Wallace, J. M. and Gutzler, D. S.: Teleconnections in the Geopotential Height Field during the Northern Hemisphere Winter, Mon. Weather Rev., 109, 784–812, 1981.

Wang, H. J. and He, S. P.: The North China/Northeastern Asia Severe Summer Drought in 2014, J. Climate, 28, 6667–6681, 2015.

Wei, J., Zhang, Q., and Tao, S.: Physical Causes of the 1999 and 2000 Summer Severe Drought in North China, Chinese J. Atmos. Sci., 28, 125–137, 2004.

Wilks, D. S.: Principal Component (EOF) Analysis, in: Statistical methods in the atmospheric sciences, Academic press, 519–562, 2011.

Wood, E. F., Schubert, S. D., Wood, A. W., Peters-Lidard, C. D., Mo, K. C., Mariotti, A., and Pulwarty, R. S.: Prospects for Advancing Drought Understanding, Monitoring, and Prediction, J.

Hydrometeorol., 16, 1636–1657, 2015.

World Meteorological Organization (WMO): Standardized Precipitation Index User Guide, Geneva, Switzerland: available at: http://www.wamis.org/agm/pubs/SPI/WMO_1090_EN.pdf (last access: 20 November 2017), 2012.

Yang, J., Gong, D. Y., Wang, W. S., Hu, M., and Mao, R.: Extreme drought event of 2009/2010 over southwestern China, Meteorol.

Atmos. Phys., 115, 173–184, 2012.

Yoon, J. H., Mo, K., and Wood, E. F.: Dynamic-Model-Based Seasonal Prediction of Meteorological Drought over the Contiguous United States, J. Hydrometeorol., 13, 463–482, 2012.

Yuan, X., Wood, E. F., Roundy, J. K., and Pan, M.: CFSv2-Based Seasonal Hydroclimatic Forecasts over the Conterminous United States, J. Climate, 26, 4828–4847, 2013.

Crop-specific seasonal estimates of irrigation-water demand in South Asia

Hester Biemans[1], **Christian Siderius**[1,2], **Ashok Mishra**[3], **and Bashir Ahmad**[4]

[1]Alterra, Wageningen University and Research Centre, Wageningen, the Netherlands

[2]Environmental Economics and Natural Resources Group, Wageningen University, Wageningen, the Netherlands

[3]Agricultural and Food Engineering Department, IIT Kharagpur, Kharagpur, India

[4]Pakistan Agricultural Research Council, Islamabad, Pakistan

Correspondence to: Hester Biemans (hester.biemans@wur.nl)

Abstract. Especially in the Himalayan headwaters of the main rivers in South Asia, shifts in runoff are expected as a result of a rapidly changing climate. In recent years, our insight into these shifts and their impact on water availability has increased. However, a similar detailed understanding of the seasonal pattern in water demand is surprisingly absent. This hampers a proper assessment of water stress and ways to cope and adapt. In this study, the seasonal pattern of irrigation-water demand resulting from the typical practice of multiple cropping in South Asia was accounted for by introducing double cropping with monsoon-dependent planting dates in a hydrology and vegetation model. Crop yields were calibrated to the latest state-level statistics of India, Pakistan, Bangladesh and Nepal. The improvements in seasonal land use and cropping periods lead to lower estimates of irrigation-water demand compared to previous model-based studies, despite the net irrigated area being higher. Crop irrigation-water demand differs sharply between seasons and regions; in Pakistan, winter (rabi) and monsoon summer (kharif) irrigation demands are almost equal, whereas in Bangladesh the rabi demand is ~ 100 times higher. Moreover, the relative importance of irrigation supply versus rain decreases sharply from west to east. Given the size and importance of South Asia improved regional estimates of food production and its irrigation-water demand will also affect global estimates. In models used for global water resources and food-security assessments, processes like multiple cropping and monsoon-dependent planting dates should not be ignored.

1 Introduction

As global demand for food increases, water resources – one of the main resources for producing food – are becoming increasingly stressed. South Asia, home to $\sim 25\%$ of the world population, is often identified as one of the future water-stress hotspots (Kummu et al., 2014; Wada et al., 2011). Excess food production in recent years has obscured this bleak future; increases in both agricultural productivity and cropland extension have made the region food self-sufficient in its staple crops in recent decades. But the resources that supported this increase – surface- and groundwater extracted for irrigation, land converted into cropland, increased use of nutrients and pesticides – are not unlimited. Groundwater levels are already falling rapidly in large parts of South Asia due to overexploitation (Rodell et al., 2009; Tiwari et al., 2009) and surface-water irrigation is reaching its limits (Biemans, 2012), costly river interlinking schemes aside (Bagla, 2014; Gupta and Deshpande, 2004). In addition, higher temperatures and an expected higher variability in climate due to global warming further jeopardizes future food production in this region (Krishna Kumar et al., 2004; Mall et al., 2006; Moors et al., 2011).

In order to understand if, when and where water availability to sustain crop production becomes critical, a more thorough understanding of the potential mismatch between seasonal water availability and demand is required. In recent years, our insight into the seasonal pattern of water availability has increased due to a better understanding of fluctuations in monsoon onset (Goswami et al., 2010; Kajikawa et al., 2012; Ren and Hu, 2014), and the variation in the active–break cycle of the monsoon, which governs intra-seasonal droughts (Joseph and Sabin, 2008), both influenced by large-scale phenomena like El Niño (Joseph et al., 1994). Effort has also gone into quantifying the seasonal availability of snowmelt and glacier melt runoff on the regional scale (Bookhagen and Burbank, 2010; Siderius et al., 2013a), with intra-annual shifts in runoff expected in the future due to climate change (Immerzeel et al., 2013; Lutz et al., 2014; Mathison et al., 2015; Rees and Collins, 2006). When it comes to estimating water demand, however, a similar detailed understanding of the seasonal pattern is surprisingly absent.

Two essential and well-known agricultural characteristics that distinguish South Asia from most other large food-producing regions in the world govern this water demand. First, South Asia's agriculture is characterized by a high degree of multiple cropping. A first crop during the monsoon season (kharif) is often succeeded by a second crop during the dry season (rabi) (Portmann et al., 2010). Planting dates for the kharif crop are determined primarily by the onset of the monsoon rather than by an accumulation of degree days. High maximum temperatures form a constraint for crop production during the rabi season, favouring planting as early as possible. Second, with rainfall highly concentrated during June–September and significant moisture deficits occurring during the other months of the year, crop production is to a very large extent supported by a combination of canal and groundwater irrigation, especially in the dry winter season (rabi) (GoI, 2012).

Many models that are used for global to regional water resources assessments still lack representation of multiple cropping (e.g. Arnold and Fohrer, 2005; Best et al., 2011; Gerten et al., 2004; Liang et al., 1994). Typically, a single cropping period per year is simulated with a degree-day-based or predefined single planting date (see, e.g., Elliott et al., 2014; Kummu et al., 2014). Exceptions are the model by Wada et al. (2011), who applied multiple cropping in their estimation of water stress, but in a simplified aggregated form without distinguishing between different crops and the models of Alcamo et al. (2003) and Hanasaki et al. (2008), who applied multiple-cropping seasons using optimized planting dates. However, Hanasaki et al. (2008) noted that their optimization mainly reacted to cold spells and was performed under rainfed conditions, which does not lead to optimal planting dates for the South Asia region. The study of Hoogeveen et al. (2015) accounted for multiple cropping by incorporating national level FAO cropping calenders, but only present total mean annual irrigation demands for South Asia (Table 1). Siebert and Döll (2010) also took multiple cropping into account by using MIRCA land use data (as in the present study; see Sect. 2.2) and cropping calenders (Portmann et al., 2010). They showed results for global seasonal irrigation demands, but not for South Asia specifically. As a result, crop-specific seasonal estimates of irrigation-water demand in South Asia are still lacking.

Table 1. Seasonal and total net and gross irrigation-water demand estimates (BCM yr⁻¹) and groundwater contribution to irrigation-water supply for individual countries and South Asia as a whole (India, Pakistan, Nepal and Bangladesh).

	Net irrigation demand (consumption)					Percentage groundwater irrigation					Gross irrigation demand (withdrawal)				
	Kharif (M6–10)	Rabi (M11–3)	Summer (M4–5)	Total	Other estimates Total	Kharif (M6–10)	Rabi (M11–3)	Summer (M4–5)	Total	Other estimates Total	Kharif (M6–10)	Rabi (M11–3)	Summer (M4–5)	Total	Other estimates Total
Nepal	0.1	10	0.2	14	4.4[d]	19 %	62 %	34 %	54 %	20 %[d]	0.3	2.0	0.5	2.7	10[e]
Pakistan	38	42	16	96	10[d]	25 %	68 %	25 %	44 %	33 %[d]	110	86	47	243	200.2[h], 162.7[b], 117–120[c], 187.8[g]
India	59	148	31	235	317[d]	27 %	79 %	63 %	64 %	64 %[d]	136	249	58	443	575.9[h], 541[a], 558.4[b], 710–715[c]
Bangladesh	0.1	11	0.3	12	24[d]	10 %	43 %	2 %	41 %	76 %[d]	0.2	24	0.8	25	31[e]
South Asia	97	200	48	346	480[i], 532[j]	26 %	74 %	50 %	58 %		247	361	106	714	985[f], 910[j]

[a] GoI (2006); *Water Data Complete Book*, Central Water Commission, Ministry of Water Resources, Government of India; [b] AQUASTAT (http://www.fao.org/nr/water/aquastat/main/index.stm); [c] Rost et al. (2008); [d] Siebert et al. (2010); [e] AQUASTAT with reference to 2008 for Bangladesh and 2005 for Nepal: approximately 79 % and 21 % of the total water withdrawal comes from groundwater (Nepal and Bangladesh, respectively); [f] Rosegrant and Cai (2002); 1995 estimate using a basin efficiency of 0.54; [g] Water Resources Section, Ministry of Planning and Development in Ahmed et al. (2007); [h] Biemans et al. (2013); [i] Siebert and Döll (2010); [j] Hoogeveen et al. (2015).

In this paper, we aim to provide such spatially explicit, crop-specific seasonal estimates of water demand and crop production, using a revised version of the Lund-Potsdam-Jena managed Land (LPJmL) hydrology and vegetation model (Gerten et al., 2004), adjusted for the region. We distinguish two main South Asian cropping periods, kharif and rabi, and introduce zone-specific, monsoon-onset-determined planting dates for 12 major crop types, both rainfed and irrigated. We calibrate the improved model against the latest sub-national statistics on seasonal crop yields from four different countries – India, Pakistan, Nepal and Bangladesh – and explicitly evaluate the irrigation-water demand and crop production for the two cropping seasons.

2 Methodology

2.1 LPJmL

We used the LPJmL global hydrology and vegetation model for bio- and agro-spheres (Bondeau et al., 2007; Sitch et al., 2003), but developed a version that contains more spatial and temporal detail for South Asia. The LPJmL model has been widely applied to study the effects of climate change on water availability and requirements for food production at a global scale (Gerten et al., 2011; Falkenmark et al., 2009) and the potential of rainfed water-management options for raising global crop yields (Rost et al., 2009). For South Asia, the model has been applied to study the adaptation potential of increased dam capacity and improved irrigation efficiency in light of climate change (Biemans et al., 2013). LPJmL physically links the terrestrial hydrological cycle to the carbon cycle, making it a suitable tool for studying the relationship between water availability and crop production. The model includes algorithms to account for human influences on the hydrological cycle, e.g. irrigation extractions and supply (Rost et al., 2008). Production and water use for 12 different crops, both rainfed and irrigated are simulated. LPJmL is a grid-based model, run at a resolution of 0.5°, and at a daily time step.

Net irrigation-water demand (consumption) for irrigated crops is calculated daily in each grid cell as the minimum amount of additional water needed to fill the soil to field capacity and the amount needed to fulfil the atmospheric evaporative demand (Rost et al., 2008). Subsequently, the gross irrigation demand (withdrawal) accounts for application and conveyance losses, and is calculated by multiplying the net irrigation-water demand with a country-specific efficiency factor (Rohwer et al., 2007), which is different for surface-water irrigation and groundwater irrigation (as in Biemans et al., 2013). Irrigation efficiency for canal water is estimated at 37.5 % in India, Bangladesh, Nepal and 30 % in Pakistan (Rohwer et al., 2007); efficiency of groundwater irrigation is estimated at 70 % for all countries (following Gupta and Deshpande, 2004).

Surface water is defined as the water available in local rivers, lakes and reservoirs and is calculated by a daily routing algorithm (Biemans et al., 2009). Irrigation-water demand is assumed to be withdrawn from available surface water first. If surface water is unavailable, it is assumed to be withdrawn from groundwater (Rost et al., 2008).

Crop growth is simulated based on daily assimilation of carbon in four pools: leaves, stems, roots and harvestable storage organs. Carbon allocated to those pools depends on crop phenology and is adjusted in case of water stress on the plants. Crops are harvested when either maturity or the maximum number of growing days is reached (Bondeau et al., 2007; Fader et al., 2010).

To improve the understanding of spatial and temporal heterogeneity in irrigation-water demand and crop production in South Asia, we made some adjustments to the version of LPJmL that is used for global studies. First of all, we introduced the simulation of two cropping cycles per year by developing two different land use maps for kharif and rabi. Second, we applied zone-specific sowing dates related to monsoon patterns. Third, we accounted for regional differences in crop management by performing a calibration of crop yields at the subnational level. In the next three sections, those adjustments to LPJmL are explained in more detail.

In our experimental set-up, LPJmL is forced with daily precipitation, daily mean temperature, net longwave and downward shortwave radiation derived from the WFDEI data set (WATCH Forcing Data methodology applied to ERA-Interim reanalysis data) (Weedon et al., 2014). Using this data set, all LPJmL simulations were done for the period 1979–2009 after a 1000 year spin-up period to bring carbon and water pools into equilibrium. The calibration and all analysis presented in this paper uses the simulation results of the period 2003–2008 for comparison with available statistics. Kharif and rabi irrigation-water demand and crop production are estimated by performing two simulations using different land use input and sowing-date input data sets. Those two runs are subsequently combined to attain the seasonal pattern for irrigation-water demand and crop production.

For comparison and to show model improvements, LPJmL is also run with the single cropping land use input as in previous model studies (Biemans et al., 2013) for which sowing dates are determined based on climate as in Waha et al. (2012).

2.2 Development of land use maps for kharif and rabi seasons

To derive land use input for two separate cropping seasons for South Asia, we used the MIRCA2000 database (MIRCA, version 1.1; Portmann et al., 2010) on a 5 min resolution. MIRCA is a global spatially explicit data set on irrigated and rainfed monthly crop areas for 26 crop classes around the year 2000. On an annual basis, MIRCA is consistent with other gridded data sets for total cropland extent (Ramankutty et al., 2008), total harvested area (Monfreda et al., 2008), and area equipped for irrigation (Siebert et al., 2007), but has more temporal detail. For India, MIRCA2000 includes subnational (i.e. state-level) information on the start and end of cropping periods. The data set explicitly includes multiple cropping.

Crop classes in MIRCA2000 were first aggregated to the crop classes available in the LPJmL model, which are fewer (12, irrigated and non-irrigated, plus one class with "other perennial crops", versus 26 in MIRCA) but include the most important food crops for South Asia (see Fig. 2 for distinguished crops). The exact period of monsoon (kharif) and dry season (rabi) cropping differs according to region. In India, kharif sowing is strongly related to the onset of the monsoon, whereas in large parts of Pakistan – where the monsoon is less pronounced – sowing can happen earlier or later because other factors like water availability for irrigation are more important. From the monthly MIRCA cropping calendars we decided to define the cropped area of the kharif season as the area under cultivation per crop as in September and that of rabi as the area per crop as in January. Perennial crops were only included in the kharif land use map.

Next, a few adjustments to the obtained data were made. First, MIRCA specifies three rotations of rice in northern India, two during summer and one during winter months. We merged the two summer rotations to the kharif rice area and allocated one to the rabi rice area, accepting a potential minor mismatch between data sets. Second, we corrected wheat and rice areas, both of which MIRCA equally divides over rabi and kharif. In reality, rice is mainly cropped during the kharif season and wheat is only cropped during the rabi (winter) season, when temperatures are lower and heat stress is avoided. We shifted all irrigated wheat to the rabi season and made compensations where possible by shifting an equal amount of irrigated rice area to the kharif season. Third, we shifted 45 % of area cropped with pulses from the rabi to kharif season to comply with the latest agricultural statistics (GoI, 2012). In this way, consistency with other data sets was largely maintained (i.e. total cultivated area, cultivated area per crop, area irrigated), while at the same time a better match with crop phenology and regional agricultural practices was achieved.

Finally, we updated the area irrigated to the latest statistics. MIRCA represents land use and irrigated area for the period 1998–2002. Over the past 10 years, irrigated area has further increased in India alone from 76 to 86 million ha (gross irrigated area), to 44 % of the total area. Statistics for India show (GoI, 2012) that the increase in irrigated area occurred for all crops. By shifting 10 % of rainfed area to irrigated area, while keeping the overall cropped area the same, we achieved an increase in gross irrigated area. We assumed that the all-India trend is mirrored in the neighbouring counties. Cropped area was then aggregated to 0.5° grids for both kharif and rabi, which formed the input into the LPJmL model. The resulting

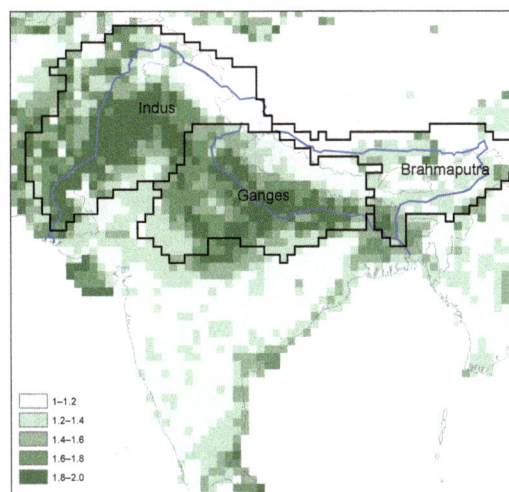

Figure 1. Cropping intensity in South Asia (land use data sets derived for this study based on MIRCA2000. Average cropping intensity is defined here as the total annual harvested area (kharif and rabi) divided by the maximum cropped area of the two cropping seasons. Study-basin delineations are indicated in black.

land use input is in good agreement with subnational statistics on cropping areas in kharif and rabi (see Supplement, Figs. S1–S6).

Figure 1 shows the cropping intensity in the study region according to this newly compiled data set, as well as the delineation of the river basins for which we will present our results. Figure 2 shows the total cropped area during the kharif and rabi seasons for all major crops in South Asia (India, Pakistan, Nepal and Bangladesh) according to the input data compiled here and compared to the agricultural statistics (GoI, 2014; Statistics, 2014).

2.3 Adjusted planting dates for kharif and rabi crops

Sowing dates for kharif crops are closely related to the onset of the monsoon as farmers start (trans)planting rice or other crops when the first rains have arrived. Normal onset dates of the monsoon over South Asia are determined by the India Meteorological Department, at 5- to 15-day intervals (IMD, 2015) (Fig. 3). The onset of the monsoon starts in Kerala in southern India around the first of June (Julian day 152) and arrives in western Pakistan around mid-July (Julian day 197). For the model simulations in this study, sowing dates for kharif crops were set to five days after the onset of the monsoon, because several days of rain are needed before a crop is (trans)planted (Fig. 3). Inter-annual variations in the onset of the monsoon were not taken into account in this study. The perennial crop sugarcane is assumed to be planted on this date as well.

In general, the kharif season ends by the end of October and the sowing of rabi crops starts early – from mid-November to early January, depending on local temperatures

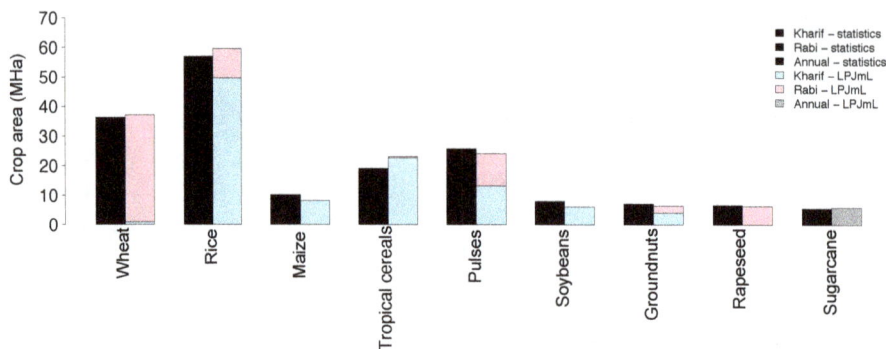

Figure 2. Total crop area in South Asia (India, Pakistan, Nepal and Bangladesh) for different crops in the two dominant growing seasons. National statistics (average of 2003–2008) versus LPJmL input data derived from MIRCA as described in Sect. 2.2. For the spatial distribution of crops between states and provinces of India and Pakistan, Nepal and Bangladesh, see Annex. Temperate and tropical roots and sunflower are not shown because they occupy relatively small areas; other perennial crops are not shown because there are no statistics available.

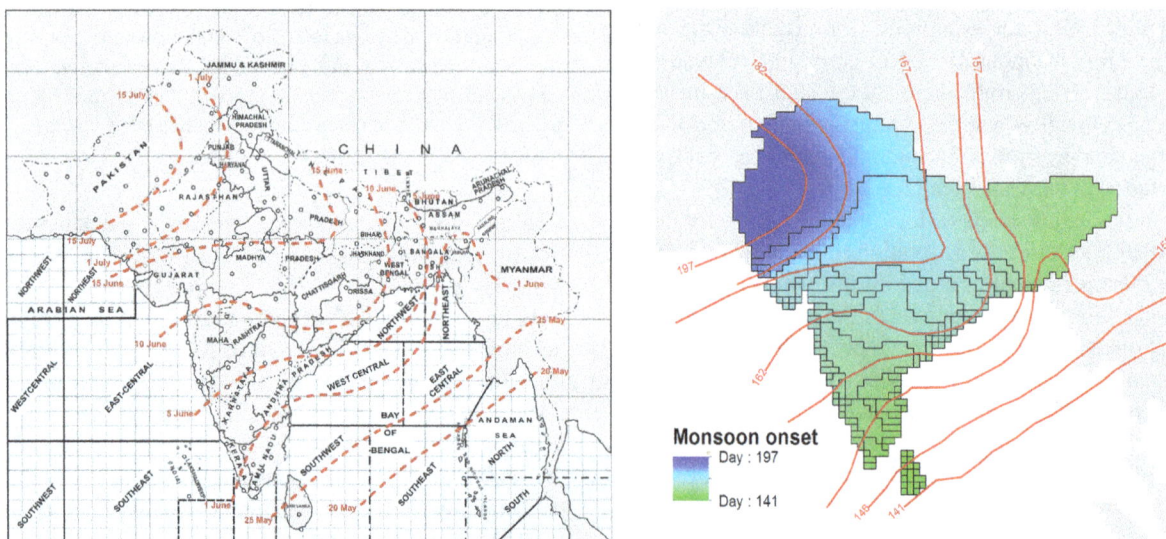

Figure 3. Normal dates for the onset of the south-west monsoon as presented by the Indian Meteorological Department (left panel) and interpolated over South Asia (right panel) derive input data for LPJmL, red numbers indicating Julian days, grey lines showing basin boundaries.

during winter and water availability in spring. As the exact date is difficult to determine, we set the first of November as the single sowing date for the rabi crops over the whole study area. Because the rabi crops are generally harvested by the end of March, the irrigation-water demand in the warm pre-monsoon summer months of April and May can almost entirely be attributed to perennial crops. In the analysis of seasonal irrigation demand, we therefore distinguish three seasons: kharif, from June until October; rabi, from November until March; and a dry "summer" season from April to May. This dry pre-monsoon summer season is sometimes also called Zaid season.

2.4 Calibration of crop yields

Crop yields in LPJmL are calibrated by varying management intensity, which is represented by three coupled parameters:

maximum leaf area index, maximum harvest index, and a parameter that scales leaf-level biomass production to plot level (Fader et al., 2010). The three parameters are related to crop density, crop varieties and the occurrence of poor soils, pests and diseases, respectively. "Plot level" in this context means the total area of the crop within the grid cell: a plot shares the same climate, soil and land use. "Scale" means that a yield reduction has been applied to translate from biomass production of individual plants to plot level. Fader et al. (2010) explain this as follows: "The assumption is that intensively managed crop stands (LAImax = 7) have little or no areas with reduced productivity due e.g. to poor soil conditions or pests and diseases ($\alpha - a = 1.0$), while such areas are more common in extensively managed crop stands (LAImax = 1; $\alpha - a = 0.4$)" (for a detailed description of the calibration procedure, see Fader et al., 2010).

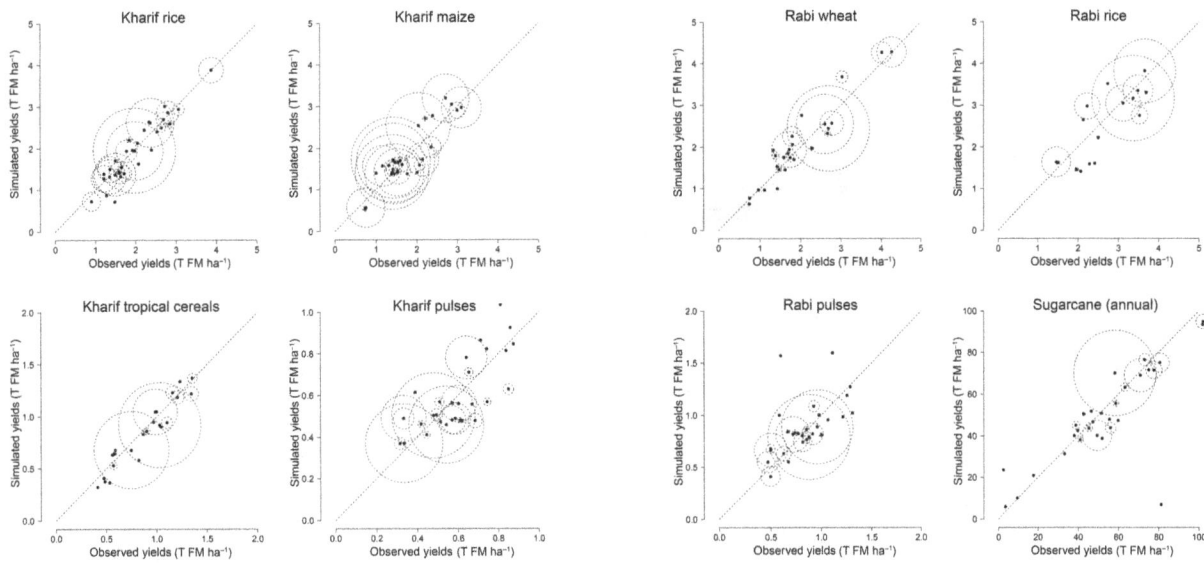

Figure 4. Observed vs. simulated (calibrated) crop yields for the most important crops in the different cropping seasons in tons of fresh matter per hectare (T FM ha^{-1}). Each dot represents one state (India), province (Pakistan) or country (Nepal, Bangladesh). Size of the circle represents the relative area under that crop (for areas, see Figs. S1–S6).

The value of these management factors affects the estimated water demand, because a poorly developed crop with little leaf area will evaporate less and therefore demands less (irrigation) water and vice versa.

The calibration is performed for each crop individually, and management factors are usually determined at the country level in global applications of LPJmL. For this model version, we calibrated crop yields for kharif and rabi separately, as they are differentiated in the agricultural statistics. Moreover, we calibrated the management parameters at the sub-national level for India and Pakistan (state and province level, respectively) and at the national level for Nepal and Bangladesh. By calibrating at the sub-national level, existing spatial heterogeneity in management and crop yields between regions could be better represented. We used 5-year average yield statistics, for 2003–2004 till 2007–2008, the most recent period for which consistent records are available from different national agricultural statistics (India: GoI, 2012; Pakistan: http://www.pbs.gov.pk/content/agricultural-statistics-pakistan-2010-11, last visited 1 July 2014; Bangladesh for the years from 2003–2004 till 2005–2006 form http://www.moa.gov.bd/statistics/statistics.htm#3 and for 2007–2008 in the 2011 yearbook http://www.bbs.gov.bd/PageWebMenuContent.aspx?MenuKey=234; Nepal; GoN, 2012). After calibration, the model is able to simulate the heterogeneity of (mean annual) yields between states and regions (illustrated in Fig. 4). Kharif rice and kharif maize crops show the highest variation between states and provinces. Overall, yields during the kharif season are lower than yields during the rabi season, when a higher percentage of the area cropped is irrigated, and temperatures are more favourable. Interannual variations in crop yields are shown and discussed by Siderius et al. (2016).

3 Results

3.1 Seasonality in agricultural water demand

Table 1 shows estimates of seasonal net (consumption) and gross (withdrawal from surface and groundwater) irrigation-water demand between the four countries. India and Pakistan have the largest water demand, both in terms of consumption and withdrawal. While Pakistan's net irrigation demand is almost equally divided over the kharif and rabi seasons, India's demand is skewed towards the rabi season; almost three-quarters of net irrigation demand in India occurs in this dry season (including pre-monsoon summer). This difference between kharif and rabi is less pronounced for gross irrigation demand, i.e. water withdrawals, which include application and conveyance losses. In the rabi season a much higher proportion of the irrigation water is supplied from groundwater (Table 1), which has a higher overall efficiency than surface-water irrigation from canals.

The seasonal distribution of irrigation-water demand is a result of rainfall patterns in the region. In Bangladesh and Nepal, monsoon rainfall is abundant for sustaining crop production during the kharif season and irrigation is therefore concentrated in the dry rabi season. Groundwater irrigation, modelled as the resultant of demand minus surface-water availability, provides most water resources

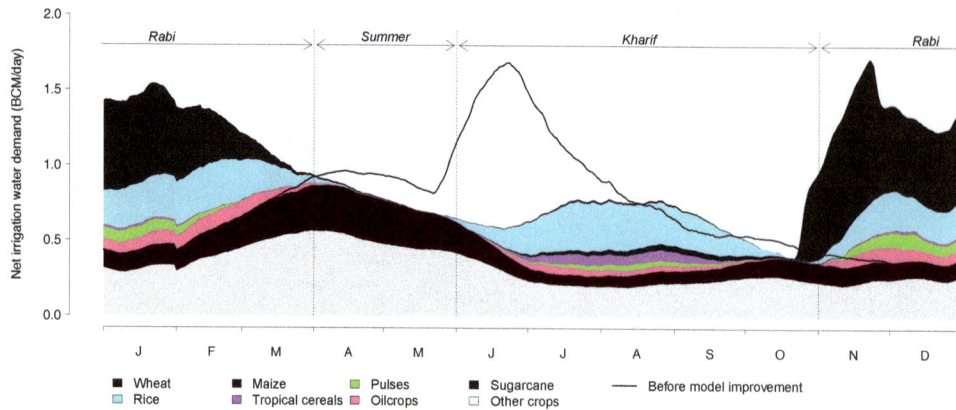

Figure 5. Mean annual cycle of net irrigation requirements for main agricultural crops in South Asia in BCM/day (30-day moving average). For comparison, the mean annual cycle of net irrigation requirements before model improvements (with single cropping season and climate driven sowing dates determination) is added in black.

during the rabi season in all countries, especially in India. In Pakistan, the Indus provides annually approximately 120 BCM yr^{-1} of utilizable runoff, of which approximately two-thirds is used during the kharif (Randhawa, 2002). Our estimate of mean annual groundwater withdrawal in Pakistan is at 60 BCM yr^{-1}, of which three-quarters occurs during the rabi season and pre-monsoon summer. This is somewhat higher than previous estimates of groundwater withdrawal, which were in the range of at 47 to 55 BCM yr^{-1} (Ahmed et al., 2007; Qureshi et al., 2003; Wada et al., 2010) but still lower than the estimated total potential of 68 BCM yr^{-1} (Randhawa, 2002). For India, the exact distribution of surface-water and groundwater withdrawal between the kharif and rabi seasons is not well documented. Our model estimate of 217 BCM yr^{-1} of groundwater withdrawal, mainly occurring during the rabi season, is in agreement with earlier groundwater studies with estimates ranging from 190 ($\pm$37) BCM yr^{-1} by Wada et al. (2010) to 212.5 BCM yr^{-1} (GoI, 2006).

Overall, our estimates of national total net and gross irrigation-water demand are in line with earlier studies and statistics, but at the lower end of the range for India. Accounting for monsoon-dependent planting dates, and thereby a more effective use of rainfall during the main kharif cropping season, reduced our estimate of total agricultural water demand compared to earlier regional studies, e.g. with the LPJmL model (Biemans et al., 2013). For Pakistan, our estimates are on the high side compared to other studies. Especially for the rabi season, we estimate a high additional demand from cash crops like cotton. This demand has to be met largely by groundwater abstractions, because runoff from the Indus and its tributaries is low during these months.

Evaluating the mean annual cycle of irrigation-water demand per crop reveals the reason behind seasonal differences in demand (Fig. 5). The single peak in net water demand for wheat during the rabi season stands out, while rice peaks in

both rabi and kharif seasons. The moderating effect of monsoon rainfall during the kharif season is obvious, with net irrigation-water demand during the kharif season only accounting for about 30 % of the annual net irrigation-water demand (Table 1). So while water-use efficiency improvements in rice receive much attention, paddy fields being the epitome of excessive water consumption, rice is actually not the most water-demanding crop in the region. Because rice is grown mainly during the kharif season in most states, its water demand is lower than for wheat and sugarcane, which are grown during the dry rabi season. Those crops therefore depend much more on groundwater availability (see also Table 1 and Fig. 6 for contribution of groundwater irrigation per cropping season). Additionally, sugarcane has an atypical demand in time, caused by its very long cultivation period of about 12 months; it requires large amounts of irrigation water in the hot dry months of March, April and May, a period when rainfall is scarce and most other fields are left fallow.

The mean annual cycle of irrigation demand as calculated with single cropping and sowing dates determined based on climate (before model improvement) are also shown in Fig. 5.

3.2 Seasonal patterns of water demand for different basins

As a result of varying climatological conditions and availability of spring and summer runoff from snow- and glacier-fed rivers, cropping patterns and thereby seasonal water demand patterns differ greatly between the major river basins (Figs. 6 and 7). The Indus basin shows a relatively stable irrigation-water demand during the year, which is primarily fed by groundwater in winter and melt runoff in summer (Fig. 7). Downstream, monsoon rainfall contributes little to crop water needs. In the Ganges basin, a more seasonal pattern can be seen with demand for irrigation water being lower during the monsoon, when rainfall is sufficient over large

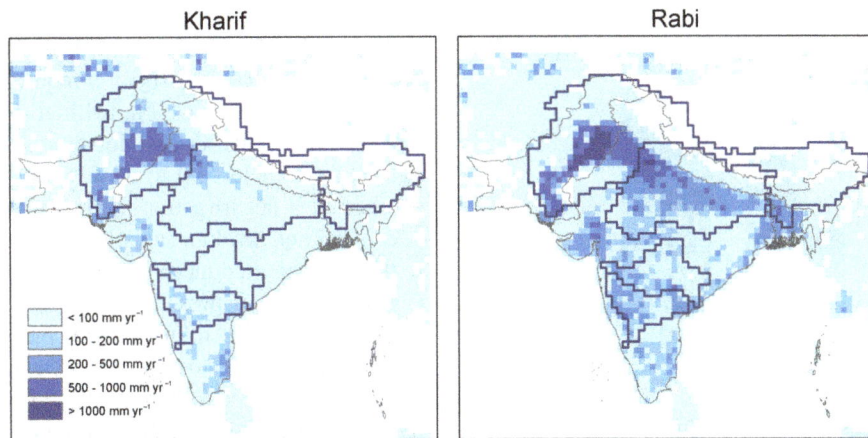

Figure 6. Gross irrigation-water demand for kharif (June to October) and rabi (November to March) cropping seasons, with selected river basins (Indus, Ganges, Brahmaputra).

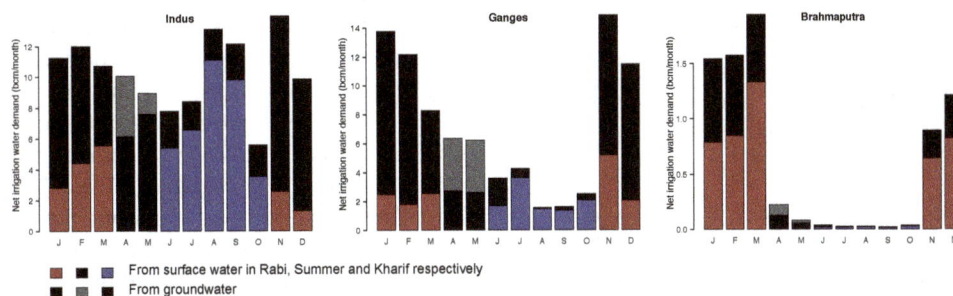

Figure 7. Monthly net irrigation-water demand for three river basins. Colours indicate the different seasons (red – kharif, grey – summer, blue – rabi) and the dark areas the source for supplying the irrigation water (dark – surface water, light – groundwater).

parts of the basin, and no additional irrigation is needed. The same pattern can be seen to be even stronger in the Brahmaputra basin.

3.3 Food production in South Asia during the kharif- and rabi-cropping seasons

Figure 8 shows the total seasonal production of only the five most important food crops (wheat, rice, maize, tropical cereals and pulses), both for the region as a whole as for the individual basins. The total area irrigated to grow these food crops is smaller in kharif than rabi (35 Mha vs. 46 Mha total for the four counties), but total (rainfed plus irrigated) area used to grow these food crops is much larger in kharif than rabi (95 Mha vs. 57 Mha). While the percentage of area under irrigation, productivity per hectare and sources of water used greatly differ between the kharif and rabi seasons, total regional food-crop production is remarkably similar in the two seasons. A lower cropped area during the rabi season is compensated for by higher yields. Of the total production of food crops in South Asia during the kharif season, ∼ 50 % is supported by irrigation (Fig. 8). In the rabi season up to ∼ 95 % of food-crop production is supported by irrigation.

These estimates agree with the recent study of Smilovic et al. (2015), who focus on rice (kharif and rabi) and wheat (rabi) production in India only. They show that during kharif 68 % of rice production is produced on irrigated lands, which is only 56 % of the rice area sown. During rabi this percentage is much higher: 96 % of the rice was irrigated (on 89 % of the sown area) and 97 % of the wheat production was irrigated (on 93 % of the sown area) (Smilovic et al., 2015).

We also calculated the potential rainfed yield on those areas currently irrigated. Absence of irrigation would reduce the kharif food-crop production with ∼ 15 % (dark blue bar in Fig. 8), against a reduction of almost 60 % in rabi. This stresses the importance of sufficient irrigation-water supply for achieving food security in this region.

A closer look into the seasonal food production in the different river basins shows clear differences. The Indus and the Ganges have a much higher annual production of food crops than the Brahmaputra.

Rabi is the most important season for the production of food crops in the Indus. The same is true for the Ganges, although the production levels between the seasons are closer to each other. The rainfed production is much larger in the Ganges than in the Indus. In the Brahmaputra basin, the ma-

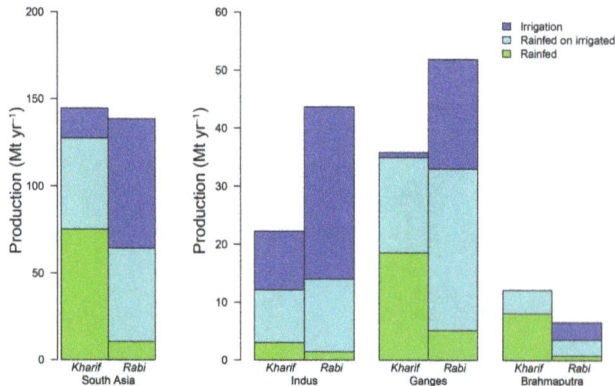

Figure 8. Mean annual seasonal irrigated (blue) and rainfed (green) production of food crops (sum of wheat, rice, maize, tropical cereals and pulses) in South Asia (Nepal, Pakistan, India and Bangladesh) and individual river basins. Light blue corresponds to potential rainfed production on irrigated land, i.e. dark blue corresponds to the increase in production due to irrigation.

jority of food-crop production takes place during the kharif season.

4   Discussion

The seasonal estimates presented here on food production and related irrigation-water demand in South Asia form a new baseline estimate of South Asian seasonal water demand and food-crop production, as they provide more spatial, temporal and crop-specific details than previous estimates.

Incorporating seasonal cropping patterns in more detail leads to improved estimation of the timing of water demand. Figure 5 shows that the simulated timing of water demand is very different compared to a simulation with old settings – thus single cropping season and calculated sowing dates. This difference shows the importance of including multiple cropping in the simulation of irrigation-water demand. Especially in this region with a very strong seasonal variability in both water availability and demand, an improved understanding of the (changes in) timing of both water availability and demand is essential to understand current and future water stressed regions. Therefore, the effect of multiple cropping on patterns of irrigation-water demand should not be neglected. We show that seasonal water demand is a factor of crop-specific seasonal consumption, availability of rainfall and different sources of water supply, i.e. groundwater or surface water, and the irrigation efficiencies connected to these sources. Despite these improvements, when modelling such large basins with complex hydrology and high diversity in agricultural and water-management practices, inevitably simplifications and local inaccuracies remain.

Our estimate of the net irrigation requirement (consumption) is influenced by the performed calibration and resulting management factors. Generally, regions with high manage-

ment factors will show higher yields and higher transpiration, but lower soil evaporation. The effect of the calibration on our estimate of net irrigation requirements was tested by making two model runs: one with all management parameters set to the lowest possible value and one with all management parameters set to the highest possible values. This resulted in a net irrigation requirement for South Asia between 307 and 389 km^3, a variation of about 10 % compared to the here reported mean annual value of 346 km^3.

Our estimate of gross irrigation demand, the water withdrawal, is strongly influenced by the water use efficiency value used, which is determined by a variety of factors like local irrigation practices, scale of analysis and source of water use. We used the most commonly reported values for the region, similar to other model-based studies in order to be able to compare results. Inclusion of regional, more application- and water-source-specific water use efficiency values in models would improve the estimation of gross water demand. Such detail is also necessary to gain better insight into the adaptation potential of different measures like drip irrigation and alternate wetting and drying.

More attention to seasonal cropping patterns and their water demand opens the scope for further model improvement. Double cropping was evaluated by combining two seasonal model runs, one for kharif and one for rabi. Use of residual soil moisture from one season to the other was not incorporated in this way, nor could the continued depletion of groundwater be accurately modelled. An integrated double-cropping routine, with proper calibrated crop-specific planting dates and yields, would provide such necessary analysis in a region where groundwater depletion is of serious concern.

Next, estimation of planting dates should be further improved, using detailed information on local agricultural practices and local water availability. Further, the sowing dates were kept constant during the whole simulation period and was based on average data of monsoon onset, although actual onsets vary year by year. In reality a farmer might decide year to year to sow earlier or later, which introduces an uncertainty in our calculations. Ample information is available in the irrigation domain but it will require a form of cooperation between experts at the local to national level and the water resources modelling community. Sharing of input data might reduce costs and time expenditure, will increase its uptake and improve overall quality of water resources assessments.

Finally, cropped area and sources of irrigation used are not constants or slowly evolving properties, but can be highly variable on inter-annual timescales in response to climate variability (Siderius et al., 2013b). These fluctuations were not assessed in the current study but are of high importance to individual farmers and the overall profitability of agriculture in regions with a variable climate. Combining an improved baseline of seasonal water demand with the inter-annual fluctuations in cropped area will lead to a more realistic assess ment of both water demand and crop production, of high relevance in today's world with its volatile food commodity markets.

This paper highlights crop-specific periods of peak water demand that can form critical moments in agricultural production. Such better understanding of the size of water demand during critical moments, the crops that are responsible for this water demand, and its relative importance for food production is essential to guide sustainable development of climate adaptation measures. This analysis can support the selection of promising options to decrease irrigation-water demand. When combined with information on the (un)availability of surface water and the resulting pressure on groundwater resources (Fig. 7), it improves our understanding of the causes of water shortages and groundwater depletion. Finally, insight into the yield gap between rainfed and irrigated agriculture in specific regions, and between regions, can help target investments to improve irrigation practices or to increase productivity of rainfed agriculture.

5 Conclusions

Introducing seasonal crop rotation with monsoon-dependent planting dates in a global vegetation–hydrological model leads to better seasonal estimates of irrigation-water demand. Irrigation-water demand between the two main cropping seasons differs sharply both in terms of source and magnitude; gross irrigation demand during the rabi season is $\sim 30\%$ higher than during the kharif season, the traditional cropping season, when monsoon rainfall reduces the amount of supplemental irrigation water needed. Our estimate of total annual water demand is lower than that of previous studies (Biemans et al., 2013), despite the net irrigated area being higher. Overall, gross annual irrigation demand is estimated at $714\,\mathrm{BCM\,yr^{-1}}$; $247\,\mathrm{BCM\,yr^{-1}}$ during the kharif monsoon season, $361\,\mathrm{BCM\,yr^{-1}}$ during rabi and $106\,\mathrm{BCM\,yr^{-1}}$ during the dry summer months of April and May.

Seasonal estimates of agricultural water demand better highlight crop-specific differences in peak water demand. Such increased temporal detail is needed for properly evaluating the impact of expected shifts in supply of water as a result of a rapidly changing climate, especially in the Himalayan headwaters of some of the main rivers in South Asia. With temperatures rising and total precipitation fairly constant, increased melt from glaciers combined with an early melt of the snow cover is expected to shift the peak in spring runoff to early in the season (Immerzeel et al., 2010; Lutz et al., 2014). Whether this shift will affect critical moments for irrigation or the ecosystem as a whole is to be assessed.

Our study has thereby more than regional relevance. Given the size and importance of South Asia, in terms of population and food production, improved regional estimates of production and its water demand will also affect global estimates.

In models used for global water resources and food-security assessments, processes like multiple cropping and monsoon-dependent planting dates should not be ignored.

Acknowledgements. This work was carried out by the Himalayan Adaptation, Water and Resilience (HI-AWARE) consortium under the Collaborative Adaptation Research Initiative in Africa and Asia (CARIAA), with financial support from the UK Government's Department for International Development and the International Development Research Centre, Ottawa, Canada. We acknowledge the Potsdam Institute for Climate Impact Research for their support in using the LPJmL model and computational facilities.

Disclaimer. The views expressed in this work are those of the creators and do not necessarily represent those of the UK Government's Department for International Development, the International Development Research Centre, Canada or its Board of Governors.

Edited by: C. Stamm

References

Ahmed, A., Iftikhar, H., and Chaudhry, G.: Water resources and conservation strategy of Pakistan, Pakistan Dev. Rev., 997–1009, 2007.

Alcamo, J., Döll, P., Henrichs, T., Kaspar, F., Lehner, B., Rösch, T., and Siebert, S.: Development and testing of the WaterGAP 2 global model of water use and availability, Hydrolog. Sci. J., 48 317–337, 2008.

Arnold, J. G. and Fohrer, N.: SWAT2000: current capabilities and research opportunities in applied watershed modelling, Hydrol. Process., 19, 563–572, 2005.

Bagla, P.: India plans the grandest of canal networks, Science, 345, 128, doi:10.1126/science.345.6193.128, 2014.

Best, M. J., Pryor, M., Clark, D. B., Rooney, G. G., Essery, R. L. H., Ménard, C. B., Edwards, J. M., Hendry, M. A., Porson, A., Gedney, N., Mercado, L. M., Sitch, S., Blyth, E., Boucher, O., Cox, P. M., Grimmond, C. S. B., and Harding, R. J.: The Joint UK Land Environment Simulator (JULES), model description – Part 1: Energy and water fluxes, Geosci. Model Dev., 4, 677–699, doi:10.5194/gmd-4-677-2011, 2011.

Biemans, H.: Water constraints on future food production, Wageningen UR, Wageningen, 168 pp., 2012.

Biemans, H., Hutjes, R. W. A., Kabat, P., Strengers, B. J., Gerten, D., and Rost, S.: Effects of precipitation uncertainty on discharge calculations for main river basins, J. Hydrometeorol., 10, 1011–1025, 2009.

Biemans, H., Speelman, L., Ludwig, F., Moors, E., Wiltshire, A., Kumar, P., Gerten, D., and Kabat, P.: Future water resources for food production in five South Asian river basins and potential for adaptation – A modeling study, Sci. Total Environ., 468, S117–S131, 2013.

Bondeau, A., Smith, P. C., Zaehle, S., Schaphoff, S., Lucht, W., Cramer, W., and Gerten, D.: Modelling the role of agriculture for the 20th century global terrestrial carbon balance, Global Change Biol., 13, 679–706, 2007.

Bookhagen, B. and Burbank, D. W.: Toward a complete Himalayan hydrological budget: Spatiotemporal distribution of snowmelt and rainfall and their impact on river discharge, J. Geophys. Res., 115, 1–25, 2010.

Elliott, J., Deryng, D., Müller, C., Frieler, K., Konzmann, M., Gerten, D., Glotter, M., Flörke, M., Wada, Y., Best, N., Eisner, S., Fekete, B. M., Folberth, C., Foster, I., Gosling, S. N., Haddeland, I., Khabarov, N., Ludwig, F., Masaki, Y., Olin, S., Rosenzweig, C., Ruane, A. C., Satoh, Y., Schmid, E., Stacke, T., Tang, Q., and Wisser, D.: Constraints and potentials of future irrigation water availability on agricultural production under climate change, P. Natl. Acad. Sci., 111, 3239–3244, 2014.

Fader, M., Rost, S., Müller, C., Bondeau, A., and Gerten, D.: Virtual water content of temperate cereals and maize: Present and potential future patterns, J. Hydrol., 384, 218–231, 2010.

Falkenmark, M., Rockstrom, J., and Karlberg, L.: Present and future water requirements for feeding humanity, Food Secur., 1, 59–69, 2009.

Gerten, D., Schaphoff, S., Haberlandt, U., Lucht, W., and Sitch, S.: Terrestrial vegetation and water balance – hydrological evaluation of a dynamic global vegetation model, J. Hydrol., 286, 249–270, 2004.

Gerten, D., Heinke, J., Hoff, H., Biemans, H., Fader, M., and Waha, K.: Global water availability and requirements for future food production, J. Hydrometeorol., 12, 885–899, doi:10.1175/2011jhm1328.1, 2011.

GoI: Dynamic groundwater resources of India (as of March, 2004), Central Ground Water Board, Ministry of Water Resources, Government of India, Faridabad, 2006.

GoI: Agricultural Statistics at a glance 2012, Government of India, Ministry of Agriculture, New Delhi, 2012.

GoI: http://eands.dacnet.nic.in/, last access: 22 July 2014, Directorate of economics and statistics, Department of Agriculture and Cooperation, Ministery of Agriculture, Government of India, 2014.

GoN: Statistical information on Nepalese Agriculture 2011/2012, Government of Nepal, Ministry of Agriculture, Kathmandu, 2012.

Goswami, B. N., Kulkarni, J. R., Mujumdar, V. R., and Chattopadhyay, R.: On factors responsible for recent secular trend in the onset phase of monsoon intraseasonal oscillations, Int. J. Climatol., 30, 2240–2246, 2010.

Gupta, S. K. and Deshpande, R. D.: Water for India in 2050: first-order assessment of available options, Current Science, 86, 1216–1224, 2004.

Hanasaki, N., Kanae, S., Oki, T., Masuda, K., Motoya, K., Shirakawa, N., Shen, Y., and Tanaka, K.: An integrated model for the assessment of global water resources – Part 2: Applications and assessments, Hydrol. Earth Syst. Sci., 12, 1027–1037, doi:10.5194/hess-12-1027-2008, 2008.

Hoogeveen, J., Faurès, J.-M., Peiser, L., Burke, J., and van de Giesen, N.: GlobWat – a global water balance model to assess water use in irrigated agriculture, Hydrol. Earth Syst. Sci., 19, 3829–3844, doi:10.5194/hess-19-3829-2015, 2015.

IMD: http://www.imd.gov.in/doc/wxfaq.pdf, last access: 30 September 2014, Indian Meteorological Department, Pune, 2015.

Immerzeel, W. W., van Beek, L. P. H., and Bierkens, M. F. P.: Climate Change Will Affect the Asian Water Towers, Science, 328, 1382–1385, 2010.

Immerzeel, W. W., Pellicciotti, F., and Bierkens, M. F. P.: Rising river flows throughout the twenty-first century in two Himalayan glacierized watersheds, Nat. Geosci., 6, 742–745, 2013.

Joseph, P. V. and Sabin, T. P.: An ocean–atmosphere interaction mechanism for the active break cycle of the Asian summer monsoon, Clim. Dynam., 30, 553–566, 2008.

Joseph, P. V., Eischeid, J. K., and Pyle, R. J.: Interannual variability of the onset of the Indian summer monsoon and its association with atmospheric features, El Nino, and sea surface temperature anomalies, J. Climate, 7, 81–105, 1994.

Kajikawa, Y., Yasunari, T., Yoshida, S., and Fujinami, H.: Advanced Asian summer monsoon onset in recent decades, Geophys. Res. Lett., 39, L03803, doi:10.1029/2011GL050540, 2012.

Krishna Kumar, K., Rupa Kumar, K., Ashrit, R. G., Deshpande, N. R., and Hansen, J. W.: Climate impacts on Indian agriculture, Int. J. Climatol., 24, 1375–1393, 2004.

Kummu, M., Gerten, D., Heinke, J., Konzmann, M., and Varis, O.: Climate-driven interannual variability of water scarcity in food production potential: a global analysis, Hydrol. Earth Syst. Sci., 18, 447–461, doi:10.5194/hess-18-447-2014, 2014.

Liang, X., Lettenmaier, D. P., Wood, E. F., and Burges, S. J.: A Simple hydrologically Based Model of Land Surface Water and Energy Fluxes for GSMs, J. Geophys. Res., 99, 415–428, 1994.

Lutz, A., Immerzeel, W., Shrestha, A., and Bierkens, M.: Consistent increase in High Asia's runoff due to increasing glacier melt and precipitation, Nat. Clim. Change, 4, 587–592, doi:10.1038/nclimate2237, 2014.

Mall, R., Singh, R., Gupta, A., Srinivasan, G., and Rathore, L.: Impact of Climate Change on Indian Agriculture: A Review, Climatic Change, 78, 445–478, 2006.

Mathison, C., Wiltshire, A. J., Falloon, P., and Challinor, A. J.: South Asia river-flow projections and their implications for water resources, Hydrol. Earth Syst. Sci., 19, 4783–4810, doi:10.5194/hess-19-4783-2015, 2015.

Monfreda, C., Ramankutty, N., and Foley, J. A.: Farming the planet: 2. Geographic distribution of crop areas, yields, physiological types, and net primary production in the year 2000, Global Biogeochem. Cy., 22, GB1022, doi:10.1029/2007GB002947, 2008.

Moors, E. J., Groot, A., Biemans, H., van Scheltinga, C. T., Siderius, C., Stoffel, M., Huggel, C., Wiltshire, A., Mathison, C., Ridley, J., Jacob, D., Kumar, P., Bhadwal, S., Gosain, A., and Collins, D. N.: Adaptation to changing water resources in the Ganges basin, northern India, Environ. Sci. Policy, 14, 758–769, 2011.

Portmann, F. T., Siebert, S., and Döll, P.: MIRCA2000 – Global monthly irrigated and rainfed crop areas around the year 2000: A new high-resolution data set for agricultural and hydrological modeling, Global Biogeochem. Cy., 24, Gb1011, doi:10.1029/2008gb003435, 2010.

Qureshi, A. S., Shah, T., and Akhtar, M.: The groundwater economy of Pakistan, Working Paper 64, International Water Management Institute, Lahore, Pakistan, 31 pp., 2003.

Ramankutty, N., Evan, A. T., Monfreda, C., and Foley, J. A.: Farming the planet: 1. Geographic distribution of global agricultural lands in the year 2000, Global Biogeochem. Cy., 22, GB1003, doi:10.1029/2007gb002952, 2008.

Randhawa, H. A.: Water development for irrigated agriculture in

Pakistan: Past trends, returns and future requirements, Food and Agricultural Organization (FAO), FAO Corporate Document Repository, available at: www.fao.org/DOCREP/005/AC623E/ac623e0i.htm (last access: 5 July 2014), 2002.

Rees, H. G. and Collins, D. N.: Regional differences in response of flow in glacier-fed Himalayan rivers to climatic warming, Hydrol. Process., 20, 2157–2169, 2006.

Ren, R. and Hu, J.: An emerging precursor signal in the stratosphere in recent decades for the Indian summer monsoon onset, Geophys. Res. Lett., 41, 7391–7396 doi:10.1002/2014GL061633, 2014.

Rodell, M., Velicogna, I., and Famiglietti, J. S.: Satellite-based estimates of groundwater depletion in India, Nature, 460, 999–1002, 2009.

Rohwer, J., Gerten, D., and Lucht, W.: Development of Functional Irrigation Types for Improved Global Crop Modelling, No. 104, Potsdam Institute for Climate Impact Research, Postdam, 98 pp., 2007.

Rosegrant, M. W. and Cai, X.: Global Water Demand and Supply Projections, Water Int., 27, 170–182, 2002.

Rost, S., Gerten, D., Bondeau, A., Lucht, W., Rohwer, J., and Schaphoff, S.: Agricultural green and blue water consumption and its influence on the global water system, Water Resour. Res., 44, W09405, doi:10.1029/2007WR006331, 2008.

Rost, S., Gerten, D., Hoff, H., Lucht, W., Falkenmark, M., and Rockström, J.: Global potential to increase crop production through water management in rainfed agriculture, Environ. Res. Lett., 4, 044002, doi:10.1088/1748-9326/4/4/044002, 2009.

Siderius, C., Biemans, H., Walsum, P. E. V. V., Ierland, E. V., Kabat, P., and Hellegers, P.: Snowmelt contributions to discharge of the Ganges, Sci. Total Environ., 468–469, S93–S101, 2013a.

Siderius, C., Hellegers, P. J. G. J., Mishra, A., van Ierland, E. C., and Kabat, P.: Sensitivity of the agroecosystem in the Ganges basin to inter-annual rainfall variability and associated changes in land use, Int. J. Climatol., 34, 3066–3077, doi:10.1002/joc.3894, 2013b.

Siderius, C., Biemans, H., van Walsum, P. E., van Ierland, E. C., Kabat, P., and Hellegers, P. J.: Flexible Strategies for Coping with Rainfall Variability: Seasonal Adjustments in Cropped Area in the Ganges Basin, Plos One, 11, e0149397, doi:10.1371/journal.pone.0149397, 2016.

Siebert, S. and Döll, P.: Quantifying blue and green virtual water contents in global crop production as well as potential production losses without irrigation, J. Hydrol., 384, 198–217, doi:10.1016/j.jhydrol.2009.07.031, 2010.

Siebert, S., Döll, P., Feick, S., Hoogeveen, J., and Frenken, K.: Global map of irrigation areas version 4.0.1, Johann Wolfgang Goethe University, Frankfurt am Main, Germany/Food and Agriculture Organization of the United Nations, Rome, Italy, 2007.

Siebert, S., Burke, J., Faures, J. M., Frenken, K., Hoogeveen, J., Döll, P., and Portmann, F. T.: Groundwater use for irrigation – a global inventory, Hydrol. Earth Syst. Sci., 14, 1863–1880, doi:10.5194/hess-14-1863-2010, 2010.

Sitch, S., Smith, B., Prentice, I. C., Arneth, A., Bondeau, A., Cramer, W., Kaplan, J. O., Levis, S., Lucht, W., Sykes, M. T., Thonicke, K., and Venevsky, S.: Evaluation of ecosystem dynamics, plant geography and terrestrial carbon cycling in the LPJ dynamic global vegetation model, Global Change Biol,, 9, 161–185, 2003.

Smilovic, M., Gleeson, T., and Siebert, S.: The limits of increasing food production with irrigation in India, Food Secur., 7, 835–856, 2015.

Statistics: P.B.o., http://www.pbs.gov.pk/content/agriculture-statistics-pakistan-2010-11, last access: 22 July 2014.

Tiwari, V. M., Wahr, J., and Swenson, S.: Dwindling groundwater resources in northern India, from satellite gravity observations, Geophys. Res. Lett., 36, L18401, doi:10.1029/2009GL039401, 2009.

Wada, Y., van Beek, L. P. H., van Kempen, C. M., Reckman, J. W. T. M., Vasak, S., and Bierkens, M. F. P.: Global depletion of groundwater resources, Geophys. Res. Lett., 37, L20402, doi:10.1029/2010GL044571, 2010.

Wada, Y., Van Beek, L. P. H., Viviroli, D., Dürr, H. H., Weingartner, R., and Bierkens, M. F. P.: Global monthly water stress: 2. Water demand and severity of water stress, Water Resour. Res., 47, doi:10.1029/2010WR009792, 2011.

Waha, K., van Bussel, L. G. J., Muller, C., and Bondeau, A.: Climate-driven simulation of global crop sowing dates, Global Ecol. Biogeogr., 21, 247–259, doi:10.1111/j.1466-8238.2011.00678.x, 2012.

Weedon, G. P., Balsamo, G., Bellouin, N., Gomes, S., Best, M. J., and Viterbo, P.: The WFDEI meteorological forcing data set: WATCH Forcing Data methodology applied to ERA-Interim reanalysis data, Water Resour. Res., 50, 7505–7514, 2014.

4

Pairing FLUXNET sites to validate model representations of land-use/land-cover change

Liang Chen[1]**, Paul A. Dirmeyer**[1]**, Zhichang Guo**[1]**, and Natalie M. Schultz**[2]

[1]Center for Ocean-Land-Atmosphere Studies, George Mason University, Fairfax, Virginia, USA

[2]School of Forestry and Environmental Studies, Yale University, New Haven, Connecticut, USA

Correspondence: Liang Chen (lchen15@gmu.edu)

Abstract. Land surface energy and water fluxes play an important role in land–atmosphere interactions, especially for the climatic feedback effects driven by land-use/land-cover change (LULCC). These have long been documented in model-based studies, but the performance of land surface models in representing LULCC-induced responses has not been investigated well. In this study, measurements from proximate paired (open versus forest) flux tower sites are used to represent observed deforestation-induced changes in surface fluxes, which are compared with simulations from the Community Land Model (CLM) and the Noah Multi-Parameterization (Noah-MP) land model. Point-scale simulations suggest the CLM can represent the observed diurnal and seasonal changes in net radiation (R_{net}) and ground heat flux (G), but difficulties remain in the energy partitioning between latent (LE) and sensible (H) heat flux. The CLM does not capture the observed decreased daytime LE, and overestimates the increased H during summer. These deficiencies are mainly associated with models' greater biases over forest land-cover types and the parameterization of soil evaporation. Global gridded simulations with the CLM show uncertainties in the estimation of LE and H at the grid level for regional and global simulations. Noah-MP exhibits a similar ability to simulate the surface flux changes, but with larger biases in H, G, and R_{net} change during late winter and early spring, which are related to a deficiency in estimating albedo. Differences in meteorological conditions between paired sites is not a factor in these results. Attention needs to be devoted to improving the representation of surface heat flux processes in land models to increase confidence in LULCC simulations.

1 Introduction

Earth system models (ESMs) have long been used to investigate the climatic impacts of land-use/land-cover change (LULCC) (cf. Pielke et al., 2011; Mahmood et al., 2014). Results from sensitivity studies largely depend on the land surface model (LSM) that is coupled to the atmospheric model within ESMs. In the context of the Land-Use and Climate, Identification of Robust Impacts (LUCID) project, Pitman et al. (2009) found disagreement among the LSMs in simulating the LULCC-induced changes in summer latent heat flux over the Northern Hemisphere. de Noblet-Ducoudré et al. (2012) and Boiser et al. (2012) argued that the inter-model spread of LULCC sensitivity (especially regarding the partitioning of available energy between latent and sensible heat fluxes within the different land-cover types) highlights an urgent need for a rigorous evaluation of LSMs. From Phase 5 of the Coupled Model Intercomparison Project (CMIP5), Brovkin et al. (2013) also found different climatic responses to LULCC among the participating models, and the diverse responses are associated with different parameterizations of land surface processes among ESMs. To deal with the uncertainties in LULCC sensitivity among models, the Land Use Model Intercomparison Project (LUMIP) has been planned, with a goal to develop metrics and diagnostic protocols that quantify LSM performance and related sensitivities with respect to LULCC (Lawrence et al., 2016).

However, a paucity of useful observations has hindered the assessment of the simulated impacts of LULCC and limited the understanding of the discrepancies among models. In situ and satellite observations make it possible to quantify the im-

Table 1. Information about the variables used from FLUXNET2015. The marginal distribution sampling (MDS) filling method is based on Reichstein et al. (2005), and the ERA-Interim filling method can be found in Vuichard and Papale (2015).

Name	Gap-filling	Description
SW_IN_F	MDS and ERA-Interim	downwelling shortwave radiation
LW_IN_F	MDS and ERA-Interim	downwelling longwave radiation
PA_F	MDS and ERA-Interim	atmospheric pressure
TA_F	MDS and ERA-Interim	air temperature
VPD_F	MDS and ERA-Interim	vapor pressure deficit
P_F	ERA-Interim	precipitation
WS_F	ERA-Interim	wind speed
LE_F_MDS	MDS	latent heat flux
H_F_MDS	MDS	sensible heat flux
G_F_MDS	MDS	ground heat flux
NETRAD	n/a	net radiation
LE_CORR	n/a	corrected LE_F_MDS by energy balance closure correction factors. LE_CORR_25, LE_CORR, and LE_CORR_75 are calculated based on the 25th, 50th, and 75th percentiles of the factors, respectively.
H_CORR	n/a	corrected H_F_MDS by energy balance closure correction factors. H_CORR_25, H_CORR, and H_CORR_75 are calculated based on the 25th, 50th, and 75th percentiles of the factors, respectively.

pacts of LULCC on land surface variables. Satellite-derived datasets have been used to explore the albedo, evapotranspiration (ET), and land surface temperature changes due to historical LULCC (Boisier et al., 2013, 2014) and the climatic effects of forest (Li et al., 2015).

Meanwhile, the development of FLUXNET (Baldocchi et al., 2001) enables the study of land surface responses to different land-cover types based on paired field observations from neighboring flux towers over forest and open land (Juang et al., 2007; Lee et al., 2011; Luyssaert et al., 2014; Teuling et al., 2010; Williams et al., 2012). In terms of LSM evaluation, the paired site observations have been mainly used to simulated impacts of LULCC on land surface temperature (Chen and Dirmeyer 2016; Lejeune et al., 2016; Vanden Broucke et al., 2015). However, a more fundamental question, "whether a model can represent the observed LULCC-induced changes in surface energy fluxes well", has not been thoroughly investigated, even though we know that the turbulent fluxes are tightly associated with both energy and water exchange between the land surface and atmosphere.

In this study, we evaluate the performance of the Community Land Model (CLM) version 4.5 and the Noah Multi-Parameterization (Noah-MP) LSM in simulating the impacts of LULCC on surface energy fluxes based on observations from FLUXNET sites. The CLM and Noah-MP represent perhaps the two most readily available and widely used state-of-the-art community land models developed in the US. The CLM is chosen because, as the land component of the Community Earth System Model (CESM), it prioritizes the simulation of biogeophysical and biogeochemical processes for climate applications (Oleson et al., 2013). Much effort has gone into improving the representation of the land–atmosphere interactions among different biomes (Bonan et al., 2011), and the model itself has been used for many LULCC sensitivity studies (e.g., Chen and Dirmeyer, 2016, 2017; Schultz et al., 2016; Lejeune et al., 2017; Lawrence et al., 2012). Noah-MP has found use mainly in shorter

timescale, limited area applications, such as weather and hydrologic forecasting, and as a LSM run at very high resolution coupled to mesoscale models (e.g., WRF-Hydro, Gochis et al., 2015). It is intended to become the LSM used in global weather and seasonal forecasting applications at the National Centers for Environmental Prediction (NCEP). Its performance over varying land-cover types has direct consequences for its use in forecast models.

The rest of this paper is structured as follows. Section 2 describes the datasets used in the study and experimental design. Section 3 presents a comparison between observations and model simulations in surface latent and sensible heat flux, ground heat flux, and net radiation. Section 4 shows the uncertainties within the FLUXNET pairs and model simulations. Sections 5 and 6 include discussion and conclusions, respectively.

2 Methodology

2.1 Observational data

We use half-hourly observations from 24 selected pairs of flux sites from the FLUXNET2015 Tier 1 dataset (http://fluxnet.fluxdata.org/data/fluxnet2015-dataset) and 4 pairs from the AmeriFlux dataset (Baldocchi et al., 2001). These observations include meteorological forcings for the LSM, and surface flux measurements for model validation, which include latent heat flux (LE), sensible heat flux (H), ground heat flux (G), and net radiation (R_{net}). All of these variables have been gap-filled (Reichstein et al., 2005; Vuichard and Papale, 2015). Table 1 shows the variable names and gap-filling algorithms used in FLUXNET2015. Because there is no directly measured humidity variable reported, which is needed as a meteorological forcing for the LSMs, relative humidity is calculated based on the reported vapor pressure

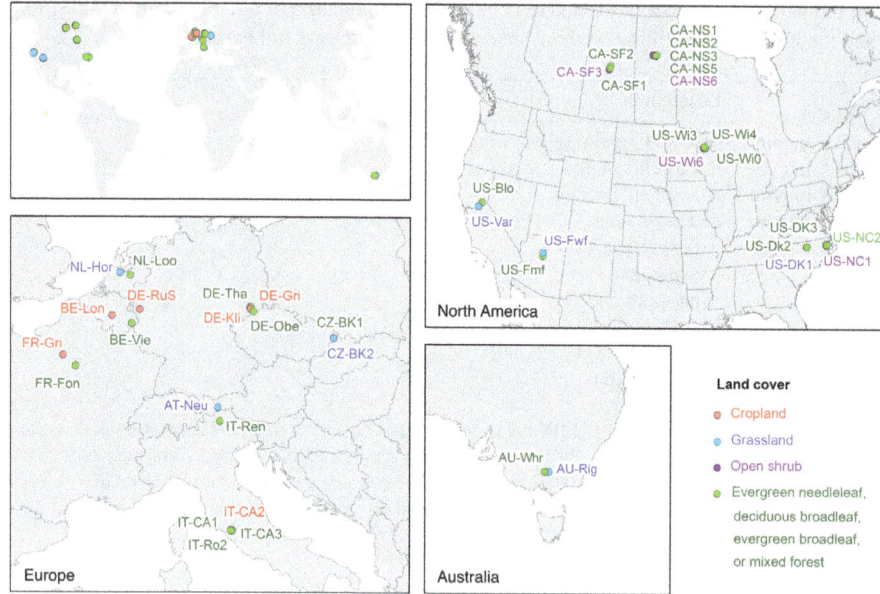

Figure 1. Location and land-cover type of the paired sites. The land-cover type of each site is based on the reported land cover in FLUXNET database.

deficit and surface air temperature (Eqs. 1 and 2).

$$e_s = 6.11 \exp\left(17.26938818 \frac{T_a}{237.3 + T_a}\right), \quad (1)$$

$$\text{RH} = \left(1 - \frac{\text{VPD}}{e_s}\right) \times 100, \quad (2)$$

in which T_a is air temperature (°C), e_s is saturation vapor pressure (hPa), VPD is vapor pressure deficit (hPa), and RH is relative humidity (%). Additionally, for the turbulent flux measurements over 18 pairs, FLUXNET2015 provides "corrected" fluxes based on an energy balance closure correction factor, which is calculated for each half-hour as $(R_{net} - G)/(H + LE)$. More details about the data processing can be found on the FLUXNET2015 website (http://fluxnet.fluxdata.org/data/fluxnet2015-dataset/data-processing/).

To simulate local land-cover change for each pair, one flux tower is located in forest (deciduous, evergreen, or mixed; broadleaf or needleleaf) and the other is in a nearby open land-cover type (grassland, cropland, or open shrub). Figure 1 shows the locations of the paired sites. Their general characteristics are listed in Table S1. The median linear distance between the paired sites is 21.6 km, and the median elevation difference is 20.0 m. Because of their proximities, the paired sites share similar atmospheric background conditions; however, they are not identical (Chen and Dirmeyer, 2016). Below we show that the differences in meteorology are usually small and not likely a dominant factor in simulated surface flux differences in most of the pairs. We consider the differences (open minus forest) in observed surface fluxes to be representative of the effects of LULCC (deforestation in this case).

2.2   Model simulations

We have run the offline version of CLM4.5 and Noah-MP at the point scale for individual sites. The forcing data, described below, include downwelling longwave radiation (W m^{-2}), downwelling shortwave radiation (W m^{-2}), air temperature (K), precipitation (mm s^{-1}), relative humidity (%), surface pressure (Pa), and wind speed (m s^{-1}) at half-hourly time steps. The plant functional type (PFT) in the CLM for each site is identified based on its reported land-cover type (Table S1) with prescribed climatological satellite phenology (Lawrence and Chase, 2010). Because of the focus on biogeophysical impacts of LULCC in this study, the biogeochemistry carbon–nitrogen module has been disabled in our simulations. The initial conditions for each site are generated by cycling through available atmospheric forcings for about 40 years until soil moisture and temperature reach quasi-equilibrium.

The differences in simulated surface fluxes between the paired sites are compared against the observations, so that the performance of the CLM in representing LULCC-induced surface flux changes can be evaluated. In the single-point simulations, two types of forcing data are used for each site: (1) measurements at this site; (2) measurements at the neighboring paired site. Consequently, three types of differences in simulated surface fluxes can be calculated: (1) the difference derived from individual forcings; (2) the difference from identical "forest forcings" (both of the paired sites use the same forcings measured at the forest site); (3) the difference from identical "open forcings" (both of the paired sites use the same forcings measured at the open sites). Such an experimental design can eliminate well the influence from the uncertainties of forcing data and the difference in the atmospheric background of the paired sites.

The ultimate goal of evaluating the CLM's performance at single-point scale is to assess its ability to be used in global LULCC sensitivity simulations in both offline and coupled modes. The paired sites are close enough that they are typically located within a single grid cell of the CESM. Moreover, the sub-grid heterogeneity of the CLM allows the biogeophysical processes to be calculated at the individual PFT level (15 PFTs available), and makes it possible to output surface fluxes for individual land-cover types. The paired sites can be presented as paired PFTs within a single grid of the CESM. They then share the same atmospheric forcings, and their differences can be considered as the impacts of LULCC. It should be noted that the PFT-level calculation is independent of the percentage of individual PFTs in the grid cell. Therefore, the coverage of the PFTs in the shared grid cell does not influence the flux difference between the paired PFTs in the global simulations.

We run the CLM offline, globally driven by the CRUNCEP forcings from 1991 to 2010 (Viovy, 2011) and present land-cover conditions (Lawrence et al., 2012) at a horizontal resolution of $0.9° \times 1.25°$. The paired PFTs are identified based on the locations and land-cover types of the FLUXNET paired sites, to ensure the single-point and global simulations are comparable.

Schultz et al. (2016) found the shared-soil-column configuration for vegetated land units in the CLM caused issues with PFT-level ground heat fluxes. They propose an individual-soil-column scheme (PFTCOL) to better represent the PFT-level energy fluxes, so we also extract and examine the output for the paired PFTs from the PFTCOL model configuration. Details about the PFTCOL simulations can be found in Schultz et al. (2016). Additionally, a coupled simulation with the Community Atmosphere Model (CAM) has also been conducted. It shows very similar results to the offline simulations, because the paired PFTs in a single model grid box always share the same atmospheric forcings no matter if the CLM is run offline or coupled with the CAM. Therefore, results from the coupled simulation are not included in this study.

Furthermore, we compare the performance of the CLM with Noah-MP (Niu et al., 2011), which serves as a participant model in Land Data Assimilation Systems (LDAS, Cai et al., 2014). Single-point Noah-MP simulations are conducted in the same way as CLM simulations to ensure their comparability. The monthly leaf area index (LAI) of each site is identical to the prescribed satellite-based LAI in the corresponding CLM simulation. Table S2 shows selected options for various physical processes in Noah-MP. Information about all model simulations is summarized in Table 2.

3 Surface energy fluxes and their changes

First, we analyze the diurnal and seasonal cycles of surface energy fluxes and the LULCC-induced changes. The diurnal cycle analysis is primarily focused on summer (DJF for the two austral sites and JJA for the other sites). The seasonal cycle for the austral sites is shifted by 6 months to keep summer in the middle of the time series when comparing or compositing with the Northern Hemisphere sites. The results shown below are composites averaged over all open (or forest) sites or open-forest pairs. Not all sites have energy-balance corrected fluxes available; exclusion of those sites shows very similar results for uncorrected fluxes to the average over all sites (or pairs, not shown). There are also some pairs with relatively large changes in surface fluxes. Exclusion of those pairs shows very consistent patterns with the results including all sites, even though there is a slight influence on the magnitude of the changes (Fig. S1). Therefore, all sites are included in our analyses for each variable.

3.1 Latent heat flux (LE)

Figure 2a–b shows the diurnal cycle of LE averaged over all the open sites and forest sites during summer. Compared with the observations without energy-balance correction, single-point CLM simulations overestimate LE for the open sites with both their actual meteorological forcings and the nearby forest forcings, but underestimate LE over the forest sites. The extracted PFT-level output from the global simulations also exhibit similar biases. Relative to the CLM, Noah-MP simulations show better agreement with observations over the open sites, but a greater underestimation over forest. The energy-balance correction tends to increase the values of LE. Therefore, both the CLM and Noah-MP have negative biases compared to the corrected fluxes (except LE_CORR_25 over the open sites).

Figure 2c shows the difference in the diurnal cycle of LE due to LULCC (deforestation). It should be noted that there is a substantial spread among the pairs in model simulations and especially observations, indicating the diverse geographical backgrounds and specific vegetation changes of these paired sites. The observations suggest an overall lower summer daytime LE over the open land compared to forest. In spite of the considerable spread among the energy-balance corrected LE observations (Fig. 2a, b), the differences between the forest and open lands show consistent signals. However, both CLM and Noah-MP single-point simulations fail to represent the observed decreased daytime LE as a result of deforestation. The simulated LE over the open land is usually slightly greater than the forest from 10:00 to 16:00 at local time. Such a discrepancy may be attributed to the large underestimation of daytime forest LE in the models. Meanwhile, simulations by different forcings of the paired sites show robust signals, implying that the bias of the simulated LE sensitivity should not be attributed to the uncertain-

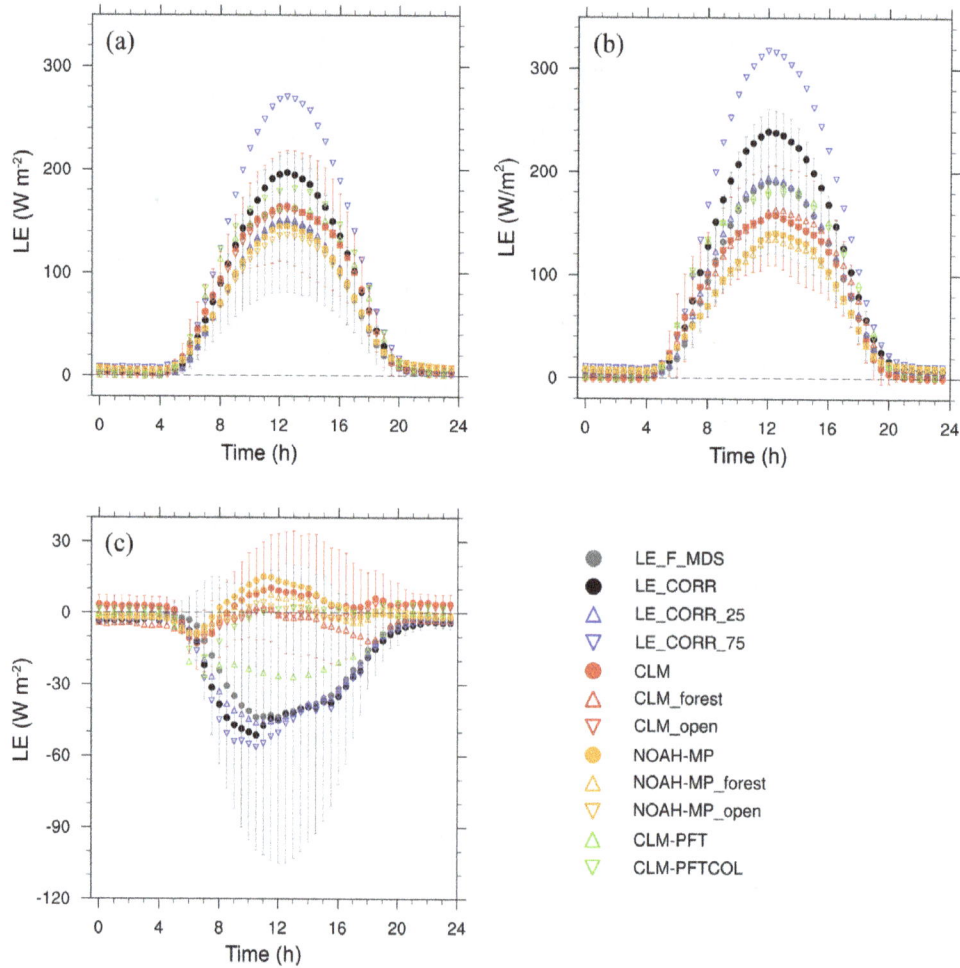

Figure 2. The diurnal cycle of LE ($W\,m^{-2}$) averaged over all the open sites (**a**) and forest sites (**b**) and their difference (open−forest, **c**) during the summer. The gray error bars indicate the standard deviation of the observed LE (MDS) among the sites; the red error bars are for the simulated LE in the CLM case. Details about the four types of FLUXNET observations can be found in Table 1. Information about model simulations in the CLM and Noah is given in Table 2.

Table 2. Information about model simulations. "Nearby" observations indicate that the paired sites have the identical forcings either from the companion forest or open sites.

Name	Forcings	Description
CLM	observations from individual sites	single-point CLM simulations with its own observations
CLM_forest	observations only from forest sites	single-point CLM simulations with the (nearby) forest observations
CLM_open	observations only from open sites	single-point CLM simulations with the (nearby) open land observations
CLM-PFT	CRUNCEP	global CLM simulations with default soil-column scheme with PFT-level output
CLM-PFTCOL	CRUNCEP	global CLM simulations with default individual-soil-column scheme scheme with PFT-level output
NOAH-MP	observations from individual sites	single-point NOAH-MP simulations with its own observations
NOAH-MP_forest	observations only from forest sites	single-point NOAH-MP simulations with the (nearby) forest observations
NOAH-MP_open	observations only from open sites	single-point NOAH-MP simulations with the (nearby) open land observations

ties of the forcing data. For the CLM global simulations, the PFTCOL case exhibits a similar diurnal pattern to the single-point simulations, while decreased daytime LE is found consistently only in the PFT simulations. As CLM-PFT is less physically realistic than CLM-PFTCOL from a soil hydrologic perspective, its superior performance needs further investigation.

To explore the mechanism of the LE changes within the CLM, we examine the changes in the three components of evapotranspiration, namely canopy evaporation, canopy transpiration, and ground evaporation (Fig. 3). Unfortunately, these separate components are not measured and cannot be directly validated. The CLM, PFT, and PFTCOL simulations show an agreement in decreased canopy evaporation after deforestation, with the greatest decrease during the early morning. There is also an agreement in an overall decreased canopy transpiration, but CLM simulations do not exhibit an obvious change during the morning, when greatly decreased canopy transpiration can be found in the PFT and PFTCOL simulations. The main discrepancy among model versions is found in ground evaporation, which increases after deforestation in the CLM and PFTCOL simulations. The increased ground evaporation has exceeded the decreased canopy evaporation and transpiration, resulting in slightly increased LE (Fig. 2c). Interestingly, the PFT simulations, which have known issues with PFT-level ground heat flux (Schultz et al., 2016), show decreased daytime ground evaporation. Along with decreased canopy evaporation, transpiration, and ground evaporation, the total LE decreases sharply after deforestation in the PFT simulations, which agrees better with the observations than other simulations (Fig. 2c). However, the decreased ground evaporation may be associated with a problematic soil-column scheme at sub-grid scale, which undermines the credibility of the agreement between the observations and PFT simulations.

Figure 4 shows the changes in monthly LE after deforestation across the annual cycle. There is clear and consistent seasonality in the LE changes from the observations. The four types of observations show decreased LE (up to $-24.0\,\mathrm{W\,m^{-2}}$) during local summer. There is little change in LE in the uncorrected observations during the winter season. However, there is significantly increased LE (up to $+17.9\,\mathrm{W\,m^{-2}}$) in the energy-balance corrected observations in late winter and early spring. Neither the CLM nor Noah-MP captures the observed seasonality of LE change. As found in the change in the diurnal cycle of the LE, the PFT-COL simulations exhibit a similar pattern to the single-point simulations, while the PFT simulations show decreased LE throughout the year, with the maximum from May to August, and the best correlation ($R = 0.81$, $P < 0.01$) with observations.

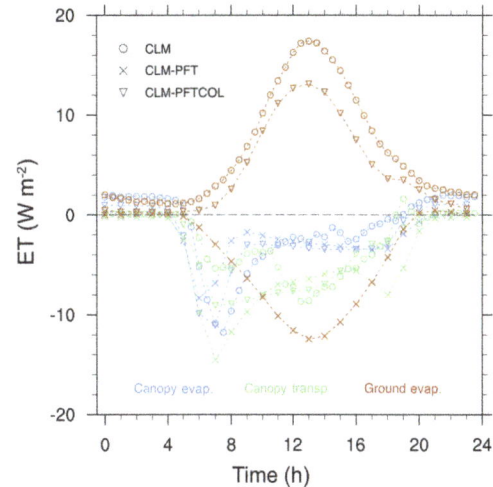

Figure 3. Change in the diurnal cycle of components (colors) of evapotranspiration (canopy evaporation, canopy transpiration, and ground evaporation) due to LULCC from forest to open land (open−forest).

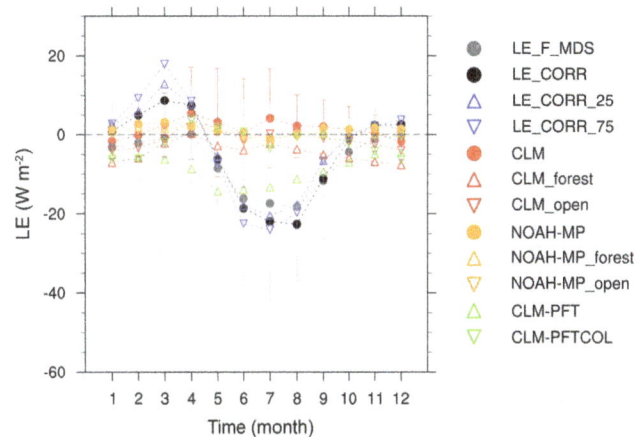

Figure 4. Change in the seasonal cycle of LE ($\mathrm{W\,m^{-2}}$) due to LULCC from forest to open land (open−forest).

3.2 Sensible heat flux (H)

Figure 5a–b show the diurnal cycle of H averaged over all open and forest sites during local summer. Generally, the models overestimate H throughout the day, with the largest positive bias during midday. Compared with the observations without energy-balance correction, the overestimation can be up to $86.5\,\mathrm{W\,m^{-2}}$ from the CLM over the forest during noon and $46.4\,\mathrm{W\,m^{-2}}$ over the open sites. The difference in H between the forest and open sites is shown in Fig. 5c. Robust signals are found among the four types of observations, so results from the energy-balance corrected observations are not included hereafter, but are shown in Fig. S2. Both observations and models exhibit a clear diurnal pattern of change in H after deforestation – a small nighttime increase and a large

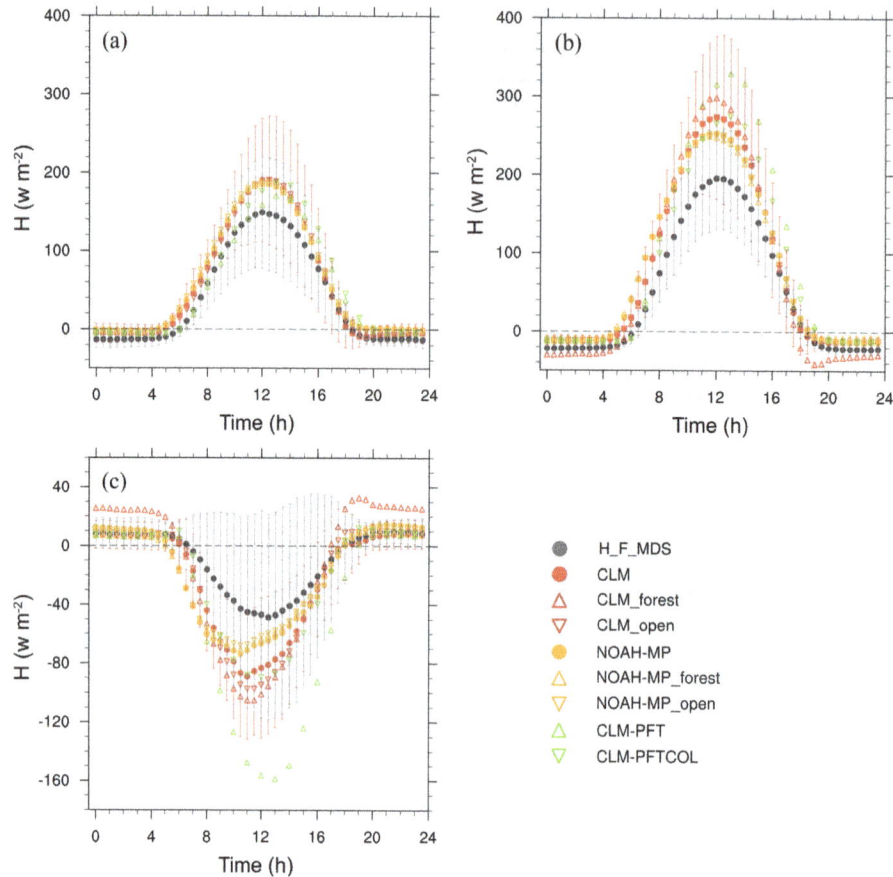

Figure 5. The diurnal cycle of H ($W\,m^{-2}$) averaged over all the open sites (**a**) and forest sites (**b**) and their difference (open−forest, **c**) during the summer. The gray error bars indicate the standard deviation of the observed H among the sites; the red error bars are for the simulated H in the CLM case.

daytime decrease. Observations show a large spread among the 28 pairs, which is much greater than that from the CLM simulations, indicating uncertainties and variability among the observed fluxes and the robustness of simulated H sensitivity to LULCC in the LSM. Compared with the observations, the CLM shows a greater H decrease, which is twice as much as in the observations. The overestimated H decrease may be related to the large positive bias in H over the forest sites (Fig. 5b). Additionally, the PFT simulations show the largest H decrease, which may be associated with the ground heat issues in the shared-soil-column scheme.

Seasonally, decreased H is found throughout the year after deforestation in both observations and models (except for the same-forest-forcing CLM simulations in winter, Fig. 6). The greatest decrease is observed during spring, when both of the single-point CLM and PFTCOL simulations show good agreement. However, CLM and Noah-MP simulations also show a large decrease during summer, which has not been observed in the FLUXNET dataset. Again, the PFT simulations show the greatest H decrease among the simulations and the largest bias compared with the observations during the warm season.

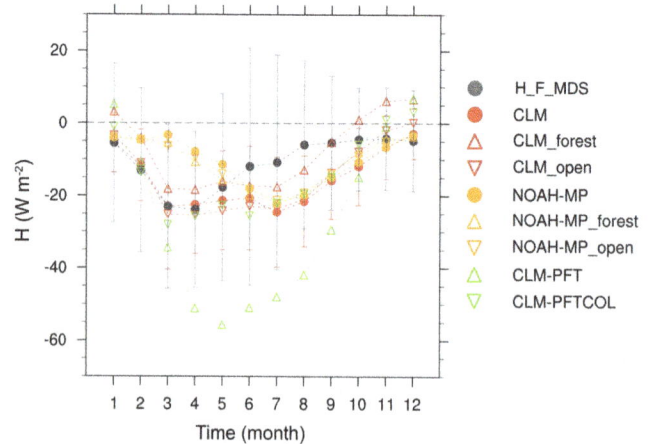

Figure 6. Change in the seasonal cycle of H ($W\,m^{-2}$) due to LULCC from forest to open land (open−forest).

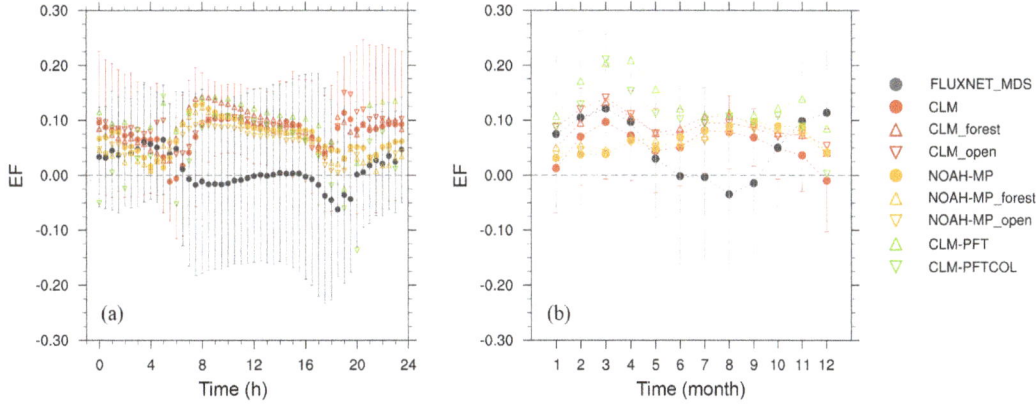

Figure 7. Change in the summer diurnal (**a**) and seasonal (**b**) cycles of EF (unitless) due to LULCC from forest to open land (open−forest). The observed EF (FLUXNET_MDS) is calculated based on the MDS gap-filled LE (LE_F_MDS) and H (H_F_MDS).

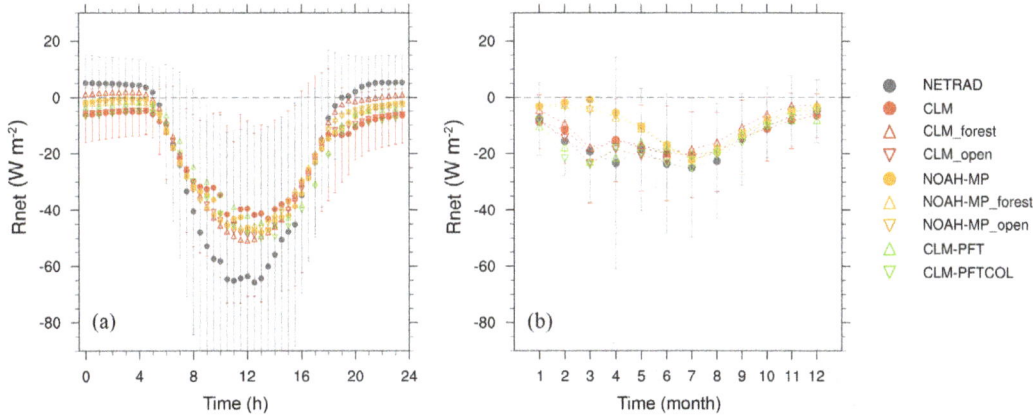

Figure 8. Change in the summer diurnal (**a**) and seasonal (**b**) cycles of G (W m^{-2}) due to LULCC from forest to open land (open−forest). It should be noted that the changes in the CLM-PFT simulation are much further from the observations than the other simulations. Some of its values are beyond the limit of the figure (**b**). The smallest value is −11.2 W m^{-2} in January, while the largest value is 52.9 W m^{-2} in May.

Additionally, evaporative fraction (EF), which is defined as the ratio of LE to the available energy (LE + H), is a useful diagnostic of the surface energy balance (Gentine et al., 2011). Meanwhile, most of the correction methods to solve the imbalance issue of the surface energy budget assume that the Bowen ratios for small- and large-scale eddies are similar or even equal (Wilson et al., 2002; Foken, 2008; Zhou and Wang, 2016). Under such an assumption, EF can be independent of the energy closure issue, because EF is related to the Bowen ratio (B) as

$$EF = (1 + B)^{-1}. \qquad (3)$$

Figure 7 shows the change in the diurnal (summer only) and seasonal cycle of EF due to LULCC from forest to open land. During summer, there are small changes in observed daytime EF (Fig. 7a) because of the decreases in both LE and H. However, both CLM and Noah-MP show increased daytime EF due to the decreased H and slightly increased LE after deforestation. Seasonally, the models show year-around increased EF, however, which is not observed in FLUXNET from June to September, further demonstrating the models' deficiencies in representing energy partitioning during summer.

3.3 Diurnal and seasonal cycle of ground heat flux (G) and net radiation (R_{net})

Figure 8a shows the change in the diurnal cycle of G after deforestation. Both the observations and models exhibit increased G during the day and decreased G during the night. However, models overestimate the magnitude of the G change, and discrepancies also exist in the timing of maximum change. The greatest increase in G is observed during the early afternoon, while the greatest increase in simulated G occurs at noon in the CLM (single-point and PFTCOL) and during the morning in Noah-MP. Because G is strongly

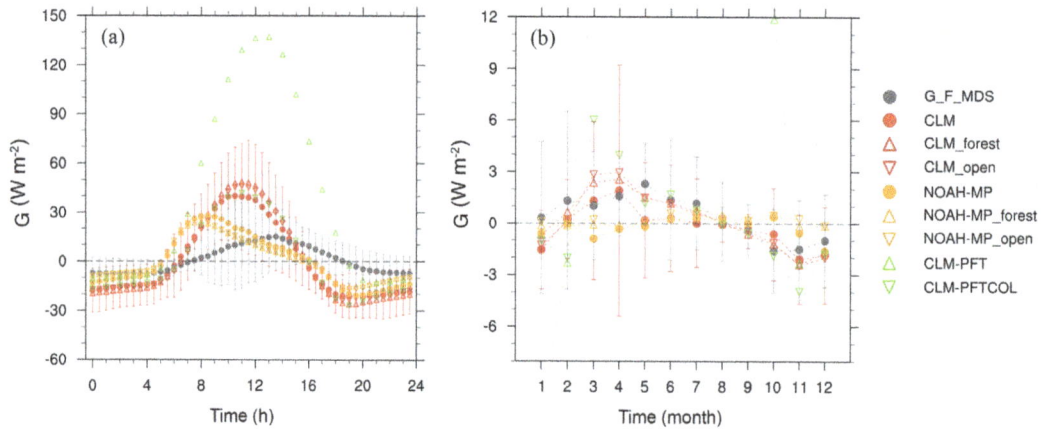

Figure 9. Change in the summer diurnal (**a**) and seasonal (**b**) cycles of R_{net} (W m^{-2}) due to LULCC from forest to open land (open−forest).

correlated with R_{net} (Santanello and Friedl, 2003), we examine the timing of maximum observed G and R_{net} during summer. There are some sites showing about a 1 h lag between maximum R_{net} and G (not shown). Therefore, the lag between simulated and observed peaks in G change can be partially attributed to the uncertainties in G measurements that are commonly estimated with heat flux plates installed at some depth (e.g., 5–10 cm) below the surface (Wang and Bou-Zeid, 2012), while the LSM simulated G is calculated at the surface. Meanwhile, the G changes (in both the diurnal and seasonal cycles) in the PFT simulations are further from the observations than the other simulations. Such disagreement further confirms the issues with the sub-grid soil-column scheme in the CLM, which is discussed in the following section. The changes in observed G also have a clear seasonal pattern – an increase during the warm season and a decrease during the cold season (Fig. 8b). This seasonality is captured well by the CLM simulations (especially the simulations with identical forcings for the paired sites) in both magnitude and timing, but is not evident in Noah-MP simulations.

After exploring the three flux components of the surface energy balance, it is worthwhile examining the change in R_{net} after deforestation. During summer, the observations show that R_{net} slightly increases during the night, and decreases considerably (up to -65.7 W m^{-2}) during the day, which can be attributed to the increased albedo after deforestation (Fig. 9a). Decreased daytime R_{net} is also found in the CLM simulations, but with a slightly smaller magnitude. Seasonally, there is a good agreement between the observations and CLM simulations, showing a large R_{net} decrease during spring and summer but a relatively small decrease during autumn and winter (Fig. 9b). The Noah-MP simulations are comparable to the CLM, but with a notable deficiency in simulating the R_{net} change during late winter and early spring.

4 Uncertainty analysis

4.1 Uncertainties among the FLUXNET pairs

The results discussed above are based on composites averaged over all forest and open sites. It is worthwhile examining the uncertainties in surface flux changes among different paired sites. Figure 10a shows the changes in summer daytime (08:00–16:00) LE from the observations and model simulations across the 28 pairs. This time period is chosen because it is the time of the greatest differences in surface energy fluxes (Figs. 2c, 5c, 7a, 8a). The observations show decreased LE associated with deforestation over 23 pairs, among which the pairs of evergreen needleleaf forest and open shrub (nos. 16–25) exhibit consistent decreases and the pairs of deciduous broadleaf forest and crops (nos. 1–4) show the overall greatest decrease. However, both the CLM and Noah-MP show relatively weak increases over most of the pairs, which further demonstrate their deficiency in simulating LE change. Additionally, for both the CLM or Noah, the choice of forcing does not exert much influence on the simulated change in summer daytime LE.

The changes in R_{net} over individual pairs are shown in Fig. 10b. There are 27 pairs (all except number 21) showing decreased R_{net} after deforestation, with the greatest decreases over the pairs of evergreen needleleaf forest and grassland. Both the CLM and Noah-MP capture the observed decreases in R_{net} well over most of the pairs.

It should be noted that pair 15 shows large LE and R_{net} changes in Fig. 10. This pair consists of a site over valley grassland and the other site over mountain evergreen needleleaf forest with 60.29 km separation and 1186 m elevation difference. There are significantly different air temperature and downwelling longwave radiation measurements between the sites (Fig. S3). Such large differences in LE and R_{net} here are likely associated with the distinct although proximate geographical sites. Even though the exclusion of this site does not make a significant change to the composite anal-

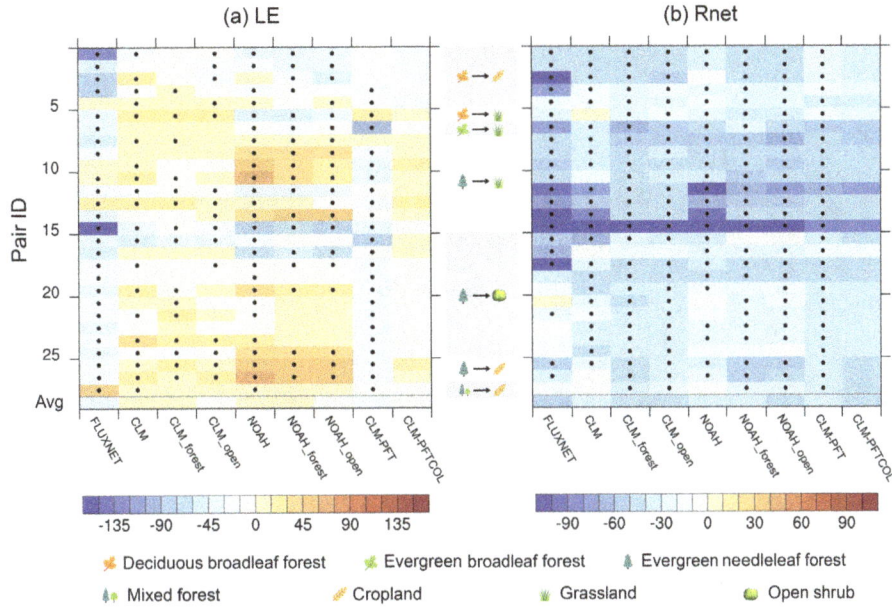

Figure 10. Change (open−forest) in observed and simulated LE **(a)** and R_{net} **(b)** during summer daytime (averaged during the period 08:00–16:00) over individual pairs and their averages. The vertical labels show the pair IDs from 1 to 28 based on Table S1. The pairs are grouped based on the type of LULCC (shown as the icons in the middle). The bottom row is the average over all pairs. The Student's t test is performed on the daily (daytime average) time series for each pair. Dots indicate statistically significant changes at the 95 % confidence level. No significant test is carried out for the CLM-PFTCOL simulation (the last column), because we only have long-term averaged hourly output for each month.

ysis in Sect. 3 (not shown), it may raise another question of whether the simulated sensitivity of surface energy fluxes is associated with the inconsistencies of atmospheric forcings of LSMs at the single pair level.

4.2 Uncertainties within the forcings for LSMs

Based on the composite analysis in Sect. 3, we have found that the simulated changes in surface energy fluxes with identical forcings (either from forest or open sites) are consistent with the simulations with individual forcings, demonstrating that the overall sensitivities of surface energy fluxes are robust among the choices of different forcings. In this subsection, we explore the uncertainties of the simulated surface flux changes due to the different forcings for individual pairs, especially with the focus on the roles of separation and elevation difference in the simulated sensitivity of surface energy fluxes.

Since we have simulations with identical forcings for the paired sites, the difference in surface flux changes between "forest forcings" and "open forcings" can be considered as the simulated sensitivity of surface energy fluxes to variation in the atmospheric forcings. Figure 11 shows the relationship with separation and elevation difference for individual pairs. Overall, the flux changes are not associated with the separation and elevation difference between the paired sites, further confirming the robustness of simulated signals from paired-site simulations. Nevertheless, some "outliers" are identified.

In the CLM simulations, only pair 15 shows large differences in LE and H change. However, pairs 3, 7, and 12 also exhibit large differences in Noah-MP simulations. The uncertainties in pairs 12 and 15 may be attributed to their large elevation differences. For pair 7 in Australia, Noah-MP shows greater sensitivity of H and R_{net} to atmospheric forcings over the evergreen broadleaf forest than grassland (not shown), leading to large differences in the surface flux changes. However, this is the only pair with evergreen broadleaf forest, and its behavior in Noah-MP needs further investigation. Even though the pair 3 sites are close with small elevation differences, we found considerably different downwelling shortwave and longwave radiation between the two sites (not shown), which may explain the uncertainties in the Noah-MP simulations.

5 Discussion

This study has examined simulated changes in the surface energy budget in response to local land-cover change based on paired proximate FLUXNET sites with differing land cover. Our results suggest that the CLM represents the observed changes in R_{net} and G well, but there remain issues in simulating the energy partitioning between LE and H, which also further confirms the large uncertainties in simulated ET responses to LULCC revealed in several recent studies (e.g., Pitman et al., 2009; Boisier et al., 2012, 2014; de Noblet-Ducoudré et al., 2012; Vanden Broucke et al., 2015). Based

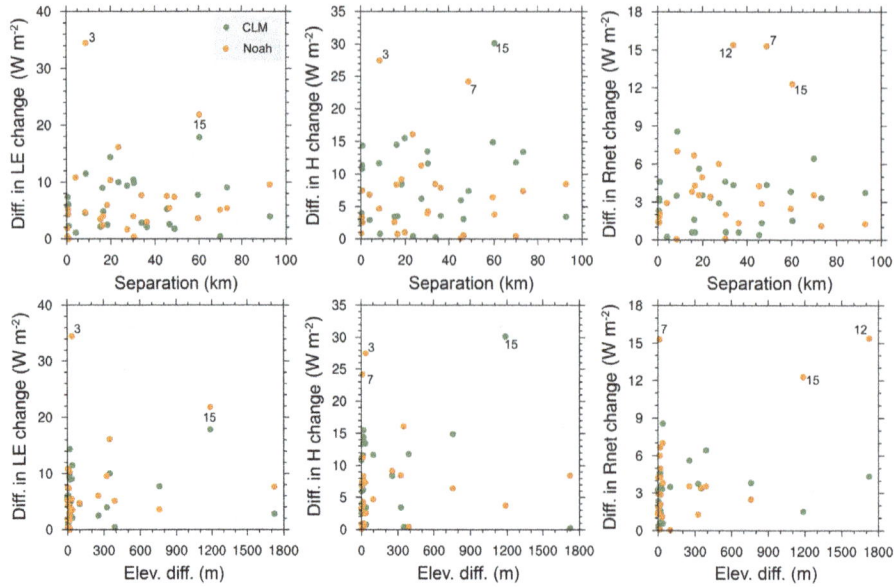

Figure 11. Sensitivity of differences in simulated surface energy flux changes (left column: LE, middle: H, and right: R_{net}) between "forest forcing" and "open forcing" simulations to site separation (top) and elevation difference (bottom) between the forest and open sites in individual pairs. The pair nos. 3, 7, 12, and 15 are labeled because of the greatest differences in surface flux changes.

on the observations, deforestation generally leads to a decrease in summer daytime R_{net}, accompanied by decreased LE and H. On the one hand, the CLM captures the observed signal of H change, but overestimates the decrease due to its large overestimation of H over the forest. On the other hand, the model underestimates the LE over the forest, leading to an opposite signal (a slight increase) of LE change compared to the observations. Simulations in Noah-MP show similar biases. Therefore, uncertainties in current LULCC sensitivity studies may persist specifically in the representation of turbulent fluxes over forest land-cover types.

Scrutinizing the three components of ET suggests that the simulated increase in summer daytime LE is mainly attributable to a large increase in ground evaporation, which counteracts the decreased canopy evaporation and transpiration. This may raise another issue about the soil resistance parameterization in CLM4.5. Previous studies indicate that the model generates excessive ground evaporation when the canopy is sparse or absent (Swenson and Lawrence, 2014; Tang et al., 2015). If there is overestimated ground evaporation over the open land, such a bias can also contribute to the disagreement in the LULCC-induced ET changes. Swenson and Lawrence (2014) have implemented a dry surface layer for the soil resistance parameterization to solve this issue for the upcoming CLM5. An extension of the evaluation with CLM5 would be useful to examine whether the issue within the soil resistance parameterization is responsible for the uncertainties in ET changes.

Besides the uncertainties in estimating turbulent fluxes over different land-cover types, the simulations show that differences in the meteorological forcings between nearby

paired sites seem to have little impact on the simulation of surface flux changes due to LULCC. Many LSMs besides the CLM employ a sub-grid tiling parameterization where multiple land surface types exist within a single grid box, each maintaining a separate set of surface balances and returning a weighted average set of fluxes to the atmosphere based on areal coverage of each surface type. In this arrangement, each land surface type within a grid box receives the same meteorological forcing from the overlying atmospheric model. It appears from our forcing-sensitivity studies that this arrangement does not significantly impact the simulation of surface flux changes associated with LULCC on the grid scale.

That said, the sub-grid comparison between different land-cover types may yet be problematic due to the shared soil-column issue for vegetated land units in the CLM (Schultz et al., 2016). Both the single-point observations and simulations show significant differences in surface soil moisture between most of the paired sites, even though no clear drying or wetting pattern is found (Fig. S4). The differences between the paired sites suggests that the shared soil column for vegetated land in the CLM may not represent soil moisture and temperature well at the sub-grid scale, which may influence the simulations of land surface energy and water fluxes. We find an unreasonably large change in PFT-level G between forest and open land especially for the seasonal cycle in PFT simulations, while both observations and single-point and PFTCOL simulations show a seasonal change with a very small range (within $\pm 3\,W\,m^{-2}$). As G is calculated as the residual of the surface energy budget in the CLM (Oleson et al., 2013), this sub-grid G issue may cast even more uncertainties on the calculation of LE and H at the PFT level, as

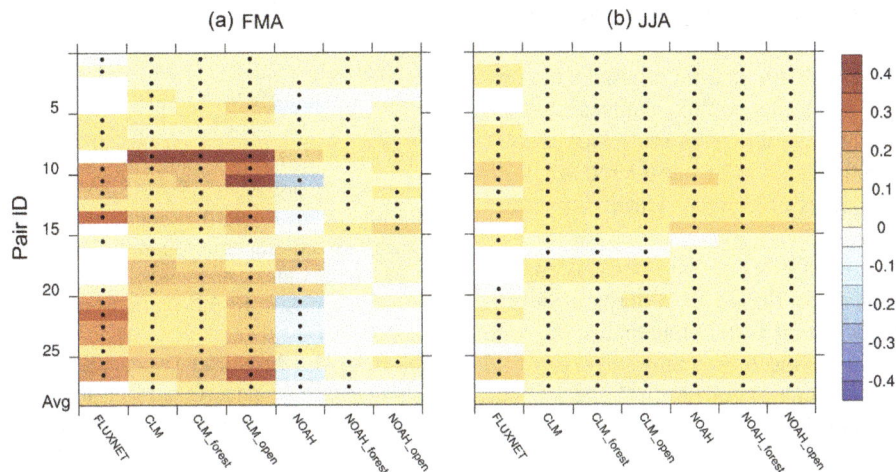

Figure 12. Change (open−forest) in observed and simulated daytime albedo during late winter/early spring (FMA, **a**) and summer (JJA, **b**). White areas indicate missing observations.

well as their aggregated values at the grid level for regional or global simulations. Therefore, caution should be taken when examining the LULCC sensitivity which involves sub-grid PFT changes.

Compared with the CLM, Noah-MP exhibits a similar ability to simulate surface flux changes, except for a deficiency in simulating H and R_{net} changes during late winter and early spring. We have examined the daytime albedo change after deforestation, calculated from available shortwave radiation terms, from observations and model simulations during local late winter/early spring (February–April, FMA) and summer (Fig. 12). Both the CLM and Noah-MP agree with the observations during summer. However, Noah-MP does not capture the observed albedo increase over nearly half of the pairs during late winter/early spring. Greater disagreement is also found during the local winter season (DJF, not shown), suggesting a deficiency in snowmelt timing or snow albedo sensitivity to LULCC, despite improvement in the snow surface albedo simulations by implementation of the Canadian Land Surface Scheme (CLASS; Verseghy, 1991) in Noah-MP (Niu et al., 2011).

Finally, it should be recognized that the observational data are not perfect. In particular, there may be systematic biases or even trends in specific instruments that contribute to the perceived differences between paired sites (e.g., site 3). Ideally, redundant instrumentation at sites, or in this case the rotation of an extra set of instruments among nearby paired sites, could be used to identify, quantify, and account for significant systematic biases in measurements for suspicious variables. Furthermore, footprints of the flux towers may bias the comparison of surface fluxes between the open and forest sites (Baker et al., 2003; Griebel et al., 2016). In other words, the observed differences between sites can only be partially attributed to LULCC because their environmental conditions may also be different. As most of the current stud-

ies used paired sites to represent LULCC, we have assumed that the paired sites share similar background atmospheric conditions, and any observed differences in surface climate conditions can be attributed to LULCC (e.g., Lejeune et al., 2017; Luyssaert et al., 2014; Teuling et al., 2010; Vanden Broucke et al., 2015). Meanwhile, model simulations with the different forcings can effectively examine the effects of the local environment of individual sites, because their footprints can also be taken by the meteorological measurements. Our results show robust signals of LULCC-induced changes in surface fluxes, implying that impacts of footprints at individual sites are probably trivial.

6   Conclusions

This study has evaluated the performance of two state-of-the-art LSMs in simulating the LULCC-induced changes in surface energy fluxes. Observations from 28 FLUXNET pairs (open versus forest) are used to represent the observed flux changes following deforestation, which are compared with the LSM simulations forced with meteorological data from the observation sites. Diurnal and seasonal cycles of the flux changes have been investigated.

The single-point simulations in the CLM and Noah-MP show the greatest bias in simulating LE change. Significantly decreased daytime LE is observed during local summer, but is not captured by the models. The observed LE changes also exhibit an evident seasonality, which is not represented in the model. The energy partitioning between LE and H might be a common issue within the LSMs. Other studies have noted problems in the simulation of surface fluxes by LSMs, including poor performance relative to non-physical statistical models (Best et al., 2015; Haughton et al., 2016).

The sub-grid comparison from the global simulations in the CLM yields unrealistic changes in G and H when the soil column is shared among vegetated land units, even though there is a better agreement in LE change with the observations. The individual-soil-column scheme improves the representation of the PFT-level energy flux changes, but uncertainties still remain as with the point-scale simulations. Therefore, these uncertainties must be considered when interpreting global experiments of LULCC sensitivity studies with current LSMs.

Consistent aggregate performance across many paired sites suggests the problems in these LSMs may not lie primarily with parameter selection at individual sites, but with more fundamental issues of the representation of physical processes in LSMs. The simulation of LULCC may or may not have become more consistent among models since LU-CID (de Noblet-Ducoudré et al., 2012), but consistency with observed biophysical responses appears to be lacking. LU-MIP (Lawrence et al., 2016) will be a step toward better LSM simulation of LULCC responses, and ultimately better simulations of the response of climate to LULCC.

Acknowledgements. This study was supported by the National Science Foundation (AGS-1419445). This work used eddy covariance data acquired and shared by the FLUXNET community, including these networks: AmeriFlux, CarboEuropeIP, CarboItaly, CarboMont, Fluxnet-Canada, GreenGrass, ICOS, and OzFlux-TERN. The ERA-Interim reanalysis data are provided by ECMWF and processed by LSCE. The FLUXNET eddy covariance data processing and harmonization were carried out by the European Fluxes Database Cluster, AmeriFlux Management Project, and Fluxdata project of FLUXNET, with the support of CDIAC and the ICOS Ecosystem Thematic Center, and the OzFlux, ChinaFlux and AsiaFlux offices. We thank all site investigators and flux networks for their work in making our model evaluation possible. The authors wish to thank Ahmed Tawfik at the National Center for Atmospheric Research for his assistance in preparing forcing datasets for the AmeriFlux sites. Computing resources for the CLM and Noah-MP experiments were provided by the NSF/CISL/Yellowstone supercomputing facility.

Edited by: Shraddhanand Shukla

References

Baker, I., Denning, A. S., Hanan, N., Prihodko, L., Uliasz, M., Vidale, P., Davis, K., and Bakwin, P.: Simulated and observed fluxes of sensible and latent heat and CO_2 at the WLEF-TV tower using SiB2.5, Global Change Biol., 9, 1262–1277, https://doi.org/10.1046/j.1365-2486.2003.00671.x, 2003.

Baldocchi, D., Falge, E., Gu, L., Olson, R., Hollinger, D., Running, S., Anthoni, P., Bernhofer, C., Davis, K., Evans, R., Fuentes, J., Goldstein, A., Katul, G., Law, B., Lee, X., Malhi, Y., Meyers, T., Munger, W., Oechel, W., Paw, K. T., Pilegaard, K., Schmid, H. P., Valentini, R., Verma, S., Vesala, T., Wilson, K., and Wofsy, S.: FLUXNET: A New Tool to Study

the Temporal and Spatial Variability of Ecosystem–Scale Carbon Dioxide, Water Vapor, and Energy Flux Densities, B. Am. Meteorol. Soc., 82, 2415–2434, https://doi.org/10.1175/1520-0477(2001)082<2415:FANTTS>2.3.CO;2, 2001.

Betts, A. K.: Understanding Hydrometeorology Using Global Models, B. Am. Meteorol. Soc., 85, 1673–1688, https://doi.org/10.1175/BAMS-85-11-1673, 2004.

Boisier, J. P., de Noblet-Ducoudré, N., and Ciais, P.: Historical land-use-induced evapotranspiration changes estimated from present-day observations and reconstructed land-cover maps, Hydrol. Earth Syst. Sci., 18, 3571–3590, https://doi.org/10.5194/hess-18-3571-2014, 2014.

Boisier, J. P., de Noblet-Ducoudré, N., and Ciais, P.: Inferring past land use-induced changes in surface albedo from satellite observations: a useful tool to evaluate model simulations, Biogeosciences, 10, 1501–1516, https://doi.org/10.5194/bg-10-1501-2013, 2013.

Boisier, J. P., de Noblet-Ducoudré, N., Pitman, A. J., Cruz, F. T., Delire, C., van den Hurk, B. J. J. M., van der Molen, M. K., Müller, C., and Voldoire, A.: Attributing the impacts of land-cover changes in temperate regions on surface temperature and heat fluxes to specific causes: Results from the first LU-CID set of simulations, J. Geophys. Res.-Atmos., 117, D12116, https://doi.org/10.1029/2011JD017106, 2012.

Bright, R. M., Zhao, K., Jackson, R. B., and Cherubini, F.: Quantifying surface albedo and other direct biogeophysical climate forcings of forestry activities, Glob. Change Biol., 21, 3246–3266, https://doi.org/10.1111/gcb.12951, 2015.

Brovkin, V., Boysen, L., Arora, V. K., Boisier, J. P., Cadule, P., Chini, L., Claussen, M., Friedlingstein, P., Gayler, V., van den Hurk, B. J., Hurtt, G. C., Jones, C. D., Kato, E., de Noblet-Ducoudré, N., Pacifico, F., Pongratz, J., and Weiss, M.: Effect of Anthropogenic Land-Use and Land-Cover Changes on Climate and Land Carbon Storage in CMIP5 Projections for the Twenty-First Century, J. Climate, 26, 6859–6881, https://doi.org/10.1175/JCLI-D-12-00623.1, 2013.

Cai, X., Yang, Z., Xia, Y., Huang, M., Wei, H., Leung, L. R., and Ek, M. B.: Assessment of simulated water balance from Noah, Noah-MP, CLM, and VIC over CONUS using the NLDAS test bed, J. Geophys. Res.-Atmos., 119, 13751–13770, https://doi.org/10.1002/2014JD022113, 2014.

Chen, L. and Dirmeyer, P. A.: Adapting observationally based metrics of biogeophysical feedbacks from land cover/land use change to climate modeling, Environ. Res. Lett., 11, 034002, https://doi.org/10.1088/1748-9326/11/3/034002, 2016.

Chen, L. and Dirmeyer, P. A.: Impacts of Land Use/Land Cover Change on Afternoon Precipitation over North America, J. Climate, 30, 2121–2140, https://doi.org/10.1175/JCLI-D-16-0589.1, 2016.

de Noblet-Ducoudré, N., Boisier, J., Pitman, A., Bonan, G. B., Brovkin, V., Cruz, F., Delire, C., Gayler, V., van den Hurk, B. J., Lawrence, P. J., van der Molen, M. K., Müller, C., Reick, C. H., Strengers, B. J., and Voldoire, A.: Determining Robust Impacts of Land-Use-Induced Land Cover Changes on Surface Climate over North America and Eurasia: Results from the First Set of LUCID Experiments, J. Climate, 25, 3261–3281, https://doi.org/10.1175/JCLI-D-11-00338.1, 2012.

Dirmeyer, P. A., Koster, R. D., and Guo, Z.: Do Global Models Properly Represent the Feedback between Land and Atmosphere?, J. Hydrometeor., 7, 1177–1198, https://doi.org/10.1175/JHM532.1, 2006.

Foken, T.: The Energy Balance Closure Problem: An Overview, Ecol. Appl., 18, 1351–1367, https://doi.org/10.1890/06-0922.1, 2008.

Gochis, D. J., Yu, W., and Yates, D. N.: The WRF-Hydro Model Technical Description and User's Guide, Version 3.0, NCAR Technical Document, 120 pp., available at: https://www.ral.ucar.edu/sites/default/files/public/images/project/WRF_Hydro_User_Guide_v3.0.pdf, 2015.

Griebel, A., Bennett, L. T., Metzen, D., Cleverly, J., Burba, G., and Arndt, S. K.: Effects of inhomogeneities within the flux footprint on the interpretation of seasonal, annual, and interannual ecosystem carbon exchange, Agr. Forest Meteorol., 221, 50–60, https://doi.org/10.1016/j.agrformet.2016.02.002, 2016.

Janowiak, J. E., Kousky, V. E., and Joyce, R. J.: Diurnal cycle of precipitation determined from the CMORPH high spatial and temporal resolution global precipitation analyses, J. Geophys. Res.-Atmos., 110, D23105, https://doi.org/10.1029/2005JD006156, 2005.

Juang, J., Katul, G., Siqueira, M., Stoy, P., and Novick, K.: Separating the effects of albedo from eco-physiological changes on surface temperature along a successional chronosequence in the southeastern United States, Geophys. Res. Lett., 34, L21408, https://doi.org/10.1029/2007GL031296, 2007.

Lawrence, D. M., Hurtt, G. C., Arneth, A., Brovkin, V., Calvin, K. V., Jones, A. D., Jones, C. D., Lawrence, P. J., de Noblet-Ducoudré, N., Pongratz, J., Seneviratne, S. I., and Shevliakova, E.: The Land Use Model Intercomparison Project (LUMIP) contribution to CMIP6: rationale and experimental design, Geosci. Model Dev., 9, 2973–2998, https://doi.org/10.5194/gmd-9-2973-2016, 2016.

Lawrence, D. and Vandecar, K.: Effects of tropical deforestation on climate and agriculture, Nat. Clim. Change, 5, 27–36, 2015.

Lawrence, P. J. and Chase, T. N.: Investigating the climate impacts of global land cover change in the community climate system model, Int. J. Climatol., 30, 2066–2087, https://doi.org/10.1002/joc.2061, 2010.

Lawrence, P. J., Feddema, J. J., Bonan, G. B., Meehl, G. A., O'Neill, B. C., Oleson, K. W., Levis, S., Lawrence, D. M., Kluzek, E., Lindsay, K., and Thornton, P. E.: Simulating the Biogeochemical and Biogeophysical Impacts of Transient Land Cover Change and Wood Harvest in the Community Climate System Model (CCSM4) from 1850 to 2100, J. Climate, 25, 3071–3095, https://doi.org/10.1175/JCLI-D-11-00256.1, 2012.

Lee, X., Goulden, M. L., Hollinger, D. Y., Barr, A., Black, T. A., Bohrer, G., Bracho, R., Drake, B., Goldstein, A., Gu, L., Katul, G., Kolb, T., Law, B. E., Margolis, H., Meyers, T., Monson, R., Munger, W., Oren, R., Paw U, K. T., Richardson, A. D., Schmid, H. P., Staebler, R., Wofsy, S., and Zhao, L.: Observed increase in local cooling effect of deforestation at higher latitudes, Nature, 479, 384–387, https://doi.org/10.1038/nature10588, 2011.

Lejeune, Q., Seneviratne, S. I., and Davin, E. L.: Historical land-cover change impacts on climate: comparative assessment of LU-CID and CMIP5 multi-model experiments, J. Climate, 30, 1439–1459, https://doi.org/10.1175/JCLI-D-16-0213.1, 2016.

Li, Y., Zhao, M., Motesharrei, S., Mu, Q., Kalnay, E., and Li, S.: Local cooling and warming effects of forests based on satellite observations, Nat. Comm., 6, 6603, https://doi.org/10.1038/ncomms7603, 2015.

Luyssaert, S., Jammet, M., Stoy, P. C., et al.: Land management and land-cover change have impacts of similar magni-

tude on surface temperature, Nat. Clim. Change, 4, 389–393, https://doi.org/10.1038/nclimate2196, 2014.

Mahmood, R., Pielke, R. A., Hubbard, K. G., Niyogi, D., Dirmeyer, P. A., McAlpine, C., Carleton, A. M., Hale, R., Gameda, S., Beltrán-Przekurat, A., Baker, B., McNider, R., Legates, D. R., Shepherd, M., Du, J., Blanken, P. D., Frauenfeld, O. W., Nair, U. S., and Fall, S.: Land cover changes and their biogeophysical effects on climate, Int. J. Climatol., 34, 929–953, https://doi.org/10.1002/joc.3736, 2014.

Niu, G.-Y., Yang, Z.-L., Mitchell, K. E., Chen, F., Ek, M. B., Barlage, M., Kumar, A., Manning, K., Niyogi, D., Rosero, E., Tewari, M., and Xia, Y.: The community Noah land surface model with multiparameterization options (Noah-MP): 1. Model description and evaluation with local-scale measurements, J. Geophys. Res.-Atmos., 116, D12109, https://doi.org/10.1029/2010JD015139, 2011.

Oleson, K. W., Lawrence, D. M., Bonan, G. B., Drewniak, B., Huang, M., Koven, C. D., Levis, S., Li, F., Riley, W. J., Subin, Z. M., Swenson, S. C., Thornton, P. E., Bozbiyik, A., Fisher, R., Heald, C. L., Kluzek, E., Lamarque, J.-F., Lawrence, P. J., Leung, L. R., Lipscomb, W., Muszala, S., Ricciuto, D. M., Sacks, W., Sun, Y., Tang, J., Yang, Z.-L.: Technical Description of version 4.5 of the Community Land Model (CLM), NCAR Technical Note, TN-503+STR, National Center for Atmospheric Research, Boulder, CO, USA, 434 pp., available at: http://www.cesm.ucar.edu/models/cesm1.2/clm/CLM45_Tech_Note.pdf, 2013.

Pielke, R. A., Pitman, A., Niyogi, D., Mahmood, R., McAlpine, C., Hossain, F., Goldewijk, K. K., Nair, U., Betts, R., Fall, S., Reichstein, M., Kabat, P., and de Noblet, N.: Land use/land cover changes and climate: modeling analysis and observational evidence, Wiley Interdisciplinary Reviews: Climate Change, 2, 828–850, https://doi.org/10.1002/wcc.144, 2011.

Pitman, A. J., de Noblet-Ducoudré, N., Cruz, F. T., Davin, E. L., Bonan, G. B., Brovkin, V., Claussen, M., Delire, C., Ganzeveld, L., Gayler, V., van den Hurk, B. J. J. M., Lawrence, P. J., van der Molen, M. K., Müller, C., Reick, C. H., Seneviratne, S. I., Strengers, B. J., and Voldoire, A.: Uncertainties in climate responses to past land cover change: First results from the LU-CID intercomparison study, Geophys. Res. Lett., 36, L14814, https://doi.org/10.1029/2009GL039076, 2009.

Portmann, F. T., Siebert, S., and Döll, P.: MIRCA2000—Global monthly irrigated and rainfed crop areas around the year 2000: A new high-resolution data set for agricultural and hydrological modeling, Global Biogeochem. Cy., 24, L14814, https://doi.org/10.1029/2008GB003435, 2010.

Reichstein, M., Falge, E., Baldocchi, D., Papale, D., Aubinet, M., Berbigier, P., Bernhofer, C., Buchmann, N., Gilmanov, T., Granier, A., Grünwald, T., Havránková, K., Ilvesniemi, H., Janous, D., Knohl, A., Laurila, T., Lohila, A., Loustau, D., Matteucci, G., Meyers, T., Miglietta, F., Ourcival, J.-M., Pumpanen, J., Rambal, S., Rotenberg, E., Sanz, M., Tenhunen, J., Seufert, G., Vaccari, F., Vesala, T., Yakir, D., and Valentini, R.: On the separation of net ecosystem exchange into assimilation and ecosystem respiration: review and improved algorithm, Global Change Biol., 11, 1424–1439, https://doi.org/10.1111/j.1365-2486.2005.001002.x, 2005.

Santanello, J. A. and Friedl, M. A.: Diurnal Covariation in Soil Heat Flux and Net Radiation, J. Appl. Meteor., 42, 851–862, https://doi.org/10.1175/1520-0450(2003)042<0851:DCISHF>2.0.CO;2, 2003.

Schultz, N. M., Lee, X., Lawrence, P. J., Lawrence, D. M., and Zhao, L.: Assessing the use of subgrid land model output to study impacts of land cover change, J. Geophys. Res.-Atmos., 121, 6133–6147, https://doi.org/10.1002/2016JD025094, 2016.

Swenson, S. C. and Lawrence, D. M.: Assessing a dry surface layer-based soil resistance parameterization for the Community Land Model using GRACE and FLUXNET-MTE data, J. Geophys. Res.-Atmos., 119, 10299–10312, https://doi.org/10.1002/2014JD022314, 2014.

Tang, J., Riley, W. J., and Niu, J.: Incorporating root hydraulic redistribution in CLM4.5: Effects on predicted site and global evapotranspiration, soil moisture, and water storage, J. Adv. Model Earth Sy., 7, 1828–1848, https://doi.org/10.1002/2015MS000484, 2015.

Teuling, A. J., Seneviratne, S. I., Stöckli, R., Reichstein, M., Moors, E., Ciais, P., Luyssaert, S., van den Hurk, B., Ammann, C., Bernhofer, C., Dellwik, E., Gianelle, D., Gielen, B., Grünwald, T., Klumpp, K., Montagnani, L., Moureaux, C., Sottocornola, M., and Wohlfahrt, G.: Contrasting response of European forest and grassland energy exchange to heatwaves, Nat. Geosci., 3, 722–727, https://doi.org/10.1038/ngeo950, 2010.

Vanden Broucke, S., Luyssaert, S., Davin, E. L., Janssens, I., and van Lipzig, N.: New insights in the capability of climate models to simulate the impact of LUC based on temperature decomposition of paired site observations, J. Geophys. Res.-Atmos., 120, 5417–5436, https://doi.org/10.1002/2015JD023095, 2015.

Verseghy, D. L.: Class – A Canadian land surface scheme for GCMS, I. Soil model, Int. J. Climatol., 11, 111–133, https://doi.org/10.1002/joc.3370110102, 1991.

Viovy, N.: CRUNCEP data set for 1901–2008, available at: https://www.earthsystemgrid.org/dataset/ucar.cgd.ccsm4.CRUNCEP.v4.html, 2011.

Vuichard, N. and Papale, D.: Filling the gaps in meteorological continuous data measured at FLUXNET sites with ERA-Interim reanalysis, Earth Syst. Sci. Data, 7, 157–171, https://doi.org/10.5194/essd-7-157-2015, 2015.

Williams, C. A., Reichstein, M., Buchmann, N., Baldocchi, D., Beer, C., Schwalm, C., Wohlfahrt, G., Hasler, N., Bernhofer, C., Foken, T., Papale, D., Schymanski, S., and Schaefer, K.: Climate and vegetation controls on the surface water balance: Synthesis of evapotranspiration measured across a global network of flux towers, Water Resour. Res., 48, W06523, https://doi.org/10.1029/2011WR011586, 2012.

Wilson, K., Goldstein, A., Falge, E., Aubinet, M., Baldocchi, D., Berbigier, P., Bernhofer, C., Ceulemans, R., Dolman, H., Field, C., Grelle, A., Ibrom, A., Law, B. E., Kowalski, A., Meyers, T., Moncrieff, J., Monson, R., Oechel, W., Tenhunen, J., Valentini, R., and Verma, S.: Energy balance closure at FLUXNET sites, Agr. Forest Meteorol., 113, 223–243, https://doi.org/10.1016/S0168-1923(02)00109-0, 2002.

Xu, Z., Mahmood, R., Yang, Z., Fu, C., and Su, H.: Investigating diurnal and seasonal climatic response to land use and land cover change over monsoon Asia with the Community Earth System Model, J. Geophys. Res.-Atmos., 120, 1137–1152, https://doi.org/10.1002/2014JD022479, 2015.

Zhang, L., Mao, J., Shi, X., Ricciuto, D., He, H., Thornton, P., Yu, G., Li, P., Liu, M., Ren, X., Han, S., Li, Y., Yan, J., Hao, Y., and Wang, H.: Evaluation of the Community Land Model simulated carbon and water fluxes against observations over ChinaFLUX sites, Agr. Forest Meteorol., 226–227, 174–185, https://doi.org/10.1016/j.agrformet.2016.05.018, 2016.

Zhou, C. and Wang, K.: Biological and Environmental Controls on Evaporative Fractions at AmeriFlux Sites, J. Appl. Meteor. Climatol., 55, 145–161, https://doi.org/10.1175/JAMC-D-15-0126.1, 2016.

5

Temperature signal in suspended sediment export from an Alpine catchment

Anna Costa[1], Peter Molnar[1], Laura Stutenbecker[2], Maarten Bakker[3], Tiago A. Silva[4], Fritz Schlunegger[5], Stuart N. Lane[3], Jean-Luc Loizeau[4], and Stéphanie Girardclos[6]

[1]Institute of Environmental Engineering, ETH Zurich, 8093 Zurich, Switzerland

[2]Institute of Applied Geosciences, Technische Universität Darmstadt, Darmstadt, Germany

[3]Institute of Earth Surface Dynamics, University of Lausanne, 1015 Lausanne, Switzerland

[4]Department F.-A. Forel for Environmental and Aquatic Sciences, University of Geneva, 1211 Geneva, Switzerland

[5]Institute of Geological Sciences, University of Bern, 3012 Bern, Switzerland

[6]Department of Earth Sciences and Institute for Environmental Sciences, University of Geneva, 1205 Geneva, Switzerland

Correspondence: Anna Costa (costa@ifu.baug.ethz.ch)

Abstract. Suspended sediment export from large Alpine catchments (> 1000 km^2) over decadal timescales is sensitive to a number of factors, including long-term variations in climate, the activation–deactivation of different sediment sources (proglacial areas, hillslopes, etc.), transport through the fluvial system, and potential anthropogenic impacts on the sediment flux (e.g. through impoundments and flow regulation). Here, we report on a marked increase in suspended sediment concentrations observed near the outlet of the upper Rhône River Basin in the mid-1980s. This increase coincides with a statistically significant step-like increase in basin-wide mean air temperature. We explore the possible explanations of the suspended sediment rise in terms of changes in water discharge (transport capacity), and the activation of different potential sources of fine sediment (sediment supply) in the catchment by hydroclimatic forcing. Time series of precipitation and temperature-driven snowmelt, snow cover, and ice melt simulated with a spatially distributed degree-day model, together with erosive rainfall on snow-free surfaces, are tested to explore possible reasons for the rise in suspended sediment concentration. We show that the abrupt change in air temperature reduced snow cover and the contribution of snowmelt, and enhanced ice melt. The results of statistical tests show that the onset of increased ice melt was likely to play a dominant role in the suspended sediment concentration rise in the mid-1980s. Temperature-driven enhanced melting of glaciers, which cover about 10 % of the catchment surface, can increase suspended sediment yields through an increased contribution of sediment-rich glacial meltwater, increased sediment availability due to glacier recession, and increased runoff from sediment-rich proglacial areas. The reduced extent and duration of snow cover in the catchment are also potential contributors to the rise in suspended sediment concentration through hillslope erosion by rainfall on snow-free surfaces, and increased meltwater production on snow-free glacier surfaces. Despite the rise in air temperature, changes in mean discharge in the mid-1980s were not statistically significant, and their interpretation is complicated by hydropower reservoir management and the flushing operations at intakes. Overall, the results show that to explain changes in suspended sediment transport from large Alpine catchments it is necessary to include an understanding of the multitude of sediment sources involved together with the hydroclimatic conditioning of their activation (e.g. changes in precipitation, runoff, air temperature). In addition, this study points out that climate signals in suspended sediment dynamics may be visible even in highly regulated and human-impacted systems. This is particularly relevant for quantifying climate change and hydropower impacts on streamflow and sediment budgets in Alpine catchments.

1 Introduction

Erosion processes and sediment dynamics in Alpine catchments are determined by geological, climatic, and anthropogenic factors. Geological forcing is one of the main drivers of sediment production and landscape development, through crustal thickening, deformation and isostatic uplift, and glacier inheritance (e.g. England and Molnar, 1990; Schlunegger and Hinderer, 2001; Vernon et al., 2008). Glacier inheritance influences sediment production and transport as demonstrated by a strong spatial association between sediment yield and past and current glacial cover (Hinderer et al., 2013; Delunel et al., 2014). Almost continuous temperature-driven glacier recession in the European Alps since the late 19th century (Paul et al., 2004, 2007; Haeberli et al., 2007) has maintained large parts of the landscape in early stages of the paraglacial phase, where unstable or metastable sediment sources (Ballantyne, 2002; Hornung et al., 2010) can maintain high sediment supply rates. Anthropogenic impacts on sediment yields are more recent, and on a global scale largely related to land-cover change through intensified agriculture and the trapping of sediment in reservoirs (e.g. Syvitski et al., 2005). Land-use changes mainly impact fine-sediment production (e.g. Foster et al., 2003; Wick et al., 2003), while river channelization, flow regulation, water abstraction, and sediment extraction have caused a general reduction in sediment yield and consequently led to sediment-starved rivers worldwide (Kondolf et al., 2014). In Alpine catchments, in addition to trapping in reservoirs, sediment transfer is also disturbed by flow abstraction at hydropower intakes. The reduction of sediment transport capacity downstream of intakes and the periodic flushing of locally trapped sediment has severe impacts on the sediment budget (e.g. Anselmetti et al., 2007) and downstream river ecology (e.g. Gabbud and Lane, 2016).

Here we focus on the dominant role of climate in sediment production and transfer in Alpine environments (e.g. Huggel et al., 2012; Zerathe et al., 2014; Micheletti et al., 2015; Palazón and Navas, 2016; Wood et al., 2016). The premise behind this work is that to explain impacts of changes in climate on Alpine catchment suspended sediment yield, it is necessary to consider both transport capacity and sediment supply. Sediment supply depends on many factors, most importantly the spatial location of sediment sources (e.g. lithology, distance to outlet, connectivity) and the specific processes of sediment production (e.g. hillslope erosion, glacial erosion, release of subglacially stored sediment, channel bed and bank erosion, mass wasting events) and transport (e.g. hysteresis).

In this study we look at specific sediment sources and the hydroclimatic conditioning of their activation (e.g. precipitation, runoff, and air temperature) with a process-based perspective with the aim to infer the possible effects of changes in hydroclimate, such as increases in temperature and/or precipitation intensity on suspended sediment dynamics. We identify four main sediment sources typical of Alpine environments: glacial erosion, hillslope erosion, channel bed or bank erosion, and mass wasting events (e.g. rockfalls, debris flows). Climatic conditions, specifically precipitation and air temperature, contribute to the activation of these four sediment sources through different processes and at different rates. Erosive processes of abrasion, bed-rock fracturing, and plucking at the base of glaciers provide proglacial areas with large amounts of sediment (Boulton, 1974). Due to glacial erosion, discharge from subglacial channels has high suspended sediment concentrations (e.g. Aas and Bogen, 1988). Temperature-driven snow and ice melt in spring and summer, as well as intense rainfall on snow-free surfaces, may lead to entrainment from proglacial areas provided they are connected to the river network (Lane et al., 2016). Hillslope erosion driven by overland flow and rainfall erosivity may be exacerbated in Alpine catchments by permanently or partially frozen ground (Quinton and Carey, 2008). In summer and autumn, when Alpine catchments are largely free of snow, intense rainfall may erode large amounts of sediment and transport it in rills and gullies to the river network. Intense rainfall is also responsible for triggering mass wasting events, such as debris flows and landslides, where a large mass of sediment is delivered to the channel network instantaneously (e.g. Bennett et al., 2012). Flow conditions (e.g. shear stress, stream power) then determine the sediment transport capacity and in-stream sediment mobilization along rivers, and hence its transfer to downstream locations.

The close link between precipitation, air temperature, runoff, and the activation–deactivation of sediment sources in Alpine catchments becomes critical in the context of climate change. Alpine regions represent a sensitive environment in relation to current rapid warming. In Switzerland, together with glacier recession, a reduction in snow-cover duration and mean snow depth has been observed during the last 30 years (e.g. Beniston, 1997; Laternser and Schneebeli, 2003; Scherrer et al., 2004; Marty, 2008; Scherrer and Appenzeller, 2006). Although current effects of climate change are less clear for precipitation (Brönnimann et al., 2014) than for temperature, a sharp reduction in the number of snowfall days has been observed at many meteorological stations in Switzerland (Serquet et al., 2011).

The upper Rhône River Basin draining into Lake Geneva in Switzerland is at the centre of our investigation. The basin has experienced a rise in air temperature that coincided with a rise in suspended sediment concentrations in the mid-1980s. Our main objective is to explore the presence of the signal of a warmer climate in the suspended sediment dynamics of this regulated and human-impacted Alpine catchment. In this work, we refer to fine sediment as the sediment transported in suspension. To investigate the potential causes of the observed increase in suspended sediment concentration, we conceptualize the upper Rhône Basin as a series of spatially distributed sediment sources that are activated or deactivated by hydroclimatic forcing. In addition

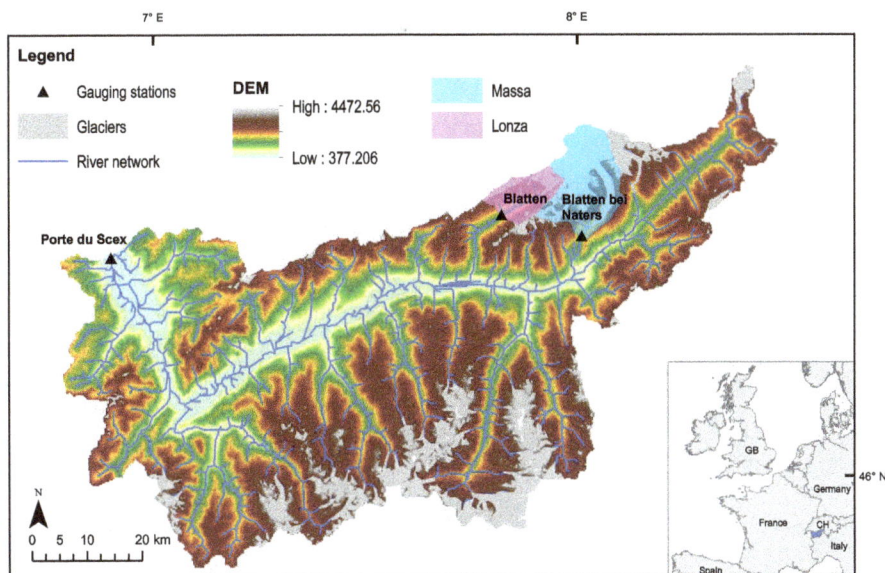

Figure 1. Map of the upper Rhône Basin with topography (DEM resolution is 250×250 m), glacierized areas, and river network. The inset shows the position of the upper Rhône Basin in Europe (blue). Locations of gauging stations used in this analysis are shown as triangles. Massa and Lonza sub-basins used in the calibration and validation of the ice melting component are highlighted.

to discharge (transport capacity), we consider four main hydroclimatic variables: (a) ice melt runoff (IM), which evacuates accumulated fine sediment, the product of glacial erosion, through subglacial channels (e.g. Swift et al., 2005); (b) snow-cover fraction (SCF), which influences ice melting onset, impacts ice-melt efficiency through albedo, and may result in more rapid erosion and sediment production through an increased glacier basal velocity (e.g. Herman et al., 2015); (c) snowmelt runoff (SM) from snow-covered areas, which may generate downstream hillslope erosion and channel erosion (e.g. Lenzi et al., 2003); and (d) effective rainfall (ER), defined as liquid precipitation over snow-free areas, which leads to hillslope erosion, mass wasting, and also, due to enhanced discharge, channel erosion (e.g. Bennet et al., 2012; Meusburger and Alewell, 2014). Our aims are the following: (a) to estimate daily basin-wide ice melt, snow-cover fraction, snowmelt, and effective rainfall over the Rhône Basin for the last 40 years; and (b) to analyse these variables with the goal to provide statistical evidence for possible reasons for the rise in suspended sediment concentrations in the mid-1980s.

2 Study site description

The upper Rhône Basin is located in the southwestern part of Switzerland, in the central Swiss Alps (Fig. 1). It has a total surface area of 5338 km^2 and an altitudinal range of 372 to 4634 m a.s.l. About 10 % of the surface is covered by glaciers, which are mostly located in the eastern and southeastern part of the catchment (Stutenbecker et al., 2016). The Rhône River originates at the Rhône Glacier and flows for

about 160 km through the Rhône valley before entering Lake Geneva a few kilometres downstream of the gauging station located at Porte du Scex (Fig. 1). Basin-wide mean annual precipitation is about 1400 mm yr^{-1} and shows strong spatial variability driven mostly by orography and the orientation of the main valley. The hydrological regime of the catchment, typical of Alpine environments, is strongly influenced by snow and ice melt with highest discharge in summer and lowest in winter. Mean annual discharge is 180 m^3 s^{-1}, which corresponds to about 1060 mm yr^{-1} and an annual runoff coefficient of 75 %.

The catchment has been strongly affected by anthropogenic impacts during the last century. The main course of the Rhône River has been extensively channelized for the purposes of flood protection: levees were constructed and the channel was narrowed and deepened in the periods 1863–1894 and 1930–1960 (first and second Rhône corrections). Due to the residual flood risk that affects the main valley, a third project was started in 2009 with the main objectives to increase channel conveyance capacity and river ecological rehabilitation (Oliver et al., 2009). In addition, significant gravel mining operations are carried out along the main channel and many tributaries. Since the 1960s, several large hydropower dams have been built in the main tributaries of the Rhône River. The total storage capacity of these reservoirs corresponds to about 20 % of the mean annual streamflow (Loizeau and Dominik, 2000). Flow impoundment, water abstraction, and diversion through complex networks of intakes, tunnels, and pumping stations have significantly impacted the flow and sediment regime of the catchment. Flow regulation due to hydropower production has resulted in a

Table 1. List of the variables analysed: observed SSC and hydroclimatic variables originating from measurements (Q), spatial interpolation of measurements (T, P), simulations of the snow and ice-melt model (SCF, SM, IM), or a combination thereof (ER). Information on the source and the spatial and temporal resolution are reported for each variable.

Variable	Data source	Resolution
T	Daily mean temperature (°C) on a $\sim 2\,\mathrm{km} \times 2\,\mathrm{km}$ grid provided by MeteoSwiss	basin-averaged, daily, 1975–2015
P	Daily total precipitation ($\mathrm{mm\,day^{-1}}$) on a $\sim 2\,\mathrm{km} \times 2\,\mathrm{km}$ grid provided by MeteoSwiss	basin-averaged, daily, 1975–2015
Q	Daily mean discharge ($\mathrm{m^3\,s^{-1}}$) at three stations (Porte du Scex, Blatten, Blatten bei Naters) provided by FOEN	daily, 1975–2015
SSC	Suspended sediment concentration ($\mathrm{mg\,L^{-1}}$) at Porte du Scex provided by FOEN	2 times per week, 1975–2012
SCF	Snow-cover fraction [0–1] simulated by the Snowmelt model on a $250\,\mathrm{m} \times 250\,\mathrm{m}$ grid, and calibrated with MODIS satellite data for the period 2000–2009	basin-averaged, daily, 1975–2012
SM	Snowmelt rate ($\mathrm{mm\,day^{-1}}$) simulated by the Snowmelt model on a $250\,\mathrm{m} \times 250\,\mathrm{m}$ grid	basin-averaged, daily, 1975–2012
IM	Ice-melt rate ($\mathrm{mm\,day^{-1}}$) simulated by the ice-melt model on a $250\,\mathrm{m} \times 250\,\mathrm{m}$ grid, and calibrated at Blatten and Blatten bei Naters	basin-averaged, daily, 1975–2012
ER	Effective rainfall ($\mathrm{mm\,day^{-1}}$) (rainfall on snow-free pixels), estimated from P, T and SCF on a $250\,\mathrm{m} \times 250\,\mathrm{m}$ grid	basin-averaged, daily, 1975–2012

considerable decrease in discharge in summer and increase in winter (Loizeau and Dominik, 2000). That said, the construction of dams and the start of hydropower operation has coincided with a drop in the suspended sediment load of the main Rhône River measured at Porte du Scex in the 1960s (Loizeau et al., 1997; Loizeau and Dominik, 2000).

Two sub-catchments of the upper Rhône Basin are used for the calibration and validation of the ice-melt model: the Massa and the Lonza sub-catchments (Fig. 1). The Massa is a medium-sized basin ($195\,\mathrm{km^2}$) with a mean elevation of 2945 m a.s.l. More than 60 % of the surface is glacierized, and the remaining surface is classified mostly as rock and firn (Boscarello et al., 2014). The basin includes the Aletsch Glacier, which is the largest glacier in the European Alps with a length (1973) of around 23.2 km and a surface area (1973) of approximately $86\,\mathrm{km^2}$ (Haeberli and Holzhauser, 2003). The Lonza is a relatively small basin located to the west of the Massa with an average elevation of 2630 m a.s.l. It has a total drainage area of roughly $77.8\,\mathrm{km^2}$ and almost 36 % of its surface (1991) is covered by glaciers.

3 Methods

Our objective is to explore the potential effect of climate on suspended sediment dynamics in the upper Rhône Basin during the period 1975–2015. To this end, we analyse observed and simulated hydroclimatic and sediment-transport variables as listed in Table 1: mean daily temperature T, total daily precipitation P, mean daily discharge Q, suspended sediment concentration SSC, daily snow-cover fraction SCF, snowmelt SM, ice melt IM, and effective rainfall over snow-free areas ER. Some variables originate from observations

(Q, SSC) or spatial interpolation of observations (T, P), others from simulations by spatially distributed snow and ice-melt models (SCF, SM, IM), or from a combination thereof (ER). The snowmelt model is described in Sect. 3.1, the ice-melt model in Sect. 3.2, and their calibration in Sect. 3.3. We first interpolate the input datasets of precipitation and temperature on a $250\,\mathrm{m} \times 250\,\mathrm{m}$ grid resolution by the nearest-neighbour interpolation method. Second we run the snow and ice-melt model on the daily timescale over the period 1975–2015. In a third step, we analyse the variables (Table 1) as mean monthly and mean annual values averaged over the basin area. To quantify changes in the hydroclimatic variables and in suspended sediment concentration, we apply standard statistical tests for change detection described in Sect. 3.4. A description of all datasets used in this analysis is reported in more detail in Sect. 4.

3.1 Snowmelt model

We use a snowmelt model to predict SM and SCF over the entire basin, because snow station measurements are sparsely and irregularly distributed and a physical consistency between precipitation and air temperature as climatic driving forces and snowmelt and snow cover as response variables is needed. The spatially distributed temperature index method (degree-day model) was used due to its simplicity, low data requirements, and demonstrated success on daily temporal scales over large basins (e.g. Hock, 2003; Boscarello et al., 2014). The degree-day approach also matches the coarse spatial ($250\,\mathrm{m} \times 250\,\mathrm{m}$) and temporal (daily) resolution of our analysis and the areal averaging on the basin scale. Models based on energy balance, or enhancements of the degree-day approach, represent physical processes better and could be used when higher spatial and temporal resolution and accuracy are needed (e.g. Pellicciotti et al., 2005).

The snowmelt model includes snow accumulation and melt. On the grid scale, precipitation P (mm day^{-1}) is first partitioned into solid and liquid form based on (a) daily minimum T_{min} (°C) and maximum air temperature T_{max} (°C) and (b) a rain–snow threshold temperature T_{RS} (°C). If minimum air temperature T_{min} is above the threshold temperature T_{RS}, all precipitation falls as rainfall R; if the maximum air temperature T_{max} is below the threshold temperature T_{RS}, all precipitation falls as snow S; otherwise precipitation is a mixture of solid and liquid form, partitioned proportionally to the temperature difference

$$\begin{cases} R = c_p P \\ S = (1 - c_p) P \end{cases}, \tag{1}$$

where

$$\begin{cases} c_p = 1, & T_{min} > T_{RS} \\ c_p = 0, & T_{max} \leq T_{RS} \\ c_p = \dfrac{T_{max} - T_{RS}}{T_{max} - T_{min}}, & T_{min} \leq T_{RS} < T_{max} \end{cases}. \tag{2}$$

The daily snowmelt rate SM_i (mm day^{-1}) is estimated from a linear relation with air temperature:

$$\begin{cases} SM_i = k_{snow}(T_{mean} - T_{SM}), & T_{mean} > T_{SM} \\ SM_i = 0, & T_{mean} \leq T_{SM} \end{cases}, \tag{3}$$

where T_{mean} (°C) is the mean daily air temperature, T_{SM} (°C) is a threshold temperature for the onset of melt, and k_{snow} is a melt factor (mm day^{-1} °C^{-1}). Snow depth (SD), in millimetres snow water equivalent, for time t is then simulated from a balance between accumulation and melt at every grid cell i:

$$SD_i(t) = SD_i(t - 1) + S_i(t) - SM_i(t). \tag{4}$$

The snow-cover fraction, SCF, for a chosen area containing $i = 1, \ldots, N$ grids is

$$SCF(t) = \frac{1}{N} \sum_{i=1}^{N} H[SD_i(t)], \tag{5}$$

where H is a unit step function: $H = 0$ when $SD = 0$ and $H = 1$ when $SD > 0$. The area of integration N can be the entire catchment, sub-basins, elevation bands, etc. For the entire catchment, we estimate mean daily snowmelt SM (mm day^{-1}) as the arithmetic average over all grid melt rates:

$$SM(t) = \frac{1}{N} \sum_{i=1}^{N} SM_i(t). \tag{6}$$

The threshold temperatures for defining the precipitation type, T_{RS}, and the onset of melt, T_{SM}, depend on many factors such as atmospheric boundary layer conditions, temperature, and humidity, among others. Different parameterizations and temperature values are available in the literature (Wen et al., 2013). Depending on region, altitude, and

modelling approach, rain–snow temperature thresholds show a range of variability from -5 °C (Collins et al., 2004) to more than 6 °C (Auer, 1974). For the upper Rhône Basin we assume a constant rain–snow temperature threshold $T_{RS} = 1$ °C, resulting from a calibration and validation of the physically based fully distributed hydrological model Topkapi-ETH in the catchment (Fatichi et al., 2015). To reduce degrees of freedom, the threshold temperature for the onset of melt T_{SM} is set equal to 0 °C, which is a typical value for Alpine regions (e.g. Schaefli et al., 2005; Corbari et al., 2009; Boscarello et al., 2014). The calibration of the snowmelt model consists of estimating only the melt factor k_{snow} with methods described in Sect. 3.3. In addition, we apply a sensitivity analysis on the three parameters k_{snow}, T_{RS}, and T_{SM}, as described in the Supplement Sect. S1.

3.2 Ice-melt model

Similar to snowmelt, ice melt is also simulated with a temperature index (degree-day) model on grid cells that are identified as glacier covered. The daily ice melt IM_i (mm day^{-1}) on glacier surfaces that are snow-free is estimated as

$$\begin{cases} IM_i = k_{ice}(T_{mean} - T_{IM}) & T_{mean} > T_{IM} \\ IM_i = 0 & T_{mean} \leq T_{IM} \end{cases}, \tag{7}$$

where T_{mean} (°C) is the mean daily air temperature, T_{IM} (°C) is a threshold temperature for the onset of ice melt, and k_{ice} (mm day^{-1} °C^{-1}) is the ice-melt factor. For the entire catchment, we estimate mean daily ice melt IM (mm day^{-1}) as the arithmetic average over all ice-covered grid cells as follows:

$$IM(t) = \frac{1}{N} \sum_{i=1}^{N} IM_i(t). \tag{8}$$

The threshold temperature for glacier melting T_{IM} is set equal to 0 °C. Ice melt occurs only if the glacier cell is snow free. The snow cover simulated by the snowmelt model in Sect. 3.1 is thus essential for estimating ice melt. The calibration of the ice melt model consists of estimating only the melt factor k_{ice} as described in Sect. 3.3.

3.3 Calibration and validation of snowmelt and ice-melt models

We perform the calibration and validation of the snow and ice-melt model parameters in sequence, since the snow-covered surface is required for ice-melt estimation on glaciers. The snowmelt factor k_{snow} is calibrated based on comparisons with snow-cover maps. Snow-cover observations are split into two periods: 1 October 2000–30 September 2005 for calibration and 1 October 2005–31 December 2008 for validation, for a total number of 217 calibration and 143 validation days. Snow-cover maps at 500 m × 500 m resolution are distributed by proximal interpolation to the snowmelt model 250 m × 250 m computational grid. Maps of snow depth simulated with Eq. (4) are first transformed into simulated snow-cover fraction SCF^{sim} with Eq. (5) and are afterwards compared with snow-cover fraction derived from the observations SCF^{obs}.

The objective function for calibration is based on a combination of mean absolute error and true skill statistic. The mean absolute error MAE is estimated as

$$MAE = \frac{1}{n}\sum_{j=1}^{n}\left|SCF_j^{obs} - SCF_j^{sim}\right|, \tag{9}$$

where n is the number of MODIS image maps. MAE captures the overall ability of the model to reproduce the snow-cover fraction accurately. The true skill statistic (TSS) is a spatial statistic that measures the grid-to-grid performance of the model in capturing snow–no-snow presence. It is computed as the sum of sensitivity SE (correct snow predictions) and specificity SP (correct no-snow predictions) computed from contingency tables (e.g. Wilks, 1995; Mason and Graham, 1999; Corbari et al., 2009) in each image j and averaged over the n MODIS maps in the simulation period:

$$TSS = \frac{1}{n}\sum_{j=1}^{n}TSS_j = \frac{1}{n}\sum_{j=1}^{n}\left(SE_j + SP_j\right). \tag{10}$$

Because TSS includes both sensitivity and specificity, it captures both predictions of snow-covered and snow-free areas. It takes on values between 0 and 1, where 1 indicates perfect performance, and is a widely applied metric for assessing spatial model performance (e.g. Begueria, 2006; Allouche et al., 2006). We combine both goodness-of-fit measures (MAE and TSS) into an objective function OF, by giving more weight to MAE. Finally, we evaluate the objective function OF over $b = 5$ different elevation bands in order to better capture the topographic gradients in snowmelt distribution in the Rhône Basin:

$$OF = \sum_{b=1}^{5}OF_b = \sum_{b=1}^{5}-0.6\,MAE_b + 0.4\,TSS_b. \tag{11}$$

This objective function is maximized in calibration. The rationale of using both MAE and TSS in evaluating performance is to give weight to both basin-integrated snow cover as well as to grid-based predictions. Indeed, the same value of snow-cover fraction can result in two different spatial arrangements of snow-covered pixels, and a correct spatial distribution of snow-covered and snow-free areas is relevant for this analysis insofar as it affects the activation and deactivation of specific sediment sources. The weights assigned to MAE and TSS in Eq. (11) are the outcome of sensitivity tests with the model. After calibration, we also estimate the Nash–Sutcliffe efficiency (NS; Nash and Sutcliffe, 1970) and the mean square error MSE to quantify the performance of the

model:

$$NS = 1 - \frac{\sum_{j=1}^{n}\left(SCF_j^{obs} - SCF_j^{sim}\right)^2}{\sum_{j=1}^{n}\left(SCF_j^{obs} - \overline{SCF}\right)^2}, \tag{12}$$

$$MSE = \frac{1}{n}\sum_{j=1}^{n}\left(SCF_j^{obs} - SCF_j^{sim}\right)^2, \tag{13}$$

where $\overline{SCF}$ is the average observed snow-cover fraction during the calibration–validation period.

We calibrate the ice-melt factor k_{ice} with the data of the sub-basin of the river Massa (Fig. 1), on the basis of daily discharge measurements, and focusing only on months when the ice-melt contribution is not negligible (June–October). The gauging station is located upstream of the Gebidem dam, therefore discharge is not influenced by reservoir regulation and represents undisturbed natural flow. Calibration is performed on the period 1 January 1975–31 December 2005, while validation covers the remaining 10 years of available data, i.e. the period 1 January 2006–31 December 2015. We then validate the model on the Lonza sub-basin with the same procedures and goodness-of-fit measures.

The optimal value of k_{ice} is found by minimizing the mass balance error MBE_S computed for the period June–October:

$$MBE_S = 100\frac{\sum_{i=1}^{ny}\left(V_i^{obs} - V_i^{sim}\right)}{\sum_{i=1}^{ny}V_i^{obs}}, \tag{14}$$

where ny is the number of calibration years, V_i^{obs} and V_i^{sim} (mm year^{-1}) are the observed and simulated discharge volumes per unit area reaching the outlet of the catchment during the period June–October of each calibration year i:

$$V^{obs} = \sum_{j=1}^{nd}Q_j^{obs}, \tag{15}$$

$$V^{sim} = \sum_{j=1}^{nd}Q_j^{sim} = \sum_{j=1}^{nd}(R_j + SM_j + IM_j). \tag{16}$$

Here, nd is the number of observation days from June to October, Q_j^{obs} (mm day^{-1}) is the daily discharge per unit area observed at Blatten Bei Naters (Blatten), R_j, SM_j, IM_j are respectively the total daily rainfall, snowmelt, and ice melt aggregated over the Massa (Lonza) basin. Rainfall (R) and snowmelt (SM) are simulated with the snow accumulation and melt model in Sect. 3.1, while ice melt (IM) is simulated with the ice-melt model in Sect. 3.2.

It should be noted that in this study we neither consider glacier evolution, i.e. changes in ice thickness due to accumulation and melt, nor glacier ice flow. Neglecting glacier retreat raises the possibility that we overestimate the ice-melt contribution over the study period. To quantify the potential effect of glacier retreat, we compare our simulations with time series produced from the Global Glacier Evolution Model (GloGEM), a model accounting for both the mass balance and glacier evolution (Huss and Hock, 2015). For comparison, we use total monthly runoff generated from glacierized surfaces of the upper Rhône Basin, simulated with Glo-GEM for the period 1980–2010. GloGEM computes the mass balance for every 10 m elevation band of each glacier, by estimating snow accumulation, snow and ice melt, and refreezing of rain and melt water. The response of glaciers to changes in mass balance is modelled on the basis of an empirical equation between ice thickness changes and normalized elevation range parameterized as proposed by Huss et al. (2010). Normalized surface elevation changes Δh_r are derived for each elevation band from mass balance changes (mass conservation). Starting from initial values derived by the method of Huss and Farinotti (2012), ice thickness is updated at the end of each hydrological year by applying the relation between normalized elevation range h_r and normalized surface elevation change Δh_r. The area of each glacier is finally adjusted by a parabolic cross-sectional shape of the glacier bed (Huss and Hock, 2015). GloGEM is calibrated and validated over the period 1980–2010 with estimates of glacier mass changes by Gardner et al. (2013) and in situ measurements provided by the World Glacier Monitoring Service.

3.4 Statistical testing for change

We use the non-parametric Pettitt test (Pettitt, 1979) for the detection of the time of change (year-of-change) in the air temperature data. We then test the other variables (SSC, P, Q, SM, SCF, IM, and ER) for changes in the mean (and variance) by splitting the time series into two periods before and after the identified year-of-change, and by applying two-sample two-sided t tests for the equality of the means (and variances). The null hypothesis of no change is tested at the 5 % significance level. The t test is a parametric test commonly used in hydrology to assess the validity of the null hypothesis of two samples having equal means and unknown unequal variances. We apply the t test to all hydroclimatic variables averaged on the annual and monthly timescales with the same year-of-change to determine which hydroclimatic variables, and therefore the activation or deactivation of which sediment sources, are possibly responsible for the observed changes in suspended sediment concentration.

In our catchment, SSC is sampled intermittently (twice per week). This might have an effect on the change detection analysis of the hydroclimatic variables. We estimate this potential effect by considering the hydroclimatic variables SM, IM, ER, and Q only on days corresponding to SSC-measurement days. We compare these new time series with the original ones by estimating the cumulative distribution

functions of the variables and by testing changes of mean monthly and annual values over time. We consider only the positive (non-zero) part of the distributions. Results are reported in Sect. 5.4.

4 Data description

4.1 Precipitation and air temperature

For precipitation and air temperature we use spatially distributed datasets provided by the Swiss Federal Office of Meteorology and Climatology (MeteoSwiss). Total daily precipitation and mean, minimum, and maximum daily air temperature are available on a $\sim 2\,\mathrm{km} \times 2\,\mathrm{km}$ resolution grid for Switzerland (MeteoSwiss, 2013a, b). All four datasets are developed by spatial interpolation of quality-checked data collected at MeteoSwiss meteorological stations (Frei et al., 2006; Frei, 2014). We apply the statistical analysis of change to basin-averaged values of precipitation and temperature and not to individual grid point values, which might be potentially affected by substantial interpolation errors. Moreover, the variability in time of the number of stations involved in the spatial interpolation may induce non-homogeneities in the datasets. This is particularly relevant when analysing long-term changes as in the case of this study. Therefore, we verify the effects of potential non-homogeneities by using an experimental dataset developed by MeteoSwiss specifically for this research, based on a constant number of stations (294 for precipitation and 48 for temperature) for the period 1971–2013. We applied the statistical tests for detecting changes both on the original and the experimental datasets of P and T. Results of the statistical tests on the two datasets coincide. This confirms that temporally variable number of meteorological stations employed to build the product does not influence the changes detected in the original dataset.

4.2 Discharge and suspended sediment concentration

We use daily discharge data measured by the Swiss Federal Office for the Environment (FOEN) at three gauging stations: Porte du Scex (available since 1905), Blatten Bei Naters (available since 1931) and Blatten (available since 1956) (Fig. 1). For suspended sediment concentration, two in-stream samples per week collected by FOEN at Porte du Scex are available since October 1964 (Grasso et al., 2012).

In this work, we focus on sediment transported in suspension. Previous analysis on the grain size distribution of suspended sediment at the outlet of the upper Rhône River reports a bimodal distribution, with mode diameters equal to 13.7 μm (silt) for the finer fraction and 39.6 μm (silt) for the coarser grains (Santiago et al., 1992). The composition of grains cover a wide range of values, including clay (16.9 %), silt (64.7 %), and sand (18.4 %). The mean suspended sediment size is reported to be equal to 17.7 μm (silt), and the largest grains transported in suspension during summer high-flow conditions are in the range of coarse sand ($> 500\,\mu\mathrm{m}$) (Santiago et al., 1992).

4.3 Snow cover and glacier data

We use snow-cover maps derived from satellite imagery for the upper Rhône Basin over the period 2000–2008 processed in previous research (Fatichi et al., 2015). We use the 8-day snow-cover product MOD10A2 retrieved from the Moderate Resolution Imaging Spectroradiometer (MODIS) (Dedieu et al., 2010) for the calibration and validation of the snowmelt model. MOD10A2 is provided at a $500\,\mathrm{m} \times 500\,\mathrm{m}$ spatial resolution, where cells are classified as snow covered, snow free, inland water, or cloud covered. In order to reduce the impacts of clouds in estimating snow-cover fraction, maps with cloud cover greater than 30 % are excluded from the dataset, resulting in a total number of usable images equal to 360, i.e. on the average 40 days per year.

The surface covered by glaciers is assigned based on the GLIMS (Global Land Ice Measurements from Space) Glacier Database (Fig. 1). Ice-covered cells identified based on the GLIMS data of 1991 show that more than 10 % of the upper Rhône Basin as covered by ice with a total glacier surface of almost $620\,\mathrm{km}^2$.

4.4 Digital terrain model

We use a digital elevation model (DEM) with $250\,\mathrm{m} \times 250\,\mathrm{m}$ resolution (85 409 cells in total; Fig. 1), obtained by resampling a finer model ($25\,\mathrm{m} \times 25\,\mathrm{m}$) provided by SwissTopo in the ETH geodata portal (GeoVITe). The DEM is used as a mask for extracting climatic inputs and for elevation information in the snowmelt modelling.

5 Results

5.1 Calibration of snowmelt and ice-melt models

The snowmelt factor, calibrated following the procedure described in Sect. 3.3, is $k_{\mathrm{snow}} = 3.6\,\mathrm{mm\,day}^{-1}\,{}^\circ\mathrm{C}^{-1}$. The snowmelt model reproduces well the seasonal fluctuations of snow-cover fraction (SCF) in the basin, with Nash–Sutcliffe efficiencies (NS) close to 0.90 and low mean square errors (MSE). The model maintains good performances also in the validation period showing a slight reduction in the goodness-of-fit measures (Table 2). The temporal variability in SCF is also well simulated on the basin scale. Although the comparison between observed and simulated SCF is affected by the discontinuous nature of the MODIS data (8-day resolution), Fig. 2 shows that the model with a single constant k_{snow} for the entire catchment reproduces the snow-cover dynamics reasonably well for all of the studied elevation bands. At lower elevations, the model tends to slightly underestimate SCF in autumn and overestimate it in winter. The model performs better at higher elevation bands, even at the very

Table 2. (a) Calibrated snowmelt factor k_{snow} and goodness-of-fit measures for validation and calibration periods: Nash–Sutcliffe efficiency (NS), mean square error (MSE), true skill statistic (TSS), sensitivity (SE), and specificity (SP) for the entire upper Rhône Basin; and **(b)** calibrated ice-melt factor k_{ice} and goodness-of-fit measures: mass balance error computed on June–October (MBE$_{\mathrm{S}}$) and on the entire year (MBE$_{\mathrm{A}}$) for the Massa and Lonza sub-basins.

(a)	$k_{\mathrm{snow}} = 3.6\,\mathrm{mm\,day}^{-1}\,{}^\circ\mathrm{C}^{-1}$	
	Calibration	Validation
NS	0.88	0.86
MSE	0.01	0.01
TSS	0.54	0.46
SE	0.77	0.76
SP	0.73	0.70
(b)	$k_{\mathrm{ice}} = 6.1\,\mathrm{mm\,day}^{-1}\,{}^\circ\mathrm{C}^{-1}$	
	MBE$_{\mathrm{S}}$ (%)	MBE$_{\mathrm{A}}$ (%)
Calibration Massa	6.10	7.22
Validation Massa	6.77	9.19
Validation Lonza	11.35	10.09

highest elevations with permanent snow cover (Fig. 2; bottom). The spatial distribution of snow cover is satisfactory, with average values of sensitivity and specificity greater than 0.7 (Table 2). Goodness-of-fit measures indicate that, on average, more than 70 % of snow-covered and snow-free pixels are correctly identified. The true skill score, which combines both metrics, results in values around 0.5 (Table 2). Snow-cover-duration maps averaged over the period 2000–2008 for MODIS observations and simulations show a good spatial coherence (Fig. 3). In summary, we conclude that the snowmelt model represents the spatial and temporal dynamics of snow cover in the Rhône Basin satisfactorily. Results of the sensitivity analysis on k_{snow}, T_{RS}, T_{SM} are reported in the Supplement (Sect. S1).

The ice-melt factor, calibrated following the procedure described in Sect. 3.3, is $k_{\mathrm{ice}} = 6.1\,\mathrm{mm\,day}^{-1}\,{}^\circ\mathrm{C}^{-1}$. Calibration and validation results are summarized in Table 2. In Fig. 4 we show the seasonal pattern of basin-averaged IM, SM, and R simulated with the calibrated snow and ice-melt model, together with discharge Q observed at the outlet of the two highly glacierized sub-catchments Massa (Fig. 4a) and Lonza (Fig. 4b). The fit of the simulated (computed as $\mathrm{IM} + \mathrm{SM} + R$) to the observed discharge is good, with mass balance errors about 7 % for the Massa and 8 % for the Lonza.

Although in our hydrological model we do not include glacier evolution, the annual runoff volumes ($\mathrm{SM} + \mathrm{IM} + R$) from glacierized areas during the period 1980–2010 correlate well with the results of GloGEM (Fig. 5a). Measures of performance confirm the agreement between the two models: the correlation coefficient is equal to 0.86 and the Nash–

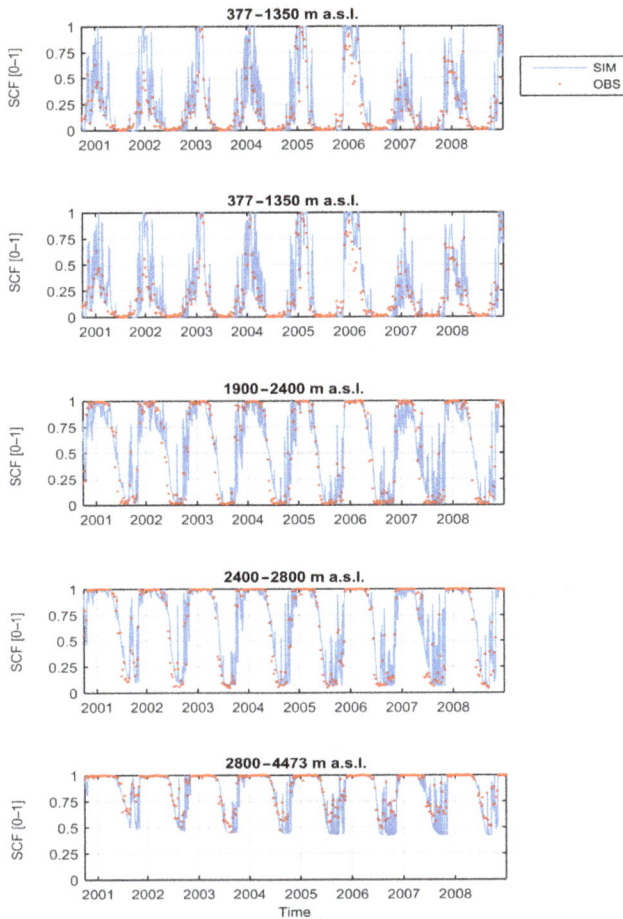

Figure 2. Comparison between observed (red circles) and simulated (light blue lines) snow-cover fraction (SCF) of the upper Rhône Basin for five different elevation bands. Simulations are computed with calibrated snowmelt factor $k_{snow} = 3.6\,\mathrm{mm\,day^{-1}\,^\circ C^{-1}}$.

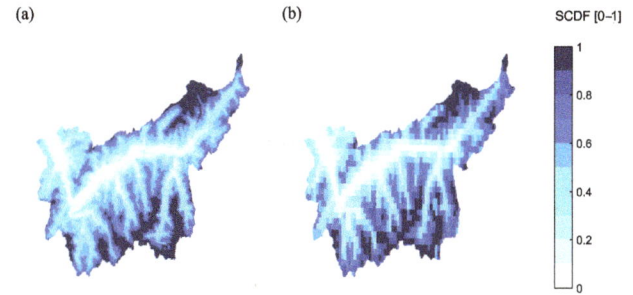

Figure 3. Map of average snow permanence during the period 2000–2008, expressed as the fraction of time in which pixels are snow-covered (snow cover duration fraction, SCDF[0–1]): **(a)** observations (MODIS) and **(b)** simulations.

Sutcliffe efficiency is equal to 0.67. We also capture quite well the seasonal pattern of runoff generated from glacierized areas (Fig. 5b). Perhaps most importantly, GloGEM simulations show that total annual runoff is increasing throughout the period and there is no evidence for decreasing ice-melt rates. This confirms that, although glaciers of the upper Rhône Basin are retreating, melt-water discharges from glacierized and proglacial areas are increasing during the 1980–2010 period. As expected, total runoff from glacierized surfaces and ice melt is highly correlated (Fig. 5a; correlation coefficient = 0.95), thus indicating that the increase in total runoff is due to an increase of the ice-melt component. Indeed, non-parametric Mann–Kendall tests indicate an increasing trend with 5 % significant level. Trend slopes, estimated with the Theil–Sen estimator, confirm the agreement between the two models: we find a total increase of runoff of ~ 27.65 millon m^3 year^{-2} with GloGEM and ~ 21.71 million m^3 year^{-2} with our model. We also computed the basin-averaged mass balance accounting for snow accumulation

and snow and ice melt for each hydrological year. The mean mass balance over the period 1980–2010 is equal to -0.78 ± 0.22 m year^{-1} of water equivalent, which is within the uncertainty range of recent studies (Fischer et al., 2015). In summary, although we do not account for glacier retreat, our model results agree well with state-of-the-art glaciological models that include glacier evolution. Both comparisons with GloGEM and our basin-averaged mass balance indicate that we are not significantly overestimating ice melt during the period 1975–2015.

5.2 Temperature, precipitation, discharge, and SSC in the Rhône Basin

Mean annual air temperature shows a clear and statistically significant increase in 1987 (p value < 0.01). A two-sample t test for equal means (p value < 0.01) confirms an increase in mean daily temperature greater than 1 °C (Fig. 6a). Statistical tests on monthly means reveal that the 1987 temperature jump is mainly in spring and summer months from March to August, while changes in the autumn and winter months are not statistically significant (Fig. 7a). For the period March–August, mean monthly temperatures have risen by about 1.7 °C on the average.

The change in air temperature around 1987 coincides with statistically significant changes in mean annual suspended sediment concentration (Fig. 6c). After the abrupt warming, mean annual suspended sediment concentrations are roughly 40 % larger than before: average values have risen from 172 ± 6.86 mg L^{-1} before 1987 up to 242 ± 14.45 mg L^{-1} after 1987, where the ranges express the standard error of the mean. This change can be ascribed to statistically significant (p value < 0.01) increases in summer (July–August) concentrations (Fig. 7c). Suspended sediment concentration is also characterized by much larger interannual variability after 1987 than before: the standard deviation of mean annual SSC increases from ~ 32 mg L^{-1} before 1987 up to ~ 78 mg L^{-1} after (Fig. 6c). A statistically significant in-

(a)   (b)

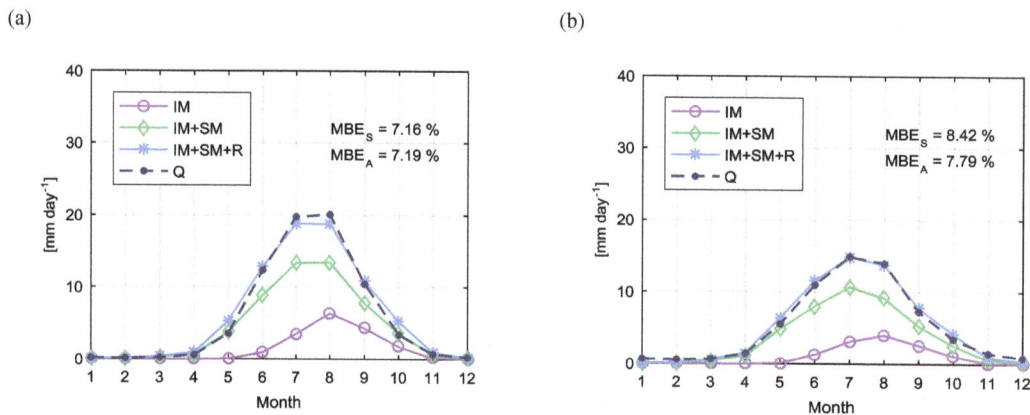

Figure 4. Comparison of mean monthly observed (dark blue) and simulated (light blue) discharge for the period 1975–2015: **(a)** Massa basin and **(b)** Lonza basin. Simulated discharge is the sum of three components: ice melt (IM), snowmelt (SM), and rainfall (R).

(a)   (b)

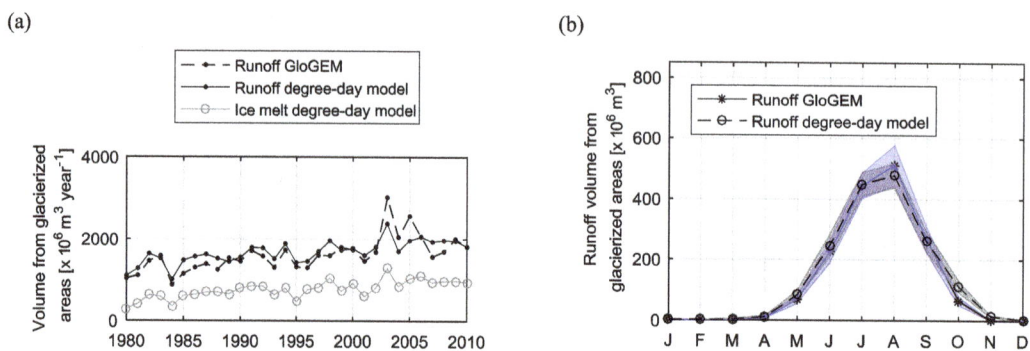

Figure 5. Runoff (snowmelt + ice melt + rainfall) generated from glacierized areas within the upper Rhône Basin, simulated with GloGEM and with the snowmelt and ice-melt models (degree-day) for the period 1980–2010: **(a)** total annual values; **(b)** mean monthly values. **(a)** also depicts the time series of total annual ice melt simulated with the ice-melt model.

crease in the variance is confirmed with a two-sample F test at 5 % significant level.

While the upper Rhône Basin underwent an abrupt warming around 1987, mean annual precipitation (Fig. 6b) and mean monthly precipitation (Fig. 7b) did not change significantly in time. Likewise, mean annual discharge does not show any statistically significant change in 1987 (Fig. 6d). Mean monthly discharge (Fig. 7d) is characterized by a small statistically significant increase in winter (November–February) runoff, most likely due to increased snowmelt and possibly changes in hydropower generation.

5.3   Hydroclimatic activation of sediment sources

Mean annual simulated snowmelt (SM) shows a decreasing tendency during the last 30 years (Fig. 8a). The reduction in snowmelt after 1987 occurs mostly in summer and early autumn (Fig. 9a) mainly due to poor snow cover (Fig. 9b). However, except July and September, the changes in all months are within the 95 % confidence interval. The increase in snowmelt in March and April is due to warmer temperatures in spring. Results are coherent with the temporal evo-

lution of simulated snow-cover fraction, which is also gradually decreasing (Fig. 8b), especially in spring and summer (Fig. 9b). Statistical analysis reveals a step-like reduction of more than 10 % for mean annual values of snow-cover fraction in 1987 (p value < 0.01).

Although mean annual and monthly precipitation were shown not to change significantly in the mid-1980s, effective rainfall (ER) on snow-free areas has increased, especially in early summer (Figs. 8d, 9d). Effective rainfall increases in conjunction with decreases in snow-cover fraction, and a statistically significant jump is identified in 1987 (p value < 0.01) (Fig. 8d). However, although snow-cover fraction is significantly lower throughout the entire melting season, only June and especially July show statistically significant increases in ER after 1987 (Fig. 9d).

Our results show that the temporal evolution of ice melt is consistent with suspended sediment concentration rise. Although the change is rather gradual on the annual scale (Fig. 8c), the step-like increase in ice melt is evident in the ice-melting season (May–September) and reaches highest magnitudes in July and August (Fig. 9c) in conjunction with

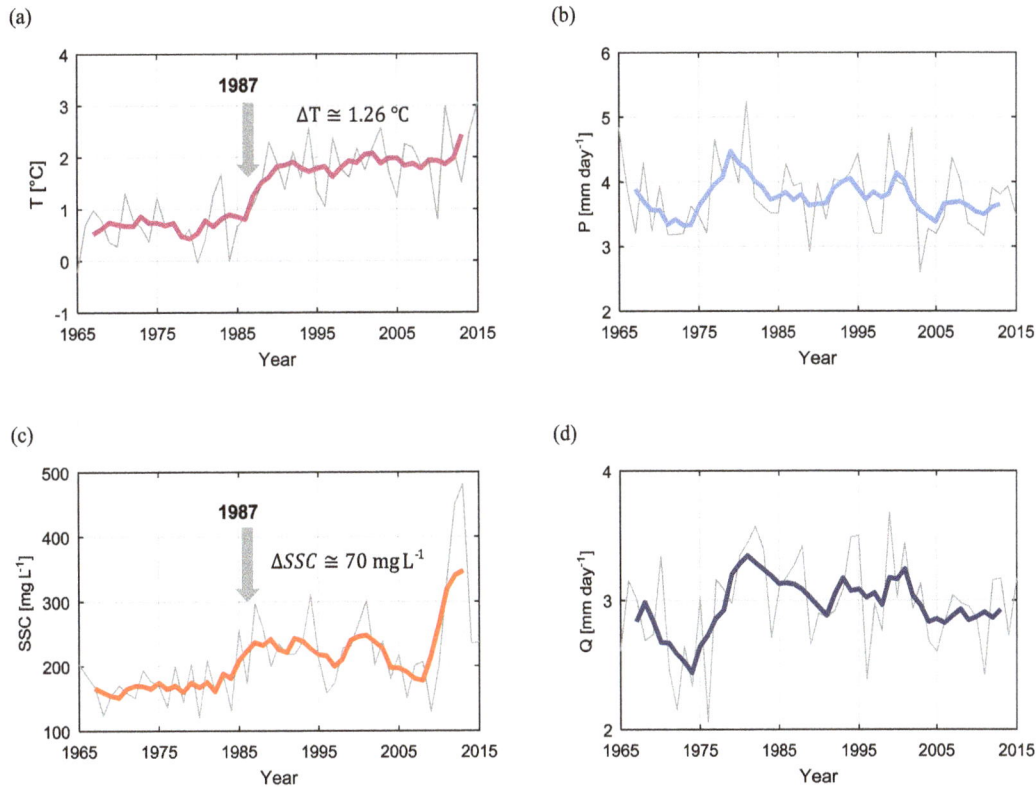

Figure 6. Observations for the period 1965–2015 of **(a)** basin-averaged air temperature, **(b)** basin-averaged daily precipitation, **(c)** suspended sediment concentration measured at the outlet of the basin, and **(d)** daily discharge per unit area measured at the outlet of the basin. Mean annual values are shown in grey and the 5-years moving average is shown with a bold line.

rises in suspended sediment concentration in those months (Fig. 7c).

The simultaneous increase in ice melt and decrease in snowmelt suggests that the abrupt warming has led to important alterations of the hydrological regime. To quantify this alteration, we compute the relative contribution of rainfall, snow, and ice melt on the sum of these three components in July and August. The average relative contribution of ice melt has almost doubled after 1987 (from ~ 12 to $\sim 22\,\%$; Fig. 10), while the relative contribution of snowmelt has reduced by more than $30\,\%$ (from ~ 52.5 to $\sim 35\,\%$; Fig. 10). This indicates the substantial effect of the sharp temperature rise on the basin hydrology.

5.4 Effect of intermittent SSC sampling

The empirical cumulative distribution functions of total daily basin-averaged SM, IM, ER, and Q, computed on all days ("all non-zero days") and only on days corresponding to SSC measurements ("SSC-measurement non-zero days") is shown in Fig. 11. Although extremely high and low values may indeed be missed by the non-continuous sampling, cumulative distributions of SM, IM, ER, and Q on "SSC-measurement non-zero days" and on "all non-zero days" are

similar. This indicates that, although SSC is measured at a fixed interval, the sampling captures accurately the process variability. In addition, results of the statistical tests on mean monthly and mean annual values of all analysed hydroclimatic variables are unchanged. We therefore conclude that our results are not significantly influenced by the discontinuous nature of the SSC sampling.

6 Discussion

6.1 Snowmelt and ice-melt models

The value of the snowmelt factor k_{snow} ($3.6\,\text{mm day}^{-1}\,^{\circ}\text{C}^{-1}$) is in agreement with previous studies carried out in this region. In the upper Rhône Basin, Boscarello et al. (2014) found a snowmelt factor equal to $4.3\,\text{mm day}^{-1}\,^{\circ}\text{C}^{-1}$ based on previous studies on the Toce basin in Italy (Corbari et al., 2009). Calibration of a semi-lumped conceptual model for the three tributary catchments of the upper Rhône Basin – Lonza, Drance, and Rhône at Gletsch – led to snowmelt factors equal to 6.1, 4.5, and $6.6\,\text{mm day}^{-1}\,^{\circ}\text{C}^{-1}$, respectively (Schaefli et al., 2005). Differences in k_{snow} between this and previous studies are attributable to the different temporal resolution of models, lengths of calibration datasets,

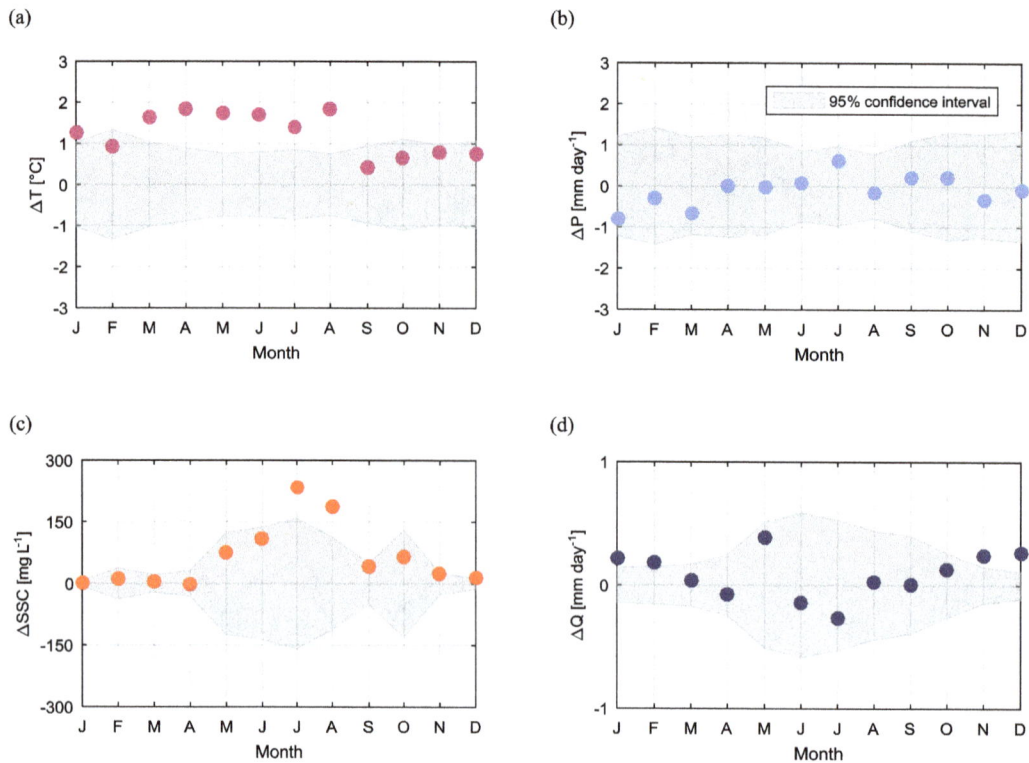

Figure 7. Monthly differences between the period after and before the year-of-change (1987–2015 and 1965–1986) of: **(a)** basin-averaged air temperature, **(b)** basin-averaged daily precipitation, **(c)** mean suspended sediment concentration measured at the outlet of the basin, and **(d)** daily discharge per unit area measured at the outlet of the basin. Points outside the confidence interval (grey shaded area) represent statistically significant (5 % significance level) changes in the monthly mean.

type and thresholds of precipitation partitioning, climatic inputs, threshold temperature for melt, and others. We highlight that the higher performance of the model in simulating snow cover at the highest elevations in our study, where most of the glaciers are located, is a prerequisite for successful ice-melt estimation. The underestimation of SCF in autumn and the overestimation in winter at lower elevations are likely related to errors in partitioning precipitation into solid and liquid form. One of the main problems of degree-day models is related to their poor performance in reproducing the spatial distribution of snow accumulation and melt in complex topography. The temperature-index approach does not take into account features that affect melting, such as topographic slope, aspect, surface roughness, and albedo (Pellicciotti et al., 2005). However, in our case, the spatial distribution of snow cover is satisfactory. Sensitivity and specificity are characterized by a strong seasonal signal. In summer, when a large part of the basin is snow-free, it is much easier for the model to capture snow-free pixels correctly than snow-covered pixels. In winter, when the basin is largely snow-covered, the situation is reversed. We account for this by computing the true skill score, which combines both sensitivity and specificity into a better representation of overall model performance.

Despite the large regional and temporal variability that characterizes ice-melt factors, comparison with previous studies confirms that the calibrated value ($7.1\,\mathrm{mm\,day^{-1}\,°C^{-1}}$) is reasonable for the Alpine environment (e.g. Schaefli et al., 2005; Boscarello et al., 2014). A range from 5 to $20\,\mathrm{mm\,day^{-1}\,°C^{-1}}$ has been reported in the literature (e.g. Hock, 2003; Schaefli, 2005). It should be noted that, when calibrating the ice-melt factor, we neglect evaporation (evapotranspiration). However, evaporation indeed plays a secondary role in the long-term water balance in Alpine environments compared to precipitation and snowmelt (Braun et al., 1994; Huss et al., 2008b), especially at high elevations such as in the case of the Massa and Lonza sub-catchments.

Considering that the aim of this study is to evaluate long-term changes in hydro-climatology and sediment dynamics of the upper Rhône Basin and not the short-term variability in ice melt on the daily scale, we consider the snowmelt and ice-melt model performances as satisfactory. In addition, we show that although our model does not account for glacier retreat, it does not overestimate the ice-melt contribution during the period 1975–2015. However, considering climate projections further into the future, and glaciers that continue to retreat, the issue of future ice-melt contribution will

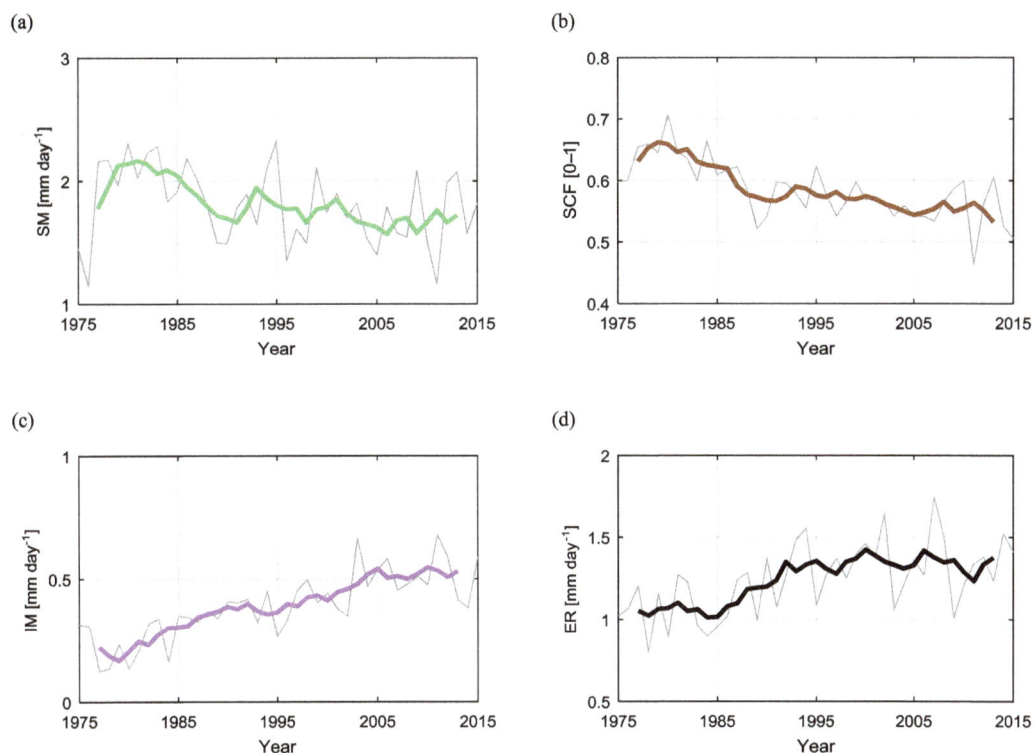

Figure 8. Simulations for the period 1975–2015 of mean annual variables: **(a)** snowmelt SM, **(b)** snow-cover fraction SCF, **(c)** ice melt IM, and **(d)** effective rainfall ER. Mean annual values are shown in grey and a 5-year moving average is shown with a thick line.

need to be revised. Under climate change, even the largest glacier in the basin, the Aletsch Glacier, is expected to shrink at a rate where its ice-melt contribution would start decreasing before 2050 (Farinotti et al., 2012; FOEN, 2012; Brönnimann et al., 2014).

6.2 Changes in hydro-climatology and SSC

Abrupt temperature jumps, such as the one we observed in the upper Rhône basin, rather than gradual changes in air temperature have been observed globally (e.g. Jones and Moberg, 2003; Rebetez and Reinhard, 2008). Observations indicate that Switzerland has experienced two main rapid warming periods in the past, with the 1940s and 1980s being the warmest decades of the last century (Beniston et al., 1994; Beniston and Rebetez, 1996). The simultaneous increase in temperature and suspended sediment concentration indicates that changes in climatic conditions may effectively impact sediment dynamics, especially in Alpine environments where temperature-driven processes, like snow and ice-melt, have a strong influence on the basin hydrology. The statistically significant change in the SSC variance supports the finding that processes related to fine sediment regime of the upper Rhône Basin have been altered by changing climatic conditions, resulting in greater concentrations and higher variability of suspended sediment reaching the outlet of the basin.

Conversely, differences in precipitation before and after 1987 are within the 95 % confidence interval and are not statistically significant. Differences in discharge are also not statistically significant except in winter, when the suspended sediment concentration does not show changes. Therefore, it is very unlikely that the abrupt increase in suspended sediment concentration around mid-1980s in July and August is caused by changes in mean precipitation and/or discharge.

6.3 Hydroclimatic activation of sediment sources

Our simulations of snow cover and melt are in agreement with snow observations across Switzerland. The decreasing tendency in snow cover after the mid- or late 1980s has been demonstrated for the Swiss Alps (Beniston, 1997; Laternser and Schneebeli, 2003; Scherrer et al., 2004; Scherrer and Appenzeller, 2006; Marty, 2008). Snow depth, number of snowfall days, and snow cover show similar patterns during the last century: a gradual increase until the early 1980s, interrupted in late 1950s and early 1970s, and a statistically significant decrease afterwards (Beniston, 1997; Laternser and Schneebeli, 2003). Previous analyses also state that the reduction in snow cover after mid-1980s is characterized more by an abrupt shift than by a gradual decrease (Marty, 2008), in agreement with our simulations. The reduction in snow-cover duration, which is observed to be stronger at lower and mid-altitudes than at higher elevations, is mainly the re-

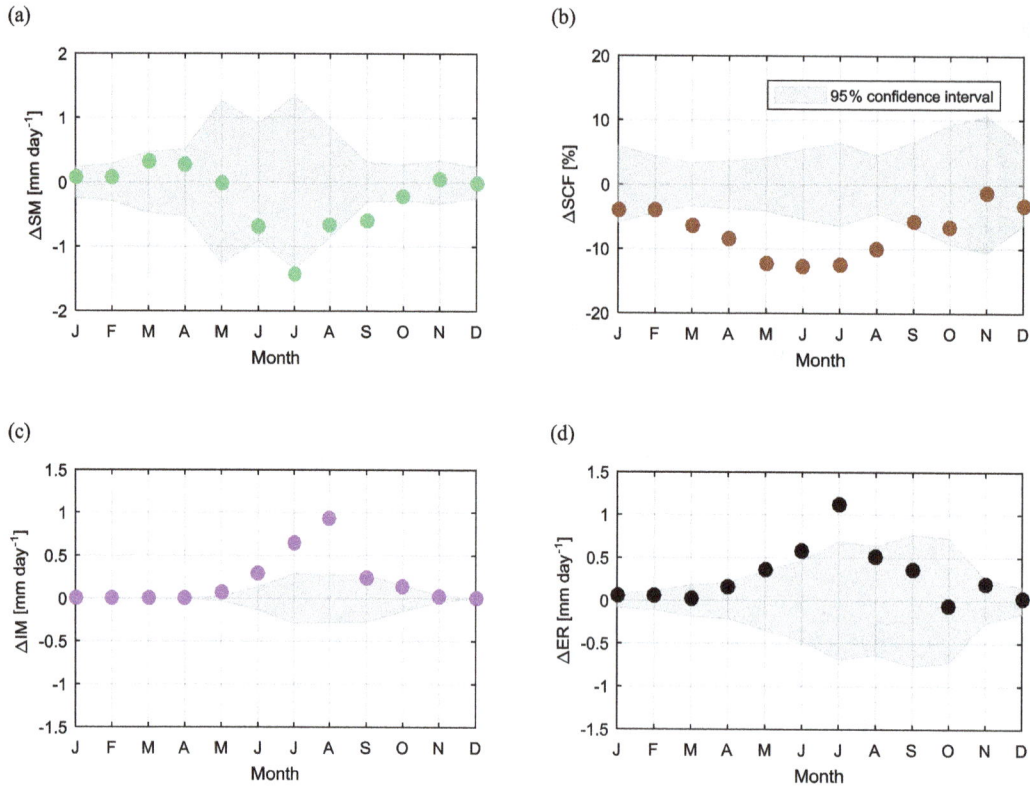

Figure 9. Mean monthly differences (between 1987–2015 and 1975–1986) in variables **(a)** snowmelt SM, **(b)** snow-cover fraction SCF, **(c)** ice melt IM, and **(d)** effective rainfall ER. Points outside the confidence interval (grey shaded area) represent statistically significant (5 % significance level) changes in the monthly mean.

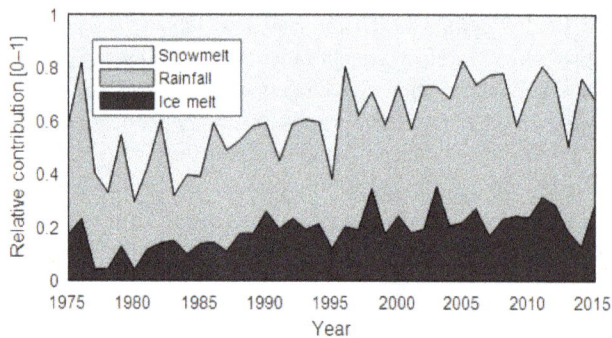

Figure 10. Relative contribution of snowmelt (SM), rainfall (R), and ice melt (IM) for the summer months July–August, computed as the ratio between each component and their sum. Rainfall is extracted from observed precipitation by using a rain–snow temperature threshold, snow and ice melt are simulated with spatially distributed temperature-index models.

sult of earlier snow melting in spring due to warmer temperatures (Beniston, 1997; Laternser and Schneebeli, 2003; Marty, 2008). Moreover, by analysing 76 meteorological stations in Switzerland, Serquet et al. (2011) demonstrated a sharp decline in snowfall days relative to precipitation days, both for winter and early spring, showing the impact of

higher temperature on reduced snowfall, independently of variability in precipitation frequency and intensity. Therefore, despite the high complexity that characterizes snow dynamics in the Alps (Scherrer et al., 2006, 2013), the dominant effect of temperature rise on snow-cover decline after the late 1980s has been clearly shown (Beniston, 1997; Marty, 2008; Serquet et al., 2011; Scherrer et al., 2004; Scherrer and Appenzeller, 2006).

The increase in potentially erosive rainfall is partially confirmed by recent observations. Rainfall erosivity, expressed by the R-factor of the Revised Soil Loss Equation (Wischmeier and Smith, 1978; Brown and Foster, 1987), computed on the basis of 10 min resolution precipitation data, was recently analysed for Switzerland. Although the upper Rhône Basin together with the eastern part of Switzerland was found to have relatively low rainfall erosivity (low R-factor) compared to the rest of the country, due to a lower frequency of thunderstorms and convective events (Schmidt et al., 2016), there is evidence of an increasing trend for the R-factor from May to October during the last 22 years (1989–2010; Meusburger et al., 2012). This suggests that the increase in effective rainfall on snow-free surfaces may have contributed to suspended sediment concentration rise, through a combination of reduced snow-cover fraction, increased rainfall–snowfall ratio, and possible increases in

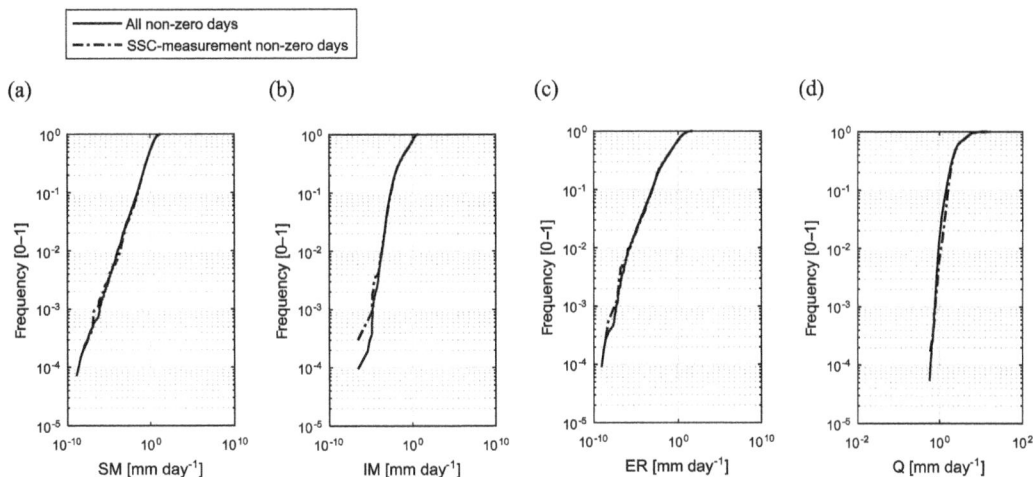

Figure 11. Empirical cumulative distribution functions of total daily basin-averaged SM **(a)**, IM **(b)**, ER **(c)**, and Q **(d)**, computed on all days and only on days corresponding to SSC measurements. Only non-zero values of SM, IM and ER, are included.

rainfall intensity on a sub-daily scale. However, simulations show a statistically significant jump in effective rainfall in June and July, while SSC is significantly larger in July and August. Therefore, we argue that erosive rainfall alone is unlikely to explain the abrupt jump in suspended sediment concentration observed around mid-1980s.

Enhanced ice melt is coherent with the observed acceleration of Alpine glacier retreat after the mid-1980s. Ground-based and satellite observations, combined with mass balance analysis, reveal that current rates of glacier retreat are consistently greater than long-term averages (Paul et al., 2004, 2007; Haeberli et al., 2007). Estimations of glacier area reduction rates indicate a loss rate for the period 1985–1999, which is 7 times greater than the decadal loss rate for the period 1850–1973 (Paul et al., 2004). Investigations with satellite data and in situ observations suggest that the volume loss of Alpine glaciers during the last 30 years is more attributable to a remarkable down-wasting rather than to a dynamic response to changed climatic conditions (Paul et al., 2004, 2007). Haeberli et al. (2007) estimated that glaciers in the European Alps lost about half of their total volume (roughly 0.5 % year^{-1}) between 1850 and 1975, another 25 % (1 % year^{-1}) between 1975 and 2000, and an additional 10–15 % (2–3 % year^{-1}) in the period 2001 to 2005. The appearance of proglacial lakes and rock outcrops with lower albedo and high thermal inertia, separation of glaciers from the accumulation area, and general albedo lowering in the European Alps (Paul et al., 2005) are among the main positive feedbacks that accelerate glacier disintegration and make it unlikely to stop in the near future (Paul et al., 2007). Although glacier dynamics are quite complex and involve many variables and feedbacks, the dominant role played by temperature rise in glacier wasting has been clearly demonstrated (e.g. Oerlemans and Reichert, 2000). The major volume loss in the recent past in Swiss Alpine glaciers is attributable

to negative mass balances during the ablation season rather than to a lower accumulation by precipitation (Huss et al., 2008a). For small high-altitude Alpine glaciers, Micheletti and Lane (2016) showed negligible ice-melt contributions to runoff between the mid-1960s and mid-1980s, after which contributions increased markedly.

Most importantly, runoff coming from glaciers is notoriously rich in sediments. Very fine silt-sized sediment resulting from glacier erosion is transported in suspension most often as wash load (Aas and Bogen, 1988). Proglacial areas generally represent rich sources of sediment due to active glacier erosive processes of abrasion, bed-rock fracturing, and plucking (Boulton, 1974; Hallet et al., 1996). Glacier retreat discloses a large amount of sediments available to be transported by proglacial streams. Moreover, change in climatic conditions and specifically temperature-driven glacier recession and permafrost degradation may initiate specific erosional processes that consequently enhance sediment supply in proglacial environments (Micheletti et al., 2015; Micheletti and Lane, 2016; Lane et al., 2016).

As shown in Sect. 5.3, the ice-melt increase is highest in July and August (Fig. 9c) in agreement with the jump in suspended sediment concentration (Fig. 7c), while ER rise occurred mainly in June and July (Fig. 9d). We then conclude that the significant increase in ice melt detected in the mid-1980s (Figs. 8c, 9c, 10) is likely to be the main cause of the sharp rise in suspended sediment concentration entering Lake Geneva; this occurs through a combination of (1) increased discharge originated in proglacial environments, which implies higher suspended sediment concentration; (2) a larger relative contribution of sediment-rich ice melt compared to snowmelt and precipitation fluxes; and (3) intensified sediment production and augmented sediment supply in proglacial areas due to rapid ice recession.

6.4 Anthropogenic factors and climate signals

The interpretation of increases in suspended sediment concentration may be complicated by anthropogenic drivers and changes in the mid-1980s. Three main anthropogenic activities may have potentially influenced the suspended sediment regime of the upper Rhône Basin: river channelization, construction of reservoirs and hydropower operations, and gravel extraction along the main stream and tributaries. However, the second and last large channelization project was completed in 1960 (Oliver et al., 2009), much earlier than the observed increase in SSC. Likewise, the largest reservoirs in the catchment have been in operation since 1975 (Loizeau and Dominik, 2000). Therefore, it is unlikely that these two anthropogenic factors have contributed to the SSC rise detected in the mid-1980s. The same holds for gravel mining activities. Annual volumes of gravel extracted from the Rhône, provided by the Valais Cantonal Authorities as differences from the average over the period 1989–2014, do not show any significant correlation with mean annual suspended sediment concentration ($R^2 = 0.08$). Although gravel mining operation may perturb SSC for short periods after river bed disturbance by causing local pulses of fine sediments, this process does not significantly affect the suspended sediment load at the outlet of the basin on seasonal and annual timescales. A possibility still remains that changes in the hydropower operation itself, i.e. the distribution of flow responding to electricity demand, and the flushing of dams have increased SSC concentrations. We currently do not have any evidence for such changes; however, we think it is unlikely that they would have long-term effects on SSC.

Our results show that even in highly human-impacted and regulated catchments such as the Rhône Basin, a strong climatic signal in hydrological and sediment dynamics can persist. This also suggests that the decrease in fine-sediment load at the outlet of the upper Rhône Basin observed in the 1960s on the basis of sediment cores recovered in the Rhône delta region and reported by Loizeau et al. (1997), could be the result of a combined effect of hydropower system development, as it has been hypothesized (Loizeau et al., 1997; Loizeau and Dominik, 2000), but also reduced ice-melt loads due to colder temperatures at the time. The cooling period, which occurred between the 1950s and late the 1970s (e.g. Beniston et al., 1994) was characterized by colder and snowy winters (e.g. Laternser and Schneebeli, 2003) and has been accompanied by reduced ice-melt rates, glacier advance, and positive glacier mass balances (Zemp et al., 2008; VAW-ETH, 2015).

The climate signal in sediment dynamics takes on particular importance in the context of climate change projections into the future. Despite the large uncertainty, future projections under different climate change scenarios show a common tendency for Switzerland, characterized by a shift from snow-dominated to rain-dominated hydrological regime, reduced summer discharge, increased winter discharge, reduced snow cover, and enhanced glacier retreat (Bavay et al., 2009; Jouvet et al., 2011; Brönnimann et al., 2014; Fatichi et al., 2015; Huss and Fischer, 2016). In contrast to these hydrological predictions, changes in sediment fluxes are highly uncertain due to the complexity and feedbacks of the processes involved, inherent stochasticity in sediment mobilization and transport, and large regional variability in sediment connectivity across the Alpine landscape (Cavalli et al., 2013; Heckmann and Schwanghart, 2013; Bracken et al., 2015; Lane et al., 2017).

7 Conclusions

The aim of this research was to analyse changes in the hydroclimatic and suspended sediment regimes of the upper Rhône Basin during the period 1975–2015. We show an abrupt increase in basin-wide mean air temperature in the mid-1980s. The simultaneous step-like increase in suspended sediment concentration at the outlet of the catchment, detected in July and August, suggests a causal link between fine sediment dynamics and climatic conditions. Two main factors link warmer climate and enhanced SSC: increased transport capacity and increased sediment supply resulting from spatial and/or temporal activation–deactivation of sediment sources. Our results show that transport capacity, through discharge, is not likely to explain the increases in SSC because no statistically significant changes in the mid-1980s are present in Rhône Basin discharge, neither on the annual nor monthly timescales. The suggestion is that the impact of warmer climatic conditions acts on fine sediment dynamics through the activation and deactivation of different sediment sources and different sediment production and transport processes.

To understand sediment supply conditions, we analyse the temporal evolution of three main sediment fluxes: (1) sediments sourced and transported by snowmelt along hillslopes and channels; (2) sediments entrained and transported by erosive rainfall events over snow-free surfaces, including hillslope, channel bank erosion, and mass wasting events; and (3) fine sediment fluxes generated by glacier ice-melt. The fluxes of snow and ice melt together with snow-cover fraction and rainfall are analysed to detect changes in time and their coherence with changes in SSC.

Our results show that while mean annual precipitation does not show any evident change between the periods before and after the SSC jump in the mid-1980s, potentially erosive rainfall clearly increases over time especially in June and July, but not in August. On the other hand, ice melt has significantly increased due to temperature-driven enhanced ablation. Statistically significant shifts in ice melt were identified for summer, with highest increases in July and August, in accordance with the rise in SSC. Concurrently to the temperature and SSC rise, the relative contribution of ice melt to total annual runoff (sum of rainfall, snow, and ice-melt) presents a significant increase in the mid-1980s, substantially altering the hydrological regime of the Rhône Basin. Based on these results, we propose that climate has an effect on fine sediment dynamics by altering the three main fluxes of suspended sediment in the Rhône Basin, and that ice melt plays a dominant role in the suspended sediment concentration rise in the mid-1980s through (1) increased flow derived from sediment-rich subglacial and proglacial areas, (2) a larger relative contribution of sediment-rich ice melt compared to snowmelt and precipitation, and (3) increased sediment supply in hydrologically connected proglacial areas due to glacier recession. While snowmelt has decreased, the reduced extent and duration of snow cover may also have contributed to the suspended sediment concentration rise through enhanced erosion by heavy rainfall events over snow-free surfaces.

Because changes in SSC are not consistent with changes in discharge and transport capacity, our work emphasizes how the inclusion of sediment sources and their activation through different processes of production and transport is necessary for attributing change. This analysis also demonstrates that climate-driven changes in suspended sediment dynamics may be significantly strong even in highly regulated and human-impacted catchments such as the upper Rhône Basin, where sediment fluxes are affected by flow regulation due to hydropower production and by grain-size dependent trapping in reservoirs. This has consequences for climate change impact assessments and projections for Alpine catchments with hydropower systems, where climate change signals are sometimes thought to be secondary to human regulation. Although at this stage we cannot reliably conclude in which direction sediment fluxes will change in the future, our paper clearly shows that a more process-based understanding of the connections between hydrological change and the activation of sediment sources will provide us with a better framework for analysing and attributing changes in sediment yields in Alpine catchments in the future.

Author contributions. AC and PM designed the methodology. AC developed the code and carried out simulations and computations. AC prepared the manuscript with contributions from all co-authors.

Competing interests. The authors declare that they have no conflict of interest.

Acknowledgements. We thank Christoph Frei (Federal Office of Meteorology and Climatology MeteoSwiss) for providing us with experimental temperature and precipitation datasets and for suggestions on the right use of MeteoSwiss gridded data and the application of statistical tests. We also thank Daniel Farinotti (Swiss Federal Institute for Forest, Snow and Landscape Research WSL, Department of Civil, Environmental and Geomatic Engineering ETH Zurich) for providing us with GloGEM simulations, for the fruitful discussion on glacier retreat and glacier dynamics and for kindly revising the manuscript. The Federal Office of the Environment (FOEN) provided discharge and suspended sediment concentration data. We thank Alessandro Grasso (FOEN) for the explanation on the SSC data collection procedures. Finally, we would like to thank the Valais Cantonal Authorities for supplying information on gravel mining extraction. This research was supported by the Swiss National Science Foundation Sinergia grant 147689 (SEDFATE).

Edited by: Laurent Pfister

References

Aas, E. and Bogen, J.: Colors of Glacier Water, Water Resour. Res., 24, 561–565, 1988.

Allouche, O., Tsoar, A., and Kadmon, R.: Assessing the accuracy of species distribution models: prevalence, kappa and the true skill statistic (TSS), J. Appl. Ecol., 43, 1223–1232, 2006.

Anselmetti, F. S., Bühler, R., Finger, D., Girardclos, S., Lancini, A., Rellstab, C., and Sturm, M.: Effects of Alpine hydropower dams on particle transport and lacustrine sedimentation, Aquat. Sci., 69, 179–198, 2007.

Auer, A. H.: The rain versus snow threshold temperatures, Weatherwise, 27, 67–67, https://doi.org/10.1080/00431672.1974.9931684, 1974.

Ballantyne, C. K.: A general model of paraglacial landscape response, Holocene, 12, 371–376, 2002.

Bavay, M., Lehning, M., Jonas, T., and Löwe, H.: Simulations of future snow cover and discharge in Alpine headwater catchments, Hydrol. Process., 23, 95–108, 2009.

Begueria, S.: Validation and Evaluation of Predictive Models in Hazard Assessment and Risk Management, Nat. Hazards, 37, 315–329, https://doi.org/10.1007/s11069-005-5182-6, 2006.

Beniston, M.: Variations of snow depth and duration in the Swiss Alps over the last 50 years: Links to changes in large-scale climatic forcings, Clim. Change, 36, 281–300, 1997.

Beniston, M. and Rebetez, M.: Regional behavior of minimum temperatures in Switzerland for the period 1979–1993, Theor. Appl. Climatol., 53, 231–243, 1996.

Beniston, M., Rebetez, M., Giorgi, F., and Marinucci, R.: An analysis of regional climate change in Switzerland, Theor. Appl. Climatol., 49, 135–159, 1994.

Bennett, G., Molnar, P., Eisenbeiss, H., and McArdell, B. W.: Erosional power in the Swiss Alps: characterization of slope failure in the Illgraben, Earth Surf. Proc. Land., 37, 1627–1640, https://doi.org/10.1002/esp.3263, 2012.

Boscarello, L., Ravazzani, G., Rabuffetti, D., and Mancini, M.: Integrating glaciers raster–based modelling in large catchments hydrological balance: the Rhône case study, Hydrol. Process., 28, 496–508, https://doi.org/10.1002/hyp.9588, 2014.

Boulton, G. S.: Processes and patterns of glacial erosion, in: Glacial Geomorphology, edited by: Coates, D. R., Springer, Dordrecht, 41–87, 1974.

Bracken, L. J., Turnbull, L., Wainwright, J., and Bogaart, P.: Sediment connectivity: a framework for understanding sediment transfer at multiple scales, Earth Surf. Proc. Land., 40, 177–188, 2015.

Braun, L. N., Aellen, M., Funk, M., Hock, R., Rohrer, M. B., Steinegger, U., Kappenberger, G., and Müller-Lemans, H.: Measurement and simulation of high alpine water balance components in the Linth- Limmern head watershed (north-eastern Switzerland), Z. Gletscherkunde Glazialgeologie, 30, 161–185, 1994.

Brown, L. C. and Foster, G. R.: Storm erosivity using idealized intensity distributions, T. ASAE, 30, 379–386, 1987.

Brönnimann, S., Appenzeller, C., Croci-Maspoli, M., Fuhrer, J., Grosjean, M., Hohmann, R., Ingold, K., Knutti, R., Liniger, M. A., Raible, C. C., Röthlisberger, R., Schär, C., Scherrer, S. C., Strassmann, K., and Thalmann. P.: Climate change in Switzerland: a review of physical, institutional, and political aspects, WIRES Clim. Change, 5, 461–481, https://doi.org/10.1002/wcc.280, 2014.

Cavalli, M., Trevisani, S., Comiti, F., and Marchi, L.: Geomorphometric assessment of spatial sediment connectivity in small Alpine catchments, Geomorphology, 188, 31–41, https://doi.org/10.1016/j.geomorph.2012.05.007, 2013.

Collins, W. D., Rasch, P. J., Boville, B. A., Hack, J. J., McCaa, J. R., Williamson, D. L., Kiehl, J. T., and Briegleb, B.: Description of the NCAR community atmosphere model (CAM3), Tech. Rep. NCAR/TN- 464+STR, 226 pp., 2004.

Corbari, C., Ravazzani, G., Martinelli, J., and Mancini, M.: Elevation based correction of snow coverage retrieved from satellite images to improve model calibration, Hydrol. Earth Syst. Sci., 13, 639–649, https://doi.org/10.5194/hess-13-639-2009, 2009.

Dedieu, J.-P., Boos, A., Kiage, W., and Pellegrini, M.: Snow cover retrieval over Rhône and Po river basins from MODIS optical satellite data (2000–2009), Geophys. Res. Abstracts, 12, 5532, EGU General Assembly 2010, 2010.

Delunel, R., van der Beek, P., Bourlès, D., Carcaillet J., and Schlunegger, F.: Transient sediment supply in a high-altitude Alpine environment evidenced through a ^{10}Be budget of the Etages catchment (French Western Alps), Earth Surf. Proc. Land., 39, 890–899, https://doi.org/10.1002/esp.3494, 2014.

England, P. and Molnar, P.: Surface uplift, uplift of rocks, and exhumation of rocks, Geology, 18, 1173–1177, 1990.

Farinotti, D., Usselmann, S., Huss, M., Bauder, A., and Funk M.: Runoff evolution in the Swiss Alps: projections for selected high-alpine catchments based on ENSEMBLES scenarios, Hydrol. Process., 26, 1909–1924, 2012.

Fatichi, S., Rimkus, S., Burlando, P., Bordoy, R., and Molnar, P.: High-resolution distributed analysis of climate and anthropogenic changes on the hydrology of an Alpine catchment, J. Hydrol., 525, 362–382, 2015.

Fischer, M., Huss, M., and Hoelzle, M.: Surface elevation and mass changes of all Swiss glaciers 1980–2010, The Cryosphere, 9, 525–540, https://doi.org/10.5194/tc-9-525-2015, 2015.

FOEN: Auswirkungen der Klimaänderung auf Wasserressourcen und Gewässer. Synthesebericht zum Projekt "Klimaänderung und Hydrologie in der Schweiz" (CCHydro), Bundesamt für Umwelt, Bern, Umwelt-Wissen, 1217, 76 pp., 2012.

Foster, G. C., Dearing, R. A., Jones, R. T., Crook, D. S., Siddle, D. J., Harvey, A. M., James, P. A., Appleby, P. G., Thompson, R., Nicholson, J., and Loizeau, J.-L.: Meteorological and land use controls on past and present hydro-geomorphic processes in the pre-alpine environment: an integrated lake-catchment study at the Petit Lac d'Annecy, France, Hydrol. Process., 17, 3287–3305, 2003.

Frei, C.: Interpolation of temperature in a mountainous region using nonlinear profiles and non-Euclidean distances, Int. J. Climatol., 34, 1585–1605, 2014.

Frei, C., Schöll, R., Fukutome, S., Schmidli, J., and Vidale, P. L.: Future change of precipitation extremes in Europe: An intercomparison of scenarios from regional climate models, J. Geophys. Res., 111, D06105, https://doi.org/10.1029/2005JD005965, 2006.

Gabbud, C. and Lane, S. N.: Ecosystem impacts of Alpine water intakes for hydropower: the challenge of sediment management, WIRES Water, 3, 41–61, https://doi.org/10.1002/wat2.1124, 2016.

Gardner, A. S., Moholdt, G., Cogley, J. G., Wouters, B., Arendt, A. A., and Wahr, J.: A reconciled estimate of glacier contributions to sea level rise: 2003 to 2009, Science, 340, 852–857, https://doi.org/10.1126/science.1234532, 2013.

Grasso, A., Bérod, D., and Hodel, H.: Messung und Analyse der Verteilung von Schwebstoffkonzentrationen im Querprofil von Fliessgewässern, Wasser Energie Luft, 107, 61–65, 2012.

Haeberli, W. and Holzhauser, H.: Alpine glacier mass changes during the past two millennia, PAGES News, 11, 13–15, 2003.

Haeberli, W., Hoelzle, M., Paul, F., and Zemp, M.: Integrated monitoring of mountain glaciers as key indicators of global climate change: the European Alps, Ann. Glaciol., 46, 150–160, 2007.

Hallett, B., Hunter, L., and Bogen, J.: Rates of erosion and sediment evacuation by glaciers: A review of field data and their implications, Global Planet. Change, 12, 213–235, 1996.

Heckmann, T. and Schwanghart, W.: Geomorphic coupling and sediment connectivity in an alpine catchment – Exploring sediment cascades using graph theory, Geomorphology, 182, 89–103, 2013.

Herman, F., Beyssac, O., Brughelli, M., Lane, S. N., Leprince, S., Adatte, T., Lin, J. Y. Y., and Avouac, J. P.: Erosion by an Alpine glacier, Science, 350, 193–195, 2015.

Hinderer, M., Kastowski, M., Kamelger, A., Bartolini, C., and Schlunegger, F.: River loads and modern denudation of the Alps – A review, Earth-Sci. Rev., 118, 11–44, 2013.

Hock, R.: Temperature index melt modelling in mountain areas, J. Hydrol., 282, 104–115, 2003.

Hornung, J., Pflanz, D., Hechler, A., Beer, A., Hinderer, M., Maisch, M., and Bieg, U.: 3-D architecture, depositional patterns and climate triggered sediment fluxes of an alpine alluvial fan (Samedan, Switzerland), Geomorphology, 115, 202–14, 2010.

Huggel, C., Clague, J. J., and Korup, O.: Is climate change responsible for changing landslide activity in high mountains?, Earth Surf. Proc. Land., 37, 77–91, 2012.

Huss, M. and Farinotti, D.: Distributed ice thickness and volume of all glaciers around the globe, J. Geophys. Res., 117, F04010, https://doi.org/10.1029/2012JF002523, 2012.

Huss, M. and Fischer, M.: Sensitivity of Very Small Glaciers in the Swiss Alps to Future Climate Change, Front. Earth Sci., 4, 34, https://doi.org/10.3389/feart.2016.00034, 2016.

Huss, M. and Hock, R.: A new model for global glacier change and sea-level rise, Front. Earth Sci., 3, 54, https://doi.org/10.3389/feart.2015.00054, 2015.

Huss, M., Bauder, A., Funk, M., and Hock, R.: Determination of the seasonal mass balance of four Alpine glaciers since 1865, J. Geophys. Res.-Atmos., 113, F01015, https://doi.org/10.1029/2007JF000803, 2008a.

Huss, M., Farinotti, D., Bauder, A., and Funk, M.: Modelling runoff from highly glacierized alpine drainage basins in a changing climate, Hydrol. Process., 22, 3888–3902, 2008b.

Huss, M., Jouvet, G., Farinotti, D., and Bauder, A.: Future high-mountain hydrology: a new parameterization of glacier retreat, Hydrol. Earth Syst. Sci., 14, 815–829, https://doi.org/10.5194/hess-14-815-2010, 2010.

Jones, P. D. and Moberg, A.: Hemispheric and large-scale surface air temperature variations: an extensive revision and an update to 2001, J. Climate, 16, 206–223, 2003.

Jouvet, G., Huss, M., Funk, M., and Blatter, H.: Modelling the retreat of Grosser Aletschgletscher, Switzerland, in a changing climate, J. Glaciol., 57, 1033–1045, 2011.

Kondolf, G. M., Gao, Y., Annandale, G. W., Morris, G. L., Jiang, E., Zhang, J., Cao, Y., Carling, P., Fu, K., Guo, Q., Hotchkiss, R., Peteuil, C., Sumi, T., Wang, H.-W., Wang, Z., Wei, Z., Wu, B., Wu, C., and Yang, C. T.: Sustainable sediment management in reservoirs and regulated rivers: Experiences from five continents, Earths Future, 2, 256–280, https://doi.org/10.1002/2013EF000184, 2014.

Lane, S. N., Bakker, M., Gabbud, C., Micheletti, N., and Saugy, J. N.: Sediment export, transient landscape response and catchment-scale connectivity following rapid climate warming and Alpine glacier recession, Geomorphology, 277, 210–227, https://doi.org/10.1016/j.geomorph.2016.02.015, 2016.

Laternser, M. and Schneebeli, M.: Long-term Snow Climate Trends of the Swiss Alps (1931–99), Int. J. Climatol., 23, 733–750, 2003.

Lenzi, M. A., Mao, L., and Comiti, F.: Interannual variation of suspended sediment load and sediment yield in an alpine catchment, Hydrolog. Sci. J., 48, 899–915, https://doi.org/10.1623/hysj.48.6.899.51425, 2003.

Loizeau, J.-L. and Dominik J.: Evolution of the Upper Rhône River discharge and suspended sediment load during the last 80 years and some implications for Lake Geneva, Aquat. Sci., 62, 54–67, https://doi.org/10.1007/s000270050075, 2000.

Loizeau, J.-L., Dominik, J., Luzzi, T., and Vernet J.-P.: Sediment Core Correlation and Mapping of Sediment Accumulation Rates in Lake Geneva (Switzerland, France) Using Volume Magnetic Susceptibility, J. Great Lakes Res., 23, 391–402, 1997.

Marty, C.: Regime shift of snow days in Switzerland, Geophys. Res. Lett., 35, L12501, https://doi.org/10.1029/2008GL033998, 2008.

Mason, S. J. and Graham, N. E.: Conditional Probabilities, Relative Operating Characteristics, and Relative Operating Levels, Weather Forecast., 14, 713–725, 1999.

Meteoswiss, Federal Office of Meteorology and Climatology: Documentation of MeteoSwiss Grid-Data Products Daily Precipitation (final analysis): RhiresD, available at: http://www.meteoswiss.admin.ch/content/dam/meteoswiss/de/service-und-publikationen/produkt/raeumliche-daten-niederschlag/doc/ProdDoc_RhiresD.pdf, 2013a.

Meteoswiss, Federal Office of Meteorology and Climatology: Documentation of MeteoSwiss Grid-Data Products Daily Mean, Minimum and Maximum Temperature: TabsD, TminD, TmaxD, available at: https://www.ethz.ch/content/dam/ethz/special-interest/baug/ifu/hydrology-dam/documents/research-data/ifu-hydrologie-data-proddoctabsd.pdf, 2013b.

Meusburger K. and Alewell C.: Soil Erosion in the Alps. Experience gained from case studies (2006–2013), Federal Office for the Environment, Bern, Environmental studies no. 1408, 116 pp., 2014.

Meusburger, K., Steel, A., Panagos, P., Montanarella, L., and Alewell, C.: Spatial and temporal variability of rainfall erosivity factor for Switzerland, Hydrol. Earth Syst. Sci., 16, 167–177, https://doi.org/10.5194/hess-16-167-2012, 2012.

Micheletti, N. and Lane, S. N.: Water yield and sediment export in small, partially glacierized Alpine watersheds in a warming climate, Water Resour. Res., 52, 4924–4943, https://doi.org/10.1002/2016WR018774, 2016.

Micheletti, N., Lambiel, C., and Lane, S. N.: Investigating decadal-scale geomorphic dynamics in an alpine mountain setting, J. Geophys. Res.-Earth, 120, 2155–2175, https://doi.org/10.1002/2015JF003656, 2015.

Nash, J. E. and Sutcliffe, J. V.: River Flow Forecasting Through Conceptual Models Part 1 – A Discussion of Principles, J. Hydrol., 10, 282–290, 1970.

Oerlemans, J. and Reichert, B. K.: Relating glacier mass balance to meteorological data using a Seasonal Sensitivity Characteristic (SSC), J. Glaciol., 46, 1–6, 2000.

Oliver, J.-M., Carrel, G., Lamouroux, N., Dole-Oliver, M.-J., Malard, F., Bravard, J.-P., and Amoros, C.: The Rhône River Basin, in: Rivers of Europe, chap. 7, Academic Press, London, 247–295, 2009.

Palazón, L. and Navas, A.: Land use sediment production response under different climatic conditions in an alpine-prealpine catchment, Catena, 137, 244–255, 2016.

Paul, F., Kääb, A., Maisch, M., Kellenberger, T., and Haeberli, W.: Rapid disintegration of Alpine glaciers observed with satellite data, Geophys. Res. Lett., 31, L21402, https://doi.org/10.1029/2004GL020816, 2004.

Paul, F., Machguth, H., and Kääb, A.: On the impact of glacier albedo under conditions of extreme glacier melt: the summer of 2003 in the Alps, EARSeL eProceedings 4, 139–149, 2005.

Paul, F., Kääb, A., and Haeberli, W.: Recent glacier changes in the Alps observed from satellite: Consequences for future monitoring strategies, Global Planet. Change, 56, 111–122, 2007.

Pellicciotti, F., Brock, B., Strasser, U., Burlando, P., Funk, M., and Corripio, J.: An enhanced temperature-index melt model including the shortwave radiation balance: development and testing for Haut Glacier d'Arolla, Switzerland, J. Glaciol., 51, 573–587, 2005.

Pettitt, A. N.: A Non-parametric Approach to the Change-point Problem, Appl. Statist., 28, 126–135, 1979.

Quinton, W. L. and Carey, S. K.: Towards an energy-based runoff generation theory for tundra landscapes, Hydrol. Process., 22, 4649–4653, 2008.

Rebetez, M. and Reinhard, M.: Monthly air temperature trends in Switzerland 1901–2000 and 1975–2004, Theor. Appl. Climatol., 91, 27–34, https://doi.org/10.1007/s00704-007-0296-2, 2008.

Santiago, S., Thomas, R. L., McCarthy, L., Loizeau, J. L., Larbaigt, G., Corvi, C., Rossel, D., Tarradellas, J., and Vernet, J. P.: Particle Size Characteristics of Suspended and Bed Sediments in The Rhône River, Hydrol. Process., 6, 227–240, 1992.

Schaefli, B., Hingray, B., Niggli, M., and Musy, A.: A conceptual glacio-hydrological model for high mountainous catchments, Hydrol. Earth Syst. Sci., 9, 95–109, https://doi.org/10.5194/hess-9-95-2005, 2005.

Scherrer, S. C. and Appenzeller, C.: Swiss Alpine snow pack variability: major patterns and links to local climate and large-scale flow, Climate Res., 32, 187–199, 2006.

Scherrer, S. C., Appenzeller C., and Laternser, M.: Trends in Swiss Alpine snow days: the role of local- and large-scale climate variability, Geophys. Res. Lett., 31, L13215, https://doi.org/10.1029/2004GL020255, 2004.

Scherrer, S. C., Wüthrich, C., Croci-Maspoli, M., Weingartner, R., and Appenzeller, C.: Snow variability in the Swiss Alps 1864–2009, Int. J. Climatol., 33, 3162–3173, https://doi.org/10.1002/joc.3653, 2013.

Schlunegger, F. and Hinderer, M.: Crustal uplift in the Alps: why the drainage pattern matters, Terra Nova, 13, 425–432, 2001.

Schmidt, S., Alewell, C., Panagos, P., and Meusburger, K.: Regionalization of monthly rainfall erosivity patterns in Switzerland, Hydrol. Earth Syst. Sci., 20, 4359–4373, https://doi.org/10.5194/hess-20-4359-2016, 2016.

Serquet, G., Christoph, M., Dulex, J. P., and Rebetez, M.: Seasonal trends and temperature dependence of the snowfall/precipitation-day ratio in Switzerland, Geophys. Res. Lett., 38, L07703, https://doi.org/10.1029/2011GL046976, 2011.

Stutenbecker, L., Costa, A., and Schlunegger, F.: Lithological control on the landscape form of the upper Rhône Basin, Central Swiss Alps, Earth Surf. Dynam., 4, 253–272, https://doi.org/10.5194/esurf-4-253-2016, 2016.

Swift, D. A., Nienow, P. W., and Hoey, T. B.: Basal sediment evacuation by subglacial meltwater: suspended sediment transport from Haut Glacier d'Arolla, Switzerland, Earth Surf. Proc. Land., 30, 867–883, 2005.

Syvitski, J. P. M., Vörösmarty, C. J., Kettner, A. J., and Green, P.: Impact of Humans on the Flux of Terrestrial Sediment to the Global Coastal Ocean, Science, 308, 376–380, 2005.

VAW-ETH: The Swiss Glaciers, Yearbooks of the Cryospheric Commission of the Swiss Academy of Sciences (SCNAT) (1881–2016), Laboratory of Hydraulics, Hydrology and Glaciology (VAW), Glaciological reports no. 1–134, available at: http://swiss-glaciers.glaciology.ethz.ch/publications.html, last access: 10 January 2018.

Vernon, A. J., Van der Beek, P. A., Sinclair, H. D., and Rahn, M. K.: Increase in Late Neogene denudation of the European Alps confirmed by analysis of a fission-track thermochronology database, Earth Planet. Sci. Lett., 270, 316–329, 2008.

Wen, L. J., Nagabhatla, N., Lü, S. H., and Wang, S. Y.: Impact of rain snow threshold temperature on snow depth simulation in land surface and regional atmospheric model, Adv. Atmos. Sci., 30, 1449–1460, 2013.

Wick, L., Van Leeuwen, J. F. N., Van der Knaap, W. O., and Lotter, A. F.: Holocene vegetation development in the catchment of Sägistalsee (1935 m a.s.l.), a small lake in the Swiss Alps, J. Paleolimnol., 30, 261–272, 2003.

Wilks, D. S.: Statistical Methods in the Atmospheric Sciences, Academic Press, 467 pp., 1995.

Wischmeier, W. H. and Smith, D. D.: Predicting Rainfall Erosion Losses – A Guide to Conservation Planning, Supersedes Agriculture Handbook, No. 537, Washington DC, 58 pp., 1978.

Wood, J. L., Harrison, S., Turkington, T. A. R., and Reinhardt, L.: Landslides and synoptic weather trends in the European Alps, Clim. Change, 136, 297–308, 2016.

Zemp, M., Paul, F., Hoelzle, M., and Haeberli, W.: Glacier fluctuations in the European Alps, 1850–2000: an overview and spatio-temporal analysis of available data, in: Darkening Peaks: Glacier Retreat, Science, and Society, edited by: Orlove, B., Wiegandt, E., and Luckman, B. H., Berkeley, US, 152–167, 2008.

Zerathe, S., Lebourg, T., Braucher, R., and Bourles, D.: Mid-Holocene cluster of large-scale landslides revealed in the Southwestern Alps by Cl-36 dating. Insight on an Alpine-scale landslide activity, Quaternary Sci. Rev., 90, 106–127, 2014.

6

Subsurface storage capacity influences climate–evapotranspiration interactions in three western United States catchments

E. S. Garcia[1] and C. L. Tague[2]

[1]Department of Atmospheric Sciences, University of Washington, Seattle, WA, USA

[2]Bren School of Environmental Science and Management, University of California, Santa Barbara, CA, USA

Correspondence to: E. S. Garcia (esgarcia@uw.edu)

Abstract. In the winter-wet, summer-dry forests of the western United States, total annual evapotranspiration (ET) varies with precipitation and temperature. Geologically mediated drainage and storage properties, however, may strongly influence these relationships between climate and ET. We use a physically based process model to evaluate how plant accessible water storage capacity (AWC) and rates of drainage influence model estimates of ET–climate relationships for three snow-dominated, mountainous catchments with differing precipitation regimes. Model estimates show that total annual precipitation is a primary control on inter-annual variation in ET across all catchments and that the timing of recharge is a second-order control. Low AWC, however, increases the sensitivity of annual ET to these climate drivers by 3 to 5 times in our two study basins with drier summers. ET–climate relationships in our Colorado basin receiving summer precipitation are more stable across subsurface drainage and storage characteristics. Climate driver–ET relationships are most sensitive to subsurface storage (AWC) and drainage parameters related to lateral redistribution in the relatively dry Sierra site that receives little summer precipitation. Our results demonstrate that uncertainty in geophysically mediated storage and drainage properties can strongly influence model estimates of watershed-scale ET responses to climate variation and climate change. This sensitivity to uncertainty in geophysical properties is particularly true for sites receiving little summer precipitation. A parallel interpretation of this parameter sensitivity is that spatial variation in storage and drainage properties are likely to lead to substantial within-watershed plot-scale differences in forest water use and drought stress.

1 Introduction

In high-elevation forested ecosystems in the western US, the majority of precipitation falls during the winter; there is often a disconnect between seasonal water availability and growing seasonal water demand. Consequently, forests in these regions are frequently water-limited, even when annual precipitation totals are high (Boisvenue and Running, 2006; Hanson and Weltzin, 2000). This disconnect between water inputs and energy demands also highlights the importance of storage of winter recharge by both snowpack and by soils. The importance of snowpack storage in these systems for hydrologic fluxes has received significant attention, particularly given their vulnerability to climate warming. Warmer temperatures are already shifting seasonal water availability in the western US through reductions in snowpack accumulation (Knowles et al., 2006) and earlier occurrence of peak snowpack (Mote et al., 2005) and shifts in streamflow timing (Stewart et al., 2005). Recently, field and modeling studies have shown that the years with greater snowpack accumulation can be a strong predictor of vegetation water use and productivity for sites in the California Sierra (Tague and Peng, 2013; Trujillo et al., 2012).

Less attention, however, has been paid to the role of subsurface storage and drainage that can influence whether or not winter precipitation or snowmelt is available for plant water use during the summer months. Previous studies have shown that plant access to stored water is a substantial contributor to summer evapotranspiration in semi-arid regions (Bales et al., 2011). Plant accessible storage includes both water stored in soil and in saprolite and bedrock layers that can be accessed by plant roots (McNamara et al., 2011). Like snowpack, the storage of water in the subsurface has the potential to act as a water reservoir, storing winter precipitation for use later in the growing season (Geroy et al., 2011). The amount of water that can be stored varies substantially in space with topography, geologic properties, and antecedent moisture conditions (Famiglietti et al., 2008; McNamara et al., 2005). If the rate of snowmelt allows for subsurface moisture stores to be replenished later in the growing season, more of the winter precipitation is made available for plant water use. If storage capacity is too shallow to capture a significant amount of runoff or if the rate of rain or snowmelt inputs exceeds the rate of infiltration, then subsurface storage will not be physically able to extend water availability. While field studies in the western US have shown that shallow soils can limit how much snowmelt is available for ecological use during the summer (Kampf et al., 2014; Smith et al., 2011), these studies cannot fully characterize the relative impact of subsurface storage on ET given inter-annual and cross-site variation in climate drivers.

In this paper, we focus on the potential for plant accessible subsurface water storage to mediate the sensitivity of ET to inter-annual variation in climate drivers, precipitation and temperature. Understanding how ET varies with climate drivers is important, both from the perspective of how ET influences downstream water supply and water availability for forests and other vegetation (Grant et al., 2013). Western US forests show substantial vulnerability to drought, with declines in productivity and increases in mortality and disturbance in drought years (Allen et al., 2010; Hicke et al., 2012; Williams et al., 2013). Understanding these ecosystems' responses to primary climate drivers is of particular concern given recent warming trends (Sterl et al., 2008) and multi-year droughts (Cook et al., 2004; Dai et al., 2004) and that these changes in water and energy demands are expected to intensify (Ashfaq et al., 2013). Increased temperatures also affect plant phenology, leading to earlier spring onset of plant water use and productivity (Cayan et al., 2001), and thus can influence water requirements and water use. However, increases in early season water use, combined with higher atmospheric moisture demand, may lead to increased soil water deficit later in the season.

Forest evapotranspiration is also a substantial component of the water budget (Post and Jones, 2001) and thus any change in forest water use will potentially have significant impacts on downstream water use. Goulden et al. (2012), for example, use flux tower and remote sensing data to argue that warming may result in an increase of up to 60 % in vegetation water use at high elevations in the Upper Kings River watershed in California's southern Sierra watershed. We note however that these projected increases depend on how subsurface storage capacity interacts with snowpack at high elevations.

This paper's primary research objective is to quantify the interaction between subsurface storage characteristics and key climate-related metrics that influence forest water availability and use in snow-dominated environments receiving

Table 1. Explanatory variables.

Abbreviation	Definition
P	Total annual precipitation
T_{AMJ}	Average daily temperature for April, May, June
R_{75}	Day of water year by which 75 % of soil water recharge occurs
AWC	Available water capacity of soil (field capacity–wilting point)

a range of summer precipitation. Heterogeneity in subsurface properties in soil, sapprolite and bedrock layers makes the characterization of subsurface storage difficult at the watershed scale. Here we use a spatially distributed process-based model, the Regional Hydro-Ecologic Simulation System (RHESSys), to quantify how uncertainty or spatial variation in subsurface storage properties might be expected to influence watershed response to these climate-related drivers. We apply RHESSys in three case study watersheds of differing precipitation regimes to investigate how climate and subsurface storage combine to control inter-annual variation in ET.

2 Methods

We apply our model at a daily time step to three watersheds located in the western Oregon Cascades (OR-CAS), central Colorado Rocky Mountains (CO-ROC) and central California Sierras (CA-SIER). All three watersheds receive a substantial fraction of precipitation as snowfall, but vary in their precipitation and temperature regimes and amount of precipitation that falls as snow (Fig. 1). We compare a humid, seasonally dry watershed (OR-CAS) to two catchments that receive half as much precipitation annually. The more water-limited catchments differ in that CO-ROC receives a significant amount of its precipitation budget during the summer growing season. We use these case studies to estimate ET sensitivity to storage and drainage properties for several different precipitation and temperature regimes common in western US mountain watersheds. For each watershed, we quantify how subsurface storage and drainage properties interact with a combination of inter-annual variation in precipitation timing and magnitude, and shifts in snowpack storage. We first establish how inter-annual variation in three primary climate-related metrics (precipitation, average spring temperature, and timing of soil moisture recharge) influences annual ET with average subsurface storage properties. We then explore how these relationships change across physically plausible storage values.

2.1 RHESSys model description

We use a physically based model (RHESSys v.5.15) to calculate vertical water, energy, and carbon fluxes in our three

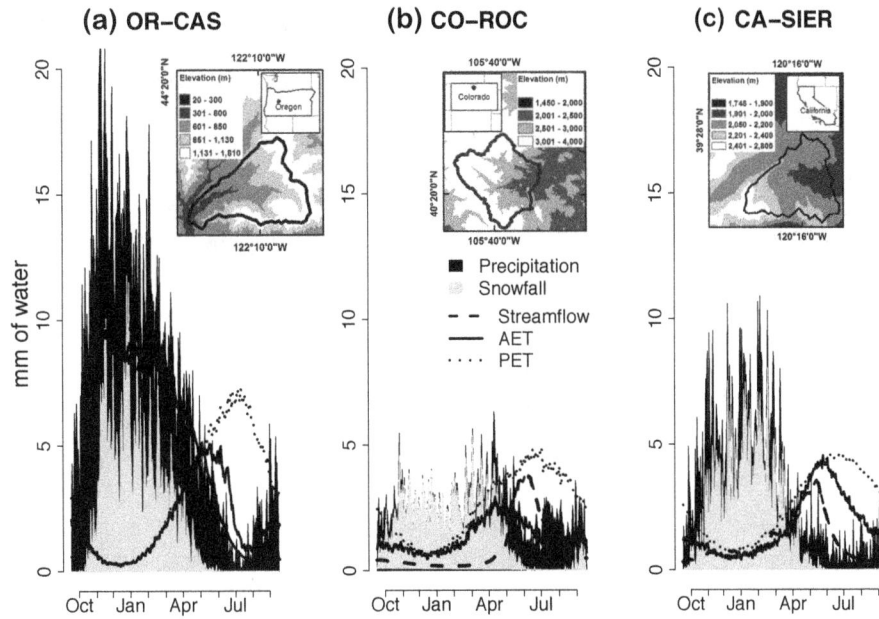

Figure 1. Locations and average daily water fluxes averaged from 1980 to 2000 for three case study watersheds located in (a) the western Oregon Cascades (OR-CAS), (b) Colorado Rockies (CO-ROC), and (c) California Sierra Nevada (CA-SIER).

Table 2. Basin topography, geology, vegetation and climate characteristics. Climate descriptions are averaged over the total available climate record (duration noted in table).

Watershed	CO-ROC	OR-CAS	CA-SIER
Location	Colorado	Oregon	California
US Geological Survey gage number	06733000	14161500	10343500
Geology	Holocene glacial till, rock; Precambrian gneiss, granite	Western Cascade basalt	Sierra granite, with Miocene andesite cap
Elevation range (m)	1470–4345	410–1630	1800–2650
Drainage area (km^2)	350	64	26
Topographic wetness index – mean (SD)	7.0 (1.9)	6.6 (1.7)	7.9 (1.8)
Climate record	1980–2008	1958–2008	1960–2000
Mean annual precipitation (mm)	1000	2250	850
Annual precipitation as snow (%)	64	29	55
Precipitation received in growing season (%)	46	21	19
Min/max winter T (JFM) (oC)	−12.1/−0.02	−0.9/5.2	−9.5/3.7
Min/max spring T (AMJ) (oC)	−2.7/10.9	4.0/14.0	−2.5/13.8
P:PET	0.9	2.3	1.2
Vegetation	Subalpine fir, aspen, meadows, shrub	Douglas fir, western hemlock	Mixed conifer, Jeffrey and lodgepole pine
Mean basin LAI	3.5	9.0	4.1
Annual NPP range for calibration (gC m^{-2} yr^{-1})	280–520	620–1100	450–800
Literature sources used to bound annual NPP range	Arthur and Fahey (1992) Bradford et al. (2008)	Grier and Logan (1977) Gholz (1982)	Hudiburg et al. (2009) Goulden et al. (2012)*

* Values reported as gross primary productivity, converted to NPP using RHESSys-calculated values of respiration.

watersheds (Tague and Band, 2004). RHESSys is a spatially explicit model that partitions the landscape into units representative of the different hydro-ecological processes modeled (Band et al., 2000). RHESSys has been used to address diverse eco-hydrologic questions across many watersheds (Baron et al., 2000; Shields and Tague, 2012; Tague and Peng, 2013). Key model processes are described below and a full account is provided in Tague and Band (2004).

RHESSys requires data describing spatial landscape characteristics and climate forcing; a digital elevation model

(DEM) and geologic and vegetation maps are used to represent the topographic, geologic, carbon and nitrogen characteristics within a watershed. RHESSys accounts for variability of climate processes within the catchment using algorithms developed for extrapolation of climate processes from point station measurements over spatially variable terrain (Running and Nemani, 1987). Hydrologic processes modeled in RHESSys include interception, evapotranspiration, infiltration, vertical and lateral subsurface drainage, and snow accumulation and melt. The Penman–Monteith formula (Monteith, 1965) is used to calculate evaporation of canopy interception, snow sublimation, evaporation from subsurface and litter stores, and transpiration by leaves. A model of stomatal conductance allows transpiration to vary with soil water availability, vapor pressure deficit, atmospheric CO_2 concentration, and radiation and temperature (Jarvis, 1976). A radiation transfer scheme that accounts for canopy overstory and understory, as well as sunlit and shaded leaves, controls energy available for transpiration. RHESSys accounts for changes in vapor pressure deficit for fractions of days that rain occurs (wet versus dry periods). Plant canopy interception and ET are also a function of leaf area index (LAI) and gappiness of the canopy such that as LAI increases and gap size decreases, plant interception capacity and transpiration potential increases. RHESSys partitions rain to snow at a daily time step based on each patch's air temperature. Snowmelt is estimated using a combination of an energy budget approach for radiation-driven melt and a temperature index-based approach for latent heat-drive melt processes. Subsurface water availability varies as a function of infiltration and water loss through transpiration, evaporation and drainage. RHESSys also routes water laterally and thus patches can receive additional moisture inputs as either re-infiltration of surface flow or through shallow subsurface flow from upslope contributing areas. Lateral subsurface drainage routes subsurface and surface water between spatial units and it is a function of topography and soil and saprolite drainage characteristics. Deep groundwater stores are drained to the stream using a simple linear reservoir representation.

Carbon and nitrogen cycling in RHESSys was modified from BIOME-BGC (Thornton, 1998) to account for dynamic rooting depth, sunlit and shaded leaves, multiple canopy layers, variable carbon allocation strategies, and drought stress mortality. The Farquhar equation is used to calculate gross primary productivity (GPP) (Farquhar et al., 1980). Plant respiration costs include both growth and maintenance respiration and are influenced by temperature following Ryan (1991). Net primary productivity (NPP) is calculated by subtracting total respiration costs from GPP.

In our three study sites, RHESSys is driven with daily records of precipitation and maximum and minimum temperature. Each basin is calibrated for seven parameters that characterize subsurface storage and drainage properties. Drainage rates are controlled by saturated hydraulic conduc-

tivity (K) and its decay with depth (m). Air-entry pressure (ϕ_{ae}), pore size index (b), and rooting depth (Z_r) control subsurface water holding capacity (Brooks and Corey, 1964). In all basins, we assume that geologic properties allow for deeper groundwater stores that are inaccessible to vegetation (Table 2). Vegetation however can access more shallow groundwater flow. These deep groundwater stores are controlled by two parameters representing the percentage of water that passes to the store (gw_1) and the rate of its release to streamflow (gw_2). Calibration is conducted with a Monte Carlo based approach, the generalized likelihood uncertainty estimation (GLUE) method (Beven and Binley, 1992). Parameter sets (1000 in total) are generated by random sampling from uniform distributions of literature-constrained estimates for the individual parameters; all calibration parameter sets are physically viable representations of soils within each basin. In other words, though a single parameter set may not meet streamflow and annual NPP calibration metrics, that particular subsurface storage capacity may still exist within the basin.

Model validation and drainage/storage parameter calibration were performed using two measures: daily streamflow statistics and annual measures of NPP. Streamflow statistics were set such that good parameters resulted in daily flow magnitude errors of less than 15 %, Nash–Sutcliffe efficiencies (NSE, a measure of hydrograph shape) greater than 0.65, and logged NSE values greater than 0.7 (a test of peak and low flows) (Nash and Sutcliffe, 1970). We select all parameter sets from these acceptable values; the total number of parameters equals 87, 246, and 47 for CA-SIER, CO-ROC, and OR-CAS, respectively. Daily hydrologic fluxes are calculated over 15 years for each soil parameter set in order to account for variability due to parameters in establishing relationships with our climate-related indices, the results of which are presented in Figs. 2–4. We verify our annual ET estimates against limited field estimates published in literature for subwatersheds of CO-ROC and OR-CAS (Baron and Denning, 1992; Webb et al., 1978). The average of our model estimated annual ET matches these limited field-based measurements and also fall within the bounds of annual ET estimated through water balance by subtracting annual streamflow from our records of annual precipitation. We assess the performance of the carbon-cycling model by comparing with published forest field measurements of annual NPP (values reported in Table 2). In our fully coupled eco-hydrologic model, accurate estimates of NPP also suggest that ET estimates are reasonable. Finally we note that RHESSys estimates of ET and NPP have been evaluated in a number of previous studies by comparison with flux tower and tree ring data, and these studies confirm that RHESSys provides reasonable estimates of ET and its sensitivity to climate drivers (Vicente-Serrano et al., 2015; Zierl et al., 2007). We quantify the sensitivity of ET–climate relationships to geologic properties by varying subsurface storage parameters (Figs. 5–6).

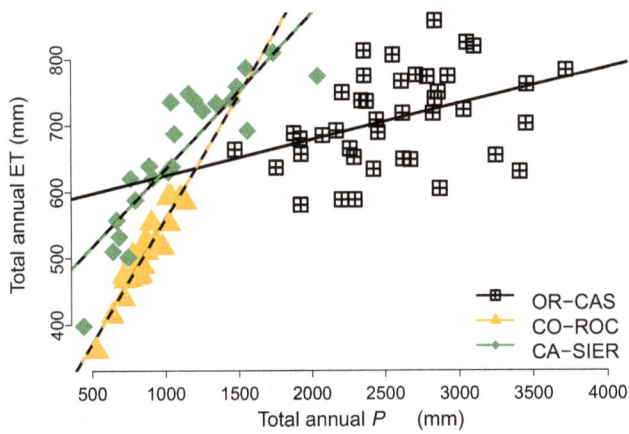

Figure 2. Total annual ET increases with total annual precipitation. Lines indicate statistically significant relationships (p value < 0.05).

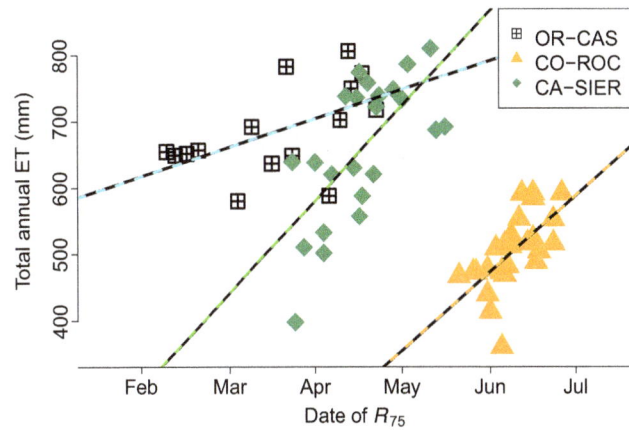

Figure 4. (a) Warmer spring temperatures are correlated with lower total annual ET in the two snow-dominated watersheds. **(b)** An earlier occurrence of soil moisture recharge is correlated with warmer temperatures in CO-ROC.

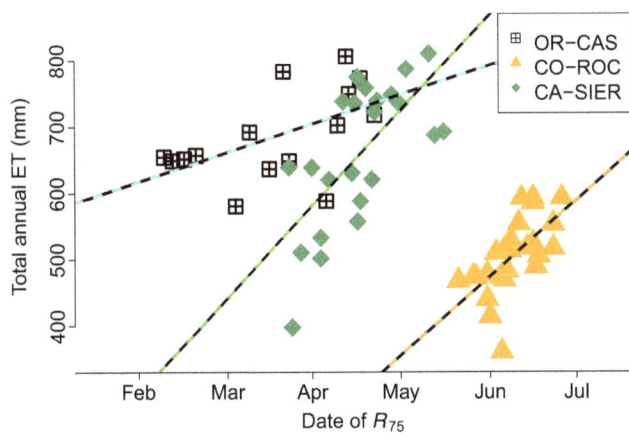

Figure 3. Later occurrence of soil moisture recharge (R_{75}) is significantly correlated with increased annual ET in all study watersheds.

2.2 Study sites

These analyses are conducted in three western US mountain catchments: Big Thompson in Colorado's Rocky Mountains (CO-ROC), Lookout Creek in Oregon's Western Cascades (OR-CAS), and Sagehen Creek Experimental Forest in California's northern Sierra Nevada (CA-SIER). Basin characteristics pertinent to modeling annual ET are listed in Table 2 and we highlight important similarities and differences here. All sites are located on steep, mountainous slopes and are dominated by forest cover. All basins have climates typical of the western US, on average receiving 54–81 % of their annual precipitation during the winter, 29–64 % of the annual P falls as snow, and they do not meet potential evaporative demand during the growing season (Fig. 1, Table 2). On average, OR-CAS is a much wetter basin and receives more than twice as much annual precipitation than CO-ROC and CA-SIER. Despite OR-CAS receiving more precipitation, a much lower fraction of that winter precipitation is received as

snow. On average, OR-CAS's peak streamflow occurs in December, 4 to 5 months earlier than CO-ROC and CA-SIER (Fig. 1). The drier watersheds, CO-ROC and CA-SIER, receive more than half of their annual precipitation as snow (Table 2). CO-ROC also experiences a summer monsoonal season and on average receives 46 % of its annual precipitation from April to September. Landscape carbon (C) and nitrogen (N) stores in general vary with total annual P across basins. For example, OR-CAS receives the most precipitation and also supports stands of large, old-growth forests; its LAI is more than twice that of either CO-ROC or CA-SIER. As presented in the model description (Sect. 2.1), we use a stable, climatic optimum for vegetation biomass for all analyses in this paper. Garcia et al. (2013) and Tague and Peng (2013) provide detailed descriptions of the geology and climate data, model vegetation, and organic soil carbon store spin-up and calibration used for model implementations of OR-CAS and CA-SIER, respectively. We note that all precipitation and temperature data were derived from daily measurements made at climate stations located within the basins and extrapolated across the terrain using MT-CLM algorithms (Running and Nemani, 1987) and 30 m resolution DEMs. Though RHESSys has previously been used in CO-ROC (Baron et al., 2000), we have made significant updates in RHESSys since that time, so we re-implemented the model as described in the next section.

2.2.1 RHESSys model development for CO-ROC

In CO-ROC, landscape topographic characteristics including elevation, slope and aspect were derived from a digital elevation model (DEM) downloaded from the US Geologic Survey (USGS) National Elevation Dataset at 1/3 arcsec resolution (http://datagateway.nrcs.usda.gov/). A stream network was then derived to accumulate surface and subsurface flow

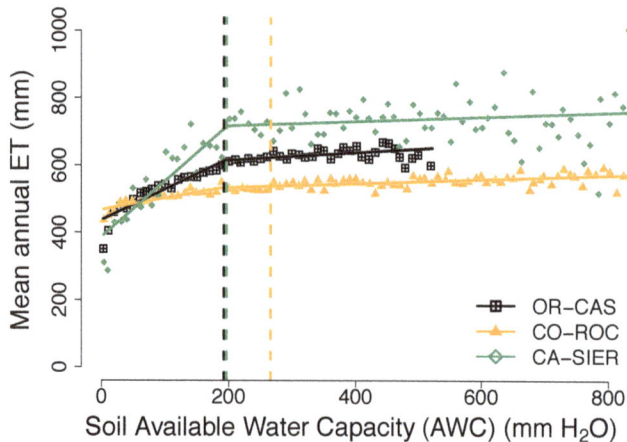

Figure 5. Each point represents the 15-year average annual ET from WY 1985 to 2000 for a physically viable mean basin soil available water capacity (AWC). Vertical lines represent the calculated break point in the nonlinear relationship between long-term ET and AWC for each basin.

at USGS gage no. 06733000. Sub-catchments were delineated using GRASS GIS's watershed basin analysis program, *r.watershed*. Terrestrial data were aggregated such that the average size of the patch units, the smallest spatial units for calculation of vertical model processes, was 3600 m^2. Soil classification data were downloaded from the Soil Survey Geographic database (SSURGO); http://sdmdataaccess. nrcs.usda.gov/ and aggregated to four primary soil types: gravelly loam, sandy loam, loamy sandy, and rock (http: //datagateway.nrcs.usda.gov/). Parameter values associated with these soil types are based on literature values (Dingman, 1994; Flock, 1978) and adjusted using model calibration, as described above. We note that these initial values are approximate and calibration permits storage values that reflect plant access to water stored in both organic soil layers and in sapprolite and rock. Vegetation land cover from the National Land Cover Database (NLCD) was aggregated to four primary vegetation types: subalpine conifer, aspen, shrubland, and meadow (Homer et al., 2007). Because a shift in precipitation patterns occurs at approximately 2700 m, we use daily records of precipitation, T_{max}, and T_{min} from two points within the watershed. RHESSys then interpolates data from these points based on MTN-CLM (Running and Nemani, 1987) to provide spatial estimates of temperature, precipitation and other meteorologic drivers for each patch. Climate data from 1980 to 2008 were downloaded from the DAYMET system for two locations – one at elevation 2460 m (latitude 40.35389, longitude −105.58361) and the second at 3448 m (latitude 40.33769, longitude −105.70315) (Thornton et al., 2012).

Plant C and N stores were initialized by converting remote-sensing-derived LAI to leaf, stem and woody carbon and nitrogen values using allometric equations appropriate to the vegetation type (http://daac.ornl.gov/MODIS/;

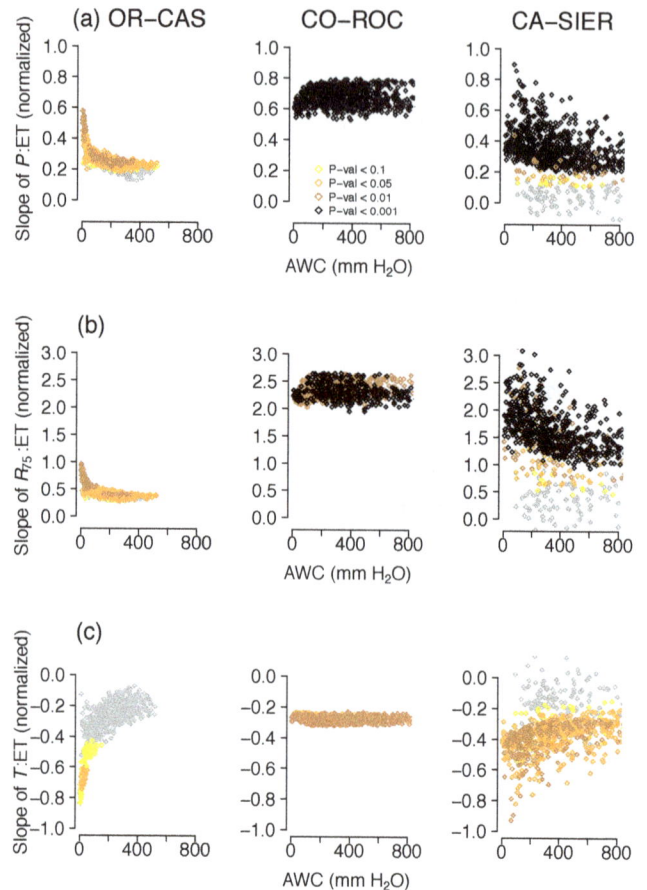

Figure 6. The impact of soil AWC on the slope of a linear regression model of annual ET as a function of climate predictors: (a) precipitation, (b) R_{75}, and (c) T_{AMJ}. The slope of the ET–climate predictor is plotted across a physically viable range of mean basin soil AWC for each climate predictor and for each study basin: OR-CAS (left column), CO-ROC (middle column), and CA-SIER (right column). The slopes are normalized to facilitate inter-basin comparison.

MOD15A2 Collection 5). In order to stabilize organic soil C and N stores relative to the LAI-derived plant C and N, we run the model repeatedly over the basin's climate record until the change in stores stabilizes (Thornton and Rosenbloom, 2005). After stabilizing soil biogeochemical processes, we remove vegetation C and N stores and then dynamically "regrow" them using daily allocation equations (Landsberg and Waring, 1997) for 160 years in order to stabilize plant and soil C and N stores with model climate drivers. For all three basins, an optimum maximum size for each vegetation type was determined using published, field-derived estimates of LAI and aboveground and total annual NPP.

2.3 Framework for primary controls on ET

In these seasonally water-limited basins, we use total annual precipitation (P) as a metric of gross climatic water input. Annual precipitation P is summed over a water year (1 October to 30 September of the following calendar year) and summer season P is summed over July, August, and September. For all climate metrics we use spatially averaged watershed values. To assess the impact of timing of soil moisture recharge (as influenced either by year-to-year variation in precipitation timing, snowmelt or rain–snow partitioning) we calculate R_{75}, the day of water year by which 75 % of the total annual recharge has occurred. Recharge is defined as liquid water (e.g., rain throughfall or snowmelt) that reaches the soil surface. For this metric, we do not differentiate between water that, upon reaching the soil surface, becomes runoff, and water that infiltrates into the soil. We treat this variable as a temporal marker of potential water availability that denotes the timing within the water year that either rain throughfall or snowmelt may potentially infiltrate the soil. To examine energy inputs, we identify a season when temperature most strongly influences estimates of annual ET modeled using historic climate. We performed linear regressions between model estimate of total annual ET and 1- and 3-month averages of daily maximum (T_{max}), minimum (T_{min}) and average temperatures ($T_{avg} = (T_{max} + T_{min})/2$) for all watersheds and for all months of the year. We test the correlation significance with a p value and set a significance threshold at 0.05; i.e., a p value greater than 0.05 is not significant. Our analysis found a 3-month average of daily T_{avg} in April, May and June (T_{AMJ}) to have the greatest explanatory power as a temperature variable for estimating inter-annual variation in annual ET under historic climate variability across our three study watersheds (results not shown). We note that the p value for T_{AMJ} in CA-SIER was greater than 0.05, so it is not reported as a significant result. The growing season is assumed to extend from 1 May to 30 September in all watersheds. For all climate metrics we use spatially averaged watershed values.

We examine the role of storage through AWC. As noted above, plants access water organic soils as well as water stored in sapprolite and rock (Schwinning, 2010). We consider an aggregate storage and do not distinguish between these layers. AWC represents the water stored after gravity drainage (field capacity) that can be extracted by plant root suction (wilting point) and is thus still viable for plant water use (Dingman, 1994, p. 236). We calculate AWC as

$$\text{AWC} = (\theta_{fc} - \theta_{wp})Z_r. \tag{1}$$

Where θ_{fc} represents the average field capacity per unit depth, θ_{wp} the average characteristic wilting point also per unit depth, and AWC is scaled by vegetation rooting depth, Z_r, a model calibration parameter. The field capacity and wilting point are calculated, respectively, as

$$\theta_{fc} = \varphi(\phi_{ae} / 0.033)^b, \tag{2}$$

$$\theta_{wp} = \varphi(\phi_{ae} / \psi_v)^{1/b}, \tag{3}$$

Where φ is average subsurface porosity, ϕ_{ae} represents the air-entry pressure (in meters), b is a pore size distribution

Table 3. Statistics for ET predictors based on linear regression models.

Watershed		CO-ROC	OR-CAS	CA-SIER
Precipitation (P)	p value	<0.001	<0.05	<0.001
	r^2	0.9	0.1	0.75
	Slope	0.4	0.1	0.2
Timing (R_{75})	p value	<0.001	<0.01	<0.001
	r^2	0.2	0.2	0.4
	Slope	3.8	1.2	4.6
Temperature T_{AMJ}	p value	<0.001	<0.05	>0.1
	r^2	0.4	0.1	−0.01
	Slope	−26.3	−25.7	15
Soil capacity (AWC)	p value	0.001	0.001	0.001
	r^2	0.43	0.53	0.11
	Slope	0.1	0.2	0.1

index that describes the moisture-characteristic curve, and ψ_v describes the pressure at which the plants' stomata close. Variables ϕ_{ae} and b are also model calibration parameters.

Larger AWC indicates that more water can be held in the subsurface and potentially interacts with climate to extend plant water availability by capturing snowmelt, one of the primary sources of water for forest ET. Our results present each watershed's average AWC; watersheds are represented by one (OR-CAS), two (CA-SIER), and five (CO-ROC) soil types and their characterizations are described in Table 2. All values of AWC calculated in the calibration represent physically feasible values for each watershed.

We use RHESSys to calculate total annual ET over the entire available climate record in each basin (28–50 years; Table 2) and use linear regression to quantify how much of the inter-annual variation in ET is related to each of the three climate metrics – P, T_{AMJ}, and R_{75}. We set a limit of less than 0.05 for p values to determine significance. We then investigate how long-term mean ET and its relationship with these climate-related indicators are influenced by AWC.

To examine how subsurface storage capacity may influence long-term average ET, we calculate average annual ET over a 15-year period (1985–2000) for a range of 1000 AWC values and linearly regress the long-term averaged ET values against AWC. We then characterize the interacting influences of AWC and each climate driver. For the 1000 values of AWC, we calculate the slope of annual ET estimates to each climate predictor (P, T_{AMJ}, R_{75}).

3 Results

3.1 Annual P vs. ET

In all watersheds higher P results in greater total annual ET (Fig. 2). This is a statistically significant relationship in all watersheds (CO-ROC and CA-SIER, correlations and p values reported in Table 3) where the years of highest annual P are correlated with the years of greatest annual ET. Of the three basins, CO-ROC's annual ET shows the greatest sensitivity to P, having the steepest slope. Annual P is the strongest explanatory variable of annual ET in both CO-ROC ($r^2 = 0.9$) and CA-SIER ($r^2 = 0.75$) (Table 3). For CO-ROC, annual P has a greater influence (steeper slope) in the drier years when P is less than 1000 mm (Fig. 2). OR-CAS has the least significant relationship between P and ET on an annual scale. OR-CAS is a relatively wet basin and on average receives more than twice the amount of winter (January–March) precipitation than CA-SIER or CO-ROC receives. High annual P in OR-CAS in most years likely diminishes the sensitivity of ET to the magnitude of P.

3.2 Timing of recharge vs. ET

For all three catchments, later R_{75} has a significant positive correlation with ET (Fig. 3). In OR-CAS and CA-SIER, R_{75} occurs between February and May. There is more scatter in the predictive power of R_{75} for annual ET when R_{75} is earlier in the water year. The earliest R_{75} is in OR-CAS, where a greater fraction of winter precipitation falls as rain. CA-SIER and CO-ROC are more sensitive to the timing of recharge than OR-CAS. Summer monsoonal pulses in CO-ROC push R_{75} to later in the water year as compared to OR-CAS or CA-SIER. The explanatory power of R_{75} for ET is greatest in CA-SIER where greater accumulation of snowpack and warmer spring temperatures can interact to increase forest water use earlier in the growing season.

3.3 Spring temperature vs. ET

Warmer spring temperature (T_{AMJ}) in all basins generally reduces annual ET (Fig. 4a) and is significantly correlated with lower ET in CO-ROC and OR-CAS. CA-SIER does not show a significant relationship between T_{AMJ} and ET. In CO-ROC and OR-CAS, increasing T_{AMJ} leads to a reduction in water availability and a decline in later season ET. The relationship between spring air temperature and snowmelt timing is demonstrated by significant correlations between T_{AMJ} and R_{75} for CO-ROC (Fig. 4b). The colder temperatures and more persistent snowpack in the CO-ROC basin are more sensitive, relative to OR-CAS, in ET response to earlier snowmelt due to temperature increases.

3.4 AWC vs. ET

Increased AWC increases the long-term average ET in all basins. Figure 5 shows a nonlinear relationship between long-term mean ET and AWC, suggesting that the effect of increasing storage diminishes for higher AWC values. Each basin reaches an approximate storage capacity above which a further increase in storage (AWC) is less important and climate (i.e., P and energy) variables limit ET. Following

Muggeo (2003), for each basin, we calculate that breakpoint value of AWC where ET is less sensitive to AWC. We find that the threshold value of AWC varies across basins and is substantially higher in CO-ROC (265 mm) as compared to CA-SIER (195 mm) and OR-CAS (190 mm) (Fig. 5). Regression of AWC against annual ET shows that a significant relationship exists in OR-CAS and CO-ROC (Table 3).

The effect of varying lateral redistribution or lateral drainage parameters can be seen in the range of slopes for a given AWC (e.g., the scatter in the slope–AWC relationship). All three watersheds show some sensitivity of climate–ET relationships to lateral redistribution parameters for a given AWC. CA-SIER shows the greatest sensitivity, followed by OR-CAS and CO-ROC. The greater sensitivity of CA-SIER to lateral drainage parameters may reflect the strong contribution of snowmelt recharge in its drier and winter precipitation dominated climate. The topography of CA-SIER is also distinctive and includes many swale-like features that concentrate drainage from upslope areas. We calculate the topographic wetness index (TWI) using a 30 m resolution DEM for each watershed (Moore et al., 1991) (Table 2). The TWI reflects the propensity of a location to develop saturated conditions under the assumption that topography controls water flow. Higher TWI values represent flatter, converging terrain and lower values reflect steep topography. The mean TWI for CA-SIER is greater than and significantly different from (Welch's t test) the mean TWI for CO-ROC and OR-CAS. Particularly for CA-SIER, changing storage parameters associated with drainage rates can alter the timing of flow into areas that concentrate flow and subsequently alter their ET rates.

3.5 Sensitivity of ET to climate drivers with AWC

We analyze the sensitivity of ET relationships with climate drivers to subsurface storage properties by plotting the slope of linear regressions between ET and P, R_{75}, and T_{AMJ}, across all storage parameter sets in Fig. 6. We note that the slope of the relationships between climate drivers and ET has been normalized by the watersheds' mean AWC in these plots to facilitate cross-site comparison.

3.5.1 Sensitivity to P with AWC

Of the climate drivers explored, ET relationships with annual precipitation P have the greatest robustness across subsurface storage parameter sets, as suggested by the number of sets that show a statistically significant relationship between annual P and annual ET (Fig. 6a). As expected, slopes are positive between P and ET across all basins. Only the drier basins CO-ROC and CA-SIER have p values less than 0.001, highlighting the strength of P as a climatic driver in these drier basins, as discussed above. The response in slope sensitivity across AWC is similar in OR-CAS and CA-SIER, where ET's sensitivity to P is highest at low AWC and decreases with increased AWC. OR-CAS has a much smaller range in sensitivities (slope varies from 0.2 to 0.6) compared to CA-SIER (slope varies from 0.0 to 0.8). Thus in CA-SIER for low values of AWC, year-to-year variation in P becomes a greater control on year-to-year variation in ET. For both OR-CAS and CA-SIER, increasing AWC becomes less important at higher values of AWC. Higher scatter in slope of annual P versus ET relationship for CA-SIER also reflects the greater sensitivity of ET to subsurface parameters that influence lateral drainage as discussed above (Sect. 3.4).

The variation of ET response to P across AWC in CO-ROC is noteworthy for two reasons. First, CO-ROC has the highest slope values (0.6–0.8), which again reflects the consistency of annual P as a control on inter-annual variation in ET in this basin. Second, unlike OR-CAS and CA-SIER, increasing AWC does not substantially reduce that sensitivity (i.e., slope) to P. Though CO-ROC's sensitivity to P does not change with AWC, the scatter in slopes (0.6–0.8) suggests that lateral drainage has a strong effect on this climate–ET relationship. We note that CO-ROC has a seasonal precipitation regime where a significant fraction of its annual precipitation is received later in the growing season as summer monsoonal pulses. When precipitation occurs during the growing season, the water available for ET is less likely to be limited by storage capacity. Instead ET is limited by the amount or intensity of precipitation. Water that does recharge the system is used relatively quickly, making variation in storage (or AWC) less important as a control on how much P can be used in CO-ROC.

3.5.2 Sensitivity to R_{75} with AWC

After precipitation, the timing of recharge (R_{75}) most significantly correlates with increased ET across all AWC and all basins (Fig. 6b). There are several similarities in the response of ET's sensitivity to R_{75} across AWC when compared to sensitivity to P (Fig. 6a). For example, the dry basins CO-ROC and CA-SIER have the highest degree of sensitivity (significant slopes > 1.0) as compared to OR-CAS (slopes < 1.0) and CA-SIER has the greatest variability in its sensitivity to AWC, with slopes ranging from 1.0 to 3.0 across variation in storage parameters. CO-ROC once again has the least variability in the ET versus R_{75} relationship, with consistently high (2.0–2.5) slopes unaffected by AWC.

3.5.3 Sensitivity to T_{AMJ} with AWC

Finally, T_{AMJ} has the fewest subsurface storage/drainage parameter sets with significant correlation with ET. None of the linear regressions of ET on T_{AMJ} have statistical significance less than 0.001 (Fig. 6c). The slopes are always negative because earlier occurrence of snowmelt results in less ET. For all basins, the sensitivity of ET to T_{AMJ} is greatest at the lowest values of AWC, though CO-ROC once again demonstrates the least variability in slopes across the entire range

of AWC (-0.2 – -0.3). At OR-CAS, T_{AMJ} is only significant for the lower AWC values. We suggest this is in part due to the small fraction of P that falls as snow. Because T_{AMJ}'s largest effect is through timing of snowmelt (Fig. 4), AWC interacts with T_{AMJ} to modulate the melt response. With relatively less snowmelt in OR-CAS, only the systems with the smallest capacities will have a significant negative interaction effect with AWC.

4 Discussion

Our model estimates show differences in the response of ET to climate-related drivers across the three watersheds, primarily due to differences in their precipitation regimes. Spatial heterogeneity in soil and geology, both within and between watersheds, substantially alters these relationships. Our model-based study provides a simplified representation of these interactions, ignoring many additional complexities. In particular, we assume no adaptation of the ecosystem structure and composition that would influence productivity, evapotranspiration and their relationship with climate (Loudermilk et al., 2013). Future work will investigate these coupled carbon cycling–hydrology interactions. In this study we focus on the energy and moisture drivers of ET and how subsurface properties influence their interaction.

The degree to which climate drivers affect ET varies with the magnitude and seasonality of basin precipitation. Total annual P is the first-order control of ET in the two drier watersheds, CO-ROC and CA-SIER. In OR-CAS, most of the inter-annual variation in precipitation is reflected in inter-annual variation in runoff rather than ET. In most years, subsurface storage is filled by this annual precipitation during the winter and spring, asynchronously to late growing season demands (Fig. 1). Our results extend findings by previous studies demonstrating that vegetation productivity and water use relate to the fraction of regional precipitation available to plants (Brooks et al., 2011; Thompson et al., 2011). The fraction of water available to plants tends to decrease with larger rainfall (given saturated soil stores, a greater proportion is lost) and with synchronicity between the timing of recharge and growing season water demands.

Our analysis highlights the timing of water availability (R_{75}) as a key predictor of total annual ET; annual ET increases when recharge occurs later in the water year, during the growing season and period of highest water demand. Previous research has shown how delayed soil moisture recharge (Tague and Peng, 2013) and snowpack dynamics (Tague and Heyn, 2009; Trujillo et al., 2012) are able to increase ET in the Sierra Nevada. In these mountain basins, the sensitivity of ET to timing of recharge is related to the fraction of precipitation received as snow. The climate metrics related to snowmelt, R_{75} and T_{AMJ}, are important secondary controls of ET, especially in the colder, snow-dominated watersheds, CA-SIER and CO-ROC. We note that CA-SIER does not show a significant relationship between T_{AMJ} and ET because the effect of temperature is strongly dependent on the amount of snowpack the basin receives in a year (Tague and Peng, 2013), which is more variable than the amount of snowpack received in CO-ROC or OR-CAS. In OR-CAS and CO-ROC, spring temperature T_{AMJ} is more strongly related to ET through its effect on snowmelt, and correlates negatively with ET. These results suggest that the dominant effect of warmer spring temperatures is earlier meltout of snowpack, which leads to more snowmelt lost as runoff and results in less net recharge. This greater loss of runoff occurs when storage capacity is exceeded. Later in the growing season, increased ET demands will have depleted subsurface stores and throughfall/snowmelt will enter the soil matrix and be available for plant water use. Previous work has shown seasonal increases in spring ET with warmer spring temperatures (Hamlet et al., 2007) that may be related to an earlier start to the vegetation growing season (Cayan et al., 2001) and an increase in vapor pressure deficits and water demand (Isaac and van Wijngaarden, 2012). Our work suggests that though early season ET may increase with warming temperatures, warmer spring temperatures may in some cases decrease total annual ET by melting the snowpack stores earlier in the water year and reducing soil moisture recharge later in the spring when energy demand is high.

The range of sensitivities of ET to climate in this study is a direct function of climatic and physical characteristics of the catchments presented in this study. For example, OR-CAS receives twice as much precipitation and spans a much lower elevation range than either CA-SIER or CO-ROC (Table 2). Because OR-CAS is considerably wetter, its sensitivity of ET to the magnitude of annual P is lessened considerably. OR-CAS' lower elevations, and related mean winter temperatures, also result in smaller average snowpacks reducing the strength of spring temperature as an explanatory variable for ET. Differences between CA-SIER and CO-ROC largely reflect seasonal distribution of precipitation, and reflect the importance of summer precipitation in CO-ROC. While climate is the dominant factor, topographic differences are also important. As discussed above, topographically driven flowpath convergence in CA-SIER tends to increase sensitivity of ET to parameters that influence lateral drainage. This effect is less evident in the other two watersheds.

Over a range of physically realistic storage characteristics, long-term averages of ET increase with greater storage (AWC) in all basins. Our analysis found the greatest sensitivity of long-term average annual ET to variation in AWC in OR-CAS (Table 3). In CO-ROC, ET ranges from 380 to 600 mm across annual P variation, and across all calibrated subsurface parameters long-term average ET ranges from 450 to 600 mm. This variation in CO-ROC's ET associated with subsurface storage characteristics is on the same order of magnitude as inter-annual variation in ET with P. Similarly, in CA-SIER, ET ranges from 400 to 800 mm across the P record and across all storage parameters, and ranges from

700 to 1000 mm long-term. There is a nonlinear relationship between ET and AWC in each basin. We suggest that below a threshold point in each basin (195–265 mm of AWC), long-term average ET is more sensitive to AWC, and above these threshold values, the effect of climate on ET is greater than an increase in subsurface storage.

The sensitivity of ET to year-to-year variability of climate drivers is also influenced by AWC. The sensitivity of ET estimates to climate drivers varies by 2 to 5 magnitudes in CA-SIER and OR-CAS across the range of plausible storage parameters. These basins receive the smallest fraction of annual P in the summer, and their annual ET estimates are most sensitive to P, R_{75}, and T_{AMJ} at low water capacity (AWC). CO-ROC has a high sensitivity to climate drivers, but this sensitivity does not change with AWC. We suggest that a strong summer P signal in CO-ROC explains the negligible change in ET's sensitivity to climate drivers across values of AWC, similar to other studies that show that summer P can offset the dependence of ET on soil replenishment or winter snowpack (Hamlet et al., 2007; Litaor et al., 2008). The relative importance of AWC to regional climate differences is apparent if we consider that a similar sensitivity to P and T_{AMJ} can be achieved for all basins by varying AWC. For example, ET at the smallest AWC values in OR-CAS is similarly sensitive (slope of 0.6) to inter-annual variation in precipitation as CO-ROC (Fig. 6a).

The two more water-limited basins demonstrate similarly high sensitivities of ET to climate drivers, but differ in the response of their sensitivity to climate across AWCs. Despite CO-ROC and CA-SIER showing similarly strong sensitivities to climate, their response across AWC differs considerably. CA-SIER's sensitivity to climate drivers is highly variable across all AWC, but still demonstrates slightly higher sensitivity at lower AWC values. Its lack of summer precipitation, like OR-CAS, gives water storage a more significant role in mediating late summer water stress. With lower AWC values there is less potential for water storage and ET becomes more sensitive to climate drivers.

In addition to the sensitivity to AWC, our results show that lateral redistribution strongly influences the sensitivity of ET to climate drivers in the drier basins; in CA-SIER and CO-ROC there is considerable scatter in the slopes for P and R_{75} across a single AWC (e.g., for an AWC of 400 mm, the P:ET ranges from 0.6 to 0.8 and 0.2 to 0.7 for CO-ROC and CA-SIER, respectively, in Fig. 6). We note that this additional sensitivity of ET–climate relationships to drainage rates, even given similar AWC or storage conditions, emphasizes the role played by lateral connections. In other words, results suggest that for the two more water-limited sites, the timing of upslope contributions to downslope areas can mediate the sensitivity of watershed-scale vegetation water use.

Our results have general implications for model-based estimates of ET in this region. Because there is substantial heterogeneity in subsurface storage characteristics within each basin (Dahlgren et al., 1997; Denning et al., 1991; McGuire et al., 2007), we might expect that the full range of AWCs can be observed when we look across individual forest stands within a basin. Thus, our estimates that show substantial changes in climate–ET relationships across subsurface parameters suggest that there may be substantial within-basin spatial heterogeneity in vegetation responses to climate variation and change. Even if model estimates are focused on basin aggregate responses such as streamflow, our results point to the importance of calibration data for defining subsurface storage and drainage properties. Estimates of subsurface parameters are often derived from readily available products such as STATSGO and SSURGO (Natural Resources Conservation Service) that provide relatively coarse-scale and imperfect information about hydrologic properties. Consequently, hydrologic models are typically calibrated to obtain estimates of storage and drainage parameters (Beven, 2011). Our results suggest that in areas where streamflow data are not available for calibration, watershed-scale estimates of ET responses to climate drivers may have substantial errors.

5 Conclusions

We demonstrate how subsurface storage and drainage properties (AWC and parameters that control lateral redistribution) interact with climate-related drivers to influence ET in three western US mountain watersheds with distinctive precipitation regimes. These watersheds reflect conditions found in many other western US snow-dominated systems, where summer water availability is influenced by the magnitude of precipitation, timing of soil moisture recharge and spring temperature and its effect on snowmelt. We found that, for our three watersheds, estimates of longer-term average (15-year) watershed-scale ET vary across a range of physically realistic storage/drainage parameters. For all watersheds, the range in long-term mean ET estimates across AWC estimates (e.g., mean ET at a high AWC versus mean ET at a low AWC) may be as large as inter-annual variation in ET, suggesting that the influence of AWC and drainage can be substantial.

Our results also point to the importance of lateral redistribution as a control on ET, particularly for CA-SIER. Only a few studies have emphasized the role of lateral redistribution in plot- to watershed-scale climate responses in the western US (Barnard et al., 2010; Tague and Peng, 2013). For the CA-SIER site, our model results suggest that there can also be interactions between AWC and hillslope to watershed-scale redistribution as controls on ET. Lateral redistribution was less important for the CO-ROC, where summer precipitation was a more important contributor to annual ET values and the least important for the wetter OR-CAS site. Results emphasize that the role of subsurface properties, including both storage and drainage, will be different for different climate regimes.

These results have important implications both for predicting ET in basins where data are not available for calibration and for understanding and predicting the spatial variability of ET within a basin. AWC also affects the sensitivity of annual ET to climate drivers, particularly in the two more seasonally water-limited basins. Although the three watersheds show different responses of annual ET to these climate drivers, there are values of AWC that would eliminate these cross-basin differences. These sensitivities highlight the need for improved information on spatial patterns of subsurface properties to contribute to the development of science-based information on forest vulnerabilities to climate change. Improved accounting for plant accessibility to moisture has improved model–data ET comparisons in previous modeling studies on regional and global scales (Hwang et al., 2009; Tang et al., 2013; Thompson et al., 2011). With expected decreases in fractional precipitation received as snow with climate change (Diffenbaugh et al., 2013; Knowles et al., 2006), we might expect soil storage to play a more important role in providing water for forests in the future. Improved understanding of how climate and subsurface storage/drainage combine to control ET can enhance our understanding of forest water stress related to increased mortality (van Mantgem et al., 2009). Western US forests show substantial vulnerability to drought, with declines in productivity and increases in mortality and disturbance in drought years (Allen et al., 2010; Hicke et al., 2012; Williams et al., 2013). Understanding these ecosystems' responses to primary climate drivers is of particular concern given recent warming trends (Sterl et al., 2008) and multi-year droughts (Cook et al., 2004; Dai et al., 2004). Identifying the physical conditions in which our ability to estimate ET is most sensitive or limited by knowledge of subsurface geologic properties helps to prioritize regional data acquisition agendas. Integrating results from recent advances in geophysical measurements and models such as those emerging from critical zone observatories in the US and elsewhere (Anderson et al., 2008) will be essential for analysis of climate ET interactions.

Acknowledgements. Data are available upon request from the author. This work was supported by funding from the US Geological Survey through the Western Mountain Initiative (award number G09AC00337) and the US National Science Foundation through the Willamette Watershed 2100 project (EAR-1039192). We also acknowledge support from the Southern Sierra Critical Zone Observatory (EAR-0725097) and the Center for Scientific Computing from the CNSI, MRL, an NSF MRSEC (DMR-1121053) and NSF CNS-0960316.

Edited by: P. Saco

References

Allen, C. D., Macalady, A. K., Chenchouni, H., Bachelet, D., Mc-Dowell, N., Vennetier, M., Kitzberger, T., Rigling, A., Bres-hears, D. D., Hogg, E. H. (Ted), Gonzalez, P., Fensham, R., Zhang, Z., Castro, J., Demidova, N., Lim, J.-H., Allard, G., Running, S. W., Semerci, A., and Cobb, N.: A global overview of drought and heat-induced tree mortality reveals emerging climate change risks for forests, For. Ecol. Manage., 259, 660–684, doi:10.1016/j.foreco.2009.09.001, 2010.

Anderson, S. P., Bales, R. C., and Duffy, C. J.: Critical Zone Observatories: Building a network to advance interdisciplinary study of Earth surface processes, Mineral. Mag., 72, 7–10, doi:10.1180/minmag.2008.072.1.7, 2008.

Arthur, M. and Fahey, T.: Biomass and nutrients in an Engelmann spruce - subalpine fir forest in north central Colorado: pools, annual production, and internal cycling, Can. J. For. Res., 22, 315–325, doi:10.1139/x92-041, 1992.

Ashfaq, M., Ghosh, S., Kao, S.-C., Bowling, L. C., Mote, P., Touma, D., Rauscher, S. A., and Diffenbaugh, N. S.: Near-term acceleration of hydroclimatic change in the western US, J. Geophys. Res. Atmos., 118, 1–18, doi:10.1002/jgrd.50816, 2013.

Bales, R., Hopmans, J., O'Green, A., Meadows, M., Hartsough, P., Kirchner, P., Hunsaker, C., and Beaudette, D.: Soil Moisture Response to Snowmelt and Rainfall in a Sierra Nevada Mixed-Conifer Forest, Vadose Zo. J., 10, 786–799, doi:10.2136/vzj2011.0001, 2011.

Band, L. E., Tague, C. L., Brun, S. E., Tenenbaum, D. E., and Fernandes, R. A.: Modelling Watersheds as Spatial Object Hierarchies: Structure and Dynamics, Trans. GIS, 4, 181–196, doi:10.1111/1467-9671.00048, 2000.

Barnard, H., Graham, C., van Verseveld, W., Brooks, J. R., Bond, B. J., and McDonnell, J. J.: Mechanistic assessment of hillslope transpiration controls of diel subsurface flow: a steady-state irrigation approach, Ecohydrology, 3, 133–142, doi:10.1002/eco.114, 2010.

Baron, J., Hartman, M., and Band, L.: Sensitivity of a high-elevation Rocky Mountain watershed to altered climate and CO_2, Water Resour. Res., 36, 89–99, 2000.

Baron, J. S. and Denning, A.: Hydrologic budget estimates, in Biogeochemistry of a Subalpine Ecosystem, edited by: Baron, J., 28–47, Springer-Verlag, New York, 1992.

Beven, K. J.: Rainfall-runoff modelling: the primer, John Wiley & Sons., 119–135, 2011.

Beven, K. J. and Binley, A.: the future of distributed models: model calibration and uncertainty prediction, Hydrol. Process., 6, 279–298, doi:10.1002/hyp.3360060305, 1992.

Boisvenue, C. and Running, S. W.: Impacts of climate change on natural forest productivity – evidence since the middle of the 20th century, Global Change Biol., 12, 862–882, doi:10.1111/j.1365-2486.2006.01134.x, 2006.

Bradford, J. B., Birdsey, R. A., Joyce, L. A., and Ryan, M. G.: Tree age, disturbance history, and carbon stocks and fluxes in subalpine Rocky Mountain forests, Global Change Biol., 14(, 2882–2897, doi:10.1111/j.1365-2486.2008.01686.x, 2008.

Brooks, P. D., Troch, P. A., Durcik, M., Gallo, E., and Schlegel, M.: Quantifying regional scale ecosystem response to changes in precipitation: Not all rain is created equal, Water Resour. Res., 47, W00J08, doi:10.1029/2010WR009762, 2011.

Brooks, R. and Corey, A.: Hydraulic properties of porous media, in Hydrology Paper 3, p. 27, Colorado State University, Fort Collins, 1964.

Cayan, D. R., Dettinger, M. D., Kammerdiener, S. A., Caprio, J. M., and Peterson, D. H.: Changes in the Onset of Spring in the Western United States, B. Am. Meteor. Soc., 82, 399–415, doi:10.1175/1520-0477(2001)082<0399:CITOOS>2.3.CO;2, 2001.

Cook, E. R., Woodhouse, C. A., Eakin, C. M., Meko, D. M., and Stahle, D. W.: Long-term aridity changes in the western United States, Science, 306, 1015–1018, doi:10.1126/science.1102586, 2004.

Dahlgren, R. A., Boettinger, J. L., Huntington, G. L., and Amundson, R. G.: Soil development along an elevational transect in the western Sierra Nevada, California, Geoderma, 78, 207–236, doi:10.1016/S0016-7061(97)00034-7, 1997.

Dai, A., Trenberth, K. E., and Qian, T.: A global dataset of Palmer Drought Severity Index for 1870–2002: Relationship with soil moisture and effects of surface warming, J. Hydrometeorol., 5, 1117–1130, 2004.

Denning, A., Baron, J., Mast, M., and Arthur, M.: Hydrologic pathways and chemical composition of runoff during snowmelt in Loch Vale watershed, Rocky Mountain National Park, Colorado, USA, Water. Air. Soil Pollut., 59, 107–123, 1991.

Diffenbaugh, N. S., Scherer, M., and Ashfaq, M.: Response of snow-dependent hydrologic extremes to continued global warming, Nat. Clim. Chang., 3, 379–384, doi:10.1038/nclimate1732, 2013.

Dingman, S. L.: Physical Hydrology, 2nd ed., Prentice Hall, Englewood Cliffs, NJ, p. 236, 1994.

Famiglietti, J. S., Ryu, D., Berg, A. A., Rodell, M., and Jackson, T. J.: Field observations of soil moisture variability across scales, Water Resour. Res., 44, W01423, doi:10.1029/2006WR005804, 2008.

Farquhar, G. D., Von Caemmerer, S., and Berry, J. A.: A biochemical model of photosynthetic CO_2 assimilation in leaves of C3 species, Planta, 149, 78–90, 1980.

Flock, J.: Lichen-Bryophyte Distribution along a Snow-Cover-Soil-Moisture Gradient, Niwot Ridge, Colorado, Arct. Alp. Res., 10, 31–47, 1978.

Garcia, E. S., Tague, C. L.. and Choate, J. S.: Method of spatial temperature estimation influences ecohydrologic modeling in the Western Oregon cascades, Water Resour. Res., 49, 1611–1624, doi:10.1002/wrcr.20140, 2013.

Geroy, I. J., Gribb, M. M., Marshall, H. P., Chandler, D. G., Benner, S. G., and McNamara, J. P.: Aspect influences on soil water retention and storage, Hydrol. Process., 25(, 3836–3842, doi:10.1002/hyp.8281, 2011.

Gholz, H. L.: Environmental limits on aboveground net primary production, leaf area, and biomass in vegetation zones of the Pacific Northwest, Ecology, 63, 469–481, 1982.

Goulden, M. L., Anderson, R. G., Bales, R. C., Kelly, A. E., Meadows, M., and Winston, G. C.: Evapotranspiration along an elevation gradient in California's Sierra Nevada, J. Geophys. Res., 117, G03028, doi:10.1029/2012JG002027, 2012.

Grant, G. E., Tague, C. L., and Allen, C. D.: Watering the forest for the trees: an emerging priority for managing water in forest landscapes, Front. Ecol. Environ., 11, 314–321, doi:10.1890/120209, 2013.

Grier, C. C. and Logan, R. S.: Old-growth Pseudotsuga menziesii communities of a western Oregon watershed: biomass distribution and production budgets, Ecol. Monogr., 47, 373–400, 1977.

Hamlet, A. F., Mote, P. W., Clark, M. P.. and Lettenmaier, D. P.: Twentieth-Century Trends in Runoff, Evapotranspiration, and Soil Moisture in the Western United States*, J. Clim., 20, 1468–1486, doi:10.1175/JCLI4051.1, 2007.

Hanson, P. and Weltzin, J.: Drought disturbance from climate change: response of United States forests, Sci. Total Environ., 262, 205–220, doi:10.1016/S0048-9697(00)00523-4, 2000.

Hicke, J. A., Allen, C. D., Desai, A. R., Dietze, M. C., Hall, R. J., Ted Hogg, E. H., Kashian, D. M., Moore, D., Raffa, K. F., Sturrock, R. N., and Vogelmann, J.: Effects of biotic disturbances on forest carbon cycling in the United States and Canada, Global Change Biol., 18, 7–34, doi:10.1111/j.1365-2486.2011.02543.x, 2012.

Homer, C., Dewitz, J., Fry, J., Coan, M., Hossain, N., Larson, C., Herold, N., Mckerrow, A., Vandriel, J. N., and Wickham, J.: Completion of the 2001 National Land Cover Database for the Conterminous United States, Photogramm. Eng. Remote Sens., 73, 337–341, 2007.

Hudiburg, T., Law, B., Turner, D. P., Campbell, J., Donato, D., and Duane, M.: Carbon dynamics of Oregon and Northern California forests and potential land-based carbon storage., Ecol. Appl., 19, 163–180, doi:10.1890/07-2006.1, 2009.

Hwang, T., Band, L., and Hales, T. C.: Ecosystem processes at the watershed scale: Extending optimality theory from plot to catchment, Water Resour. Res., 45, W11425, doi:10.1029/2009WR007775, 2009.

Isaac, V. and van Wijngaarden, W. A.: Surface Water Vapor Pressure and Temperature Trends in North America during 1948–2010, J. Clim., 25, 3599–3609, doi:10.1175/JCLI-D-11-00003.1, 2012.

Jarvis, P.: The interpretation of the variations in leaf water potential and stomatal conductance found in canopies in the field, Philos. Trans. R. Soc. London. B, Biol. Sci., 273, 593–610, doi:10.1098/rstb.1976.0035, 1976.

Kampf, S., Markus, J., Heath, J., and Moore, C.: Snowmelt runoff and soil moisture dynamics on steep subalpine hillslopes, Hydrol. Process., 29, 712–723, doi:10.1002/hyp.10179, 2014.

Knowles, N., Dettinger, M. D., and Cayan, D. R.: Trends in snowfall versus rainfall in the western United States, J. Clim., 19, 4545–4559, doi:10.1175/JCLI3850.1, 2006.

Landsberg, J. and Waring, R.: A generalised model of forest productivity using simplified concepts of radiation-use efficiency, carbon balance and partitioning, For. Ecol. Manage., 95, 209–228, 1997.

Litaor, M. I., Williams, M.. and Seastedt, T. R.: Topographic controls on snow distribution, soil moisture, and species diversity of herbaceous alpine vegetation, Niwot Ridge, Colorado, J. Geophys. Res., 113, G02008, doi:10.1029/2007JG000419, 2008.

Loudermilk, E. L., Scheller, R. M., Weisberg, P. J., Yang, J., Dilts, T. E., Karam, S. L., and Skinner, C.: Carbon dynamics in the future forest: the importance of long-term successional legacy and climate-fire interactions, Global Change Biol., 19, 3502–3515, doi:10.1111/gcb.12310, 2013.

McGuire, K. J., Weiler, M., and McDonnell, J. J.: Integrating tracer experiments with modeling to assess runoff processes and water transit times, Adv. Water Resour., 30, 824–837, doi:10.1016/j.advwatres.2006.07.004, 2007.

McNamara, J. P., Chandler, D., Seyfried, M., and Achet, S.: Soil moisture states, lateral flow, and streamflow generation in a semiarid, snowmelt-driven catchment, Hydrol. Process., 19, 4023–4038, doi:10.1002/hyp.5869, 2005.

McNamara, J. P., Tetzlaff, D., Bishop, K., Soulsby, C., Seyfried, M., Peters, N. E., Aulenbach, B. T., and Hooper, R.: Storage as a metric of catchment comparison, Hydrol. Process., 25, 3364–3371, doi:10.1002/hyp.8113, 2011.

Monteith, J.: Evaporation and Environment, in Proceedings of the 19th Symposium of the Society for Experimental Biology, Cambridge University Press, New York, 19, 205–234, 1965.

Moore, I. D., Grayson, R. B., and Ladson, A. R.: Digital terrain modelling?: A review of hydrological, geomorphological, and biological applications, Hydrol. Process., 5, 3–30, 1991.

Mote, P. W., Hamlet, A. F., Clark, M. P., and Lettenmaier, D. P.: Declining Mountain Snowpack in Western North America*, B. Am. Meteor. Soc., 86, 39–49, doi:10.1175/BAMS-86-1-39, 2005.

Muggeo, V. M. R.: Estimating regression models with unknown break-points., Stat. Med., 22, 3055–3071, doi:10.1002/sim.1545, 2003.

Nash, J. E. and Sutcliffe, J.: River flow forecasting through conceptual models part I – A discussion of principles, J. Hydrol., 10, 282–290, 1970.

Post, D. A. and Jones, J. A.: Hydrologic regimes of forested, mountainous, headwater basins in New Hampshire, North Carolina, Oregon, and Puerto Rice, Adv. Water Resour., 24, 1195–1210, doi:10.1016/S0309-1708(01)00036-7, 2001.

Running, S. and Nemani, R.: Extrapolation of synoptic meteorological data in mountainous terrain and its use for simulating forest evapotranspiration and photosynthesis, Can. J. For. Res., 17, 472–483, 1987.

Ryan, M. G.: Effects of climate change on plant respiration, Ecol. Appl., 1, 157–167, 1991.

Schwinning, S.: The ecohydrology of roots in rocks, Ecohydrology, 3, 238–245, doi:10.1002/eco.134, 2010.

Shields, C. A. and Tague, C. L.: Assessing the Role of Parameter and Input Uncertainty in Ecohydrologic Modeling: Implications for a Semi-arid and Urbanizing Coastal California Catchment, Ecosystems, 15, 775–791, doi:10.1007/s10021-012-9545-z, 2012.

Smith, T. J., McNamara, J. P., Flores, A. N., Gribb, M. M., Aishlin, P. S., and Benner, S. G.: Small soil storage capacity limits benefit of winter snowpack to upland vegetation, Hydrol. Process., 25, 3858–3865, doi:10.1002/hyp.8340, 2011.

Sterl, A., Severijns, C., Dijkstra, H., Hazeleger, W., Jan van Oldenborgh, G., van den Broeke, M., Burgers, G., van den Hurk, B., Jan van Leeuwen, P., and van Velthoven, P.: When can we expect extremely high surface temperatures?, Geophys. Res. Lett., 35, L14703, doi:10.1029/2008GL034071, 2008.

Stewart, I. T., Cayan, D. R., and Dettinger, M. D.: Changes toward Earlier Streamflow Timing across Western North America, J. Clim., 18, 1136–1155, doi:10.1175/JCLI3321.1, 2005.

Tague, C. and Band, L.: RHESSys: regional hydro-ecologic simulation system-an object-oriented approach to spatially distributed modeling of carbon, water, and nutrient cycling, Earth Interact., 8, 1–42, doi:10.1175/1087-3562(2004)8<1:RRHSSO>2.0.CO;2, 2004.

Tague, C. and Heyn, K.: Topographic controls on spatial patterns of conifer transpiration and net primary productivity under climate warming in mountain ecosystems, Ecohydrology, 554, 541–554, doi:10.1002/eco.88, 2009.

Tague, C. and Peng, H.: The sensitivity of forest water use to the timing of precipitation and snowmelt recharge in the California Sierra: Implications for a warming climate, J. Geophys. Res. Biogeosci., 118, 1–13, doi:10.1002/jgrg.20073, 2013.

Tang, J., Pilesjö, P., Miller, P. A., Persson, A., Yang, Z., Hanna, E., and Callaghan, T. V.: Incorporating topographic indices into dynamic ecosystem modelling using LPJ-GUESS, Ecohydrology, 7, 1147–1162, doi:10.1002/eco.1446, 2013.

Thompson, S., Harman, C., Konings, A., Sivapalan, M., Neal, A., and Troch, P.: Comparative hydrology across AmeriFlux sites: The variable roles of climate, vegetation, and groundwater, Water Resour. Res., 47, W00J07, doi:10.1029/2010WR009797, 2011.

Thornton, P. E.: Description of a numerical simulation model for predicting the dynamics of energy, water, carbon, and nitrogen in a terrestrial ecosystem, University of Montana, Missoula, MT, 1998.

Thornton, P. E. and Rosenbloom, N. A.: Ecosystem model spin-up: Estimating steady state conditions in a coupled terrestrial carbon and nitrogen cycle model, Ecol. Modell., 189, 25–48, doi:10.1016/j.ecolmodel.2005.04.008, 2005.

Thornton, P., Thornton, M., Mayer, B., Wilhelmi, N., Wei, Y., and Cook, R.: Daymet: Daily surface weather on a 1 km grid for North America, 1980–2012, Oak Ridge Natl. Lab. Distrib. Act. Arch. Cent., doi:10.3334/ORNLDAAC/Daymet_V2, 2012.

Trujillo, E., Molotch, N. P., Goulden, M. L., Kelly, A. E., and Bales, R. C.: Elevation-dependent influence of snow accumulation on forest greening, Nat. Geosci., 5, 705–709, doi:10.1038/ngeo1571, 2012.

van Mantgem, P. J., Stephenson, N. L., Byrne, J. C., Daniels, L. D., Franklin, J. F., Fulé, P. Z., Harmon, M. E., Larson, A. J., Smith, J. M., Taylor, A. H., and Veblen, T. T.: Widespread increase of tree mortality rates in the western United States, Science, 323, 521–524, doi:10.1126/science.1165000, 2009.

Vicente-Serrano, S. M., Camarero, J. J., Zabalza, J., Sangüesa-Barreda, G., López-Moreno, J. I., and Tague, C. L.: Evapotranspiration deficit controls net primary production and growth of silver fir: Implications for Circum-Mediterranean forests under forecasted warmer and drier conditions, Agr. For. Meteorol., 206, 45–54, doi:10.1016/j.agrformet.2015.02.017, 2015.

Webb, W., Szarek, S., Lauenroth, W., Kinerson, R., and Smith, M.: Primary Productivity and Water Use in Native Forest, Grassland, and Desert Ecosystems, Ecology, 59, 1239–1247, doi:10.2307/1938237, 1978.

Williams, A., Allen, C., Macalady, A., Griffin, D., Woodhouse, C., Meko, D., Swetnam, T. W., Rauscher, S. A., and Seager, R.: Temperature as a potent driver of regional forest drought stress and tree mortality, Nat. Clim. Chang., 3(September), 292–297, doi:10.1038/NCLIMATE1693, 2013.

Zierl, B., Bugmann, H., and Tague, C.: Water and carbon fluxes of European ecosystems: An evaluation of the ecohydrological model RHESSys, Hydrol. Process., 21, 3328–3339, doi:10.1002/hyp.6540, 2007.

7

Quantifying human impacts on hydrological drought using a combined modelling approach in a tropical river basin in central Vietnam

A. B. M. Firoz[1], **Alexandra Nauditt**[1], **Manfred Fink**[2], **and Lars Ribbe**[1]

[1]Institute for Technology and Resources Management in the Tropics and Subtropics (ITT), TH Köln, 50679 Cologne, Germany

[2]Chair of Geographic Information Science, Department of Geography, Friedrich Schiller University Jena, 07743 Jena, Germany

Correspondence: A. B. M. Firoz (abm.firoz@th-koeln.de)

Abstract. Hydrological droughts are one of the most damaging disasters in terms of economic loss in central Vietnam and other regions of South-east Asia, severely affecting agricultural production and drinking water supply. Their increasing frequency and severity can be attributed to extended dry spells and increasing water abstractions for e.g. irrigation and hydropower development to meet the demand of dynamic socioeconomic development. Based on hydro-climatic data for the period from 1980 to 2013 and reservoir operation data, the impacts of recent hydropower development and other alterations of the hydrological network on downstream streamflow and drought risk were assessed for a mesoscale basin of steep topography in central Vietnam, the Vu Gia Thu Bon (VGTB) River basin. The Just Another Modelling System (JAMS)/J2000 was calibrated for the VGTB River basin to simulate reservoir inflow and the naturalized discharge time series for the downstream gauging stations. The HEC-ResSim reservoir operation model simulated reservoir outflow from eight major hydropower stations as well as the reconstructed streamflow for the main river branches Vu Gia and Thu Bon. Drought duration, severity, and frequency were analysed for different timescales for the naturalized and reconstructed streamflow by applying the daily varying threshold method.

Efficiency statistics for both models show good results. A strong impact of reservoir operation on downstream discharge at the daily, monthly, seasonal, and annual scales was detected for four discharge stations relevant for downstream water allocation. We found a stronger hydrological drought risk for the Vu Gia river supplying water to the city of Da Nang and large irrigation systems especially in the dry season. We conclude that the calibrated model set-up provides a valuable tool to quantify the different origins of drought to support cross-sectorial water management and planning in a suitable way to be transferred to similar river basins.

1 Introduction

River basins and their hydrological systems play a key role in providing freshwater to downstream deltaic systems, for irrigation and domestic water supply and to regulate salt water intrusion (Ribbe et al., 2017). The patterns of timing and magnitude of streamflow essentially depend on climatic variables such as precipitation (Zhang et al., 2007; Min et al., 2011; Souvignet et al., 2013; Ahn and Merwade, 2014), temperature, and the resulting altered evapotranspiration rates (Vörösmarty et al., 2000; Santer et al., 2011; Trenberth, 2011; Ahn and Merwade, 2014), as well as on the modification of the hydrological systems by humans introducing water infrastructure such as reservoirs and damming, inter-basin water transfers, and construction of weirs.

Hydrological droughts are becoming more frequent disasters worldwide, which can also be attributed to both hydro-climatic and anthropogenic changes (AghaKouchak et al., 2015; van Loon et al., 2016; van Lanen et al., 2016). Regional studies show that larger changes in streamflow have been observed in anthropogenically modified river basins, in particular those altered by hydropower development and operation, than in hydrological systems which are only affected by climate variability and change (Arrigoni et al., 2010; Ahn and Merwade, 2014; Tang et al., 2014). Such alterations of the hydrological system often negatively affect downstream discharge patterns and communities dependent on the provision of freshwater for irrigation and domestic water supply (Rossi et al., 2009; Zhou et al., 2012; Song et al., 2015). Therefore, seasonal impacts of reservoir operation on low-flow patterns and trends need to be quantified in order to separate them from natural drought propagation and to inform downstream water users to properly manage water supply for irrigation, industry, and domestic water supply.

The effects of reservoir operation on streamflow have been assessed for instance in the Lena, Yenisei, and Ob' river basins of the Arctic Eurasian river system, on a seasonal and annual basis revealing that reservoir operation accounts for most of the seasonal changes in the three river basins, ranging from 60 to 100 %, particularly in winter and early spring. Reservoir operation was found to have little effect on annual trends (Ye et al., 2003; Adam et al., 2007; Adam and Lettenmaier, 2008). Räsänen et al. (2012) quantified hydrological changes in the upper Mekong basin due to hydropower operation in China, which showed that discharge increased by 34–155 % from December to May and decreased by 29–36 % from July to September. The impacts on streamflow of the Three Gorges Reservoir were quantified by Zhang et al. (2015), who assessed streamflow at three outlets on the southern bank of the Jingjiang River (a Yangtze tributary), providing evidence that the reservoir impacts were largely responsible for major droughts downstream.

Positive impacts of reservoir operation on downstream hydrological regimes have been reported for Chinese catchments (Song et al., 2015), suggesting a decreasing frequency of flood events in the Sanchahe River basin and by Tang et al. (2014), who showed an increasing surface runoff during the dry season at the upper Mekong/Lancang River in China.

Various approaches have been used to quantify and separate anthropogenic and climate change impacts on streamflow. The most commonly used approaches are streamflow time series analyses looking at seasonal and frequency patterns to assess impacts of human alterations on discharge. Wang and Hejazi (2011) used Budyko curves (Budyko, 1974) to detect human-induced changes in streamflow, investigating their deviation from the initial relationships between mean annual precipitation, evaporation, and potential evaporation as defined by the Budyko curves. Double mass curves (DMCs) are applied to compare the cumulative distribution of precipitation and discharge time series before and after human alterations (Wang et al., 2015) as well as linear regression to establish the relationship between discharge and different climatic variables (Johnson et al., 1991; Wang et al., 2012; Hu et al., 2015). However, although such relatively simple statistical analyses of hydro-climatic time series

might give a first insight into system behaviour, they might not capture the non-linear nature of hydrological systems.

Several studies have applied hydrological models to assess the different causes of streamflow changes (Zhang et al., 2012; Bao et al., 2012; Tesfa et al., 2014; Chang et al., 2015), providing simulations of naturalized and reconstructed discharge time series to quantify and separate the different impacts. Alternatively, the paired basin approach has been used to model the impact of human-induced land cover changes on streamflow by comparing simulations in catchments of very similar characteristics (Bonell and Bruijnzeel, 2005; Seibert and McDonnell, 2010).

The coupled modelling approach, which incorporates hydrological modelling information into reservoir simulation models, appears to be a promising approach which has been recently used to investigate effects of reservoir operations on hydrological systems. For example, López-Moreno et al. (2014) applied a regional hydrological model (RHESSys model) combined with a reservoir simulation model to predict the changes in flow due to reservoir operation as well as climate and land use changes in the Aragón River, Spanish Pyrenees. Reservoir operation effects on downstream flow in the Lena, Yenisei, and Ob' river basins were evaluated using a reservoir routing model coupled offline to the Variable Infiltration Capacity (VIC) land surface hydrology model (Adam et al., 2007). Estimated changes in streamflow due to reservoir operation in the Greater Alpine Region were computed using a parsimonious rainfall–runoff model combined with a hydropower simulation model (Wagner et al., 2017). The coupled approach was also used at a global scale to identify the impact of human water consumption on the intensity and frequency of hydrological drought worldwide (Wada et al., 2013).

The studies described above focussed on the evaluation of either human impacts on general streamflow behaviour or on flood risk. The implication of reservoir operation and other human alterations of the hydrological system for drought severity, duration, and frequency length have not been addressed in such studies. Also, hydrological drought risk is usually looked at on a monthly, seasonal, annual, or longterm scale. Hydro-climatic dynamics in the tropics, however, are fast and water-management-related decisions need to be made based on daily information (e.g. to avoid salt water intrusion into the irrigation and drinking water supply systems) (Nauditt et al., 2017).

The overall aim of this study was therefore to quantify and separate the impact of hydropower reservoir operation on hydrological drought in the VGTB River basin. Its specific objectives were to (1) simulate discharge to obtain naturalized streamflow time series by applying a distributed hydrological response unit (HRU) (Pfenning et al., 2009) based rainfall–runoff model – J2000 (Krause, 2002; Fink et al., 2013); (2) model reservoir storage and operation for eight major hy

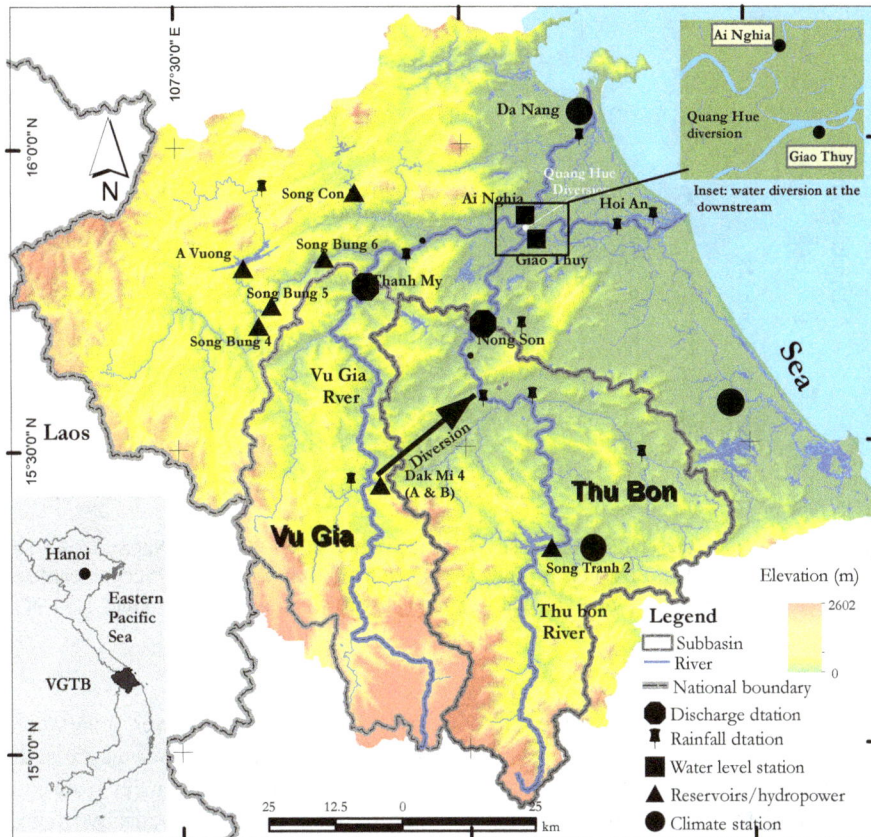

Figure 1. Topographical map of the VGTB River basin showing the hydrology, hydro-meteorological monitoring network, and eight major hydropower reservoirs as well as the diversion (in red colour) from VuGia to Thu Bon at Dak Mi 4 hydropower plant.

dropower reservoirs in order to simulate daily release rates, hydropower production, and storage using the HEC-ResSim model (USACE, 2007); (3) simulate reservoir-impacted reconstructed streamflow for downstream stations at the two main river branches; and (4) quantify to which extent hydrological drought duration and severity can be attributed to hydropower reservoir operation or climate variability by applying the variable threshold method approach (Tallaksen et al., 2009; Sung and Chung, 2014) to reconstructed and naturalized streamflow time series.

The combined assessment approach developed in this study enables us to assess the interactions between climate, catchment, and reservoir operation on the one hand and water and energy demand on the other. Furthermore, it provides us with a tool to determine drought risk on a daily scale to support water management for irrigation and drinking water supply. The results of this research provide a detailed insight into the current and potential impacts of reservoir operation on the downstream water availability, which we provided to the water managers, the reservoir operating agencies, and other decision-makers.

2 Study area and data

2.1 Study area: Vu Gia Thu Bon River basin (VGTB)

The Vu Gia Thu Bon River basin (VGTB) is located in central Vietnam (6°55′–14°55′ N and 107°15′–108°24′ E) and covers a total area of approximately 12 577 km² (Fig. 1). The main provinces in the VGTB are Quang Nam and Da Nang. It is characterized by a steep topography and the altitude ranges from 0 m at the coast to 2598 m in elevation in the South Truong Son Mountains in the west, and by the Kon Tum mountain mass in the south (Viet et al., 2017). Almost half of the land area is covered by forest (47 %), followed by cropland (26 %) and grassland (20 %) (Avitabile et al., 2016). Paddy rice cultivation and livestock farming are the two main agricultural activities in the basin. Two crops of paddy rice are planted per year in the lowlands and areas along the major rivers, yielding 5.05 t ha⁻¹ in 2013 (Quangnam Statistical Office, 2014). The VGTB is home to approximately 2.5 million inhabitants (2013), 80 % of whom live in the coastal lowlands, and 45 % of whom live in the urban areas (General Statistics Office, 2014). The VGTB river system is formed by two major rivers, the Vu Gia and the Thu Bon, which origi-

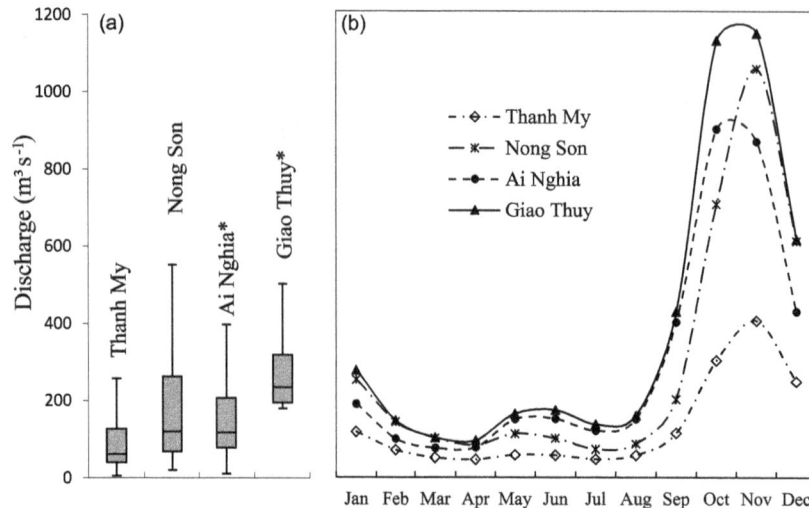

Figure 2. Mean monthly discharge at the four stations under study for the period 1979–2013 **(b)**. Naturalized flow data for Ai Nghia and Giao Thuy stations were simulated with J2000. Box plots **(a)** indicate the 25th, 50th (median), and 75th percentiles of the daily streamflow time series. Outliers have been removed from the plots. The whiskers are defined as the first quartile minus $1.5 \times$ IQR and the third quartile plus $1.5 \times$ IQR.

nate in the highlands and flow into the ocean near the cities of Da Nang and Hoi An.

The climate in the VGTB basin is characterized by a strong wet season with typhoons lasting from September to December and an extended dry season (Souvignet et al., 2013). Next to the two major seasons – which we here term the "dry" and "wet" seasons – there are four minor seasons observed in this region and referred to in this study as summer – June, July, August (JJA); autumn – September, October, November (SON); winter – December, January, February (DJF); and spring – March, April, May (MAM) (Souvignet et al., 2013). Rainfall during the wet season accounts for 65–80 % of the total annual rainfall, with 40–50 % of the annual rainfall occurring in October and November, and this high rainfall regularly causes severe floods (Souvignet et al., 2013). The long dry season lasts from January to August and is frequently accompanied by droughts (e.g. in 1982, 1983, 1988, 1990, 1998, 2005, 2012, and 2013) (Nauditt et al., 2017). February to April were considered the driest months – a period accounting for only 3–5 % of the total annual rainfall, resulting in severe water shortages and problems with saline intrusion at the coast (Souvignet et al., 2013).

The basin area of Vu Gia until reaching Ai Nghia station is approximately 5453 km^2, and the area of Thu Bon until Giao Thuy station is 3532 km^2. Around 3 km beyond Giao Thuy station, the river enters the tide-affected area and the hydrological regime of the river behaves under the interaction of tidal and upstream inflow. At two hydrological stations – Nong Son (Thu Bon River) and Thanh My (Vu Gia River) – discharge has been measured since 1976 (Fig. 1).

Water resources in the Vu Gia Thu Bon River basin (VGTB) have been intensively developed for a variety of

uses, including hydropower generation, large rice irrigation systems in the delta, and domestic and industrial water supply. Inter-basin water transfer from the Vu Gia to Thu Bon sub-basins to generate electricity from Dak Mi 4 hydropower plant is causing significant changes in the respective flow regimes. Paddy rice is the dominant crop, as it accounts for approximately 70 % of irrigated agricultural area (Pedroso et al., 2016). Water stress during drought periods is a major constraint on agricultural production in the region. Figure 2 shows mean monthly inter-annual discharge for the four gauging stations addressed in this study (two discharges and two water-level stations).

2.2 Hydro-meteorological data

Hydro-climatic records were purchased at the Regional Centre for Hydro-meteorology (RCHM) within the scope of German Ministry of Education and Research (BMBF) funded research project "Land Use and Climate Change Interaction in Central Vietnam (LUCCi)" (www.lucci-vietnam.info). A detailed description of the spatial (e.g. soil, vegetation, digital elevation model, land use, geology) and hydro-climatic data used for the hydrological model J2000 was described in Fink et al. (2013, p. 1828) and Souvignet et al. (2013). At two hydrological stations, Nong Son (Thu Bon River) and Thanh My (Vu Gia River), discharge has been measured since 1976. Rainfall data at the 17 stations and climate data at the 3 stations are completely available from 1980 onwards. Based on the data availability, this study considers the time frame 1980–2013, which covers a suitable time frame (> 30 years) for most of the available stations. Two water-level stations further downstream, Ai Nghia (Vu Gia River) and Giao Thuy

Table 1. Reservoirs in the VGTB River basin (MOIT, 2015a, b; ICEM, 2008).

Item	Unit	A Vuong	Song Tranh 2	Dak Mi 4 A	Dak Mi 4 B	Song Bung 4	Song Bung 5	Song Bung 6	Song Con 2
First year of operation		2008	2011	2011	2011	2015	2014	2014	2009
River system		VuGia	ThuBon	Vu Gia	Vu Gia	VuGia	VuGia	VuGia	VuGia
Catchment area	km^2	682	1100	1125	29	1448	2369	2386	250.1
Mean annual flow	$m^3 s^{-1}$	39.8	106	67.80	1.1	73.7	118	119	13.2
Full supply level (FSL)	m a.s.l	380	175	258	106	222.5	60	31.8	275
Minimum operation level (MOL)	m a.s.l	340	138	240	105	195	58.5	30.0	274
Reservoir area at FSL	km^2	9.1	21.5	10.4	0.45	15.65	1.68	0.398	0.13
Reservoir area at MOL	km^2	4.3	9.3	7	0.4	7.8	1.68	0.398	0.12
Reservoir total storage	$10^6 m^3$	343.6	733.4	310	2.6	510.8	20.27	3.29	1.2
Reservoir active storage	$10^6 m^3$	266.5	521.1	158	0.6	233.99	17.82	3.29	0.7
Spillway design flood	$m^3 s^{-1}$	5730	11 069	7864	642	15 427	16 780	17 011	3217
Maximum tail water level	m a.s.l	86.6	87.5	108	71.5	121.3	32.33	15.5	29.7
Normal tail water level	m a.s.l	58	71	106	67.5	101.6	30.7	12	18
Design head	m	300	88.3	135	37.5	112.4	27	13.4	246
Total turbine design discharge	$m^3 s^{-1}$	78.4	209.7	121	122	172.7	239.24	243.2	22.8
Installed capacity	MW	210	162	141	39	156	57	29	46
Annual average energy potential	GWh	825	620.7	582	161	618	220	151	168

(Thu Bon River), are also included to capture the downstream impact of hydropower. They are strongly influenced by tide (Giao Thuy) and tend to be flooded during the rainy season (Ai Nghia).

Data uncertainties

Aside from the uncertainties related to hydro-climatic data described in Fink et al. (2013) and Souvignet et al. (2013), there are no discharge time series for the downstream irrigation region. We therefore developed our methodology based on the following assumptions: before Ai Nghia station in the Vu Gia delta region, water is diverted from Vu Gia to Thu Bon via the Quang Hue channel throughout the year. Due to the strong seasonality and tidal influences, it is difficult to predict the actual amount diverted towards the Thu Bon River. There are no data on quantities of water released from the reservoirs, but we rely on the routing rules of water diverted from Vu Gia to Thu Bon through the Quang Hue chan-

nel (see Table S1 in the Supplement) (Ministry of the Environment, MONRE). To avoid complexity, we assumed in the study that Ai Nghia station is located upstream of the diversion of the Quang Hue channel. We found that the proxy station can accurately capture the influences of reservoir impact on the downstream without leading to potential errors, as it accounts for the overall water balance.

2.3 Hydropower and reservoir data

From 2008 until 2014, eight large hydropower reservoirs and plants were constructed, which have a cumulative storage capacity of more than $2 km^3$ (Table 1). The Dak Mi 4 (A & B) dam was built on the Vu Gia sub-catchment, but the water is diverted at its outflow to the Thu Bon River basin, since the turbines are located in the Thu Bon River basin (Fig. 1). The reservoir information is summarized in Table 1. The classification of the reservoirs is based on the Vietnamese description of large, medium, and small reservoirs (MOIT,

Table 2. Performance of efficiency statistics for the J2000 hydrological model: E^2, Nash–Sutcliffe efficiency; $\log E^2$, Nash–Sutcliffe efficiency with logarithmic values; R^2, coefficient of determination; and P_{bias}, relative volume error in percent.

Station	Thu Bon (Nong Son)		Vu Gia (Thanh My)
Time frame	Calibration 1 Nov 1996–31 Oct 2000	Validation 1 Nov 2000–31 Oct 2005	Validation 1 Nov 1996–31 Oct 2005
E^2	0.856	0.869	0.610
$\log E^2$	0.863	0.856	0.776
R^2	0.869	0.870	0.774
P_{bias}	−10.6	−5.37	8.59

2015a). Reservoirs which have an installed capacity of more than 29 megawatts (MW) of energy are considered large hydropower plants, while the medium and smaller plants are in the range of 10 to 29 MW. The remaining plants produce less than 10 MW (PPC, 2006). For this study we have considered all eight hydropower plants, but to evaluate the model results, we have used the hydropower release data from four of the eight reservoirs: A Vuong (February 2009 to August 2012), Dak Mi 4 A (January 2012 to December 2013), Song Con 2 (September 2010 to June 2012), and Song Tranh 2 (February 2011 to December 2013), for which the outlet data at the turbine discharge are available. Please note that A Vuong started its operation in September 2008 and that Dak Mi4 reservoir started its operation in September 2011 (Table 1). Three of the remaining four reservoirs have only been operational since 2013 (Song Bung 4, 5, and 6), and the data were not available. The last reservoir, Dak Mi 4 B, is considered a runoff reservoir, and therefore it was not necessary to account for its outflow in this study. Operational rules and rule curves were collected from the technical documents of each reservoir from the Department of Investment and Trade (DOIT) belonging to the national Ministry of Investment and Trade (MOIT) of Quang Nam Province, Vietnam (see details in Table S2).

3   Methods

3.1   JAMS/J2000 HRU-based rainfall–runoff model

The J2000 is a physically based distributed and process-oriented model, which is suitable for simulating the hydrological processes of meso- and macro-scale catchments (Kralisch and Krause, 2006; Fink et al., 2007). The model describes the hydrological processes as encapsulated or independent process modules. The model utilizes the HRU approach for the discretization of the basin, consisting of an overlay of land use, soil, geology, and the relief parameters topographic wetness index (Böhner et al., 2002), mass balance index, and solar radiation index (McCune and Dylan, 2002; Pfennig et al., 2009). Modules are described in more detail by Nepal et al. (2014) and in the on-

line documentation (http://ilms.uni-jena.de/ilmswiki/index. php/Hydrological_Model_J2000). The J2000 model was calibrated and validated for the Nong Son gauging station for the period of 1996–2005 (calibration and validation), an undisturbed period before the reservoirs were constructed in 2009.

The calibration was conducted manually and automatically using the multi-objective NSGA2 algorithm (Deb et al., 2002). The model efficiency was tested by using different efficiency criteria, which include (1) the coefficient of determination (R^2) to show the goodness of fit for the general model dynamics, (2) the Nash–Sutcliffe (E^2) efficiency to judge the goodness of fit with a focus on peak flow and simulated volumes, and (3) the Nash–Sutcliffe efficiency ($\log E^2$) with logarithmic values to achieve a stronger focus on the low-flow periods (Krause et al., 2005). As an indicator of the overall simulated volumes, we used the percent bias (P_{bias}) (Table 2). Further information about the utilized objective functions is given in Krause et al. (2005).

3.2   HEC-ResSim reservoir operation model

We applied the HEC-ResSim reservoir system simulation model (USACE, 2007) to simulate reservoir release, hydropower production, and storage in the individual reservoirs of the VGTB at a daily time step. HEC-ResSim allows the development of simulations of single or multiple reservoirs in a hydrological network, based on the available hydrological (inflow) data, the physical reservoir characteristics, and the operating rules. The model is comprehensively documented in Klipsch and Hurst (2013). J2000-simulated inflow time series (compare locations in Fig. 3a) were introduced and routed, with reservoirs altering the routed flow based on physical constraints and operating rules (Fig. 3b). Based on the technical document provided by the MOIT (more details are provided in the Supplement for the operational rules of individual reservoirs; see Fig. S1), the reservoirs were first modelled individually, calibrated, and evaluated based on the available observed outflow at their outlets. For this study, we have used the hydropower release data from four of the eight reservoirs, A Vuong, Dak Mi 4, Song Con 2, and Song Tranh 2, for which the outlet data for the turbine discharges are

Observation data	**Climate soil and land use**	**Reservoir operation and inflow data**	**Observed discharge (1980–2013)**
	↓	↓	
Modelling	J2000 Distributed hydrological model	→ Hec ResSim Reservoir simulation model	
	↓	↓	
Simulated time series	Naturalized flow (1980–2013)	Reconstructed flow (1980–2013)	Observed flow (1980–2013)
	↓	↓	↓
Drought analysis	Analysing the impact of reservoir operation on streamflow drought by applying the daily variable threshold approach		
Results	Hydrological drought severity, duration and frequency on a daily, monthly, seasonal, and annual scale		

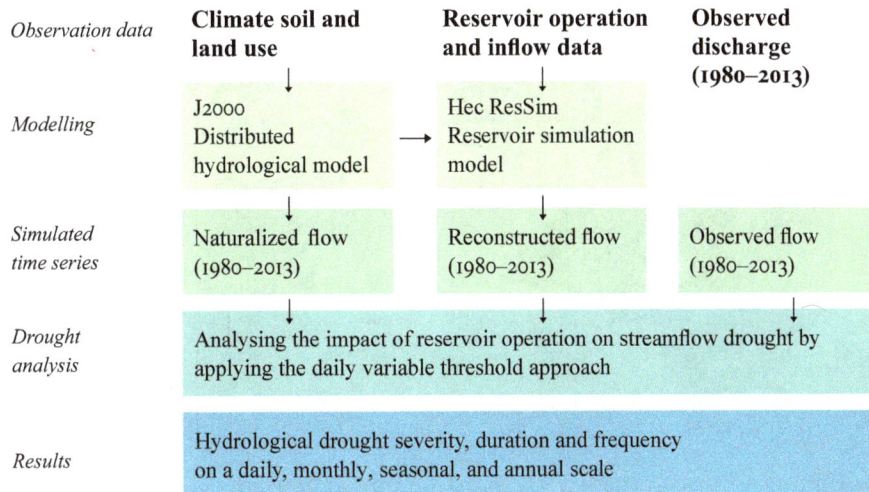

Figure 3. Drought assessment framework: (1) the Distributed Hydrological Model (J2000) (Krause, 2002) provides the simulated inflow data at various nodes and naturalized streamflow, (2) HEC-ResSim simulates reconstructed streamflow for the entire observation period, and (3) streamflow deficiency analysis through threshold-level methods provides information about the drought duration and extent. The reservoir impacts on the downstream flow have been assessed based on the reconstructed and naturalized streamflow differences.

available. Three of the remaining four reservoirs have only been operational since 2013 (Song Bung 4, 5, and 6), and the data were not available. The final reservoir, Song Con 1, is considered a runoff reservoir, and therefore it was not necessary to account for its outflow in this study.

At VGTB, the reservoirs were operated based on a defined management season, namely "*Flood season*" (from 16 September to 31 December), and "*Dry Season*" (from 1 January to 15 September) (MOIT, 2011). During the flood season, the first considerations are dam safety and spill discharge. If the inflow is greater than the maximum hydropower discharge capacity and the water level is above the flood control zone, then water is first diverted to its full capacity to produce hydropower and the excess water within that day will be released through spill discharge to ensure flood control. During the dry season, the guide curve will determine how the release of water from the reservoir will be managed. However, for each reservoir there is a monthly power production target, also controlled by the upper and lower limits of the reservoir level. Generally, if the water level is close to the upper limit of the guide curve, then energy production will be maximized, and if it is close to the lower limit, a limited amount of water will be released for hydropower production, and release rates are made considering the environmental flow.

3.3 The combined modelling–drought assessment framework

To analyse and quantify the impacts of reservoir operation on downstream low flows and to separate them from other impacts, longer time series for both the "pristine" and "impacted" periods are needed. We termed them "naturalized"

and "reconstructed" discharge, respectively. The J2000 hydrological model was utilized to simulate daily discharge for upstream HRU outlets of the VGTB River basin system as input streamflow time series to the reservoirs and to provide time series for the "naturalized" flow for the four downstream stations addressed in this study (Fig. 1). Impacts of hydropower operation on downstream low flows were assessed by using the HEC-ResSim reservoir routing model coupled offline to the J2000 for the VGTB River basin (Fig. 3). The output of this integrated model is referred to here as "reconstructed streamflow". This provides the estimated streamflow at the two existing gauging stations (Nong Son and Thanh My) and at the two additional locations further downstream of the mouth of the two reaches (Ai Nghia and Giao Thuy), to capture the influences of reservoirs located further downstream (Fig. 1). In our analysis the observed discharge data were only used for evaluating the simulated results. In the modelling process, we assumed that all eight reservoirs came into operation in 1980, and then used the reservoir model to produce the synthetic streamflow termed here as reconstructed flow. This gave us the opportunity to evaluate the long-term influences of the reservoirs on streamflow. A drought analysis was then performed for the reconstructed (reservoir-impacted) and naturalized (pristine) streamflow simulations. Figure 4 provides an overview of the applied methods.

3.4 Hydrological drought assessment

The threshold approach (Zelenhasić and Salvai, 1987) is widely used to determine hydrological drought in temperate regions, where the discharge is usually greater than zero (Tallaksen et al., 2009; van Huijgevoort et al., 2012; van Loon

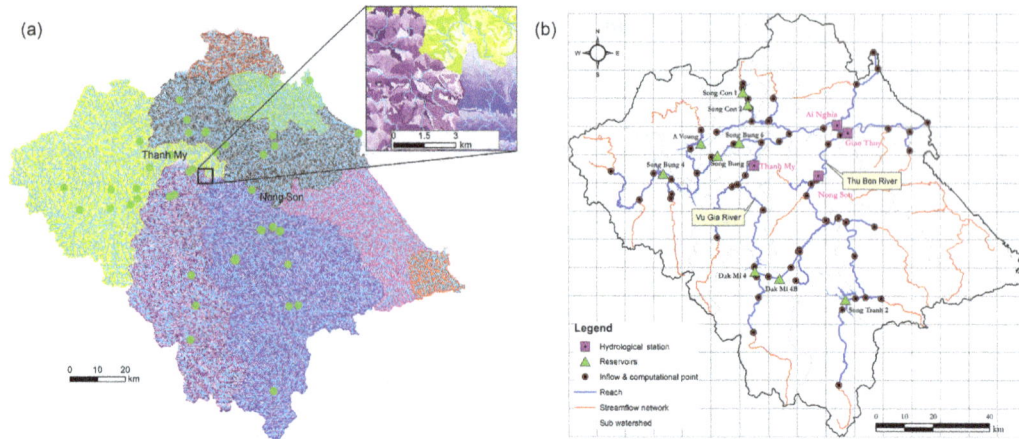

Figure 4. Coupling of the J2000 model with the HEC-ResSim model. **(a)** HRU of the J2000 model along the major sub-basin, virtual discharge stations (green points) for which J2000 simulated time series for the reservoir inflow, and relevant abstraction points in the downstream area. **(b)** The HEC-ResSim model node network, J2000 inflow discharge points (brown dots), and the locations of the reservoirs that have been incorporated within the reservoir model.

and van Lanen, 2012; Sung and Chung, 2014). It defines drought events based on a threshold value and provides information about its onset, duration, and severity (Stahl, 2011; Hisdal et al., 2004).

The daily variable threshold approach (Hisdal et al., 2004) based on flow duration curves (FDCs) has been applied to determine hydrological drought periods. We used the 90th percentile (Q_{90}) of the FDC as the daily variable threshold, which is obtained from the antecedent 365 daily streamflow values. This threshold has been selected to study the drought which has a severe impact on the livelihood of the downstream population, particularly the irrigation sectors within the VGTB River basin, and also has been used in various drought-related studies (e.g. Fleig et al., 2006; Wanders et al., 2015). Q_{90} is defined as follows: for a given day of the hydrological year d (in this study, 1 September is considered the start of the hydrological year), the daily varying $Q_{90}(d)$ is calculated based on a moving average of 30 days centred on day d (i.e. 15 days either side), starting from the first day of the hydrological year (Prudhomme et al., 2011; Van Loon et al., 2015). Due to strong seasonality within the study region, we further introduce the break-days concept to calculate the threshold level for both dry and wet season separately. Here the break days are 1 September and 1 January, which are the starting dates of the wet and dry seasons, respectively. Furthermore, lower than average flow in wet seasons contributed to the development of drought in the following season (Sung and Chung, 2014). A binary approach has been considered to identify whether it is a dry day or normal day based on the daily low-flow varying threshold. Finally, the streamflow deficits of the naturalized and reconstructed streamflow are compared to quantify the impact of reservoirs on streamflow drought.

4   Results

4.1   J2000 hydrological model calibration to simulate naturalized discharge

The J2000 model was manually calibrated and validated for the Nong Son discharge station for the period of 1996–2005. We also performed an automatic calibration using the multi-objective NSGA2 algorithm (Deb et al., 2002), which yielded similar results using the same objective functions as for the manual calibration (Table 2). The second available gauging station (Thanh My) was not separately calibrated, but tested using the same parameter set calibrated for Nong Son (see details in Table S3 for the estimation of parameters). This was done to check the ability of the model to simulate discharge for those parts of the basin where no calibration was possible due to the lack of discharge data.

Table 2 shows the efficiencies for each objective function used for the calibration and validation period (1996–2005). It is worth noting that if the model is calibrated using the first half of the time series (1996–2000; E^2, Nash–Sutcliffe efficiency, of 0.856), the runoff for the second half (2000–2005) is reasonably well simulated (E^2 of 0.869), including the low flows during the drought period in 2005 (see details in Sect. S4 for the observed and simulated discharge plots). The average of the three efficiency criteria for Nong Son station resulted in 0.865 and 0.72 for Nong Son and Thanh My, respectively, when validated for the time period from 2000 to 2005 (Table 2). Following the classification of Nash–Sutcliffe efficiency criteria proposed by Moriasi et al. (2007), most of the calibrated models are rated as "good" ($> 65\,\%$) or "very good" ($> 75\,\%$). The objective functions $\log E^2$ and R^2 show that the low-flow periods and the overall dynamics are well represented. For the calculation methods and fur-

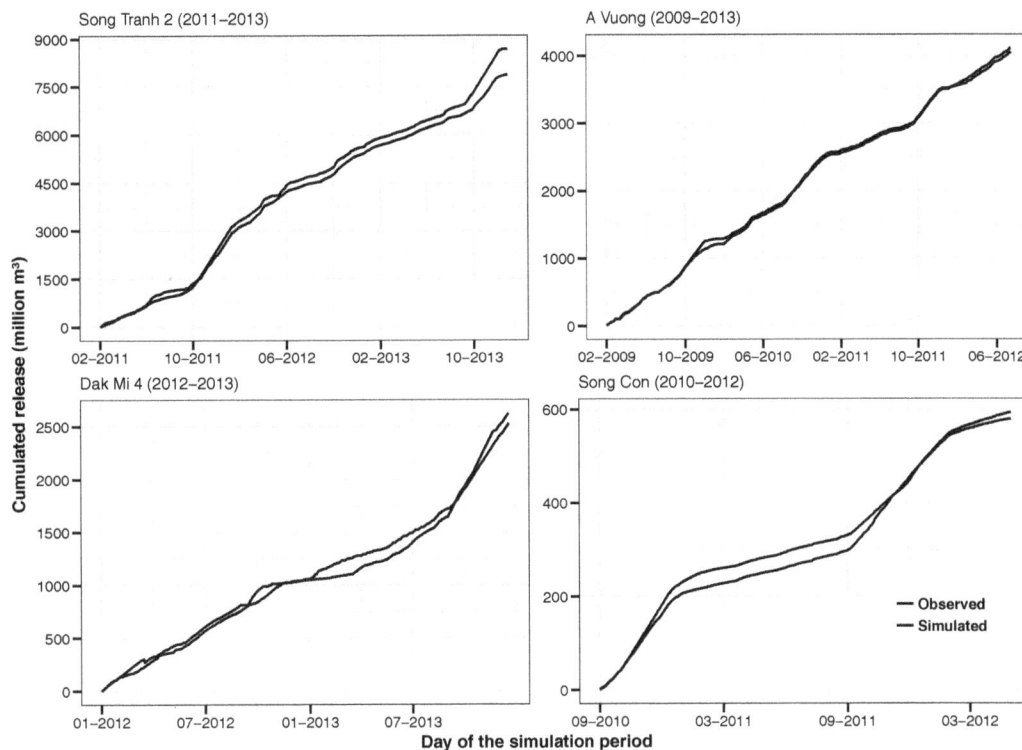

Figure 5. Simulated and observed cumulated daily release of the individual reservoirs.

ther information about the utilized objective functions, refer to Krause et al. (2005). Owing to the HRU concept of J2000 as well as the JAMS modelling framework, it is possible to generate hydrological state variables for each point in time and space. This facilitates the transfer of flow data at a daily time step at selected points along the river segments to the reservoir model (Fink et al., 2013), for example the points representing the reservoir inflow discharges.

4.2 Simulation of hydropower reservoir release discharge

We applied the Hec ResSim model to simulate reservoir release discharge for each individual reservoir in the VGTB at a daily time step. Inflow time series from J2000 hydrological models were introduced and routed at inflow locations (Fig. 4). The individual reservoir simulation results are presented in Fig. 5.

The simulation period varied for each of the reservoirs, depending on their year of construction and availability of the discharge data from the turbine. In the case of A Vuong, we compared the observed release data from February 2009 to August 2012 with the simulated cumulated daily discharge release values (Fig. 5), and there was very good agreement between the time series. There was also strong agreement at Dak Mi 4 with data from January 2012 to end of December 2012. However, for the summer period in 2013, the simulated discharge was consistently lower than the observed

discharge (Fig. 5). Simulations for Song Con 2 for the period from September 2010 until the beginning of 2011 also showed good results, while the dry season cumulative discharge for the year 2011 was underestimated but improved during the wet season. The simulation result for Song Tranh 2 was unsatisfactory for the period after January 2012 (Fig. 5).

4.3 Reconstructed streamflow simulation

Reconstructed synthetic streamflow was simulated based on the individually simulated reservoir releases (Fig. 5). We simulated the reconstructed streamflow for the period 1980–2013, incorporating varying reservoir operation options such as cascade reservoir operation and flood and dry season control. This was performed for the gauging stations Nong Son (wetter Thu Bon catchment) and Thanh My (drier Vu Gia catchment) and two downstream stations: Giao Thuy (Thu Bon) and Ai Nghia (Thanh My). These latter stations are located in the delta region where water is abstracted for rice irrigation and for the drinking water treatment plant which supplies the city of Da Nang. These simulations were needed to capture the impacts of all reservoirs on water availability in the delta area. As there are only water level but no gauging stations at Giao Thuy and Ai Nghia for calibration, we used the naturalized streamflow simulated using J2000. To evaluate the efficiency of the calibration, we applied the performance statistics for the period of 2011 to the end of 2013 (Table 3). This time frame was chosen because the Dak Mi

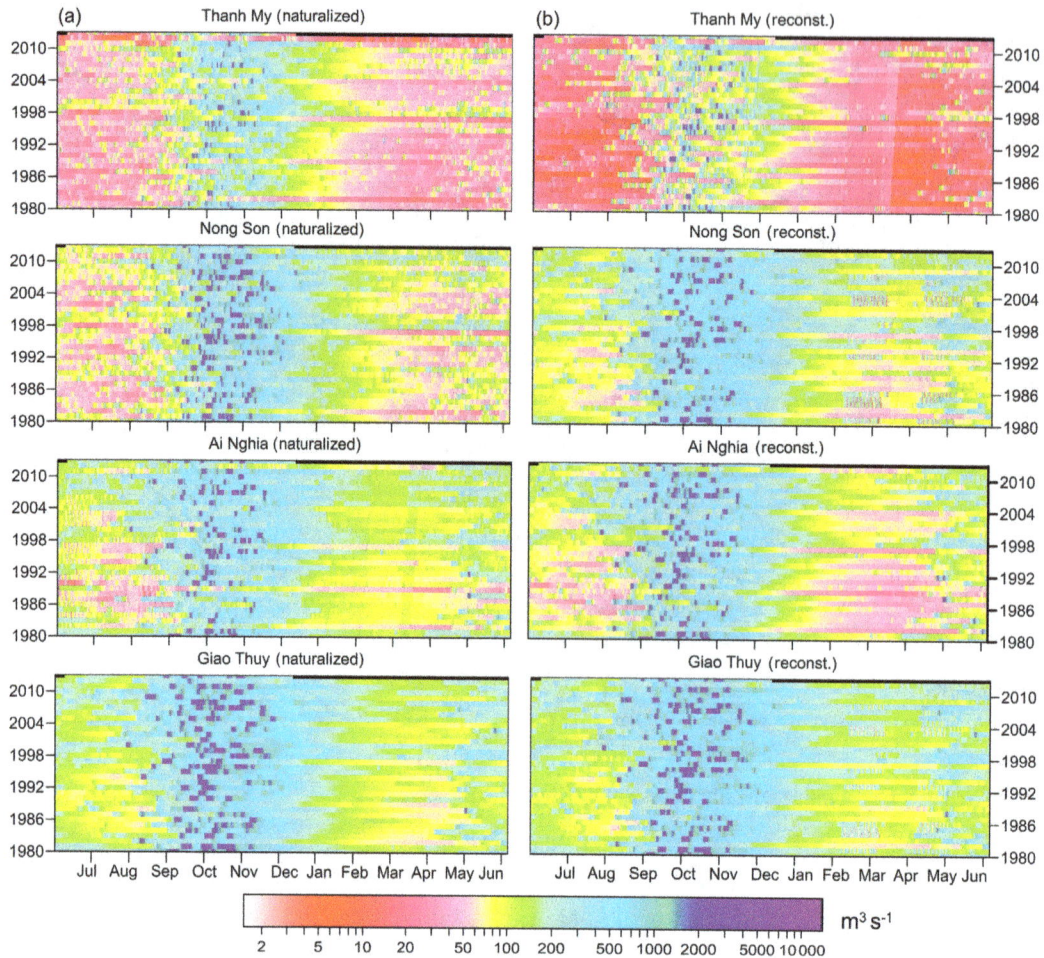

Figure 6. Daily values of discharge ($m^3 s^{-1}$) at the four discharge stations. Each pixel in the plot represents 1 day and its colour denotes discharge. The x axis represents the hydrological year, starting in July and ending in June. **(a)** shows the naturalized condition based on the J2000 model simulation. **(b)** shows the reconstructed streamflow product based on the reservoir simulation model.

4 and Song Tranh 2 hydropower plants started operation after 2011 and measured data for calibration were available for this period. The efficiency statistics show reasonable results: e.g. E^2, Nash–Sutcliffe efficiencies of 0.907 and 0.716 for Nong Son and Thanh My stations, respectively (see details in Sect. S5 for the observed and reconstructed streamflow plots for the period from 2011 to 2013). This indicates that the reconstructed streamflow is able to capture the influences of reservoir operation on streamflow. The reconstructed streamflow also shows a very good result considering the overall water balance described by the P_{bias} (relative volume error in percent). The P_{bias} values for Nong Son and Thanh My are 0.0052 and −0.077, respectively.

4.4 Daily, monthly, and seasonal effects of hydropower reservoir operation on streamflow in the subcatchments Vu Gia and Thu Bon

We compared the daily naturalized and reconstructed streamflow simulations in Fig. 6. For Thanh My station and Ai

Table 3. Performance statistics: Nash–Sutcliffe efficiency (E^2), $\log E^2$, R^2, and P_{bias} of the reservoir model (reconstructed streamflow) for the two gauging stations.

Stations	Nong Son	Thanh My
	1 Jan 2011–	1 Jan 2011–
	31 Dec 2013	31 Dec 2013
E	0.907	0.716
$\log E^2$	0.79	0.74
R^2	0.954	0.809
P_{bias}	0.0052	−0.077

Nghia station in the drier Vu Gia catchment, low flows (pink to yellow colours) during the summertime are more prominent in the reconstructed streamflow than in the naturalized streamflow. For Nong Son and Giao Thuy stations, however,

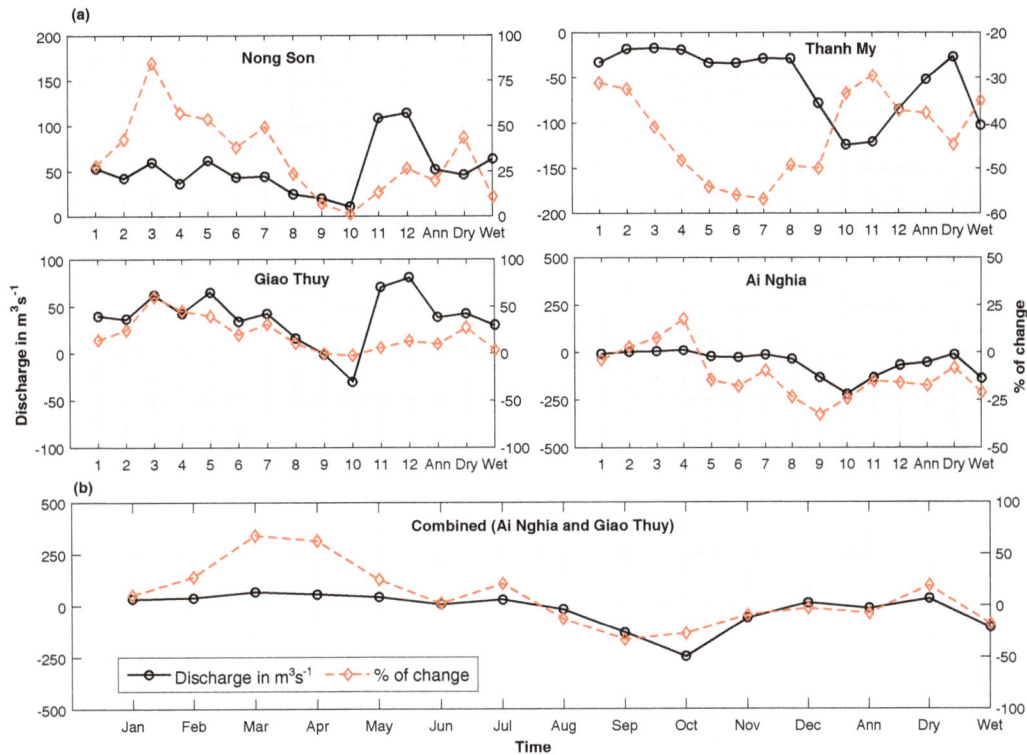

Figure 7. Reservoir impact on streamflow changes. **(a)** Mean differences of reconstructed streamflow pattern (discharge in $m^3\,s^{-1}$) and the percentage (%) of changes in streamflow from the naturalized mean flow for the period of 1980–2013. A negative value indicates a decreasing flow compared with the naturalized one. The number indicates the month starting with January referred to as 1. **(b)** Combined effect of reservoir impact for Ai Nghia and Giao Thuy, representing the overall impact on the streamflow on the VGTB basin due to reservoir construction.

fewer low flows were simulated in the reconstructed time series than in the naturalized one.

To quantify the mean monthly reservoir effects for the period from 1980 to 2013 (Fig. 7), we plotted the mean monthly values of the reconstructed streamflow against the naturalized discharges for the four stations. For Thanh My station located at the upstream of the Vu Gia River, monthly streamflow was reduced on average by approximately $51\,m^3\,s^{-1}$ (38 % of the observed flow). The impact of reservoir operation is most pronounced for the dry season (January to August), when flows decrease from 30 to 60 % compared to the naturalized mean monthly discharge. During the wet season (September to December), discharge decreased by 30 %. At Nong Son station, mean monthly streamflow increased by 24 to $62\,m^3\,s^{-1}$ (from 23 to 85 % of the observed discharge) for the period January to August. Although the mean discharge for September to December increased by 50 to $114\,m^3\,s^{-1}$, the percentage increase was rather low, varying from 1.3 % in October to 26.3 % in December (Fig. 7a). The Giao Thuy and Ai Nghia stations are located approximately 25 and 32 km downstream of the Nong Son and Thanh My stations, respectively, and exhibit a similar pattern of flow changes due to reservoir construction. Analysing the combined seasonal impact of reservoirs on water availability in

both catchments, we found that overall discharge during the wet season decreased by 2 to 38 % and increased during the dry season from January to August in which a significant increase in flow augmentation was found during March to April (62–68 %) (Fig. 7b). Figure 8 shows the annual and seasonal mean monthly hydrographs for the four stations, comparing the simulated discharge on a seasonal and an annual scale. These results show that there are strong seasonal changes in streamflow for both sub-catchments, with a significant reduction of streamflow for the Vu Gia River especially in the dry season, and an increase in water availability in the Thu Bon River.

4.5 Impacts of reservoir operation on hydrological drought

Hydrological drought occurrence, length, and severity were determined by using the daily varying threshold-level method (Q_{90}) separately applied to the dry and wet seasons (break days were 1 September and 1 January). Figure 9 shows the drought onset and duration of the naturalized and reconstructed streamflow time series to evaluate the reservoir operation impact on hydrological drought. Thanh My station (Vu Gia catchment) shows more days under drought for the

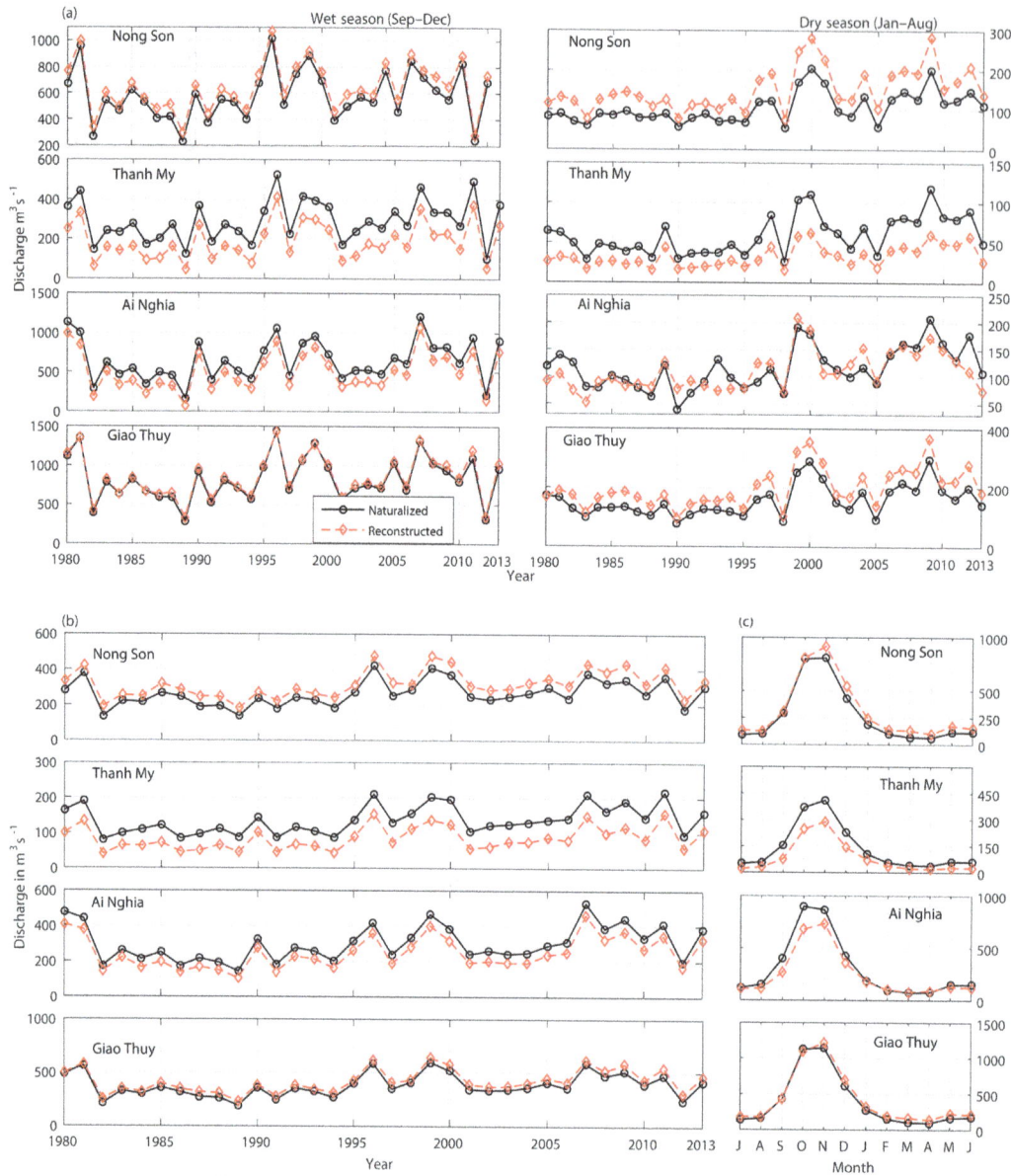

Figure 8. Comparison of mean streamflow pattern (naturalized and reconstructed streamflow): **(a)** comparison of mean seasonal flows for the dry (January to August) and wet (September to December) seasons; **(b)** comparison of mean annual streamflow; and **(c)** comparison of mean monthly streamflow (m^3 s^{-1}).

reconstructed period (1061 days) compared to the naturalized period (774 days). Similarly, an increasing number of drought days and frequency was found for the reconstructed period at Ai Nghia (1286 to 1011 days).

At Nong Son station (Thu Bon River), the analysis shows a general shift of the occurrence of drought from spring (MAM) to summer (JJA) (Fig. 9). Nong Son (upper Thu Bon River) and Giao Thuy (lower Thu Bon River) stations exhibit a decreasing number of drought days, respectively, from 821 to 680 days and from 1025 to 713 days. These reductions are due to the diversion of the Dak Mi 4 reservoir from Vu Gia to Thu Bon. The number of drought days corresponding to the

year at each of the stations are presented in the Supplement (Fig. S6).

5 Discussion

5.1 Simulating naturalized discharge with J2000 in a data-scarce environment

In the VGTB, only two discharge stations and related time series are available for calibration. Therefore, to assess changes in water availability in the delta region where water is needed for irrigation and other purposes (e.g. domestic and industrial

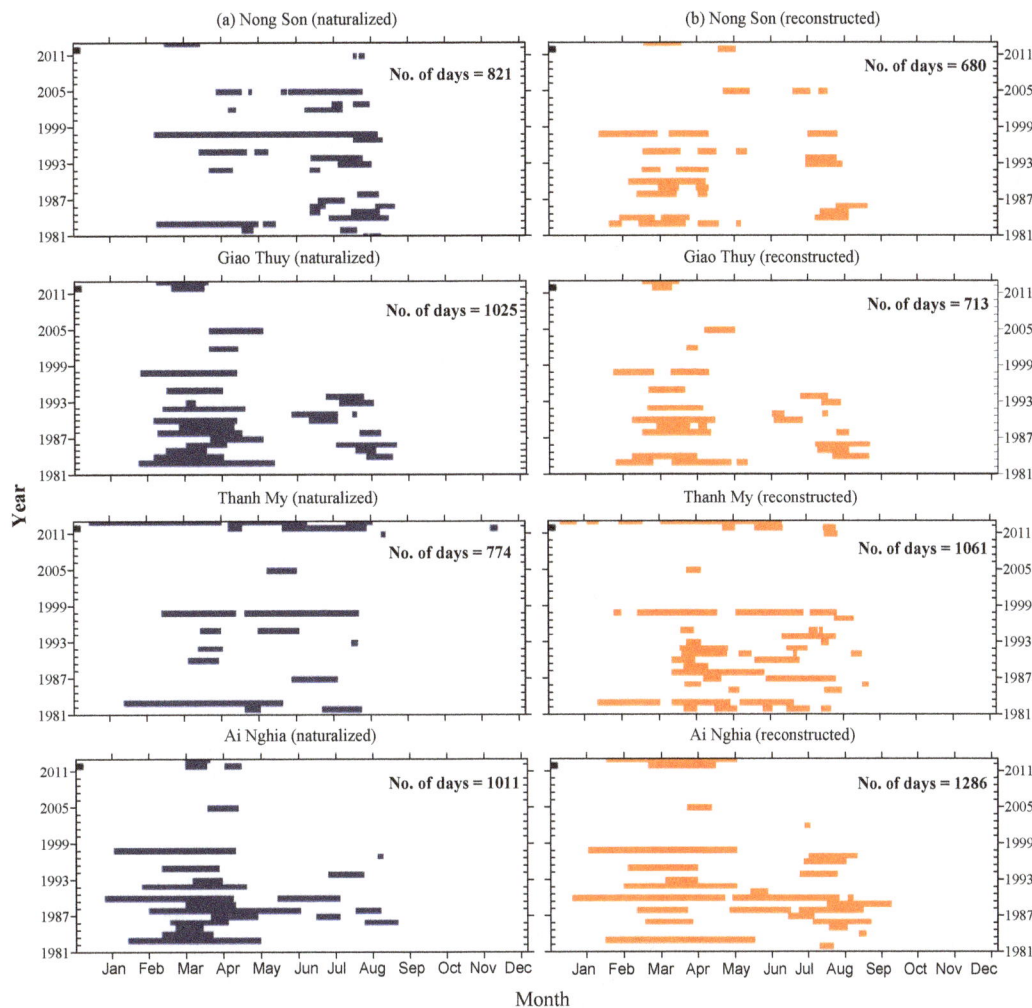

Figure 9. Number of days below the Q_{90} variable drought threshold for the VGTB at the four discharge stations (1981–2013). One day of streamflow drought is a day in which the 30-day running mean discharge is below the 10th percentile of 30-day mean discharge. The blue colour bars **(a)** show the drought onset and duration for the naturalized streamflow, whereas the orange colour bars **(b)** represent the reconstructed reservoir impacted discharge. "No." indicates the total number of drought days.

uses), we simulated discharges for locations where no validation data were available. The J2000 model was successfully calibrated and validated for the Nong Son gauging station for the period 1996–2005, an undisturbed period before the reservoirs were constructed in 2009. Results for the three applied efficiency criteria ranged from 0.72 to 0.87, which are considered very good simulation performances (Moriasi et al., 2007). The application of the Nong Son validated parameter set to Thanh My station also yielded reasonably good efficiencies (Table 2).

These results allowed us to use J2000 to simulate naturalized discharges for HRU outlets needed as reservoir inflow discharges and for the downstream delta locations Ai Nghia (Vu Gia sub-catchment) and Giao Thuy (Thu Bon sub-catchment). They can be considered valuable discharge estimations for this study. A simulation uncertainty range is presented in the Supplement (Fig. S7).

5.2 Modelling discharge release from operating hydropower reservoirs

Overall individual reservoir modelling showed good results in simulating released discharges from the turbine (Fig. 5). Available release discharge time series from operating hydropower plants for reservoir model calibration were short, and the simulation period varied for each of the reservoirs depending on their year of construction and availability of discharge data for the turbine.

The simulation results for the Song Tranh 2 reservoir were unsatisfactory for the period after January 2012 (Fig. 5) due to reservoir leakages which led to the prohibition of any storage of water in 2012–2013 to ensure dam safety. Any water entering the reservoir was sent immediately through the turbine, increasing discharge from the turbine. As a result, there was no storage functionality in the reservoir during this period. After 2013, the leakages were repaired, and the reservoir returned to its normal operating condition. Data have been available since January 2012 for Dak Mi 4, which diverts the water from Vu Gia to Thu Bon. Despite general agreement over the entire data period, the simulated discharge was lower than the observed discharge for the summer period in 2013 (Fig. 5). Furthermore, simulations for Song Con 2 underestimated dry season cumulative discharge in 2011, but improved again during the wet season. These underestimations of the simulation results can be predominantly attributed to the reservoir release constraints associated with the reservoir operation during the dry season.

5.3 Is the integrated modelling framework suitable for assessing the hydrological regime under reservoir operation?

For reservoir impact assessment, time series for either the pristine or human-impacted period are usually too short to be used for calibration. For the first time, an integrated modelling framework was applied to a data-scarce tropical mountainous mesoscale catchment to assess hydrological drought risk by using naturalized and human-impacted reconstructed streamflow and two observed discharge time series. Comparing observed, simulated, reconstructed, and naturalized discharge time series is a widely used method to assess and quantify anthropogenic impacts on streamflow (Zhang et al., 2012; Deitch et al., 2013; López-Moreno et al., 2014; Chang et al., 2015; Räsänen et al., 2017). Our softly linked model set-up shows good results in terms of statistical efficiency performances and provides reliable simulations for both reconstructed and naturalized streamflow. This applies also to the low-flow simulations and hydrological drought periods which usually pose the greatest challenges to hydrological modelling (Pilgrim et al., 1988; Nicolle et al., 2014). This method presents several advantages compared to statistics-based approaches such as Budyko curves or double mass curves. The key advantages of this approach are (1) the possibility of comparing long-term pristine and modified streamflow without relying on long-term hydropower release time series, (2) larger flexibility to account for reservoir influences at the local level, thus accurately allowing prediction of long-term influences of reservoir on streamflow, and (3) the ability to simulate and analyse scenarios dealing with changes (land use, climate, etc.) in the catchment.

Our integrated modelling approach combined with the hydrological drought analyses provided a unique and suitable set of tools to assess drought risk in a data-scarce and reservoir-impacted catchment, and can be transferred to any region where reservoirs impact downstream water availability. Existing methods are mostly able to compare the streamflow behaviour for the hydropower operations before and after their construction, especially those which were built several decades ago. Several studies used the merit of the availability of long time series data to compare before and af-

ter the construction of the hydropower reservoirs (e.g. Ye et al., 2003; Adam et al., 2007; Adam and Lettenmaier, 2008; Arrigoni et al., 2010; Ahn and Merwade, 2014; Tang et al., 2014; Zhang et al., 2015). However, without the required after-construction data, such comparative visualization and characterization of impacts become immensely challenging. Therefore, the proposed integrated model offers quantification of the impacts of newly built hydropower resources on the downstream water users and resources.

Hydropower development is growing, and as of March 2014, 3100 hydropower reservoirs with a capacity of more than 1 MW have been either planned (83 %) or are under construction (17 %) (Zarfl et al., 2015). Most of this hydropower development is concentrated in developing and emerging economies of South-east Asia, South America, and Africa, where data availability is a major issue. This method offers an opportunity to quantitatively analyse and measure the impacts of these hydropower operations at the basin scale. The understanding of our methods can be used for streamflow simulation for ensuring environmental flow of water to produce a sustainable level of food and energy production to support the growing population.

5.4 Quantification of reservoir impacts on hydrological drought

For the first time we tested the integrated hydrological modelling–drought assessment framework based on hydrological indicators, reservoir operation, and rainfall–runoff processes.

This study reveals that the intensity and frequency of hydrological drought in the entire VGTB basin are largely dependent on hydropower operation associated with the inter-basin water diversion from Dak Mi 4. Our modelling results show that drought events simulated for the human-modified catchment system are intensified by 27–37 % in the Vu Gia sub-catchment compared to those under pristine catchment conditions (Table 4). This intensification is mainly attributable to the diversion of the Vu Gia River to the Thu Bon due to Dak Mi 4 hydropower generation which controls the reservoir operation in the study region.

Part of the decreased streamflow in the Vu Gia River could be buffered by increasing reservoir release from the Dak Mi 4 reservoir. According to the technical document (MOIT, 2011), the Dak Mi 4 reservoir is required to release a minimum of $25\,\mathrm{m^3\,s^{-1}}$, a quota which has not been met throughout most of the dry season periods. Because of the high demand for energy during the dry season, some of the water needed for the minimum release towards the Vu Gia River was used for energy production and discharge to the Thu Bon River. As a result, at Nong Son and Giao Thuy stations, the drought intensity decreased by 17 and 30 %, respectively.

We found that for the entire Thu Bon catchment, there is an increasing downstream flow during the low-flow period when we consider the reservoir effects on both river dis-

Table 4. Impact of human alterations on drought intensity and changes in flow in the VGTB for the period 1980–2013 on an annual and seasonal scale. **(a)** Drought duration is calculated based on percentage changes in the number of drought days from naturalized condition to reconstructed condition (Fig. 9). **(b)** Changes in flow (%) are calculated based on the percentage changes in the mean flow between the naturalized and reconstructed streamflows for the corresponding time frame. **(c)** The changes in flow are calculated based on mean differences of reconstructed streamflow from the naturalized mean flow. The positive value indicates increasing flow or drought intensity in relation to the naturalized condition.

		Thu Bon		Vu Gia		Combined
		Nong Son	Giao Thuy	Thanh My	Ai Nghia	
(a) Drought duration (%)		−17.17	−30.43	37.08	27.20	−1.81
(b) Changes in flow (%)	Ann	19.46	10.09	−37.82	−17.41	−13.82
	Dry	43.3	27.23	−44.67	−7.91	32.54
	Wet	10.84	3.61	−35.03	−21.10	−106.53
(c) Changes in flow (in $m^3\,s^{-1}$)	Ann	51.52	38.32	−51.66	−52.14	−7.32
	Dry	45.65	42.51	−26.43	−9.97	19.31
	Wet	63.25	29.93	−102.12	−136.47	−17.48

charges (Fig. 7b and Table 4). This alleviates the general hydrological drought conditions downstream, and the seasonal amplitude of simulated streamflow tends to decrease, which also reduces downstream flood risk.

However, the impacts of reservoir operation are particularly pronounced for the more vulnerable Vu Gia River (Figs. 6–9). The Vu Gia River supplies water to the city of Da Nang and large rice irrigation systems. While Thanh My station streamflow is reduced by $51.7\,m^3\,s^{-1}$, which is 37.8 % less compared to the naturalized condition, downstream at Ai Nghia station, the streamflow reduction is less severe (17.4 % less water than the naturalized condition), as it receives water from tributaries and rainfall downstream of Thanh My (Table 4). Especially during the dry season, the damping effect of reservoirs belonging to the lower sub-basins increases (i.e. Song Bung 4, 5, 6; A Vuong and Song Con) due to their energy production during the dry season (Figs. 6 and 7a).

A further relevant impact of the reservoir operation on hydrological drought is the shift of drought occurrence from summer to spring (Figs. 7 and 9). As shown in the figures illustrating the naturalized flow simulations, low flows generally occur in spring (MAM) and extend towards summer (JJA) at all stations. Figures 7 and 9 show reconstructed flow simulations and indicate more hydrological drought periods during summer. The applied threshold-level approach (Q_{90}) was able to capture the drought events (Fig. 9) in VGTB, consistent with the observed drought events for VGTB (1982, 1983, 1988, 1990, 1998, 2005, 2012, and 2013) (Nauditt et al., 2017).

Generally, reservoir operation leads to reduced runoff volumes in the rainy season and increased runoff in the dry season, and typically serves to mitigate droughts rather than contribute to their aggravation (Wada et al., 2013; Wanders and Wada, 2015; He et al., 2017; Di Baldassarre et al., 2017). We found that the overall reservoir operation at VGTB leads

to an increased flow during the dry season of approximately $32.54\,m^3\,s^{-1}$, which is 27.23 % more than the naturalized situation, and a decreased flow during the wet season of approximately $106.53\,m^3\,s^{-1}$, which is 3.61 % less than the naturalized situation (Fig. 4). A similar pattern of streamflow changes due to hydropower operation was found in the Mekong River basin, where the dry season discharge increased by 60–90 % and the wet season discharge decreased by 17–22 % (Hoanh et al., 2010; Lauri et al., 2012; Räsänen et al., 2012).

However, due to the increased energy demand in summer, the last months of the dry season (August and September) exhibit lower streamflow values under reservoir operation than under the natural flow condition. Also, there is a lower drought risk at the beginning of the dry season, because of the additional storage in the system. At the end of the dry season, the storage is lower, which might lead to a higher likelihood of droughts. These findings on the overall impact of the reservoir operation can be transferred to other locations featuring similar climatic and topographic conditions, whereas the separate findings for the Vu Gia and Thu Bon rivers are very much influenced by the diversion at Dak Mi 4, and are therefore specific to this catchment.

5.5 Consequences of the hydrological changes

Droughts are usually assessed at a large scale and based on indices which are related to parameters such as precipitation, soil moisture, or vegetation. However, human alterations of the hydrological system and abstractions from the rivers are not incorporated into such drought analyses (Van Loon et al., 2015). A variety of anthropogenic alterations of the natural environment and river network can cause changes in downstream water availability, and these anthropogenic alterations include land cover changes, major water abstractions, and infrastructure for irrigation and drinking water supply. Nauditt et al. (2017) used varying spatial basin characteristics, such as land cover changes, to simulate low flows in the VGTB basin, and found that these only play a minor role in runoff generation processes, which are instead dominated by precipitation inputs. Therefore, it can be assumed that all the quantified changes in this study for the different temporal scales can be considered net values for reservoir operation impacts on low flow discharge.

We found that reservoirs can have multiple effects on the downstream users, particularly if they are not operated properly. In the VGTB, hydropower reservoir operation strongly alters the natural hydrological functions of the river basin. In particular, one hydropower reservoir (Dak Mi 4) generates electricity by transferring water from the drier Vu Gia sub-catchment to the wetter Thu Bon sub-catchment, due to its superior slope to produce energy (Nauditt et al., 2017). During the dry season, the combined effect of the reservoir operations at Ai Nghia and Giao Thuy (Table 4 and Fig. 7b) indicated that overall flow increases during the dry season and reduces the wet season flows. These changes resulted in dampening of VGTB's annual flood pulse. This decreased flow pattern during the flood season is expected to reduce the sediment and nutrient transport, and can affect the aquatic habitat (Pitlcik and Wilcok, 2001). The fluctuation of water supplies due to the reservoir operation degraded the river bed immediately after the turbine discharge. This degradation is typically accompanied by a coarsening of the river bed with associated loss of useable habitat for fish and benthic invertebrates (Pitlcik and Wilcok, 2001). The loss of these important habitats, combined with changes in water quality due to sediment imbalance and introduction of non-native fishes, has potentially caused long-lasting impacts on the native fish community at VGTB.

One of the major concerns is that the seasonal shift of drought occurrences, from spring (MAM) to summer (JJA), was observed at most of the stations in the VGTB. This may have impacted the VGTB's ecological productivity, which is the basis for livelihood, income, and food security for millions of people. This shift could have impacted the cropping pattern of the downstream, which relies heavily on the water during the summer season. However, the results indicated that the dry season discharge may vary considerably due to rainfall and hydropower operations. For example, in 2013, due to the low rainfall in 2012 (September–December), there was a severe shortage of water for hydropower operation during the dry season, which exacerbated the drought in the downstream for the Vu Gia catchment.

6 Conclusion

We assessed human impacts on hydrological droughts in the VGTB River basin and found that the intensity and frequency of hydrological droughts in the entire Vu Gia Thu Bon basin are largely dependent on hydropower operation associated with the Dak Mi 4 related inter-basin water diversion. Our modelling results show that drought events simulated for the human-modified catchment system were intensified by 27–37 % in the Vu Gia sub-catchment compared to the ones under pristine catchment conditions. However, when combining the overall impact of reservoir operation for the entire VGTB, we found an increase in dry season flows (ca. 27 %) and reduced flood season flows (ca. 3.5 %) compared to the naturalized condition, and a similar pattern of changes due to reservoir operation was also found in another basin in the Mekong region.

Furthermore, a seasonal shift of drought occurrence from spring (MAM) to summer (JJA) was observed, severely affecting rice cultivation as the cropping season particularly relies on the water during the spring and summer. We also identified hydropower reservoir operation impact patterns which show how energy production and demand can influence seasonality in streamflow in a tropical environment.

The multi-model framework combined with the application of a daily varying drought threshold turned out to be a suitable method to analyse human-impacted hydrological drought. To our knowledge, a distributed hydrological model such as J2000 had never been applied to such a data-scarce tropical environment. Linking the physically based model with a reservoir operation model is an effective approach to assess such a complex river system with a large number of recently built operating hydropower reservoirs and a basin transfer. In combination with the hydrological drought analysis it represents an innovative integrated framework for drought risk characterization which can be applied to any data-scarce catchment worldwide where hydropower is developed, also suitable for snowmelt-driven environments.

We conclude that the calibrated model set-up combined with the streamflow drought analysis provides a valuable tool to support cross-sectoral water management and planning in a tropical monsoon dominated region of strong seasonality.

Competing interests. The authors declare that they have no conflict of interest.

Acknowledgements. This research has been funded by the Federal Ministry of Education and Research (BMBF) in the context of the LUCCi project (grant number 01LL0908C) and by the Federal Ministry for Economic Cooperation and Development (BMZ) in the context of Higher Education Excellence in Development Cooperation (CNRD-exceed) (Grant number 57160105). The authors sincerely thank editor Hilary McMillan and three anonymous reviews for their constructive criticism of the paper. The numerous suggestions provided by the editor and reviewers have helped in the improvement of the manuscript. The authors wish to acknowledge the programming contribution by Rony Paul.

Edited by: Hilary McMillan

References

Adam, J. C., Haddeland, I., Su, F., and Lettenmaier, D. P.: Simulation of reservoir influences on annual and seasonal streamflow changes for the Lena, Yenisei, and Ob' rivers, J. Geophys. Res.-Atmos., 112, D24114, https://doi.org/10.1029/2007JD008525, 2007.

Adam, J. C. and Lettenmaier, D. P.: Application of New Precipitation and Reconstructed Streamflow Products to Streamflow Trend Attribution in Northern Eurasia, J. Climate, 21, 1807–1828, https://doi.org/10.1175/2007JCLI1535.1, 2008.

AghaKouchak, A., Feldman, D., Hoerling, M., Huxman, T., and Lund, J.: Water and climate: Recognize anthropogenic drought, Nature, 524, 409–411, https://doi.org/10.1038/524409a, 2015.

Ahn, K.-H. and Merwade, V.: Quantifying the relative impact of climate and human activities on streamflow, J. Hydrology, 515, 257–266, https://doi.org/10.1016/j.jhydrol.2014.04.062, 2014.

Arrigoni, A. S., Greenwood, M. C., and Moore, J. N.: Relative impact of anthropogenic modifications versus climate change on the natural flow regimes of rivers in the Northern Rocky Mountains, United States, Water Resour. Res., 46, W12542, https://doi.org/10.1029/2010WR009162, 2010.

Avitabile, V., Schultz, M., Herold, N., Bruin, S. de, Pratihast, A. K., Manh, C. P., Quang, H. V., and Herold, M.: Carbon emissions from land cover change in Central Vietnam, Carbon Manag., 7, 333–346, https://doi.org/10.1080/17583004.2016.1254009, 2016.

Bao, Z., Zhang, J., Wang, G., Fu, G., He, R., Yan, X., Jin, J., Liu, Y., and Zhang, A.: Attribution for decreasing streamflow of the Haihe River basin, northern China: Climate variability or human activities?, J. Hydrology, 460–461, 117–129, https://doi.org/10.1016/j.jhydrol.2012.06.054, 2012.

Bonell, M. and Bruijnzeel, L.: Forests Water and People in the Humid Tropics: Past, Present and Future Hydrological Research for Integrated Land and Water Management, Cambridge University Press, Cambridge, 2005.

Budyko, M. I.: Climate and life, International geophysics series, vol. 18, Academic Press, New York, 508 pp., 1974.

Chang, J., Zhang, H., Wang, Y., and Zhu, Y.: Assessing the impact of climate variability and human activity to streamflow variation, Hydrol. Earth Syst. Sci. Discuss., https://doi.org/10.5194/hessd-12-5251-2015, 2015.

Deb, K., Korhonen, P., and Wallenius, J.: A fast and elitist multi-objective genetic algorithm: NSGA-II, IEEE T. Evolut. Comput., 6, 182–197, 2002.

Deitch, M. J., Merenlender, A. M., and Feirer, S.: Cumulative Effects of Small Reservoirs on Streamflow in Northern Coastal California Catchments, Water Resour. Manag., 27, 5101–5118, https://doi.org/10.1007/s11269-013-0455-4, 2013.

Fink, M., Fischer, N., Frührer, N., Firoz, A., Viet, T. Q., Laux, P., and Flügel, W. A.: Distributive hydrological modeling of a monsoon dominated river system in central Vietnam, in: MODSIM2013, 20th International Congress on Modelling and Simulatio, edited by: Piantadosi, J. and Anderssen, R. S., MODSIM 2013, Australia, December 2013, 1826–1832, 2013.

Fink, M., Krause, P., Kralisch, S., Bende-Michl, U., and Flügel, W.-A.: Development and application of the modelling system J2000-S for the EU-water framework directive, Adv. Geosci., 11, 123–130, https://doi.org/10.5194/adgeo-11-123-2007, 2007.

Fleig, A. K., Tallaksen, L. M., Hisdal, H., and Demuth, S.: A global evaluation of streamflow drought characteristics, Hydrol. Earth Syst. Sci., 10, 535–552, https://doi.org/10.5194/hess-10-535-2006, 2006.

General Statistics Office: Statistical Yearbook of Vietnam, Statistical Publishing House, Hanoi, Vietnam, 2014.

Hisdal, H., Tallaksen, M. L., Clausen, B., Peters, E., and Gustard, A.: Hydrological Drought Characteristics, chap. 5, in: Hydrological Droughts: Process and Estimation Methods for Streamflow and Groundwater, Developments in Water Science, Elsevier, Amsterdam, 139–198, 2004.

Hu, Z., Wang, L., Wang, Z., Hong, Y., and Zheng, H.: Quantitative assessment of climate and human impacts on surface water resources in a typical semi-arid watershed in the middle reaches of the Yellow River from 1985 to 2006, Int. J. Climatol., 35, 97–113, https://doi.org/10.1002/joc.3965, 2015.

ICEM: Strategic Environmental Assessment of the Quang Nam Province Hydropower Plan for the Vu Gia-Thu Bon River Basin, Prepared for the ADB, MONRE, MOITT & EVN, Hanoi, Vietnam, 205 pp., 2008.

Johnson, S. A., Stedinger, J. R., and Staschus, K.: Heuristic operating policies for reservoir system simulation, Water Resour. Res., 27, 673–685, https://doi.org/10.1029/91WR00320, 1991.

Klipsch, J. D. and Hurst, M. B.: HEC-ResSim Reservoir System Simulation Version 3.1: User's Manual, US Army Corps of Engineers, Institute for Water Resources, Davis, CA, 2013.

Kralisch, P. and Krause, P.: JAMS – A Framework for Natural Resource Model Development and Application, Proceedings of the iEMSs Third Biannual Meeting, edited by: Voinov, A., Jakeman, A., and Rizzoli, A. E., iEMSs Third Biannual Meeting, Burlington, USA, IAHS, 1–4, 2006.

Krause, P.: Quantifying the impact of land use changes on the water balance of large catchments using the J2000 model, Phys. Chem. Earth, 27, 663–673, https://doi.org/10.1016/S1474-7065(02)00051-7, 2002.

Krause, P., Boyle, D. P., and Bäse, F.: Comparison of different efficiency criteria for hydrological model assessment, Adv. Geosci., 5, 89–97, https://doi.org/10.5194/adgeo-5-89-2005, 2005.

López-Moreno, J. I., Zabalza, J., Vicente-Serrano, S. M., Revuelto, J., Gilaberte, M., Azorin-Molina, C., Morán-Tejeda, E., García-Ruiz, J. M., and Tague, C.: Impact of climate and land use change on water availability and reservoir management: scenarios in the Upper Aragón River, Spanish Pyrenees, Sci. Total Environ., 493, 1222–1231, https://doi.org/10.1016/j.scitotenv.2013.09.031, 2014.

Min, S.-K., Zhang, X., Zwiers, F. W., and Hegerl, G. C.: Human contribution to more-intense precipitation extremes, Nature, 470, 378–381, https://doi.org/10.1038/nature09763, 2011.

MOIT: Decision for Hydropower Plant Operation: Technical Document, Ministry of Investment and Trade, Socialist Republic of Vietnam, 2015a.

MOIT: Decision for Hydropower Plant Operation: Technical Document, Ministry of Investment and Trade, Socialist Republic of Vietnam, 2015b.

Nauditt, A., Firoz, A., Viet, T. Q., Fink, M., Stolpe, H., and Ribbe, L.: Hydrological drought risk assessment in an anthropogenically impacted tropical catchment, in: Land Use and Climate Change Interactions in Central Vietnam: LUCCi, edited by: Nauditt, A.

and Ribbe, L., Water Resources Management and Development, Springer Book Series, 2017.

Nepal, S., Krause, P., Flügel, W.-A., Fink, M., and Fischer, C.: Understanding the hydrological system dynamics of a glaciated alpine catchment in the Himalayan region using the J2000 hydrological model, Hydrol. Process., 28, 1329–1344, https://doi.org/10.1002/hyp.9627, 2014.

Nicolle, P., Pushpalatha, R., Perrin, C., François, D., Thiéry, D., Mathevet, T., Le Lay, M., Besson, F., Soubeyroux, J.-M., Viel, C., Regimbeau, F., Andréassian, V., Maugis, P., Augeard, B., and Morice, E.: Benchmarking hydrological models for low-flow simulation and forecasting on French catchments, Hydrol. Earth Syst. Sci., 18, 2829–2857, https://doi.org/10.5194/hess-18-2829-2014, 2014.

Pedroso, R., Tran, D. H., Thi, M. H. N., van Le, A., Ribbe, L., Dang, K. T., and Le, K. P.: Cropping systems in the Vu Gia Thu Bon river basin, Central Vietnam: On farmers' stubborn persistence in predominantly cultivating rice, NJAS – Wageningen J. Life Sci., 80, 1–13, https://doi.org/10.1016/j.njas.2016.11.001, 2016.

Pfenning, B., Kipka, H., Fink, M., Krause, P., and Flügel, W. A.: Development of an extended spatially distributed routing scheme and its impact on process oriented hydrological modelling results, in: Joint IAHS & IAH International Convention, Hyderabad, India, 37–43, 2009.

Pilgrim, D. H., Chapman, T. G., and Doran, D. G.: Problems of rainfall-runoff modelling in arid and semiarid regions, Hydrol. Sci. J., 33, 379–400, https://doi.org/10.1080/02626668809491261, 1988.

PPC: Master Plan for Electricity Development in Quang Nam Province, Period of 2006–2010 Towards 2015, Provincial Peoples Committee, Quangnam, 2006.

Prudhomme, C., Parry, S., Hannaford, J., Clark, D. B., Hagemann, S., and Voss, F.: How Well Do Large-Scale Models Reproduce Regional Hydrological Extremes in Europe?, J. Hydrometeor, 12, 1181–1204, https://doi.org/10.1175/2011JHM1387.1, 2011.

Quangnam Statistical Office: Statistical Yearbook of Quang Nam 2010–2014, Statistical Publishing House, Hanoi, Vietnam, 2014.

Räsänen, T. A., Koponen, J., Lauri, H., and Kummu, M.: Downstream Hydrological Impacts of Hydropower Development in the Upper Mekong Basin, Water Resour. Manag., 26, 3495–3513, https://doi.org/10.1007/s11269-012-0087-0, 2012.

Räsänen, T. A., Someth, P., Lauri, H., Koponen, J., Sarkkula, J., and Kummu, M.: Observed river discharge changes due to hydropower operations in the Upper Mekong Basin, J. Hydrol., 545, 28–41, https://doi.org/10.1016/j.jhydrol.2016.12.023, 2017.

Ribbe, L., Viet, T. C., Firoz, A., Nguyen, A. T., Nguyen, U., and Nauditt, A.: Integrated River Basin Management in the Vu Gia Thu Bon Basin, in: Land Use and Climate Change Interactions in Central Vietnam: LUCCi, edited by: Nauditt, A. and Ribbe, L., Water Resources Management and Development, Springer Book Series, 2017.

Rossi, A., Massei, N., Laignel, B., Sebag, D., and Copard, Y.: The response of the Mississippi River to climate fluctuations and reservoir construction as indicated by wavelet analysis of streamflow and suspended-sediment load, 1950–1975, J. Hydrol., 377, 237–244, https://doi.org/10.1016/j.jhydrol.2009.08.032, 2009.

Santer, B. D., Mears, C., Doutriaux, C., Caldwell, P., Gleckler, P. J., Wigley, T. M. L., Solomon, S., Gillett, N. P., Ivanova, D., Karl, T. R., Lanzante, J. R., Meehl, G. A., Stott, P. A., Taylor, K. E., Thorne, P. W., Wehner, M. F., and Wentz, F. J.: Separating signal and noise in atmospheric temperature changes: The importance of timescale, J. Geophys. Res., 116, D22105, https://doi.org/10.1029/2011JD016263, 2011.

Seibert, J. and McDonnell, J. J.: Land-cover impacts on streamflow: A change-detection modelling approach that incorporates parameter uncertainty, Hydrol. Sci. J., 55, 316–332, https://doi.org/10.1080/02626661003683264, 2010.

Song, W.-Z., Jiang, Y.-Z., Lei, X.-H., Wang, H., and Shu, D.-C.: Annual runoff and flood regime trend analysis and the relation with reservoirs in the Sanchahe River Basin, China, Quatern. Int., 380–381, 197–206, https://doi.org/10.1016/j.quaint.2015.01.049, 2015.

Souvignet, M., Laux, P., Freer, J., Cloke, H., Thinh, D. Q., Thuc, T., Cullmann, J., Nauditt, A., Flügel, W.-A., Kunstmann, H., and Ribbe, L.: Recent climatic trends and linkages to river discharge in Central Vietnam, Hydrol. Process., 28, 1587–1601, https://doi.org/10.1002/hyp.9693, 2013.

Sung, J. H. and Chung, E.-S.: Development of streamflow drought severity-duration-frequency curves using the threshold level method, Hydrol. Earth Syst. Sci., 18, 3341–3351, https://doi.org/10.5194/hess-18-3341-2014, 2014.

Tallaksen, M. L., Madsen, H., and Clausen, B.: On the definition and modelling of streamflow drought duration and deficit volume, Hydrol. Sci. J., 42, 15–33, https://doi.org/10.1080/02626669709492003, 2009.

Tang, J., Yin, X.-A., Yang, P., and Yang, Z.: Assessment of Contributions of Climatic Variation and Human Activities to Streamflow Changes in the Lancang River, China, Water Resour. Manag., 28, 2953–2966, https://doi.org/10.1007/s11269-014-0648-5, 2014.

Tesfa, T. K., Li, H.-Y., Leung, L. R., Huang, M., Ke, Y., Sun, Y., and Liu, Y.: A subbasin-based framework to represent land surface processes in an Earth system model, Geosci. Model Dev., 7, 947–963, https://doi.org/10.5194/gmd-7-947-2014, 2014.

Trenberth, K. E.: Attribution of climate variations and trends to human influences and natural variability, WIREs Clim. Change, 2, 925–930, https://doi.org/10.1002/wcc.142, 2011.

van Huijgevoort, M. H. J., Hazenberg, P., van Lanen, H. A. J., and Uijlenhoet, R.: A generic method for hydrological drought identification across different climate regions, Hydrol. Earth Syst. Sci., 16, 2437–2451, https://doi.org/10.5194/hess-16-2437-2012, 2012.

van Lanen, H. A., Laaha, G., Kingston, D. G., Gauster, T., Ionita, M., Vidal, J.-P., Vlnas, R., Tallaksen, L. M., Stahl, K., Hannaford, J., Delus, C., Fendekova, M., Mediero, L., Prudhomme, C., Rets, E., Romanowicz, R. J., Gailliez, S., Wong, W. K., Adler, M.-J., Blauhut, V., Caillouet, L., Chelcea, S., Frolova, N., Gudmundsson, L., Hanel, M., Haslinger, K., Kireeva, M., Osuch, M., Sauquet, E., Stagge, J. H., and van Loon, A. F.: Hydrology needed to manage droughts: The 2015 European case, Hydrol. Process., 30, 3097–3104, https://doi.org/10.1002/hyp.10838, 2016.

van Loon, A. F., Gleeson, T., Clark, J., van Dijk, A. I. J. M., Stahl, K., Hannaford, J., Di Baldassarre, G., Teuling, A. J., Tallaksen, L. M., Uijlenhoet, R., Hannah, D. M., Sheffield, J., Svoboda, M., Verbeiren, B., Wagener, T., Rangecroft, S., Wanders, N., and van Lanen, H. A. J.: Drought in the Anthropocene, Nat. Geosci., 9, 89–91, https://doi.org/10.1038/ngeo2646, 2016.

Van Loon, A. F. and Van Lanen, H. A. J.: A process-based typology of hydrological drought, Hydrol. Earth Syst. Sci., 16, 1915–1946, https://doi.org/10.5194/hess-16-1915-2012, 2012.

Vörösmarty, C. J., Green, P., Salisbury, J., and Lammers, R. B.: Global Water Resources: Vulnerability from Climate Change and Population Growth, Science, 289, 284–288, https://doi.org/10.1126/science.289.5477.284, 2000.

Wada, Y., van Beek, L. P. H., Wanders, N., and Bierkens, M. F. P.: Human water consumption intensifies hydrological drought worldwide, Environ. Res. Lett., 8, 34036, https://doi.org/10.1088/1748-9326/8/3/034036, 2013.

Wagner, T., Themeßl, M., Schüppel, A., Gobiet, A., Stigler, H., and Birk, S.: Impacts of climate change on stream flow and hydro power generation in the Alpine region, Environ. Earth Sci., 76, 1–22, https://doi.org/10.1007/s12665-016-6318-6, 2017.

Wang, D. and Hejazi, M.: Quantifying the relative contribution of the climate and direct human impacts on mean annual streamflow in the contiguous United States, Water Resour. Res., 47, W00J12, https://doi.org/10.1029/2010WR010283, 2011.

Wang, H., Chen, L., and Yu, X.: Distinguishing human and climate influences on streamflow changes in Luan River basin in China, CATENA, 136, 182–188, https://doi.org/10.1016/j.catena.2015.02.013, 2015.

Wang, S., Yan, M., Yan, Y., Shi, C., and He, L.: Contributions of climate change and human activities to the changes in runoff increment in different sections of the Yellow River, Quatern. Int., 282, 66–77, https://doi.org/10.1016/j.quaint.2012.07.011, 2012.

Ye, B., Yang, D., and Kane, D. L.: Changes in Lena River streamflow hydrology: Human impacts versus natural variations, Water Resour. Res., 39, 1200, https://doi.org/10.1029/2003WR001991, 2003.

Zarfl, C., Lumsdon, A. E., Berlekamp, J., Tydecks, L., and Tockner, K.: A global boom in hydropower dam construction, Aquat. Sci., 77, 161–170, https://doi.org/10.1007/s00027-014-0377-0, 2015.

Zhang, A., Zhang, C., Fu, G., Wang, B., Bao, Z., and Zheng, H.: Assessments of Impacts of Climate Change and Human Activities on Runoff with SWAT for the Huifa River Basin, Northeast China, Water Resour. Manag., 26, 2199–2217, https://doi.org/10.1007/s11269-012-0010-8, 2012.

Zhang, R., Zhang, S.-H., Xu, W., Wang, B.-D., and Wang, H.: Flow regime of the three outlets on the south bank of Jingjiang River, China: An impact assessment of the Three Gorges Reservoir for 2003–2010, Stoch. Environ. Res. Risk Assess., 29, 2047–2060, https://doi.org/10.1007/s00477-015-1121-6, 2015.

Zhang, X., Zwiers, F. W., Hegerl, G. C., Lambert, F. H., Gillett, N. P., Solomon, S., Stott, P. A., and Nozawa, T.: Detection of human influence on twentieth-century precipitation trends, Nature, 448, 461–465, https://doi.org/10.1038/nature06025, 2007.

Zhou, Y., Zhang, Q., Li, K., and Chen, X.: Hydrological effects of water reservoirs on hydrological processes in the East River (China) basin: Complexity evaluations based on the multi-scale entropy analysis, Hydrol. Process., 26, 3253–3262, https://doi.org/10.1002/hyp.8406, 2012.

8

An adaptive two-stage analog/regression model for probabilistic prediction of small-scale precipitation in France

Jérémy Chardon, Benoit Hingray, and Anne-Catherine Favre

Univ. Grenoble Alpes, CNRS, IRD, Grenoble INP, IGE, 38000 Grenoble, France

Correspondence: Benoit Hingray (benoit.hingray@univ-grenoble-alpes.fr)

Abstract. Statistical downscaling models (SDMs) are often used to produce local weather scenarios from large-scale atmospheric information. SDMs include transfer functions which are based on a statistical link identified from observations between local weather and a set of large-scale predictors. As physical processes driving surface weather vary in time, the most relevant predictors and the regression link are likely to vary in time too. This is well known for precipitation for instance and the link is thus often estimated after some seasonal stratification of the data. In this study, we present a two-stage analog/regression model where the regression link is estimated from atmospheric analogs of the current prediction day. Atmospheric analogs are identified from fields of geopotential heights at 1000 and 500 hPa. For the regression stage, two generalized linear models are further used to model the probability of precipitation occurrence and the distribution of non-zero precipitation amounts, respectively. The two-stage model is evaluated for the probabilistic prediction of small-scale precipitation over France. It noticeably improves the skill of the prediction for both precipitation occurrence and amount. As the analog days vary from one prediction day to another, the atmospheric predictors selected in the regression stage and the value of the corresponding regression coefficients can vary from one prediction day to another. The model allows thus for a day-to-day adaptive and tailored downscaling. It can also reveal specific predictors for peculiar and non-frequent weather configurations.

1 Introduction

Statistical downscaling models (SDMs) have been widely used to generate local weather scenarios for past or future climates from outputs of climate models (e.g., Wilby et al., 1999; Hanssen-Bauer et al., 2005; Boé et al., 2007; Lafaysse et al., 2014) and to produce local weather forecasts from outputs of numerical weather prediction models (e.g., Obled et al., 2002; Gangopadhyay et al., 2005; Marty et al., 2012; Ben Daoud et al., 2016). For recent years, they have also been used to reconstruct past weather conditions from atmospheric reanalysis data (e.g., Kuentz et al., 2015; Caillouet et al., 2016).

Among the different SDM approaches presented over the last decades (see Maraun et al., 2010, for a review), perfect prognosis SDMs make use of the physical relationships that exist between some large-scale atmospheric parameters and local weather variables. Local weather scenarios can then be produced for any prediction day, conditional on the large-scale atmospheric configuration observed or simulated for this day, where the "prediction day" refers here to some future, past, or present simulation day, depending on the application context at hand (e.g., forecasting, simulation, reconstruction). Perfect prognosis SDMs include transfer functions, weather-type-based models, and methods based on atmospheric analogs. In the latter case, atmospheric analog days of the current prediction day are searched for on the basis of some atmospheric similarity criterion in the historical database. The weather variables observed for the most similar day, for one similar day chosen randomly, or for a selection of the k-most similar days are then used as a weather scenario for the prediction day of interest (e.g., Dayon et al., 2015; Raynaud et al., 2016).

Transfer functions mainly consist of regression models where the expected value of the predictand for time t is expressed as a linear or non-linear function of a set of predictors. For precipitation, the regression can be achieved with multiple linear regressions or generalized linear models (GLMs) which extend the linear regression to non-Gaussian data (e.g., Nelder and Wedderburn, 1972; Stern and Coe, 1984; Asong et al., 2016). Transfer functions can also make use of classification and regression tree (CART) (e.g., Gaitan et al., 2014) artificial neural networks or least squares support vector machines (e.g., Campozano et al., 2016).

The downscaling relationship used in transfer functions is usually established empirically between a selection of large-scale predictors and the predictand (e.g., precipitation occurrence) from a set of observations available for recent decades. As physical processes driving surface weather vary in time, the most relevant predictors and the downscaling link are however expected to vary in time too. When inferred from all observations available for a given period, the downscaling relationship – which is thus likely inferred from a heterogeneous ensemble of weather configurations – is consequently likely to be sub-optimal. To reduce this potential limitation, the parameterization of the relationship is often estimated after some data stratification. In the usual calendar stratification, one parameter set is for instance optimized for each calendar month or season (e.g., Nasseri et al., 2013). The stratification can also be based on some weather-type information. In this case, a set of parameters is usually estimated for each weather type of a given pre-established weather-type classification (e.g., Enke and Spegat, 1997). Often applied, this weather-type-based approach is expected to allow for a better identification of the most important driving large-scale variables and consequently for a more relevant downscaling. An obvious limitation however remains for prediction days that do not clearly belong to one specific weather type (e.g., prediction days that are close to the "weather frontiers" delimiting two or more weather types). Those days are indeed likely to be rather dissimilar to the weather configurations that each weather type is expected to characterize, making the downscaling relationships to be used not suited anymore or, at least, sub-optimal.

A smoother weather-type-like approach consists in defining the weather type from all atmospheric situations that are similar to the situation of the prediction day. The ensemble of days from which the downscaling link can be identified is thus expected to be rather homogeneous and to rather well inform the large- to small-scale link sought for the considered prediction day. This is in turn expected to make the link stronger and to improve the prediction (e.g., Woodcock, 1980). Such an approach can actually be achieved by iden-

tifying for each prediction day the transfer function from atmospheric analogs of that day. To our knowledge, this two-stage downscaling approach, combining in turn the two popular analog and transfer function methods, has only been explored in a few previous studies. In Ribalaygua et al. (2013), it was found to improve the probabilistic prediction of local surface temperature in the Spanish Iberian Peninsula. The multiple linear regression of the regression stage, estimated from the 150 most similar atmospheric analogs of the prediction day of interest, uses forward and backward stepwise selection of predictors from a set of four potential predictors (thickness of the air column and three temperature indexes of previous days). For precipitation, the authors did not test the potential of the two-stage combination, building directly the predictions from the precipitation observations of the 30 most similar atmospheric analogs. In the deterministic approach presented by Ibarra-Berastegi et al. (2011), incorporating the regression stage (with 79 potential atmospheric predictors) was found to allow a clear though not overwhelming improvement of precipitation prediction over the simple analog-based predictions. A multiple linear regression model was also applied here for the regression stage.

In the present study, we present a two-stage analog/regression downscaling model for the probabilistic prediction of small-scale daily precipitation: for each prediction day, the statistical downscaling link between some large-scale atmospheric predictors and small-scale precipitation is estimated from large-scale and local-scale observations available from an ensemble of days which are atmospheric analogs to the prediction day. The analog model (AM) used for the analog stage is based on developments from different studies initially focusing on the probabilistic quantitative precipitation forecasts in southern France (e.g., Bontron and Obled, 2005; Marty et al., 2012) and extended to the prediction of precipitation on larger spatial domains (e.g., Chardon et al., 2014). The statistical distribution of daily precipitation is strongly non-Gaussian with a non-negligible mass in zero (corresponding to the probability of a dry day), and a skewed distribution for non-zero daily amounts. For the regression stage, we thus do not use a multiple linear regression model as in previous studies, but a two-part GLM approach where the probability of precipitation occurrence and the distribution of wet-day amounts are modeled separately following Chandler and Wheater (2002) and Mezghani and Hingray (2009). Conversely to the work of Ibarra-Berastegi et al. (2011), this allows prediction of the full distribution of precipitation, including the probability of a wet day. In this two-stage analog/regression approach, the analogs change from one prediction day to the other. This makes the statistical downscaling link potentially adaptive; i.e., the predictors and the regressions parameters are likely to vary from one day to the other.

As mentioned above, SDMs are used for the simulation of local weather scenarios in different contexts, e.g., local weather forecasts, reconstructions, or climate impact studies. No specific context is considered here and the two-stage model could be further considered for either forecasting, reconstruction, or future projections. Depending on its intended use, some specific issues would obviously apply, calling for specific focused analyses and developments. For instance, the large-scale atmospheric parameters to be considered as predictors would depend on the dataset considered (e.g., atmospheric reanalyses, climate models, or numerical weather prediction models) as a result of their intrinsic quality (e.g., Caillouet et al., 2016). The development of climate projections would require one to check the temporal transferability of the model in a modified climate context and would thus likely also condition the selection of the predictors as highlighted by Dayon et al. (2015). These context-specific issues are not considered here. Our main objectives are to present the principles of the two-stage analog/regression approach developed for the prediction of small-scale precipitation, to assess its predictive power for both precipitation occurrence and amount, and to give some insight into its adaptive behavior and thus into the temporal variability of the downscaling link. For this, we explore the model skill and behavior for the prediction of daily precipitation for a large number of sites in France.

The paper is structured as follows: Sect. 2 describes the data and Sect. 3 the two-stage downscaling model. Section 4 presents the skill of the model for the prediction of both precipitation occurrence and amount. The adaptive behavior of the model is considered in Sects. 5 and 6.

2   Data

The predictand is the daily small-scale precipitation estimated for the 1982–2001 period over 8981 grid cells of $8 \times 8\,\mathrm{km}^2$ covering the continental French territory. The predictand is "local" precipitation, i.e., precipitation at a given grid cell. Each of the 8981 grid cells is thus considered in turn in the following independently of the other cells. In other words, the predictions do not target precipitation fields. Small-scale precipitation data are obtained from the SAFRAN analysis produced for several surface variables at an hourly time step by MeteoFrance (Quintana-Segui et al., 2008; Vidal et al., 2010). SAFRAN precipitation estimates are obtained each day from the closest measurement stations. They are considered as pseudo-observations.

Atmospheric predictors are taken from the European Centre for Medium-Range Weather Forecasts (ECMWF) Re-Analysis (ERA-40, Uppala et al., 2005). This global meteorological re-analysis is available on a $1.125° \times 1.125°$ grid with a 6-hourly temporal resolution.

For the analog stage, predictors are the 1000 and 500 hPa geopotential height fields over a large spatial domain (roughly lat $= 10°$, lon $= 8°$) centered on the target location. These predictors have been found to be the most informative large-scale predictors to be used in this context for France (e.g., Guilbaud and Obled, 1998; Obled et al.,

Table 1. Large-scale potential variables considered in the work. Stars: predictors obtained from the best GLMs identified for the 12 test SAFRAN grid cells (Sect. 2). Double stars: predictors used for the analog stage. Bold text: predictors retained for the SCAMP version presented and evaluated in this work. See Holton and Hakim (2012) for the definition of the variables.

Acronym	Predictor description
R_{850}	*Relative humidity at 850 hPa
R_{700}	**Relative humidity at 700 hPa**
R_{500}	Relative humidity at 500 hPa
TCW	Total column water
R_{850}TCW	*Product of R_{850} and TCW
T_{700}	**Air temperature at 700 hPa**
B_{700}	*Baroclinity at 700 hPa
Δ_z	700–1000 hPa thickness of the air column
Z_{1000}	**Geopotential height at 1000 hPa**
Z_{700}	Geopotential height at 700 hPa
Z_{500}	**Geopotential height at 500 hPa**
F_{700}	Wind speed at 700 hPa
U_{700}	Western component of wind speed at 700 hPa
V_{700}	Southern component of wind speed at 700 hPa
W_{700}	**Vertical velocity (vertical component of wind speed) at 700 hPa**
H	**Helicity of horizontal wind integrated from 1000 to 500 hPa**
PV_{400}	*Potential vorticity of the atmosphere at 400 hPa
$\Delta\theta$	*Potential temperature gradient between 925 and 700 hPa
FR_{700}	*Humidity flux at 700 hPa
FU_{700}	Western component of humidity flux at 700 hPa
FV_{700}	Southern component of humidity flux at 700 hPa
∇FR_{700}	*Divergence of FR_{700}
Occ -1	**Precipitation occurrence of the day before the prediction day**

2002; Radanovics et al., 2013). They also correspond to the best large-scale predictors of daily precipitation for different regions in Europe with contrasted meteorological regimes (Raynaud et al., 2016).

For the regression stage, 22 other predictors were also considered. The selection gathers most predictors considered in previous studies over Europe (e.g., Hanssen-Bauer et al., 2005; Wetterhall et al., 2009; Horton et al., 2012; Raynaud et al., 2016). They include predictors characterizing the thermal state of the atmosphere, its dynamics, the atmospheric water content, and its thermo-dynamical instability (see Table 1). As potential predictor, we also consider the occurrence of precipitation on the previous day. All predictors here are scalar variables. Atmospheric predictors are estimated on a daily time step (mean of the four values available at 06:00, 12:00, 18:00, and 24:00 UTC) from the four ERA-40 grid cells surrounding the prediction grid cell (inverse distance interpolation).

To avoid the multi-colinearity in the predictors for the regression, we identified a subset of uncorrelated predictors. The cross-correlations between all predictor pairs were first estimated on an annual basis from all available data. The correlation structure can however differ from one atmospheric configuration to the other. The set of uncorrelated predictors could thus differ from one prediction day to the other. We thus repeated the correlation analysis for each prediction day, using for this estimation the predictor values observed for the 100 nearest atmospheric analogs identified for this day. The main features of the inter-variable correlations were found to be roughly independent of the day (not shown). The final subset of uncorrelated predictors is highlighted in Table 1. These predictors are tested for the prediction of both precipitation occurrence and amount.

A large number of different possible predictor sets can be built from these predictors. In the present work, for the sake of robustness, we consider that a maximum of four predictors can be integrated into a given regression model. Predictors are obviously expected to be both day and location specific. In the present work, for the sake of simplicity and readability, we select them from a unique set of four potential predictors. This allows us to reduce the degrees of freedom in the model and to better highlight its skill and adaptive behavior.

For each predictand, the set of the four potential predictors was selected as follows. For 12 SAFRAN grid cells uniformly distributed over the French territory, we first identified with a standard iterative forward/backward algorithm the four-predictor set which leads to the best prediction skill for the all-days configuration. From the 12 different sets obtained, respectively, for the 12 grid cells, we finally retained the set which leads on average to the best prediction skill for the 8981 SAFRAN grid cells.

For precipitation occurrence, this best four-predictor set is constituted from the relative humidity R_{700}, the helicity H, the vertical velocity of the air at 700 hPa, W_{700}, and the precipitation occurrence $Occ - 1$ of the day before the prediction day. For precipitation amount, the best four-predictor set is similar except that the occurrence of the previous day $Occ - 1$ is replaced by the 700 hPa air temperature T_{700}. Note that the selection of predictors R_{700}, W_{700}, and T_{700} is consistent with results of several past studies in the region (e.g., Ben Daoud et al., 2016).

Predictors considered for the analog and regression stages obviously inform about different features of the atmosphere state for different scales. Geopotential fields, by their spatial extent, characterize the large-scale atmospheric circulation configuration (the spatial domain of several thousands of kilometers includes a part of the northeastern Atlantic and covers France and a part of the neighboring countries), whereas scalar predictors used in the regression stage are descriptive of a more local (and mostly thermodynamic) state of the atmosphere (the spatial domain of several hundreds of kilometers is roughly centered above the target location).

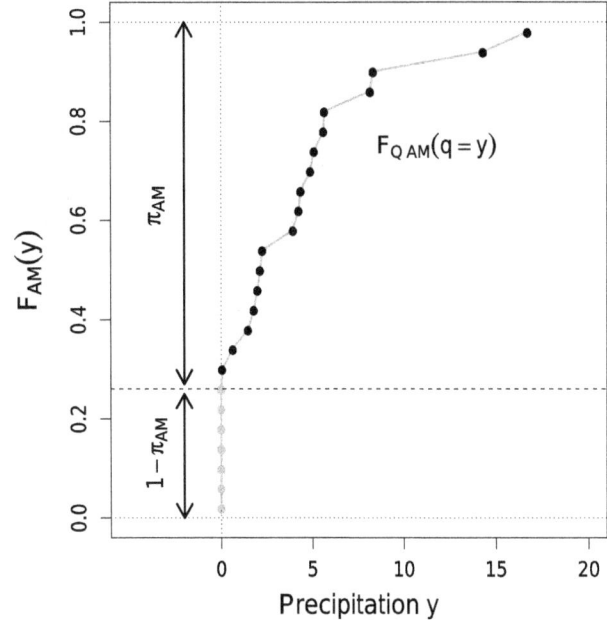

Figure 1. Cumulative distribution function (cdf) of the precipitation amount for a given prediction day (in gray) at a given grid cell. For illustration, the prediction here corresponds to the empirical cdf achieved with the analog model (AM) mentioned in Sect. 3.3. The contribution of the precipitation amount $F_{Q,\mathrm{AM}}$ cdf to the overall cdf is highlighted in black (cf. Eq. 1).

3 The two-stage analog/regression model (SCAMP)

As illustrated in Fig. 1, the cumulative distribution function (cdf) F_Y of precipitation Y at a given site (grid cell) can be expressed for any given day as the composition of the no-precipitation occurrence probability $1 - \pi$ and the cdf F_Q of the precipitation amount Q for non-zero precipitation:

$$F_Y(y) = (1 - \pi) + \pi \cdot F_Q(q = y), \tag{1}$$

where π is the precipitation occurrence probability, and y and q correspond to the precipitation value with regard to the whole precipitation distribution and to the non-zero precipitation distribution, respectively.

In the present work, the cdf of precipitation is modeled for each grid cell and each prediction day with GLMs (Coe and Stern, 1982; Stern and Coe, 1984), estimated for this specific day from atmospheric analogs of the day. The probability of precipitation occurrence and the cdf of the non-zero precipitation amount are modeled separately.

In the following, we first describe the AM used to identify atmospheric analog days (Sect. 3.1) and the GLMs applied in the regression stage (Sect. 3.2).

As discussed later, one can face prediction days where the regression stage fails, i.e., where the regression parameters are not significantly different from zero at the chosen significance level ($\alpha = 5\%$). For such days, we use the analog model as a backup prediction model. The backup model can be used for precipitation occurrence probability, for non-zero precipitation amount, or for both predictands simultaneously.

The way these different models are combined to finally give, for the current prediction day, a probabilistic prediction of precipitation, is presented in Sect. 3.3. In the following, this two-stage analog/regression model is further referred to as SCAMP (SCAMP stands for Sequential Constructive atmospheric Analogs for Multivariate weather Prediction and refers to the model presented by Raynaud et al., 2016, for the multivariate prediction of precipitation/temperature/radiation/wind).

3.1 Atmospheric analogs

The atmospheric analog days retained for the regression stage are identified with an analog model defined from the developments of several past studies in France (e.g., Obled et al., 2002; Marty et al., 2012; Chardon et al., 2014).

For any given prediction day (e.g., 31 May 2018), the analog days retained for the regression are the N_d days that are most similar to that day in terms of large-scale atmospheric circulation. The similarity is assessed using the Teweless–Wobus score (TWS, Teweless and Wobus, 1954) applied to the geopotential height at 1000 and 500 hPa at 12:00 and 24:00 UTC, respectively. The TWS compares the shapes of geopotential fields, and thus informs on the localization of low- and high-pressure systems and on the origin of air masses. Note that the N_d analog days are identified within a restricted pool of candidate days, namely all days of the archive that are included in a calendar window of ± 30 calendar days centered on the prediction day (for the prediction of 31 May 2018, the candidates are all 1 May to all 30 June from all years of the archive). The prediction day (e.g., 31 May 2018) and its 5 preceding and following days are excluded from the candidates. In the present work, the archive period corresponds to 1982–2001 (20 years), and we used the 100 nearest atmospheric analog days to estimate the GLMs in the regression stage.

Following Chardon et al. (2014), the domain considered to estimate the atmospheric similarity was optimized for each target location. A different analog model was thus considered for each of the 8981 SAFRAN grid cells. For each prediction day, the analog days thus likely differ from one SAFRAN grid cell to the next (see Chardon et al., 2014, for an illustration).

3.2 Regression stage with GLMs

The cdf of precipitation is then modeled for each prediction day with GLMs estimated for this specific day from the atmospheric analogs of the day. GLMs make the cdf, depending on some covariates, atmospheric predictors in the present case. For each prediction day, the probability of precipitation occurrence π was modeled with a GLM in the form of a lo-

gistic regression as

$$\log\left(\frac{\pi}{1-\pi}\right) = x^{oT}\boldsymbol{\beta}^o, \qquad (2)$$

where x^o is the scalar vector of the K_o predictors $(x_1^o, x_2^o, \ldots x_{K_o}^o)$ and $\boldsymbol{\beta}^o$ the scalar vector of the K_o corresponding regression coefficients $(\beta_1^o, \beta_2^o, \ldots \beta_{K_o}^o)$.

For the non-zero precipitation amount, we used a GLM with the gamma distribution and the log link function. The expected amount μ of non-zero precipitation is therefore here expressed as

$$\log(\mu) = x^{qT}\boldsymbol{\beta}^q, \qquad (3)$$

where x^q denotes the scalar vector of the K_q predictors $(x_1^q, x_2^q, \ldots x_{K_q}^q)$ and $\boldsymbol{\beta}^q$ the scalar vector of the corresponding regression coefficients $(\beta_1^q, \beta_2^q, \ldots \beta_{K_q}^q)$. The shape parameter ν of the gamma distribution is computed from the variance σ^2 of non-zero precipitation amounts estimated from Pearson's residuals (McCullagh and Nelder, 1989) as

$$\sigma^2 = \frac{1}{\{N_q - (K_q + 1)\}} \sum_{i=1}^{N_q} \frac{(q_i - \mu)^2}{\mu^2}, \qquad (4)$$

where N_q is the number of non-zero precipitation data q_i considered in the analysis. As the shape parameter ν equals the inverse of the variance $1/\sigma^2$, the gamma distribution F_Q modeling the precipitation amount thus follows a gamma distribution of this type: $\Gamma(\nu, \alpha = \mu/\nu)$.

For any given prediction day, the estimation of both GLM models practically proceeds as follows.

- The precipitation state (wet or dry), the precipitation amount, and the values of the different potential predictors are extracted for the N_d nearest analogs of the day. The precipitation state of a given day is considered to be wet if the precipitation amount for this day is higher than or equal to 0.1 mm. It is described with a binary precipitation occurrence variable **O**, set to 1 for the wet case, and 0 for the dry case.

- For occurrence probability, different sets of predictors are considered in turn. For each set, the parameters of the occurrence GLM are estimated from the predictors/occurrence values available for the N_d analogs.

- For precipitation amount, different sets of predictors are again considered in turn. For each set, the parameters of the GLM are estimated from the predictors/amount values available for the analog days which are wet (N_q, the number of days considered here for the regression, is therefore smaller than or equal to N_d, and varies a priori from one target day to another).

For the considered prediction day, the different sets of predictors considered in turn are built from the four potential

Table 2. Possible regressive structures (i.e., a combination of predictors) for the modeling of precipitation occurrence and amount.

Structure index	Precipitation occurrence	Precipitation amount
Str. no. 1	R_{700}	R_{700}
Str. no. 2	H	H
Str. no. 3	W_{700}	W_{700}
Str. no. 4	$\mathrm{Occ} - 1$	T_{700}
Str. no. 5	$R_{700} + H$	$R_{700} + H$
Str. no. 6	$R_{700} + W_{700}$	$R_{700} + W_{700}$
Str. no. 7	$R_{700} + \mathrm{Occ} - 1$	$R_{700} + T_{700}$
Str. no. 8	$H + W_{700}$	$H + W_{700}$
Str. no. 9	$H + \mathrm{Occ} - 1$	$H + T_{700}$
Str. no. 10	$W_{700} + \mathrm{Occ} - 1$	$W_{700} + T_{700}$
Str. no. 11	$R_{700} + H + W_{700}$	$R_{700} + H + W_{700}$
Str. no. 12	$R_{700} + H + \mathrm{Occ} - 1$	$R_{700} + H + T_{700}$
Str. no. 13	$R_{700} + W_{700} + \mathrm{Occ} - 1$	$R_{700} + W_{700} + T_{700}$
Str. no. 14	$H + W_{700} + \mathrm{Occ} - 1$	$H + W_{700} + T_{700}$
Str. no. 15	$R_{700} + H + W_{700} + \mathrm{Occ} - 1$	$R_{700} + H + W_{700} + T_{700}$

predictors identified in the preliminary work (cf. Sect. 2). For occurrence probability (or precipitation amount), the four potential predictors actually allow us to build 15 different sets of predictors, further denoted as "regressive structures" in the following (cf. list in Table 2). For each regressive structure, the regression coefficients of corresponding GLMs are estimated using the iterative re-weighted least squares algorithm (IRLS, Nelder and Wedderburn, 1972). The prediction skills of the different regressive structures are then compared and the regressive structure (predictor set) which minimizes the Bayesian information criterion is retained for the prediction (Schwarz, 1978; Akaike, 1974) (only the regressive structures for which all coefficients are significant at a 5 % level are compared; the significance is estimated with the Z test (or the Student's t test)).

The prediction of the occurrence probability (or the expected precipitation amount) for the prediction day is finally obtained from the best occurrence (or amount) GLM, using the values of the predictors observed for that prediction day. The final distribution of precipitation F_Y is obtained by combining the issued occurrence probability π and the amount distribution F_Q according to Eq. (1).

3.3 The analog model as a benchmark and backup prediction model

The N_d nearest analog days identified with the AM can also be directly used, without a further regression stage, for a probabilistic prediction. In the following, we also consider predictions obtained with the 25 nearest analog days (for the AM considered here, 25 was found to give the best prediction skill for France by Chardon et al., 2014). In this case, the precipitation cdf for the prediction day is simply the empirical distribution of the precipitation values observed for these 25 analogs. The predictions obtained with this analog

model, further called AM_{25}, are used as a benchmark to assess the prediction skill of the two-stage analog/regression approach. In addition, they were used as a backup prediction for days for which the regression stage failed in the two-stage approach. One can actually face the situation where no GLM satisfies the significance conditions required for the regression coefficients. This can occur for precipitation occurrence probability, for non-zero precipitation amount, or for both predictands simultaneously. In such cases, AM_{25} is applied as a backup prediction model.

If the significance conditions cannot be satisfied for the precipitation occurrence GLM, the occurrence probability π is set to that obtained with AM_{25}. It thus simply corresponds to the empirical probability $\pi_{\mathrm{AM}_{25}}$ of precipitation occurrence derived from the 25 analog days of AM_{25} as

$$\pi \equiv \pi_{\mathrm{AM}_{25}} = \frac{1}{25} \sum_{i=1}^{25} o_i. \tag{5}$$

Similarly, if the significance conditions cannot be satisfied for the precipitation amount GLM, the distribution F_Q is estimated with the empirical distribution $F_{Q,\mathrm{AM}_{25}}$ derived with AM_{25} as

$$F_Q(q) \equiv F_{Q,\mathrm{AM}_{25}}(q) = \frac{F_{\mathrm{AM}_{25}}(q) - \left(1 - \pi_{\mathrm{AM}_{25}}\right)}{\pi_{\mathrm{AM}_{25}}}, \tag{6}$$

where $F_{\mathrm{AM}_{25}}$ corresponds to the empirical cdf estimated from all precipitations (null and positive) related to the 25 analog days. Note also that if the number N_q of humid analog days is low ($N_q < 10$), the estimation of a GLM is not expected to be robust. When this case appears, F_Q is also set to the cdf obtained with AM_{25}.

As illustrated in Fig. 2, four prediction cases are thus achieved with the two-stage approach. They correspond, respectively, to cases where AM_{25} is used to back up the prediction of the whole precipitation distribution (case 1), where

3.4 Model evaluation

The prediction skill of the downscaling model is assessed with probabilistic scores usually used to evaluate ensemble prediction systems (EPSs). Let us consider a given EPS, denoted as $\mathcal{P}$.

The Brier score (Brier, 1950; Murphy, 1973) first evaluates the ability of EPS $\mathcal{P}$ to predict precipitation occurrence. When estimated over M prediction days, the mean Brier score $\overline{\mathrm{BS}}$ reads as

$$\overline{\mathrm{BS}} = \frac{1}{M} \sum_{i=1}^{M} [p_i - o_i]^2, \tag{7}$$

where, for a given prediction day i, p_i is the occurrence probability issued by EPS $\mathcal{P}$ and o_i is the effective precipitation occurrence for this day ($o_i = 1$ for a wet day, $= 0$ otherwise).

The ability of EPS $\mathcal{P}$ to estimate the precipitation amount is evaluated with the continuous ranked probability score (CRPS, Brown, 1974; Matheson and Winkler, 1976). When estimated over M prediction days, the mean CRPS reads as

$$\overline{\mathrm{CRPS}} = \frac{1}{M} \sum_{i=1}^{M} \int_{-\infty}^{+\infty} [F_i(x) - H_{y_i}(x)]^2 dx, \tag{8}$$

where, for a given prediction day i, H_{y_i} and F_i denote, respectively, the cdf of the observation y_i and the cdf derived from EPS $\mathcal{P}$. x denotes the predictand quantiles of the cdfs. Note that H_{y_i} corresponds to the Heaviside function where $H_{y_i} = 1$ if $x \geq y_i$ and $H_{y_i} = 0$ otherwise.

For this evaluation, the probabilistic prediction of the predictand y is described here, for each prediction day, with a discretized cdf composed of N values, with $N = 25$. When AM_{25} is used as a backup model, the N values are the precipitation observations of the 25th analog days. When the prediction is issued with SCAMP, the N values are those of the 25 percentiles $(k - 0.5)/25$, k in $[1, N]$, of the predicted cdf F_Y.

In the following, we discuss the prediction skill for precipitation occurrence and amount with the Brier skill score (BSS) and the continuous ranked probability skill score (CRPSS), respectively. Both scores normalize the prediction skill of EPS $\mathcal{P}$ with that obtained with a reference EPS $\mathcal{P}_\varphi$. $\mathcal{P}_\varphi$ is here a climatological EPS based on a calendar climatology defined for each prediction day by the precipitation distribution of all days belonging to a seasonal window (± 30 days) centered on the corresponding calendar day. In this context, the BSS and CRPSS, respectively, read as

$$\mathrm{BSS} = 1 - \frac{\overline{\mathrm{BS}}}{\overline{\mathrm{BS}}_\varphi} \tag{9}$$

and

$$\mathrm{CRPSS} = 1 - \frac{\overline{\mathrm{CRPS}}}{\overline{\mathrm{CRPS}}_{\mathcal{P}_\psi}}, \tag{10}$$

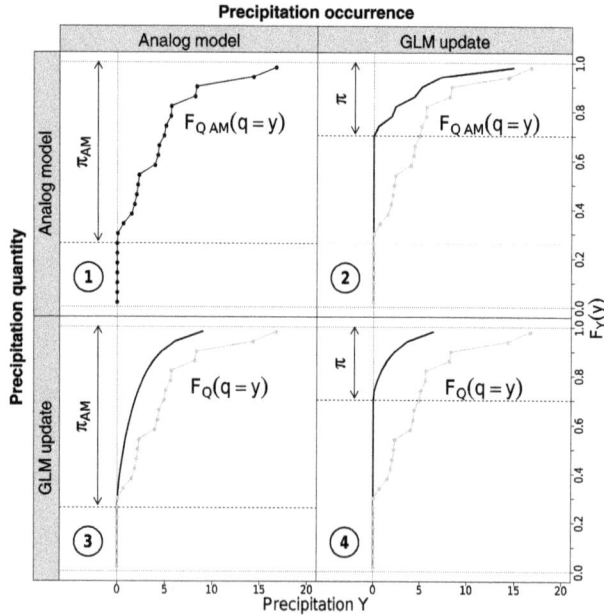

Figure 2. Illustrations of the four cases met for the issue of $F_Y(y)$ by the two-stage analog/regression model (SCAMP). Case 1: none of the occurrence and amount (quantity) GLMs could be retained during the regression stage: the backup analog model (AM_{25}) is used to predict the whole precipitation distribution. Case 2: only the occurrence GLM could be retained. It gives the estimated occurrence probability. The distribution of non-zero precipitation comes from AM_{25}. Case 3: only the amount (quantity) GLM could be retained. It gives the distribution of non-zero precipitation. The occurrence probability is the empirical occurrence probability from AM_{25}. Case 4: both occurrence and amount (quantity) GLMs could be estimated: they give, respectively, the occurrence probability and the distribution of non-zero precipitation, to be further combined for the full distribution of precipitation.

AM_{25} is applied to back up the prediction of the amount cdf (case 2), where AM_{25} is used to back up the occurrence probability prediction (case 3), and where the regression stage could be activated for both occurrence and amount (case 4).

Note that the regression stage achieved with GLMs can also be seen as a way to refine the estimation of the cdf that could have been obtained directly with the backup (and benchmark) AM_{25} analog model. The refinement leads to an update of the occurrence probability and/or the cdf of a non-zero precipitation amount.

As described previously, the two-stage analog/regression prediction process is repeated for each prediction day in turn. As the analog days vary from one prediction day to another, the predictors selected in the regression stage and the value of the corresponding regression coefficients are expected to vary from one prediction day to the other. The two-stage model SCAMP allows thus for a day-to-day adaptive and tailored downscaling.

where $\overline{BS}_\varphi$ and $\overline{CRPS}_{\mathcal{P}_\varphi}$ correspond to the mean BS and the mean CRPS obtained with EPS $\mathcal{P}_\varphi$. For both scores, a negative value indicates that the prediction obtained with EPS $\mathcal{P}$ is worse than the prediction obtained with the climatological EPS $\mathcal{P}_\varphi$. A score of 1 conversely denotes a perfect EPS $\mathcal{P}$.

In the following, to assess the added value of the two-stage SCAMP model when compared to the benchmark AM_{25} analog model, we additionally estimate the gain in prediction skill as $\Delta S = S_{SCAMP} - S_{AM_{25}}$, where S corresponds either to the BSS or the CRPSS.

4 Results

The two-stage model is used for the probabilistic prediction of small-scale precipitation over the continental French territory for each day of the 1982–2001 period. We here present the prediction skill obtained for occurrence and amount with the two predictors sets presented in Sect. 2. As discussed later in Sect. 5, the four predictors of each set are not necessarily all used; the predictors which have some predictive power for the considered predictand vary from one day to the other.

4.1 Performance of SCAMP

Figure 3a presents the BSS skill score of SCAMP for precipitation occurrence prediction. The highest BSS values – up to 0.5 – are found in the western part of the Massif Central, in the Alps and along the Atlantic coast. Lower skill (BSS from 0.45 to 0.5) is obtained in northern and western lowlands. The lowest skill (0.35) is obtained for few cells located along the Mediterranean coast.

The BSS gain obtained with SCAMP over AM_{25} is rather important (up to 0.1 BSS points) and presents a high space variability (Fig. 3b). The gain (between 0.05 and 0.1 BSS points) is high in the mountainous areas (Pyrenees, Massif Central, Alps, Vosges). The highest gains are found along the Mediterranean coast and in the southern Alps, where the BSS of SCAMP was lowest. This highlights the weakness of the AM_{25} in these regions – characterized by more frequent convective precipitation and thus a weaker link with large-scale atmospheric circulation – and the interest for thermodynamic and more local predictors. Conversely, lower gains are observed in the western part of France characterized by more frontal precipitation and thus a stronger link with large-scale circulation. Note also that the spatial distribution of ΔBSS is very close (even if it has higher values) to the one obtained by SCAMP with R_{700} as a unique predictor (not shown here).

Figure 4a shows the CRPSS obtained with SCAMP. The CRPSS values also depend on topography. The highest values, up to 0.45, are obtained in the western part of the Massif Central, the northern Alps, the Jura and the Vosges massifs. Lower values, between 0.32 and 0.45, are obtained in lowlands. The lowest skill (below 0.30) is again obtained along the Mediterranean coast.

The CRPSS gain obtained over AM_{25} is significant for most grid cells, with the highest value (up to 0.10 CRPSS points) obtained in the Rhône Valley and northeastern France (Fig. 4b). Similarly to the BSS gain, a lower CRPSS gain is also obtained here in lowlands and western France. The spatial distribution of $\Delta CRPSS$ is also very close here (even if it has higher values) to the one obtained by SCAMP with W_{700} as a unique predictor for amount (not shown here).

Despite the large dependency on regional features such as topography or proximity to the sea, adding local and thermodynamic information in SCAMP greatly improves the prediction skill over that of AM_{25}, for both precipitation occurrence and amount.

4.2 Characterization of SCAMP's behavior

As described in Sect. 3.3, the regression stage of SCAMP is equivalent to update the empirical distribution obtained from the atmospheric analogs directly. For some prediction days, the regression stage can be however only partly activated, for either occurrence or amount. It can be even not activated at all. In these cases, the prediction is fully or partly obtained from the backup model AM_{25}.

The frequency with which each activation case (cases 1 to 4) is obtained over the simulation period is given in Fig. 5. The situation where both precipitation occurrence and amount GLMs are activated (case 4) is very frequently observed. It corresponds to more than 85 % of the days except in south-eastern France, where only 60 % of the days are concerned. All in all, the regression stage of SCAMP is very often activated (more than 97 % of the days) to predict the occurrence probability (cases 2+4). In the failing full-updating cases, AM_{25} is usually applied to back up the precipitation amount prediction (cases 1+2). Case 1, where the whole prediction is backed up with AM_{25}, is finally very rare. For a large majority of the grid cells, it occurs less than 35 times in the 20-year period considered (corresponding to around 5 ‰).

Figure 6 presents the mean precipitation anomaly for each of the previous cases, i.e., the ratio between the mean amount obtained for all days belonging to the considered case and the overall mean precipitation amount. An anomaly greater (lower) than 1 indicates days that are rainier (drier) than usual. Cases 1 to 4 correspond clearly to different precipitation configurations. The mean precipitation amount of days in case 4 is close to the overall mean. Days in cases 1 and 2 are very dry. Days in case 3 are very wet, with a mean precipitation 3 times larger than the overall mean.

For a given prediction day, the precipitation state of its analog days is actually expected to be roughly similar to that of the day. This thus explains SCAMP's behavior described above. In cases 1 and 2, analog days of the prediction day are likely very dry. The number of humid analog days is thus likely small to very small, and likely too small to allow for a robust estimation of the precipitation amount GLM. Ana-

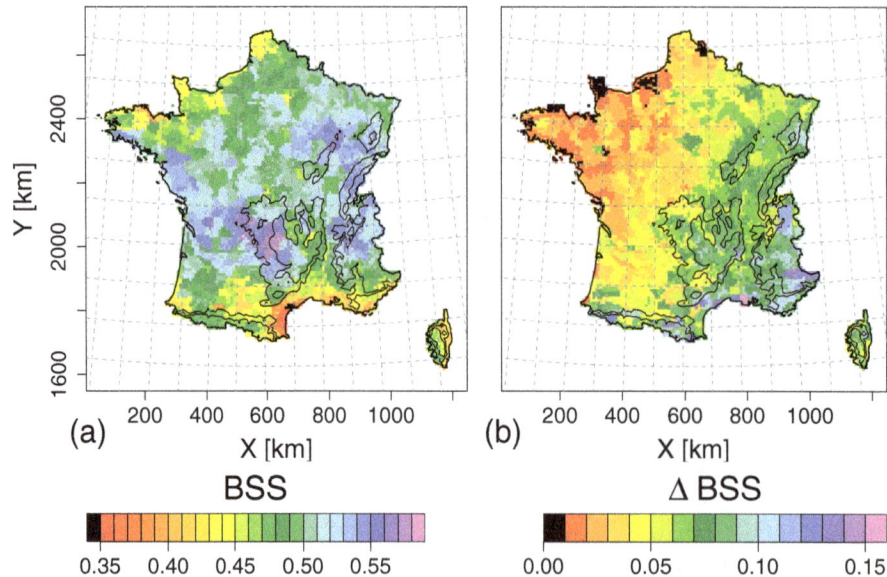

Figure 3. (a) BSS obtained with SCAMP (best possible value = 1). **(b)** BSS gain obtained with SCAMP compared to AM_{25}. Black solid lines correspond to the French borders and the contours around mountainous regions (400 and 800 m elevation), while the dashed lines show the ERA-40 grid mesh.

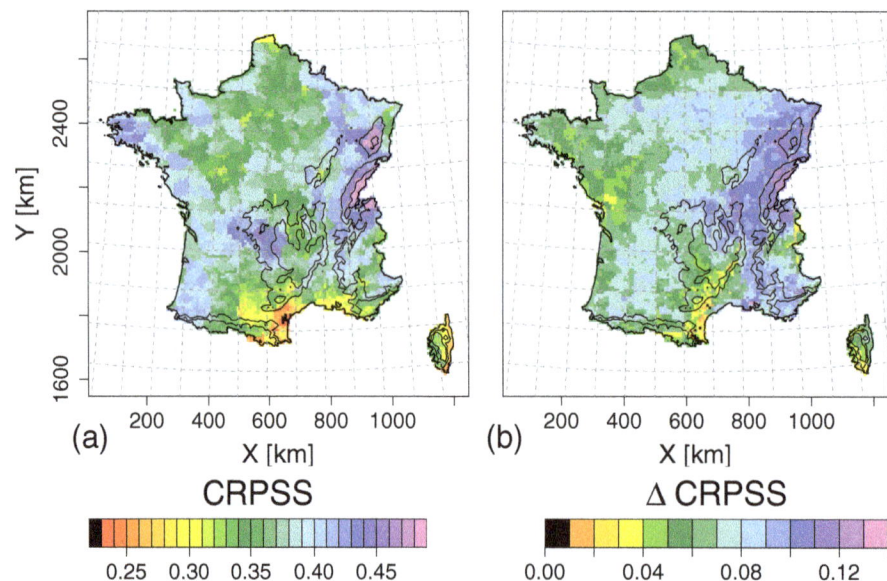

Figure 4. (a) CRPSS obtained with SCAMP (best possible value = 1). **(b)** CRPSS gain obtained with SCAMP compared to AM_{25}.

log days are conversely likely humid in case 4 or even very humid in case 3. The number of humid days in those cases is thus likely large enough to allow for a robust estimation of the precipitation amount GLM. The very humid configuration of case 3 suggests that prediction days are characterized by a very large number of humid analog days, which can in turn prevent a robust estimation of the occurrence GLM (e.g., the occurrence GLM cannot be estimated in configurations where all days are wet).

This can also explain the specific results obtained in the southeast. Case 2 is indeed activated much more often in this region (increase of 30 % percentage points) than elsewhere and, in a symmetric way, case 4 is activated much less often in this region (decrease of 30 % percentage points). The reason underlying this result is to be related to the much higher proportion of dry days in the southeast (see Fig. S1 in the Supplement). In this region, the number of wet analog days is thus likely small for a large number of prediction days. As suggested above, this is obviously not a difficulty for the estimation of the occurrence GLM. This is conversely likely for the estimation of the amount GLM. A small number of wet analogs likely prevents a robust estimation of the precipitation amount GLM. This likely explains the much lower (higher) frequency of case 4 (case 2) in the southeast.

The CRPSS gain achieved with SCAMP's results from the updated prediction of both precipitation occurrence and amount. To assess the relative effects of these updates on the gain, we further compared the following four prediction experiments.

- **Exp. 1.** The prediction of both the occurrence and the amount is achieved with AM_{25} for all prediction days. This corresponds to the results given by Chardon et al. (2014, cf. Fig. 3).

- **Exp. 2.** When possible, the precipitation occurrence probability is updated with the occurrence GLM. The non-zero precipitation amount is always predicted with AM_{25}.

- **Exp. 3.** When possible, the precipitation amount is updated with the amount GLM. The precipitation occurrence probability is always predicted with AM_{25}.

- **Exp. 4.** When possible, both precipitation occurrence probability and amount are updated with the occurrence and amount GLMs. This corresponds to the two-stage configuration already evaluated previously.

The CRPSS gains obtained between Exps. 1 and 2, between Exps. 1 and 3, and between Exps. 1 and 4 are presented in Fig. 7 (the results for Exp. 4, already presented in Fig. 4, are presented again for ease of comparison).

For a large majority of grid cells, the CRPSS gain obtained with an updated prediction of the occurrence probability (from 0 to 0.05 CRPSS points) is significantly lower than that obtained with an updated prediction of amount (from 0.03 to 0.1 CRPSS points). The CRPSS gain obtained in the latter case is additionally close to that obtained with the full two-stage model. The CRPSS gain obtained by SCAMP in Fig. 7c is thus explained in most cases by the updated prediction of precipitation amount. The scheme is somehow different in the south of France along the Mediterranean coast and in the Cevennes–Vivarais mountains. In those regions, the CRPSS gain obtained by SCAMP is mostly explained by the updated prediction of the occurrence probability. Updating only the precipitation amount leads to fairly no CRPSS gain.

5 Discussion

The sets of potential predictors used in SCAMP for the prediction of precipitation occurrence and amount have been listed in Sect. 2. For each variable, the number of potential predictors is here equal to four. All four predictors are not necessary retained for the GLM. For a given prediction day, a GLM with a single predictor or a combination of several predictors among the four can be selected. Fifteen regressive structures plus the backup AM_{25} model are possible in our context (Table 2).

For a given prediction day, the regressive structures selected by SCAMP for precipitation occurrence or for precipitation amount are supposed to include the best information for the prediction. In the following, we assess how often each structure has been selected. This allows for some insight into the atmospheric information really used for the regression stage and how this information varies in time.

Figures 8 and 9 present the percentage of times that the 15 regressive structures and the backup AM_{25} are used for the prediction of precipitation occurrence and amount, respectively. As in Fig. 5, gray cells indicate that the regression structure has been retained less than 35 times over the 20-year evaluation period. For both occurrence and amount, the selection frequency of the structures is also rather region dependent and strongly influenced by topography.

For occurrence (Fig. 8), the most often selected structure is Str. no. 1, which is only based on R_{700} (more than 25 % of the time for the whole of France). R_{700} was actually found to give the highest predictive power when used in a single predictor configuration. Another structure which is also often selected (more than 15 % for a high number of grids) is Str. no. 7 which combines R_{700} with $Occ - 1$. Secondary structures – as for example Str. no. 6 and no. 13 combining R_{700} to W_{700} and $Occ - 1$ – can be selected more than 10 % of the days for some given regions. Other structures are seldom selected and some of them (Str. no. 8, no. 11, no. 14, and no. 15) are almost never selected.

For precipitation amount, the most frequently selected structures are Str. no. 3 and Str. no. 1, both based on one single predictor, W_{700} or R_{700}, respectively. These structures are selected more than 25 % of the time (or more than 15 %) for a high number of grids. The secondary structures (Str. no. 6, no. 8, and no. 13) are selected from 5 up to 20 %, depending on the region. They always include W_{700} in combination with some other predictor (R_{700}, H, or $R_{700} + T_{700}$). Str. no. 9, no. 12, no. 14, and no. 15 are almost never selected. The others are selected less than 10 to 5 % of the time.

Note that for the selection of the best regression structure for a given prediction day, all 15 of these regressive structures have been tested in turn. The results above suggest that this systematic test is not necessary and that it could be reasonable to consider only the few structures which are frequently retained or which are retained a "reasonable" fraction of the days. However, the selection frequency of a given structure actually varies with the seasons and/or the encountered synoptic situation, and some secondary regressive structures can be retained frequently for specific situations. This is illustrated in Fig. 10 for a cell located in northwestern France. The figure presents how the selection frequency of each regression structure differs in different seasons and weather

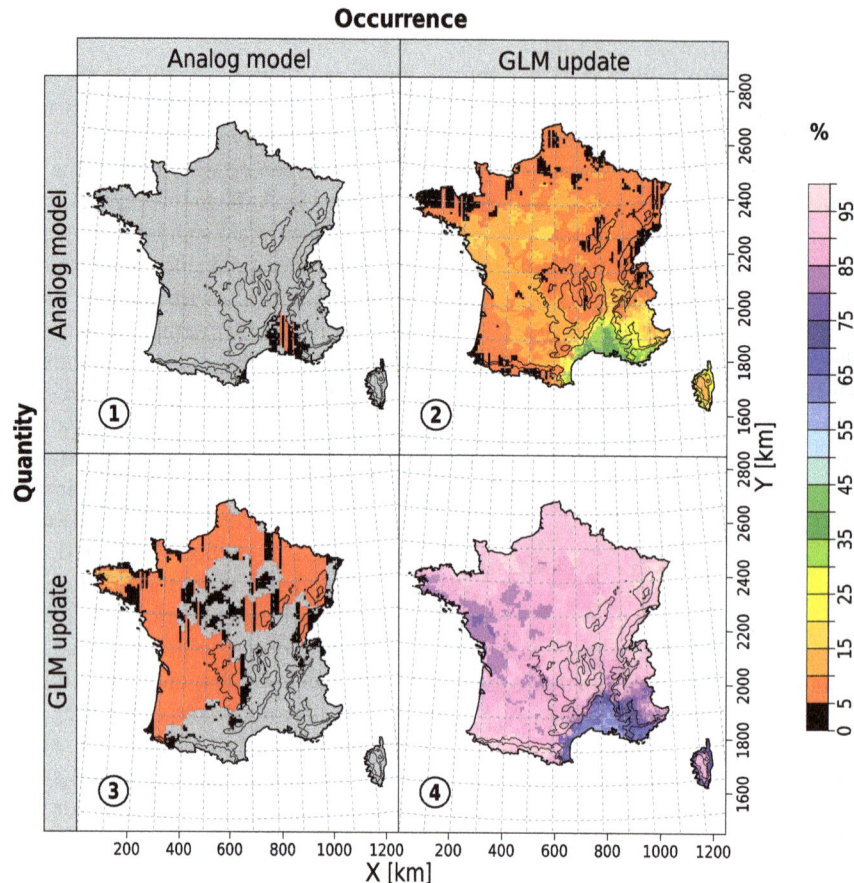

Figure 5. Percentage of days where (1) no updates are applied, (2) only the precipitation occurrence is updated, (3) only the precipitation amount is updated, and (4) the occurrence and the precipitation amount are updated. Grids with gray colors correspond to grid cells where the corresponding case has been met less than 35 times over the 20-year evaluation period.

patterns (WP, defined in Table 3, Garavaglia et al., 2010) from the selection frequency obtained for the all-days situation.

For precipitation occurrence (Fig. 10a), the selection of the main regressive structures (i.e., Str. no. 1 and no. 7, respectively, based on R_{700} and $R_{700} + \text{Occ} - 1$) is up to 15 % more frequent (less frequent) for WP3 (WP5) compared to the all-days situation. For precipitation amount (Fig. 10b), the selection frequency of the main regressive structures (Str. no. 1 and no. 3 based on R_{700} and W_{700}, respectively) can similarly change up to ±10 %. The reduced selection of a main regressive structure for a given season or WP can lead to preferential retention of some secondary regressive structure. For instance, the regressive Str. no. 8 based on $W_{700} + H$ is selected 10 % more frequently for WP2 than for the all-days situation (Fig. 10b).

The preferential (or conversely reduced) selection of some regression structures for given WTs or seasons was estimated for all grid cells of France. In most cases, the preferential (or reduced) selection was found to present a noticeable spa-

tial coherency. Different configurations are observed as illustrated in Fig. 11 and discussed below.

The preferential selection of some regression structures can first be observed over large to very large regions. As an example, the preferential selection of Str. no. 3 for the prediction of precipitation amount for days in WP7 (more than +15 % compared to usual) is obtained for all grid cells in France. Whatever the location, the vertical velocity W_{700} seems thus required in this specific weather pattern. Another example is that of WP8 which corresponds to an Anticyclonic situation. Whatever the location, no precipitation is really expected for this configuration. No predictor is thus required in addition to geopotential heights used in the analog stage. This configuration logically leads to a large preferential selection of the backup AM_{25} model.

For a given weather pattern, the preferential selection of a regressive structure can also vary from one region to the other. For WP2 for instance, the structures based on W_{700} or on W_{700} and H are selected much more often along the Atlantic coast and in the north of France. The backup AM_{25} model is conversely more selected in the southeast, on the

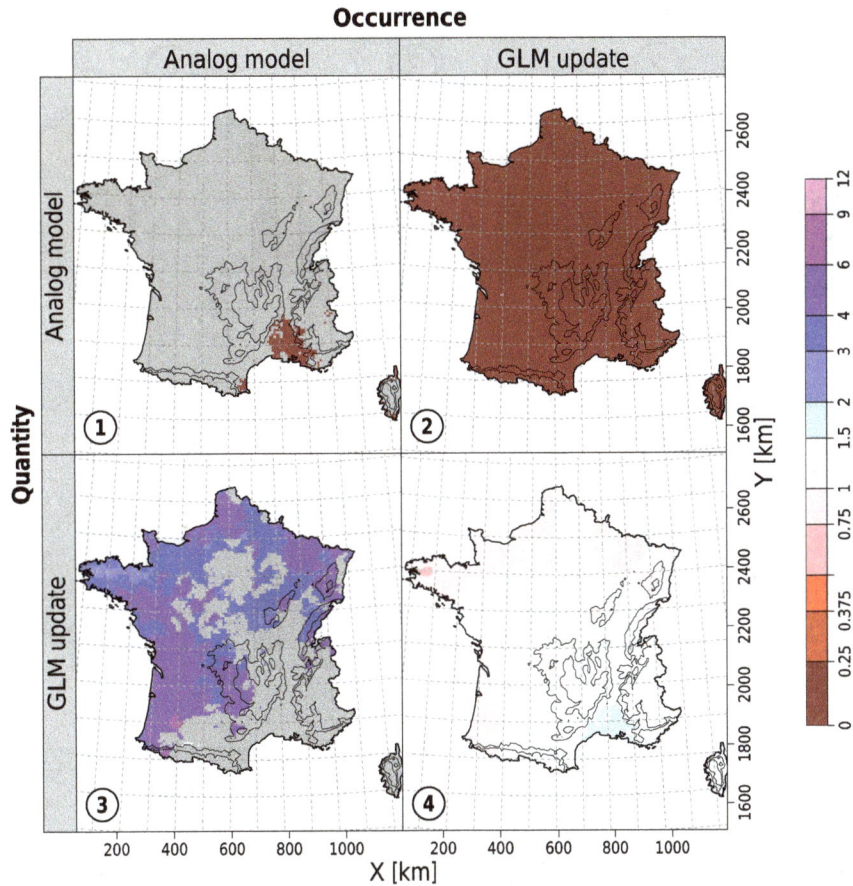

Figure 6. Ratio between the mean amount obtained for all days belonging to a given case and the overall mean precipitation amount. The four cases and gray grids: same as in Fig. 5.

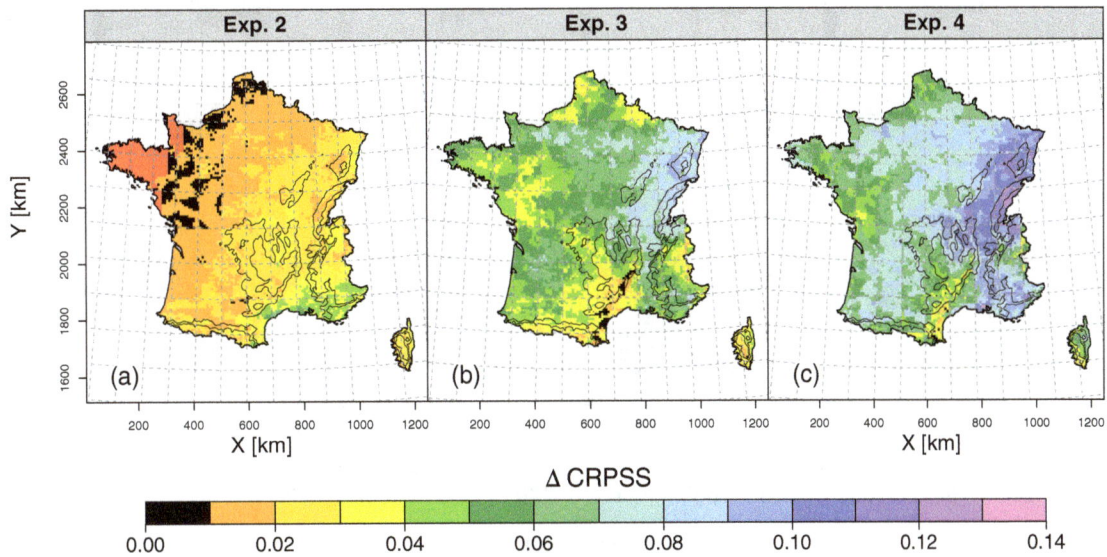

Figure 7. Gain in CRPSS for different prediction experiments (see Sect. 4.2 for details) compared to the performance of AM_{25}. (a) Exp. 2: only the precipitation occurrence probability is updated (when possible); (b) Exp. 3: only the precipitation amount is updated (when possible); (c) Exp. 4: both occurrence probability and precipitation amount are updated (when possible).

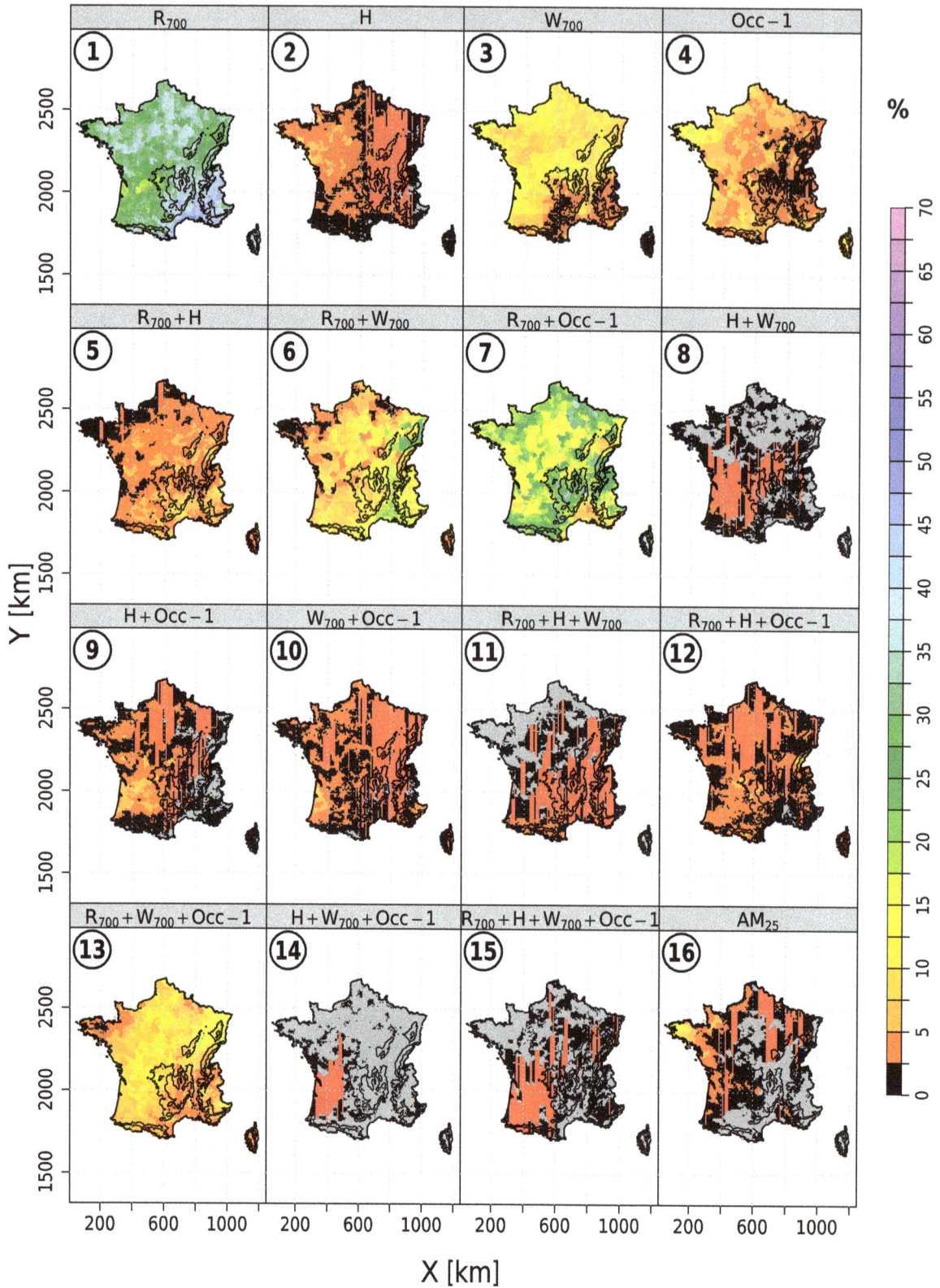

Figure 8. Prediction of occurrence probability: selection frequencies (%) of the 15 regression structures and of the backup model AM_{25}. Predictors involved are indicated in the graph headers, and the index of the regressive structure in the top left corners. Gray grids: same as in Fig. 5. The selection frequency of AM_{25} corresponds to the sum of those obtained for cases 1 and 3 in Fig. 5.

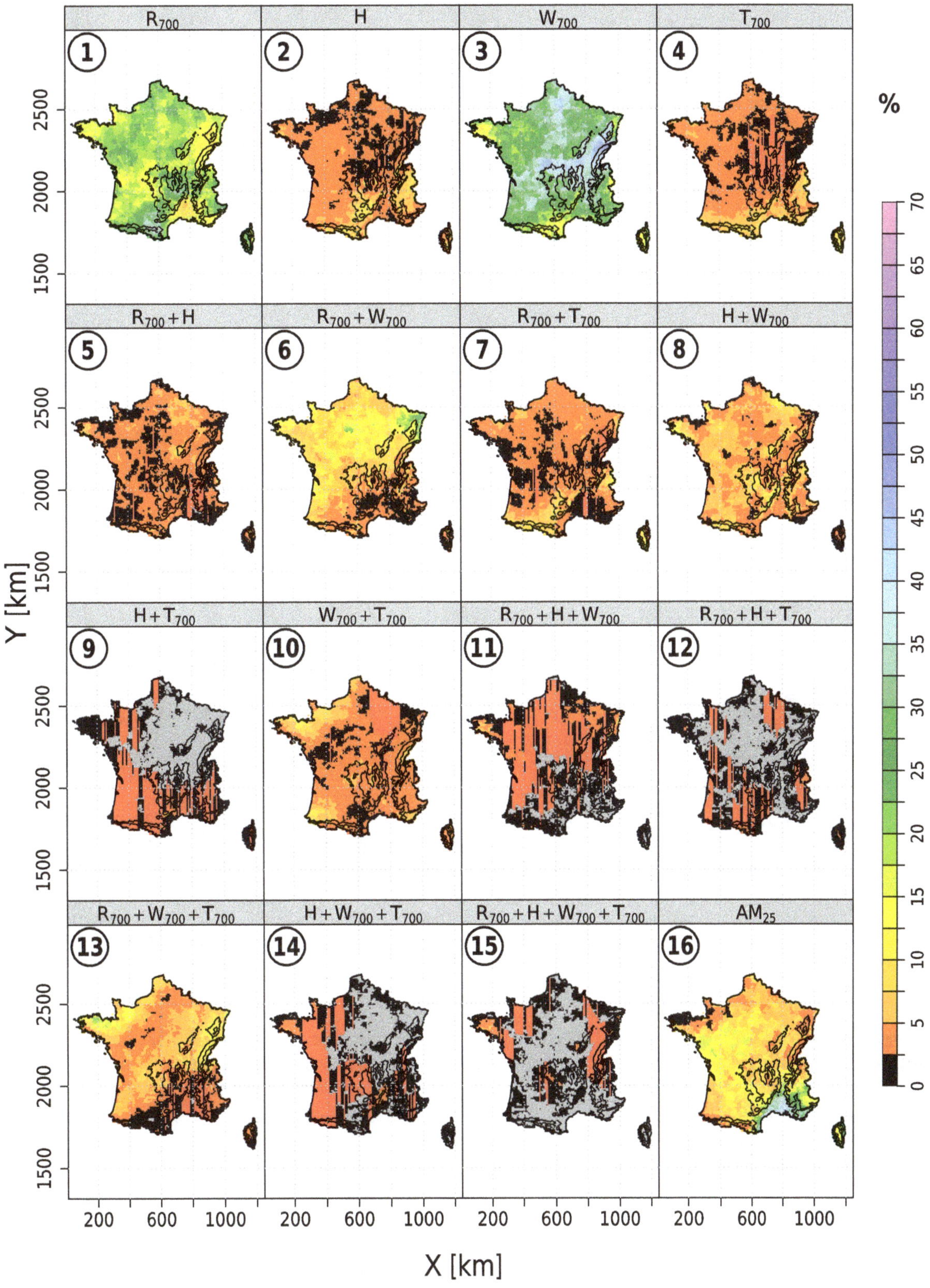

Figure 9. Same as Fig. 8 for the probabilistic prediction of precipitation amount. The selection frequency of AM_{25} corresponds to the sum of those obtained for cases 1 and 2 in Fig. 5.

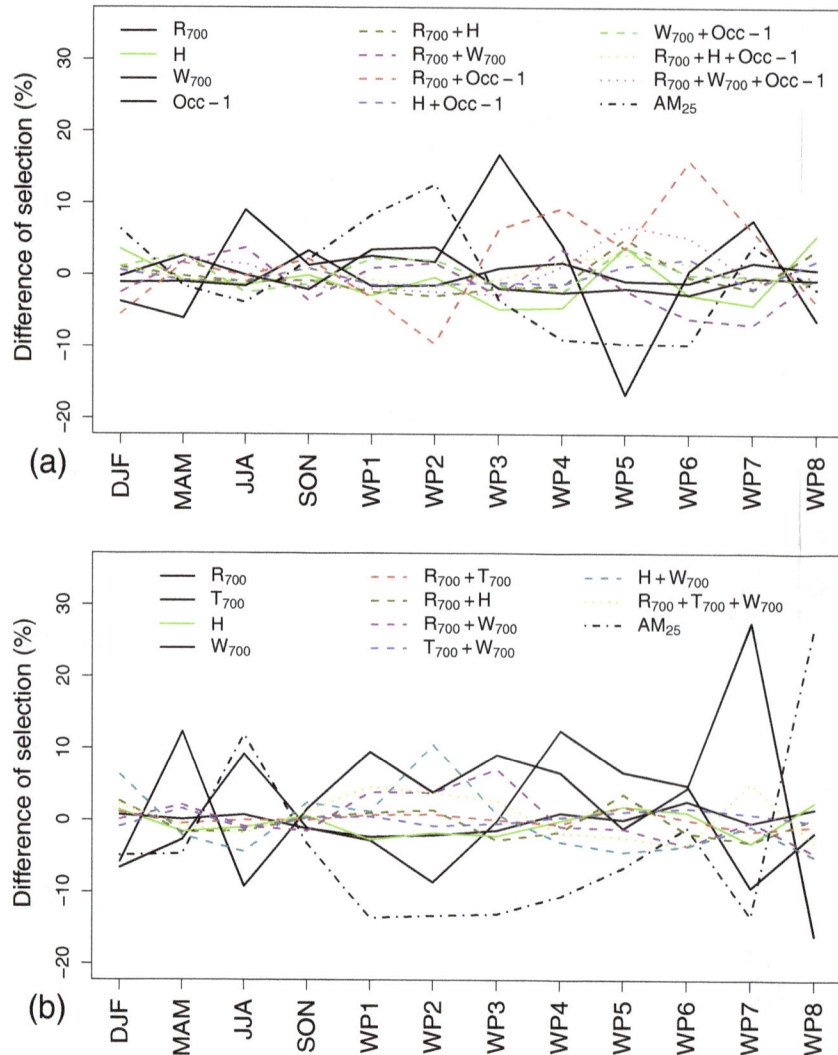

Figure 10. For each season and weather type, difference (%) in selection frequency with the all-days case for different regression structures. Results for the prediction of (a) occurrence and (b) amount. A positive difference indicates that the considered regressive structure is selected more often than for the all-day situation. Results are displayed for a grid cell located in the northwest of France. For a clearer illustration, the three or four regressive structures that are almost never selected are not displayed.

Table 3. Names of the weather patterns (WP) defined in Garavaglia et al. (2010) and the related frequency for the 1982–2001 period.

Index	Denomination	Annual frequency (%)
WP1	Atlantic wave	8
WP2	Steady oceanic	22
WP3	Southwest circulation	8
WP4	South circulation	17
WP5	Northeast circulation	6
WP6	East return	6
WP7	Central depression	4
WP8	Anticyclonic	29

Mediterranean coast especially. For this weather regime, the southeast is actually protected by the Massif Central mountain and thus usually does not receive precipitation (cf. Fig. 3 of Garavaglia et al., 2010).

The preferential selection of a regressive structure can also be obtained for rather small and specific regions. In Fig. 11b, the regressive Str. no. 8 based on $W_{700} + H$ is more frequently selected for WP7 (around $+15\,\%$) in the Cevennes–Vivarais region (southeastern part of the Massif Central) and in the pre-Alpine mountains (western part of the Alps). The combination of W_{700} and H seems thus to be very informative in those configurations for this really rare WP (4 % of the 20-year period).

Whatever the configuration, the preferential selection of regression structures presents some spatial coherency, at small or large regional scales. This obviously also suggests the spatial robustness of the informative predictors to be retained for given large-scale weather configurations.

6 Conclusions

The relevance of a two-stage analog/regression model has been explored in this study for the probabilistic prediction of precipitation over France. Atmospheric analogs of the prediction day are identified to estimate the parameters of a two-part regression model further applied for the prediction. The regression model consists of a logistic GLM for the prediction of precipitation occurrence and a logarithmic GLM for the prediction of precipitation amount. The prediction obtained with this two-stage approach updates the predictive distribution that would have been achieved directly from a one-stage analog model based on atmospheric circulation analogs. The two-stage approach makes the downscaling model adaptive: as the analog days are identified for each prediction day, the predictors and regression coefficients of the regression models can vary from one day to the other.

The regression stage allows a non-negligible prediction skill gain compared to the reference analog model (gain up to 0.1 skill score points for both the BSS and the CRPSS). The CRPSS gain is mainly achieved due to the regression model estimated for the precipitation amount. The introduction of local-scale predictors such as relative humidity is obviously crucial there. The adaptive nature of the model and thus the possibility of tailoring the downscaling relationship (both predictors and regression coefficients) to the current prediction day seems to be decisive as well. The CRPSS gain obtained with the two-stage approach is actually 2 times larger than the one obtained by Chardon et al. (2014) with a two-level analog model where a unique and same second-level analogy variable (namely humidity) is considered for all days.

The prediction skill and adaptability of this two-stage approach was illustrated for the prediction of both the precipitation occurrence and amount in a simplified configuration where four predictors, selected in a preliminary analysis from a large ensemble of potential predictors, are used in the regression stage. The predictors used for precipitation occurrence are the relative humidity and vertical velocity at 700 hPa, the helicity integrated from 1000 to 500 hPa, and the occurrence of the previous day. A similar set of predictors is used for the precipitation amount (the occurrence of the previous day is replaced by the 700 hPa temperature). Most of the time, the final regression model only includes one or two predictors. It also very often includes the relative humidity R_{700} for precipitation occurrence and R_{700} or the vertical velocity W_{700} for precipitation amount. Some combinations of predictors, almost never used in general, appear to be more frequently retained for some specific weather patterns and/or

locations in France, revealing their potential interest for these situations.

For the sake of simplicity and to limit the degrees of freedom in our analysis, we considered a unique set of four potential predictors for all SAFRAN grid cells. This obviously leads to a sub-optimal prediction configuration. The main meteorological processes driving precipitation in France obviously differ from one region to the other. The most informative predictors are thus expected to be region-dependent and the set of predictors to be considered in the regression stage could be refined on a regional basis. This is expected to improve the skill of the prediction. The same would apply for an application of SCAMP to other regions worldwide.

A number of atmospheric variables have been considered as potential predictors in similar downscaling studies. The predictors found to be of interest are most often few. They are roughly the same than those considered in the preliminary analysis of the present work. However, as in the present work, the analyses usually carried out to identify these informative predictors are potentially misleading. The selection of a variable is indeed often based on its predictive power, estimated with some prediction skill score in an all-days evaluation framework. As highlighted in the present work however, some predictors are likely to be informative for very few meteorological situations. An all-days evaluation is expected to reveal robust predictors. It however very likely misses important situation-specific predictors. The two-stage approach here estimates the statistical downscaling link from a homogeneous set of days, with respect to their large-scale atmospheric circulation configuration. Those days are moreover atmospheric analogs to the prediction day. This two-stage approach has thus the potential to reveal the predictive power of very specific predictors, suited for very specific meteorological configurations. It leaves very likely room for significant improvements of the prediction skill for such unusual configurations. It gives likely also the opportunity to better understand the atmospheric factors under play in a number of non-frequent and atypical meteorological situations. Notwithstanding the technical limitations that may hamper such analyses, a broader exploration of a much larger diversity of predictors, possibly non-conventional ones, would be thus definitively worth in this context.

Both the predictors and the regression coefficients were shown in our work to depend on the analog days identified in the analog stage. This is the reason for the adaptability of the downscaling discussed above. Besides the adaptability, we ideally expect that for a given prediction day the predictor selection and the associated regression coefficients will be robust. Further analyses should explore this issue. An interesting work would be for instance to check that the predictors and their related coefficients do not significantly change when the set of analog days considered for the estimation is modified as a result of a different setup of the analog model (e.g., when one changes the archive period or the archive length).

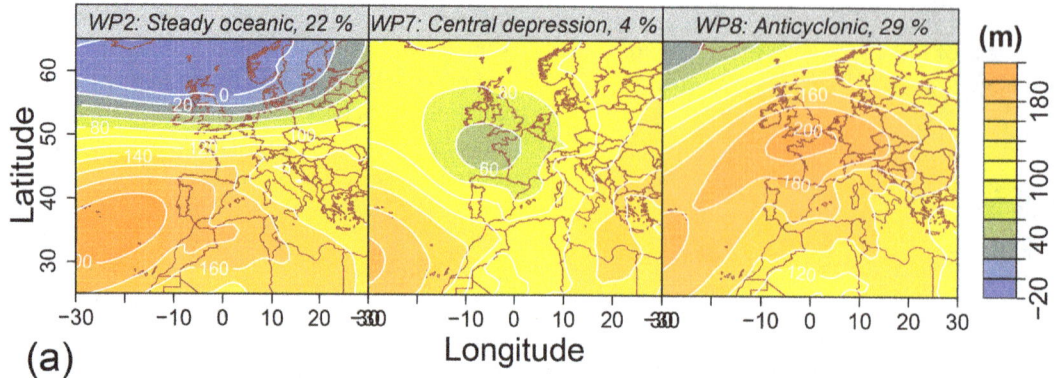

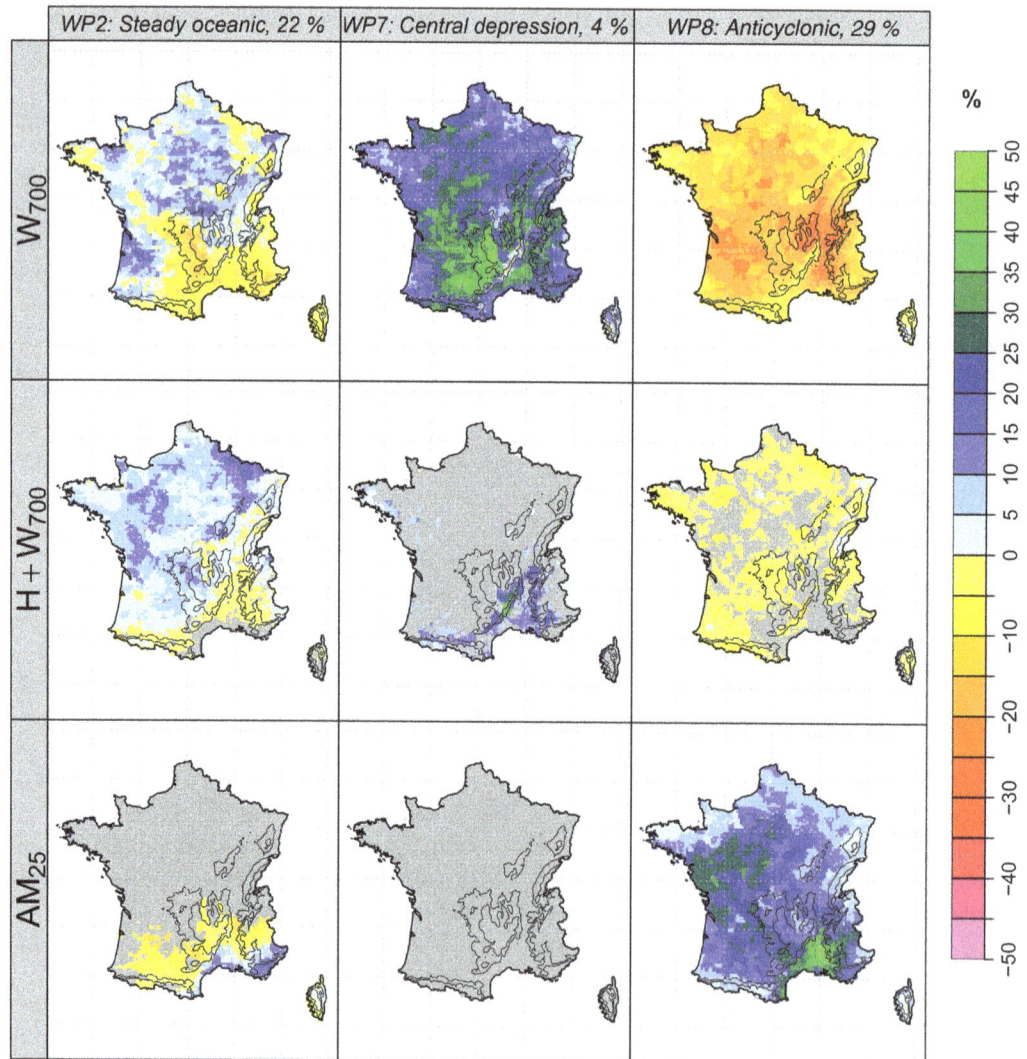

Figure 11. (a) Mean geopotential height at 1000 hPa for three WPs (Garavaglia et al., 2010). **(b)** For each WP, difference (%) in selection frequency with the all-days case. Results for two regression structures (W_{700} and $W_{700} + H$) and for AM_{25}. The predictand is the precipitation amount. A positive (negative) difference indicates an extra selection (reduced selection). Gray grids: same as in Fig. 5.

Results of our work depend on a number of choices and assumptions. They for instance likely depend on the database used for the large-scale atmospheric predictors. The day-to-day behavior of such an analog/regression approach (and the skill of the prediction) likely depends on the database and especially on the quality of the predictors. An atmospheric reanalysis with a higher spatial resolution would for instance likely allow for a better description of the shapes of geopotential fields and for a more relevant simulation of regional/local thermodynamic processes. It would likely lead in turn to higher-quality variables for some atmospheric parameters such as air instability. This may allow for a better identification of the daily specificity in the downscaling relationship and for the most informative predictors to be used each day. The reverse may occur when using lower-quality predictors, for instance lower-quality data from reanalyses available for the 20th century or lower-quality data from climate or numerical weather forecasting models. The quality of the predictors is thus obviously also an important issue to be further considered. It may lead to different informative predictors, depending on the intended use of the model (forecast, simulation, or climate impact studies).

SCAMP was used here for the prediction of small-scale precipitation at individual grid cells. The prediction of precipitation fields, obviously required for a number of impact studies, is also a challenging issue (e.g., Clark et al., 2004; Yang et al., 2005; Thober et al., 2014). Different adaptations of SCAMP would be worth investigating in this context. SCAMP could be for instance applied for the prediction of mean areal precipitation over the whole targeted spatial domain and some spatial disaggregation process could be further used to generate the required fields (e.g., Mezghani and Hingray, 2009; Rupp et al., 2012). As highlighted by Chardon et al. (2016), the prediction skill of SCAMP is expected to increase with the size of the spatial domain targeted for the prediction, which makes such an approach rather appealing. Another possible strategy for spatial predictions would be to rely on the advantages introduced by the analog stage of SCAMP. Chardon et al. (2014) indeed showed that for a given prediction day the same set of analog days can be used over rather large domains (up to a few 100s of kilometers) for a quasi-optimal prediction of local-scale precipitation. The precipitation field of each analog day (which is thus spatially coherent because already observed) could thus be used as a first-guess precipitation field scenario for the considered region. The field could be next updated at each location with day- and location-specific coefficients obtained from the regression stage of SCAMP. This spatial prediction issue will be considered in future works.

Appendix A: Acronyms

AM	Analog model
AM_{25}	Analog model (based on the 25 nearest atmospheric analogs) used as a benchmark or backup prediction model.
BS	Brier score
BSS	Brier skill score
cdf	Cumulative distribution function
CRPS	Continuous ranked probability score
CRPSS	Continuous ranked probability skill score
EPS	Ensemble prediction system
GLM	Generalized linear model
SAFRAN	8×8 km precipitation reanalysis for France from MeteoFrance
SCAMP	Two-stage analog/regression model
SDM	Statistical downscaling model
TWS	Teweless–Wobus score
WP	Weather pattern

Author contributions. This study is part of JC's PhD thesis. BH and ACF supervised the PhD. All authors contributed to the designed experiments and to the writing of the document. JC developed the model code and performed the simulations.

Competing interests. The authors declare that they have no conflict of interest.

Acknowledgements. The authors especially thank Charles Obled and Isabella Zin for fruitful discussions on the analog method. The authors also thank the Grenoble University High Performance Computing centre, CIMENT (https://ciment.ujf-grenoble.fr/wiki-pub/index.php/Welcome_to_the_CIMENT_site!), for their help and the large computing resource they provide. We would especially like to thank the associate editor and two anonymous reviewers for their relevant comments and suggestions that allowed us to improve this study and broaden its scope.

Edited by: Luis Samaniego

References

Akaike, H.: A new look at the statistical model identification, IEEE T. Automat. Contr., 19, 716–723, https://doi.org/10.1109/TAC.1974.1100705, 1974.

Asong, Z. E., Khaliq, M. N., and Wheater, H. S.: Projected changes in precipitation and temperature over the Canadian Prairie Provinces using the Generalized Linear Model statistical downscaling approach, J. Hydrol., 539, 429–446, https://doi.org/10.1016/j.jhydrol.2016.05.044, 2016.

Ben Daoud, A., Sauquet, E., Bontron, G., Obled, C., and Lang, M.: Daily quantitative precipitation forecasts based on the analogue method: improvements and application to a French large river basin, Atmos. Res., 169, Part A, 147–159, https://doi.org/10.1016/j.atmosres.2015.09.015, 2016.

Boé, J., Terray, L., Habets, F., and Martin, E.: Statistical and dynamical downscaling of the Seine basin climate for hydro-meteorological studies, Int. J. Climatol., 27, 1643–1655, https://doi.org/10.1002/joc.1602, 2007.

Bontron, G. and Obled, C.: A probabilistic adaptation of meteorological model outputs to hydrological forecasting, Houille Blanche, 23–28, 2005.

Brier, G. W.: Verification of forecasts expressed in terms of probability, Mon. Weather Rev., 78, 1–3, https://doi.org/10.1175/1520-0493(1950)078<0001:VOFEIT>2.0.CO;2, 1950.

Brown, T. A.: Admissible scoring systems for continuous distributions, Manuscript P-5235, The Rand Corporation, Santa Monica, CA, available from The Rand Corporation, 1700 Main St., Santa Monica, CA 90407-2138, 1974.

Caillouet, L., Vidal, J.-P., Sauquet, E., and Graff, B.: Probabilistic precipitation and temperature downscaling of the Twentieth Century Reanalysis over France, Clim. Past, 12, 635–662, https://doi.org/10.5194/cp-12-635-2016, 2016.

Campozano, L., Tenelanda, D., Sanchez, E., Samaniego, E., and Feyen, J.: Comparison of statistical downscaling methods for monthly total precipitation: case study for the Paute River Basin in Southern Ecuador, Adv. Meteorol., 2016, 6526341, https://doi.org/10.1155/2016/6526341, 2016.

Chandler, R. E. and Wheater, H. S.: Analysis of rainfall variability using generalized linear models: a case study from the west of Ireland, Water Resour. Res., 38, 10–1–10–11, https://doi.org/10.1029/2001WR000906, 2002.

Chardon, J., Hingray, B., Favre, A.-C., Autin, P., Gailhard, J., Zin, I., and Obled, C.: Spatial similarity and transferability of analog dates for precipitation downscaling over France, J. Climate, 27, 5056–5074, https://doi.org/10.1175/JCLI-D-13-00464.1, 2014.

Chardon, J., Favre, A.-C., and Hingray, B.: Effects of spatial aggregation on the accuracy of statistically downscaled precipitation predictions, J. Hydrometeorol., 17, 1561–1578, https://doi.org/10.1175/JHM-D-15-0031.1, 2016.

Clark, M., Gangopadhyay, S., Hay, L., Rajagopalan, B., and Wilby, R.: The Schaake shuffle: a method for reconstructing space-time variability in forecasted precipitation and temperature fields, J. Hydrometeor., 5, 243–262, https://doi.org/10.1175/1525-7541(2004)005<0243:TSSAMF>2.0.CO;2, 2004.

Coe, R. and Stern, D.: Fitting models to daily rainfall data, J. Appl. Meteorol., 21, 1024–1031, https://doi.org/10.1175/1520-0450(1982)021<1024:FMTDRD>2.0.CO;2, 1982.

Dayon, G., Boe, J., and Martin, E.: Transferability in the future climate of a statistical downscaling method for precipitation in France, J. Geophys. Res.-Atmos., 120, 1023–1043, https://doi.org/10.1002/2014JD022236, 2015.

Enke, W. and Spegat, A.: Downscaling climate model outputs into local and regional weather elements by classification and regression, Climate Res., 8, 195–207, https://doi.org/10.3354/cr008195, 1997.

Gaitan, C. F., Hsieh, W. W., and Cannon, A. J.: Comparison of statistically downscaled precipitation in terms of future climate indices and daily variability for southern Ontario and Quebec, Canada, Clim. Dynam., 43, 3201–3217, https://doi.org/10.1007/s00382-014-2098-4, 2014.

Gangopadhyay, S., Clark, M., and Rajagopalan, B.: Statistical downscaling using K-nearest neighbors, Water Resour. Res., 41, W02024, https://doi.org/10.1029/2004WR003444, 2005.

Garavaglia, F., Gailhard, J., Paquet, E., Lang, M., Garçon, R., and Bernardara, P.: Introducing a rainfall compound distribution model based on weather patterns sub-sampling, Hydrol. Earth Syst. Sci., 14, 951–964, https://doi.org/10.5194/hess-14-951-2010, 2010.

Guilbaud, S. and Obled, C.: L'approche par analogues en prévision météorologique., La Météorologie, 8, 21–35, 1998.

Hanssen-Bauer, I., Achberger, C., Benestad, R., Chen, D., and Forland, E.: Statistical downscaling of climate scenarios over Scandinavia, Climate Res., 29, 255–268, https://doi.org/10.3354/cr029255, 2005.

Holton, J. R. and Hakim, G. J.: An Introduction to Dynamic Meteorology, vol. 88, Academic press, Elsevier Ed., Amsterdam, 2012.

Horton, P., Jaboyedoff, M., Metzger, R., Obled, C., and Marty, R.: Spatial relationship between the atmospheric circulation and the precipitation measured in the western Swiss Alps by means of the analogue method, Nat. Hazards Earth Syst. Sci., 12, 777–784, https://doi.org/10.5194/nhess-12-777-2012, 2012.

Ibarra-Berastegi, G., Saénz, J., Ezcurra, A., Elías, A., Diaz Argandoña, J., and Errasti, I.: Downscaling of surface moisture flux and precipitation in the Ebro Valley (Spain) using analogues and analogues followed by random forests and multiple linear regression, Hydrol. Earth Syst. Sci., 15, 1895–1907, https://doi.org/10.5194/hess-15-1895-2011, 2011.

Kuentz, A., Mathevet, T., Gailhard, J., and Hingray, B.: Building long-term and high spatio-temporal resolution precipitation and air temperature reanalyses by mixing local observations and global atmospheric reanalyses: the ANATEM model, Hydrol. Earth Syst. Sci., 19, 2717–2736, https://doi.org/10.5194/hess-19-2717-2015, 2015.

Lafaysse, M., Hingray, B., Mezghani, A., Gailhard, J., and Terray, L.: Internal variability and model uncertainty components in a multireplicate multimodel ensemble of hydrometeorological projections : the Alpine Durance basin, Water Resour. Res., 50, 3317–3341, 2014.

Maraun, D., Wetterhall, F., Ireson, A. M., Chandler, R. E., Kendon, E. J., Widmann, M., Brienen, S., Rust, H. W., Sauter, T., Themeßl, M., Venema, V. K. C., Chun, K. P., Goodess, C. M., Jones, R. G., Onof, C., Vrac, M., and Thiele-Eich, I.: Precipitation downscaling under climate change: Recent developments to bridge the gap between dynamical models and the end user, Rev. Geophys., 48, RG3003, https://doi.org/10.1029/2009RG000314, 2010.

Marty, R., Zin, I., Obled, C., Bontron, G., and Djerboua, A.: Toward real-time daily PQPF by an analog sorting approach: application to flash-flood catchments, J. Appl. Meteorol. Clim., 51, 505–520, https://doi.org/10.1175/JAMC-D-11-011.1, 2012.

Matheson, J. E. and Winkler, R. L.: Scoring rules for continuous probability distributions, Manage. Sci., 22, 1087–1096, https://doi.org/10.1287/mnsc.22.10.1087, 1976.

McCullagh, P. and Nelder, J. A.: Generalized Linear Models, 2nd edn., Chapman and Hall/CRC, Boca Raton, 1989.

Mezghani, A. and Hingray, B.: A combined downscaling-disaggregation weather generator for stochastic generation of multisite hourly weather variables over complex terrain: development and multi-scale validation for the Upper Rhone River basin, J. Hydrol., 377, 245–260, 2009.

Murphy, A. H.: A new vector partition of the probability score, J. Appl. Meteorol. Clim., 12, 595–600, https://doi.org/10.1175/1520-0450(1973)012<0595:ANVPOT>2.0.CO;2, 1973.

Nasseri, M., Tavakol-Davani, H., and Zahraie, B.: Performance assessment of different data mining methods in statistical downscaling of daily precipitation, J. Hydrol., 492, 1–14, https://doi.org/10.1016/j.jhydrol.2013.04.017, 2013.

Nelder, J. A. and Wedderburn, R. W. M.: Generalized linear models, J. Roy. Stat. Soc. A Sta., 135, 370–384, 1972.

Obled, C., Bontron, G., and Garçon, R.: Quantitative precipitation forecasts: a statistical adaptation of model outputs through an analogues sorting approach, Atmos. Res., 63, 303–324, https://doi.org/10.1016/S0169-8095(02)00038-8, 2002.

Quintana-Segui, P., Le Moigne, P., Durand, Y., Martin, E., Habets, F., Baillon, M., Canellas, C., Franchisteguy, L., and Morel, S.: Analysis of near-surface atmospheric variables: validation of the SAFRAN analysis over France, J. Appl. Meteorol. Clim., 47, 92–107, https://doi.org/10.1175/2007JAMC1636.1, 2008.

Radanovics, S., Vidal, J.-P., Sauquet, E., Ben Daoud, A., and Bontron, G.: Optimising predictor domains for spatially coherent precipitation downscaling, Hydrol. Earth Syst. Sci., 17, 4189–4208, https://doi.org/10.5194/hess-17-4189-2013, 2013.

Raynaud, D., Hingray, B., Zin, I., Anquetin, S., Debionne, S., and Vautard, R.: Atmospheric analogues for physically consistent scenarios of surface weather in Europe and Maghreb, Int. J. Climatol., 37, 2160–2176, https://doi.org/10.1002/joc.4844, 2016.

Ribalaygua, J., Torres, L., Pórtoles, J., Monjo, R., Gaitán, E., and Pino, M. R.: Description and validation of a two-step analogue/regression downscaling method, Theor. Appl. Climatol., 114, 253–269, https://doi.org/10.1007/s00704-013-0836-x, 2013.

Rupp, D. E., Licznar, P., Adamowski, W., and Leśniewski, M.: Multiplicative cascade models for fine spatial downscaling of rainfall: parameterization with rain gauge data, Hydrol. Earth Syst. Sci., 16, 671–684, https://doi.org/10.5194/hess-16-671-2012, 2012.

Schwarz, G.: Estimating the dimension of a model, Ann. Stat., 6, 461–464, https://doi.org/10.1214/aos/1176344136, 1978.

Stern, R. D. and Coe, R.: A model fitting analysis of daily rainfall data, J. Roy. Stat. Soc. A Sta., 147, 1, https://doi.org/10.2307/2981736, 1984.

Teweless, J. and Wobus, H.: Verification of prognosis charts, B. Am. Meteorol. Soc., 35, 2599–2617, 1954.

Thober, S., Mai, J., Zink, M., and Samaniego, L.: Stochastic temporal disaggregation of monthly precipitation for regional gridded data sets, Water Resour. Res., 50, 8714–8735, https://doi.org/10.1002/2014WR015930, 2014.

Uppala, S. M., Kallberg, P. W., Simmons, A. J., Andrae, U., Bechtold, V. D., Fiorino, M., Gibson, J. K., Haseler, J., Hernandez, A., Kelly, G. A., Li, X., Onogi, K., Saarinen, S., Sokka, N., Allan, R. P., Andersson, E., Arpe, K., Balmaseda, M. A., Beljaars, A. C. M., Van De Berg, L., Bidlot, J., Bormann, N., Caires, S., Chevallier, F., Dethof, A., Dragosavac, M., Fisher, M., Fuentes, M., Hagemann, S., Holm, E., Hoskins, B. J., Isaksen, L., Janssen, P. A. E. M., Jenne, R., McNally, A. P., Mahfouf, J. F., Morcrette, J. J., Rayner, N. A., Saunders, R. W., Simon, P., Sterl, A., Trenberth, K. E., Untch, A., Vasiljevic, D., Viterbo, P., and Woollen, J.: The ERA-40 re-analysis, Q. J. Roy. Meteor. Soc., 131, 2961–3012, https://doi.org/10.1256/qj.04.176, 2005.

Vidal, J.-P., Martin, E., Franchisteguy, L., Baillon, M., and Soubeyroux, J.-M.: A 50-year high-resolution atmospheric reanalysis over France with the Safran system, Int. J. Climatol., 30, 1627–1644, https://doi.org/10.1002/joc.2003, 2010.

Wetterhall, F., Bardossy, A., Chen, D., Halldin, S., and Xu, C.-Y.: Statistical downscaling of daily precipitation over Sweden using GCM output, Theor. Appl. Climatol., 96, 95–103, https://doi.org/10.1007/s00704-008-0038-0, 2009.

Wilby, R. L., Hay, L. E., and Leavesley, G. H.: A comparison of downscaled and raw GCM output: implications for climate change scenarios in the San Juan River basin, Colorado, J. Hydrol., 225, 67–91, https://doi.org/10.1016/S0022-1694(99)00136-5, 1999.

Woodcock, F.: On the use of analogues to improve regression forecasts, Mon. Weather Rev., 108, 292–297, 1980.

Yang, C., Chandler, R., Isham, V., and Wheater, H.: Spatial-temporal rainfall simulation using generalised linear models, Water Resour. Res., 41, W11415, https://doi.org/10.1029/2004WR003739, 2005.

9

Active heat pulse sensing of 3-D-flow fields in streambeds

Eddie W. Banks[1], **Margaret A. Shanafield**[1], **Saskia Noorduijn**[1], **James McCallum**[1], **Jörg Lewandowski**[2,3], and **Okke Batelaan**[1]

[1]National Centre for Groundwater Research and Training and the College of Science and Engineering, Flinders University, Adelaide, South Australia, Australia

[2]Department Ecohydrology, IGB, Leibniz-Institute of Freshwater Ecology and Inland Fisheries, Berlin, Germany

[3]Geography Department, Humboldt University of Berlin, Berlin, Germany

Correspondence: Eddie W. Banks (eddie.banks@flinders.edu.au)

Abstract. Profiles of temperature time series are commonly used to determine hyporheic flow patterns and hydraulic dynamics in the streambed sediments. Although hyporheic flows are 3-D, past research has focused on determining the magnitude of the vertical flow component and how this varies spatially. This study used a portable 56-sensor, 3-D temperature array with three heat pulse sources to measure the flow direction and magnitude up to 200 mm below the water–sediment interface. Short, 1 min heat pulses were injected at one of the three heat sources and the temperature response was monitored over a period of 30 min. Breakthrough curves from each of the sensors were analysed using a heat transport equation. Parameter estimation and uncertainty analysis was undertaken using the differential evolution adaptive metropolis (DREAM) algorithm, an adaption of the Markov chain Monte Carlo method, to estimate the flux and its orientation. Measurements were conducted in the field and in a sand tank under an extensive range of controlled hydraulic conditions to validate the method. The use of short-duration heat pulses provided a rapid, accurate assessment technique for determining dynamic and multi-directional flow patterns in the hyporheic zone and is a basis for improved understanding of biogeochemical processes at the water–streambed interface.

1 Introduction

Application of heat as a tracer to hydrological studies has rapidly progressed in recent decades, driven by the simplicity of the methodology and low cost of sensor technology (Anderson, 2005; Rau et al., 2014). Using this method, spatial and temporal flow dynamics within the hyporheic zone, particularly hyporheic transport and exchange (e.g. longer attenuation), have been shown to enhance stream denitrification (Harvey et al., 2013; Gomez-Velez et al., 2015; Zarnetske et al., 2011), degradation of mine-pollutants (Gandy et al., 2007) and the degradation of wastewater micro-pollutants (Engelhardt et al., 2013). It is also widely used by other disciplines, e.g. ecology, where the thermal regime in river systems plays an important role in ecosystem health (Caissie, 2006; Harvey and Wagner, 2000; Brunke and Gonser, 1997; Boulton et al., 1998).

The majority of streambed heat tracer studies use vertical, ambient temperature profiles and a one-dimensional analytical solution of the heat diffusion–advection equation to estimate streambed exchange fluxes and infer hyporheic flow patterns (Constantz et al., 2002; Naranjo and Turcotte, 2015; Rau et al., 2010; Vogt et al., 2010). Series of vertical temperature profile sticks installed along transects have also be used in other studies to examine 2-D flow fields in the streambed (Constantz et al., 2013, 2016; Shanafield et al., 2010). When using ambient temperature fluctuations and one-dimensional heat transport models, several days of data are required to estimate the vertical flux. In addition, an assumption is made that the dominant exchange process is in the vertical direction only and the horizontal or lateral component of flow is considered to be negligible. There are very few investigations which have tried to capture both the vertical and horizontal component of flow, as the determination of the non-vertical component is challenging with the physical installation of sensors to measure the flow field as well as the mathematical framework to process the data (Munz et al., 2016; Briggs et al., 2012; Shanafield et al., 2016).

More recently, the suitability of active temperature sensing has been explored as an approach to characterise streambed spatial and temporal exchange dynamics in three dimensions. The injection of heat as a tracer is not new, with a number of studies using the active temperature-sensing technique to evaluate groundwater flow within wells (Sellwood et al., 2016; Read et al., 2014; Banks et al., 2014), flow in sediments (Greswell et al., 2009; Ballard, 1996; Bakker et al., 2015), surface water–groundwater exchange processes (Kurth et al., 2015) and hyporheic exchange flows in the hyporheic zone (Angermann et al., 2012a, b; Lewandowski et al., 2011).

The aim of the present study was to develop an active heat-pulse-sensing (HPS) instrument to conduct rapid assessments of the three-dimensional (3-D) flow field in the streambed from fine silt to coarse gravels across different geomorphological structures. It builds upon previous studies by Lewandowski et al. (2011) and Angermann (2012a, b), who developed an active heat pulse sensor to determine the flow direction and flow velocity in shallow sandy stream environments. In the present study we have developed a more robust field instrument and advanced the analysis of the temperature breakthrough data using the analytical solution of the heat transport equation for a 3-D array. Parameter estimation and uncertainty analysis was implemented using the differential evolution adaptive metropolis (DREAM) algorithm (Vrugt et al., 2009c), an adaption of the standard Markov chain Monte Carlo method, to determine the direction and magnitude of flow velocity patterns in the streambed at multiple depths. Laboratory tests were conducted in a sand tank using an extensive range of flow scenarios with tightly controlled hydraulic conditions to evaluate the methodology. Field tests demonstrated the active heat pulse instrument in different geomorphological structures in a small stream in the Mount Lofty Ranges, South Australia.

2 Material and methods

2.1 General design and operating principles

A 56-sensor, 3-D temperature array with three heat pulse sources (also known as the Hot Rod) was developed to measure the flow direction and magnitude up to 200 mm be-

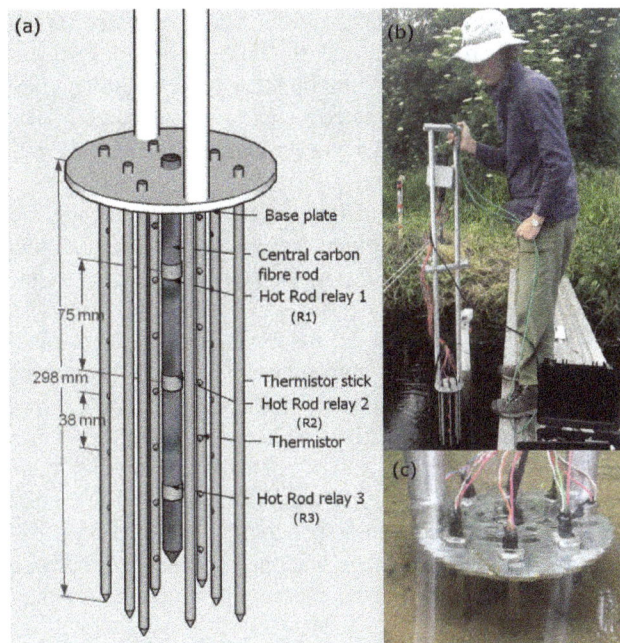

Figure 1. (a) Detailed design of the active HPS Hot Rod with three heating elements (R1, R2 and R3) on the central carbon-fibre rod surrounded by 56 temperature sensors at two distances from the central heat source (28 and 47 mm). **(b)** and **(c)** Installation of the Hot Rod in a small stream characterised by shallow bedforms.

low the water–sediment interface in the streambed (Fig. 1). The central carbon-fibre rod (260 mm long with a diameter of 12 mm) has three equally spaced, 60 W heating elements along its length at positions of 65, 140 and 215 mm below the base plate and are referred to as heat injection depth R1, R2 and R3, respectively (Fig. 1). Eight stainless steel rods (6 mm diameter and 298 mm long) housing seven equally spaced (38 mm apart) temperature thermistors (Maxim DS18B20; precision 0.06°) are arranged cylindrically around the carbon-fibre rod and at two fixed spacings of 28 and 47 mm (Table S1 in the Supplement). The central rod and thermistor sticks are fixed to a rigid circular aluminium base plate which is attached to a collapsible handle. The material, dimensions and spacing of the rods to the base plate were designed to reduce flexibility and minimise disturbance to the sediment material on insertion, as this was a limitation in previous studies. A critical design feature was to ensure that the rods stayed parallel to one another and that the spacing did not vary during installation, as it was designed to be used in a range of sedimentary environments from fine silt to coarse gravels.

A terminal program is used to communicate with the data logger and to control the sampling routine of the Hot Rod (e.g. sampling frequency, duration and power output of the active heat pulse, selected heating element used and logging period). A sealed 12 V lead acid battery together with a power supply regulator is used to maintain a constant 12 V output to the heating elements. The power delivered to the three heating elements can be adjusted in the logger program from 0 to 100 % to provide greater flexibility to the required active heat pulse. The input current from the power supply regulator is also recorded each time the temperature is measured to ensure a tight control on the actual power being delivered through the heating elements to the surrounding material and therefore reducing uncertainty in the analysis routine.

An important aspect of the design was that it could be rapidly and easily deployed to capture large spatial data sets along a reach of stream or across a pool–riffle sequence. Installation requires gently pushing (or lightly tapping with a shockless impact hammer) the device into the streambed ensuring that there is a sufficient gap between the top of the sediment and the underside of the base plate to prevent streamflow constriction (the top thermistor is set at 44 mm below the baseplate so a gap of ~ 30 mm puts the first thermistor just below the sediment–water interface). The impact of the installed device on flow velocity and direction is expected to be minimal given that the volume of the device relative to the volume of the measurement area is less than 4 %.

Once installed and equilibrated with the surrounding sediment, the logger program is executed and the ambient temperature is measured at each thermistor (T_0), directly followed by activation of the selected heat element for the chosen duration. The data logger records the temperature differential ($\Delta T = T_t - T_0$) at the chosen sample frequency to clearly discern the timing and location of the breakthrough curve at each of the thermistors. Field and laboratory tests showed that short, 1 min heat pulse injections and a 20–30 min temperature response monitoring period are appropriate for estimating the dominant direction and flux magnitude in sandy streambeds. Specific details of the analysis routine that was used are described in the following section.

2.2 Data analysis and routine outputs

2.2.1 Heat transport simulation

The magnitude and direction of the water velocity at the observation point based on the measured temperature breakthrough curves at the 56 sensors were determined using a modified version of the heat transport equation:

$$\frac{\partial T}{\partial t} = \nabla \left(D^t \nabla T \right) - \nabla \left(\frac{\rho_w c_w}{\rho c} q T \right), \tag{1}$$

where T is temperature (°C) and D^t is the thermal dispersion coefficient given as (de Marsily, 1986)

$$D_n^t = \frac{\kappa_0}{\rho c} + \beta_n \cdot \left| \frac{\rho_w c_w}{\rho c} q \right|, \tag{2}$$

where κ_0 is the bulk thermal conductivity of the water-saturated sediments (W m^{-1}°C^{-1}), ρc is the volumetric heat capacity of the water-saturated sediments (J m^{-3}°C^{-1}), β_n is the thermal dispersivity where the subscript n is

T for the transverse direction and L for the longitudinal direction, $\rho_w c_w$ is the volumetric heat capacity of water (4.1 J m^{-3}°C^{-1}) and q is the Darcy velocity (or Darcy flux) of water (m s^{-1}). Where $D_L^t \approx D_T^t$, Eq. (2) can be simplified to $D_n^t = \frac{\kappa_0}{\rho c}$.

An analogy can be made to the solute transport equation where the mean water velocity is replaced with $\frac{\rho_w c_w}{\rho c} q$ and the dispersion tensor can be replaced with D^t. Assuming that these components are constant in space and time, Eq. (1) can be reduced to

$$\frac{\partial T}{\partial t} = D^t \nabla^2 T - \frac{\rho_w c_w}{\rho c} q \nabla T. \tag{3}$$

An instantaneous injection of a thermal mass into an infinite three-dimensional sediment volume where there is only an x component of velocities can be defined as follows (Domenico and Schwartz, 1998):

$$T(x, y, z, t) = \frac{M_0}{8 \rho c D_L^t{}^{\frac{1}{2}} D_T^t (\pi t)^{\frac{3}{2}}}$$
$$\exp \left(-\frac{\left(x - \frac{\rho_w c_w}{\rho c} q t \right)^2}{4 D_L^t t} - \frac{\left(y^2 + z^2 \right)}{4 D_T^t t} \right), \tag{4}$$

where M_0 is the thermal mass (J).

The thermal mass input was not considered in previous studies as a known parameter (Angermann et al., 2012b). The Hot Rod, however, uses a known wattage, and the thermal mass term is given as

$$M_0 = F \cdot dt, \tag{5}$$

where F is the heat flux and dt is the duration of application. The thermal mass input, M_0, is measured by the data logger such that at full power the theoretical output of the 60 W heating element provides 5 A of current and an injection period of 60 s equals an energy input of 3600 J.

The requirement of Eq. (4) is that the flow component is only in the x direction, removing the non-diagonal components of the dispersion tensor. The aim of this paper is to define a flow direction and magnitude using a fixed sensor array, allowing the use of multiple sensors to constrain the thermal transport and flow properties. It is not always the case that the fluxes are oriented with the sensor array and therefore the location of the observations can be converted through rotation of the coordinate system, aligning the measurements relative to the flow direction. In this application we assume that the vector of Darcy fluxes in the x, y and z direction is defined as

$$q = \begin{bmatrix} q_x \\ q_y \\ q_z \end{bmatrix}. \tag{6}$$

The coordinate system is first rotated such that the points are orientated around the z axis and we define the angle θ

(Fig. 2), where

$$\theta = \tan^{-1}\left(\frac{q_y}{q_x}\right). \tag{7}$$

The rotational Jacobian matrix is then defined as

$$\mathbf{J} = \begin{bmatrix} \cos\theta & \sin\theta & 0 \\ -\sin\theta & \cos\theta & 0 \\ 0 & 0 & 1 \end{bmatrix}, \tag{8}$$

the coordinates as

$$x' = \mathbf{J}x, \tag{9}$$

and the flux as

$$q' = \mathbf{J}q. \tag{10}$$

Secondly, we rotate the points around the y axis. To do this we define the angle ϕ, where

$$\phi = \tan^{-1}\left(\frac{q_z'}{q_x'}\right). \tag{11}$$

The rotational Jacobian matrix is then defined as

$$\mathbf{J} = \begin{bmatrix} \cos\phi & 0 & \sin\phi \\ 0 & 1 & 0 \\ -\sin\phi & 0 & \cos\phi \end{bmatrix}, \tag{12}$$

the coordinates as

$$x'' = \mathbf{J}x', \tag{13}$$

and the flux as

$$q'' = \mathbf{J}q'. \tag{14}$$

The transformation results in the representation of the Darcy fluxes as $q_x'' = \|q\|$ (the magnitude of the non-transformed flow vector) and $q_y'' = q_z'' = 0$. The distances of the sensors are oriented relative to this new flow vector. The main advantage over previous approaches is that all sensors can be included rather than a single sensor from the array accounting for a single transverse distance (Lewandowski et al., 2011).

We then substitute these new dimensions into Eq. (4) to get

$$T(x'', y'', z'', t) = \frac{M_0}{8\rho c D_L^{t\frac{1}{2}} D_T^t (\pi t)^{\frac{3}{2}}}$$

$$\times \exp\left(-\frac{\left(x'' - \frac{\rho_w c_w}{\rho c}\|q\|t\right)^2}{4D_L^t t} - \frac{(y''^2 + z''^2)}{4D_T^t t}\right). \tag{15}$$

Equation (15) represents an impulse response function. The tests implemented used a finite pulse that did not meet this

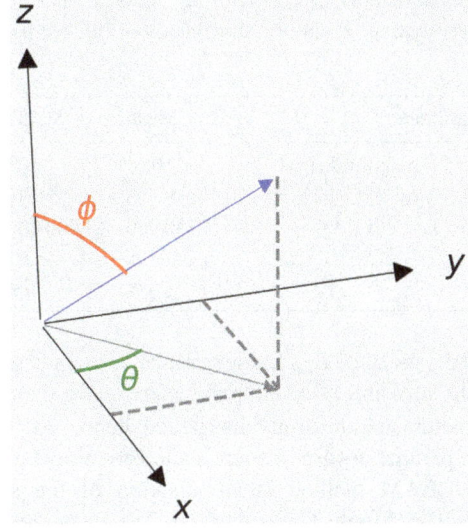

Figure 2. Schematic showing the rotation of the coordinate system to determine angles θ and ϕ.

condition. As we assume that the properties are not temperature dependent, we can treat the contribution of multiple impulse responses as additive. Hence, Eq. (15) was implemented for a series of discrete, lagged pulses to represent the actual addition of thermal mass to the system. The use of discrete pulses is implemented as

$$T_{\text{tot}}(x'', y'', z'', t) = \sum_{\tau=t_{\text{on}}}^{t_{\text{off}}} T(x'', y'', z'', t - \tau), \tag{16}$$

where T_{tot} is the total temperature response, t_{on} is the time at which the heating element was turned on and t_{off} is the time when the heating element was turned off. The operator τ represents the time lags, and for each discrete pulse $M_o = F \cdot d\tau$. The summed T term on the right-hand side is evaluated using Eq. (15) for $t \geq \tau$ and is zero otherwise. This method allows for the representation of a non-Dirac heat pulse and also for variations in the input flux from the heating elements. The interval $d\tau$ was 3 s; hence, the Dirac representation was valid on a small scale.

2.2.2 Parameter estimation

Parameter estimation and uncertainty analysis was undertaken using the DREAM algorithm (Vrugt et al., 2009a, b). The fit of the data was performed by assessing the likelihood of each individual model run. The likelihood is defined as

$$L = -\left(\sum_1^{n_{\text{obs}}} \ln\left(p_{\text{obs}}\left(T_{\text{mod}}|T_{\text{obs}}, \sigma^2\right)\right) + \sum_1^{n_{\text{pars}}} \ln\left(p_{\text{par}}(X)\right)\right), \tag{17}$$

where T_{mod} and T_{obs} represent the modelled and observed temperatures, respectively; σ^2 represents the error of the temperature observation squared; p_{obs} represents the probability of the modelled observation, assuming a normal distri-

Table 1. Initial parameter values for $\|\mathbf{q}\|$, θ, ϕ, κ_0 and ρc used in the analysis routine. A uniform distribution of the five parameters was used.

Parameter	Mean	Width
$\|\boldsymbol{q}\|$ – magnitude ($\mathrm{m\,s^{-1}}$)	10^{-5}	10^3
κ_0 ($\mathrm{W\,m^{-1}{}^{\circ}C^{-1}}$)	3	1.25
ρc ($\mathrm{J\,m^{-3}{}^{\circ}C^{-1}}$)	2 750 000	1 500 000
θ	π	2π
ϕ	π	2π

bution and a mean of T_{obs} and a variance of σ^2; and p_{par} represents the probability of the parameter X. We assume that the parameters are uniformly distributed; hence, $p_{\mathrm{par}}(X) = 1$ when the parameters are in bounds and zero elsewhere.

The DREAM method is an adaption of the standard Markov chain Monte Carlo method. The technique is initialised by specifying a number of chains. Each chain receives starting parameters by randomly sampling the parameter ranges (Table 1). After calculating an initial likelihood for the starting parameters, the algorithm selects proposed parameter values using the other chains, and the likelihood of these model parameters are also calculated. If the likelihood of these parameters is greater than the current parameters, the new parameters are accepted; however, if the likelihood of the new parameters is lower, the transition probability is calculated using a ratio of the likelihoods, and the transition is determined by generating a random number. The method is explained in greater detail in Vrugt et al. (2009c). The general outcome is that the chains spend a greater amount of time in locations of more favourable parameters, and the distribution of these parameters represents the posterior distribution of the parameter probabilities, given the model, the data and the prior knowledge of the parameter distributions.

The optimisation was undertaken using five parameters: $\|\boldsymbol{q}\|$, θ, ϕ, κ_0 and ρc. All of the parameters in the thermal dispersion term (Eq. 2) cannot be identified simultaneously in the optimisation, resulting in non-convergence of the Markov Chain approach because of the correlation between the thermal dispersivity flow term and the thermal conductivity. In the experiments that we conducted we found that the longitudinal and transverse dispersion terms, $D_L^t \approx D_T^t$ and therefore Eq. (2) can be simplified to $D_n^t = \frac{\kappa_0}{\rho c}$. Hence, κ_0 was included as an optimisation parameter and it was also physically measured in the experiments. The default parameter likelihoods and initial distributions were taken from the ranges presented in Table 1. The error of the temperature observations (σ) was taken to be 0.06 °C, the precision of the temperature sensors. Whilst the model parameters are estimated using angle and the flow magnitude, the actual flow vector can be recovered as

$$\boldsymbol{q} = \begin{bmatrix} \|\boldsymbol{q}\| \cos(\phi)\cos(\theta) \\ \|\boldsymbol{q}\| \cos(\phi)\sin(\theta) \\ \|\boldsymbol{q}\| \sin(\phi) \end{bmatrix}. \tag{18}$$

2.3 Laboratory sand tank

A laboratory sand tank was used to provide a controlled environment on the hydraulic regime to test the performance of the Hot Rod and so that the different flux calculation methods could be compared. A total of 36 combinations of flow direction, magnitude and depth of heat pulse were used. The dimensions of the sand tank for each of the scenarios varied slightly according to the fixed boundary conditions (Fig. 3). Four flow scenarios were tested: (1) horizontal flow from left to right (inflow and outflow occurred over the entire saturated cross-sectional area of the sediment volume on the left and right boundaries of the tank), (2) diagonal flow from the top left to bottom right (inflow was through a 20 mm horizontal inlet slit on the top left boundary and outflow through a 20 mm horizontal slit, 85 mm above the base of the right boundary), (3) upward flow (inflow occurred over the cross-sectional area of the base of the tank and outflow at the top of the sediment at overflow points above the sediment on the left and right boundaries), and (4) downward flow (inflow was distributed over the cross-sectional area of the sediment surface and outflow via the cross-sectional area of the tank base). A steady state flow regime for each scenario was maintained by constant heads at the inflow and outflow ports of the tank and the use of a peristaltic pump with a highly accurate ultrasonic flowmeter (Atrato ultrasonic flowmeter; 0.05 % linearity on flow less than $5\,\mathrm{L\,min^{-1}}$) to ensure a constant discharge rate. The flowmeter data were also used to determine the Darcy flux for the different flow conditions. Fine perforated mesh was used to contain the sediment and to provide the necessary flow conditions along each of the tank boundaries. The hydraulic conductivity and thermal hydraulic conductivity of the graded, saturated sand were measured using a KSAT meter (UMS GmbH, Munich, Germany) and KD2 Pro (Decagon, Washington, USA), respectively. Three different hydraulic gradients and discharge rates for the four flow scenarios were conducted to capture low- ($\sim 10^{-6}\,\mathrm{m\,s^{-1}}$), moderate- ($\sim 10^{-5}\,\mathrm{m\,s^{-1}}$) and high-flow conditions ($\sim 10^{-4}\,\mathrm{m\,s^{-1}}$). The Hot Rod was positioned in the middle of the sand tank with thermistor stick sensor number one orientated perpendicular (90°) to the flow direction for all of the scenarios. To validate the spatial arrangement of the sensors, the horizontal flow scenario was repeated with thermistor stick sensor number one rotated to be 45° to the direction of flow. In addition to these flow scenarios, a no-flow experiment was conducted to evaluate the analysis routine, where the boundary conditions meant that there was stagnant water.

2.4 Experimental field site

The Sturt River, Adelaide, Australia, is a perennial river system receiving the majority of its input from a wastewater

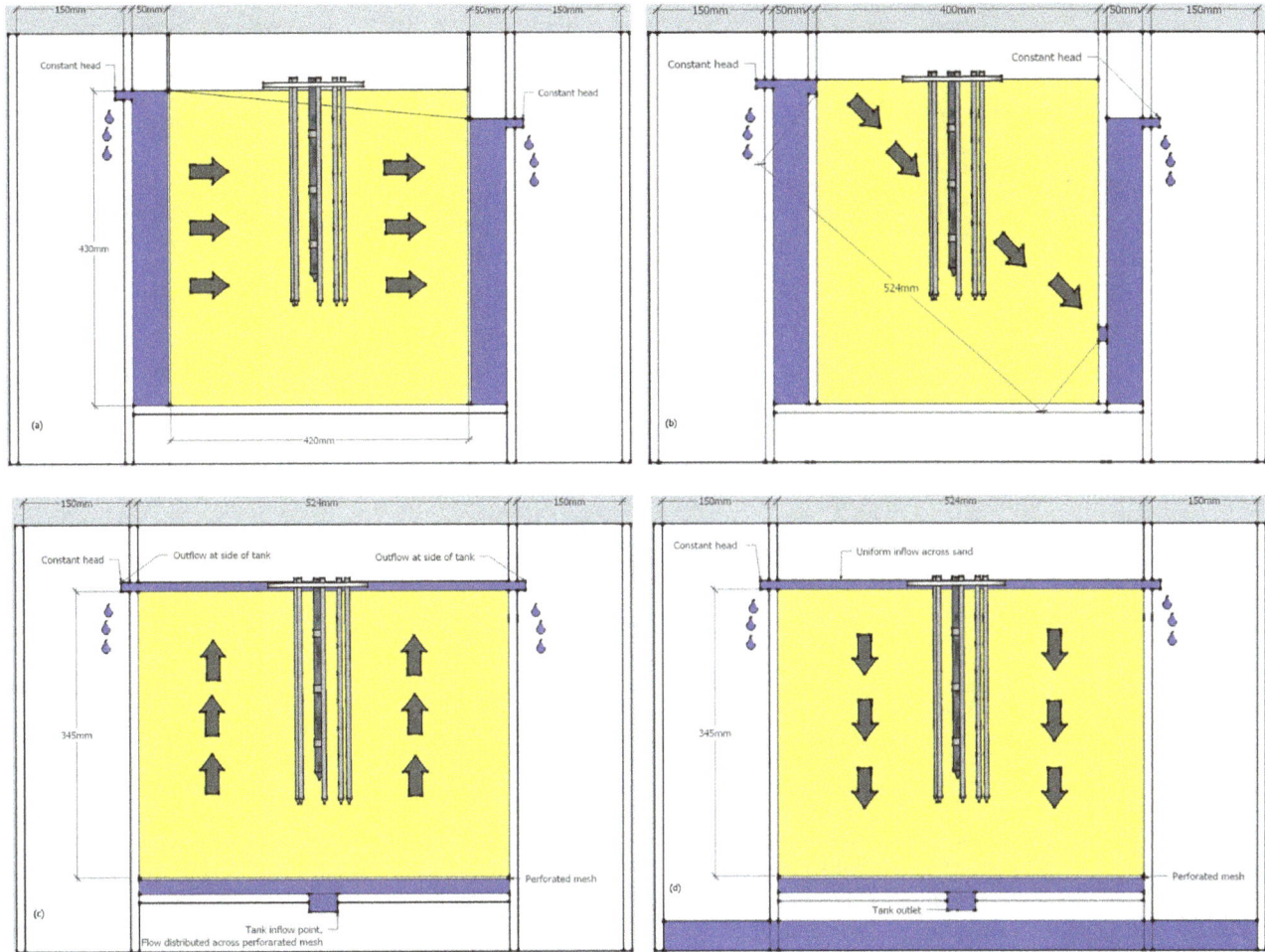

Figure 3. Laboratory sand tank dimensions for each of the flow scenarios: **(a)** horizontal flow, **(b)** diagonal flow, **(c)** upward flow and **(d)** downward flow.

treatment facility. The geomorphology of the river was characterised by a narrow channel, no more than 3 m wide with 0.3–0.5 m deep sediment ranging from fine silt to coarse gravels overlying a tight, low-permeability clay. The selection of this field site was part of another ongoing investigation looking at attenuation of micro-pollutants in the hyporheic zone. The residence time in the hyporheic zone was critical in evaluating the stream attenuation modelling.

3 Results and discussion

3.1 Laboratory sand tank

Overall, the modelled breakthrough curves closely fit the observed data from the 56-sensor array with the modelled curves capturing the rising limb, peak and tail of the measured temperature data over the sample period (Fig. 3). The variance of each parameter is included in the modelled temperature breakthrough curves; however, the uncer-

tainty is so small that it cannot be seen without zooming in on the individual curves. Selected breakthrough curve plots are shown of the fourth vertical thermistor (158 mm depth) from each radial sensor location. Four flow scenarios are presented: (a) horizontal (Fig. 4), (b) diagonal (Fig. S1-ii in the Supplement), (c) upwards (Fig. S1-iii) and (d) downwards (Fig. S1-iv). The tests presented were conducted at a moderate flow rate. Thermistor 4 was 158 mm below the base plate and at a greater depth than the heat injection depth (R2; at 140 mm). Temperature increases associated with the heat pulse were observed at the inner sensor sticks (28 mm) more quickly than at the outer (47 mm) sensor sticks in the horizontal flow scenario (see Fig. 4d and h). The inner sensors also displayed a steeper rising and falling limb compared to the outer sensors. The temperature response reflected the sensor position relative to the dominant flow direction. For example, sensor T3–4 (Fig. 4c) was directly in line and down-gradient of the heat pulse, whilst sensor T7–4 (Fig. 4g) was in line but up-gradient of the heat pulse and therefore showed a smaller response. The breakthrough

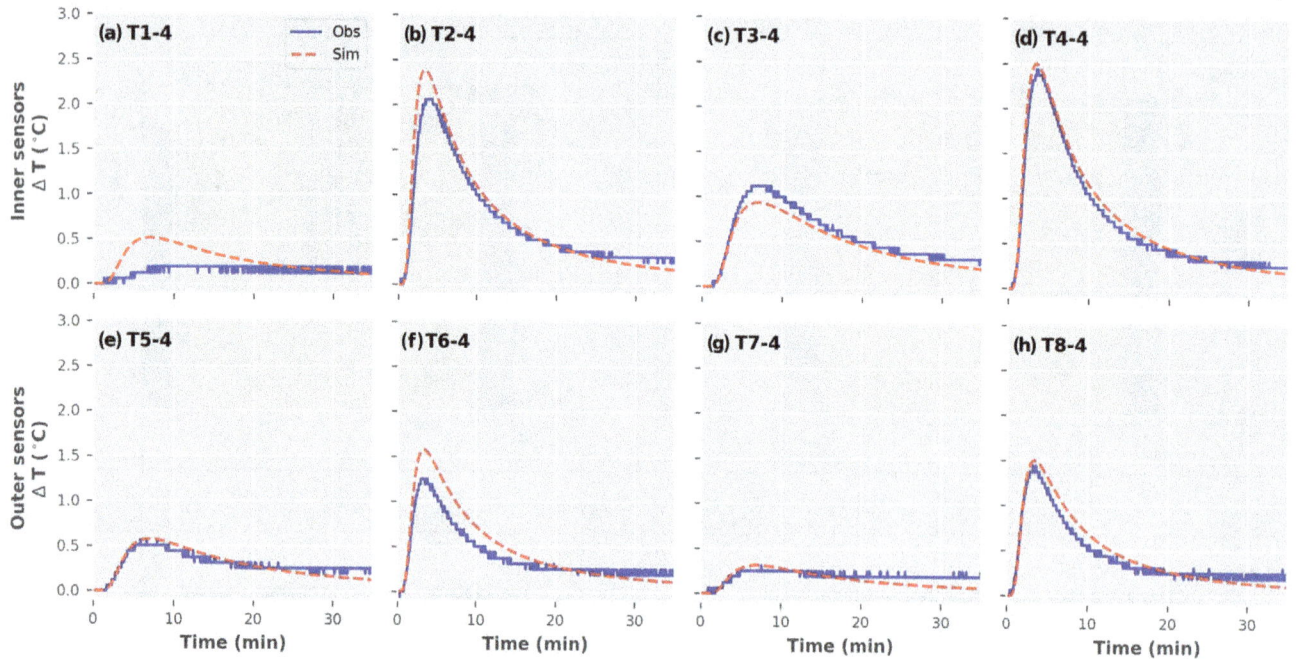

Figure 4. Breakthrough curves shown of the fourth vertical thermistor (158 mm depth) from each radial sensor location for the horizontal flow scenario from heat injection depth at relay 2. Solid lines are observed and dashed lines are modelled.

curves from the diagonal flow scenario showed a similar response in those sensors up- and down-gradient of the heat pulse (Fig. S1a). In the upward and downward flow scenarios there was very little response in the outer thermistor sensors due to the dominant vertical component of flow (Fig. S1b–c). This low response of the outer sensors was even more pronounced under higher flow conditions in the upward and downward flow scenarios. The 3-D time series videos showed clearly the migration of the heated plume vertically and highlighted the complexity of fitting multiple breakthrough curves to the most likely solution.

Overall, the 3-D flow fields calculated from the HPS Hot Rod in the laboratory sand tank for the four flux scenarios and heat injection depths (65, 140 and 215 mm) were consistent with the flow conditions established in the tank (Fig. 5). Under left to right horizontal flow conditions (Fig. 5a and b), the modelled direction of flow from each of the heat injection depths is very similar with some slight offset to the observed flow in the y direction, which was perpendicular to thermistor stick sensor 1 (90° to the flow direction). There was also a slight deviation downwards in the z direction, particularly for the shallowest heat injection depth. To refute any bias in the optimisation routine and the array configuration, the Hot Rod was rotated by 45° for the horizontal flow scenario and it showed a very similar output to when it was orientated 90° to the flow direction.

Results from the other three scenarios showed that the modelled flux direction is close to parallel to the flow con-

ditions established in the tank and the magnitude of the flux was similar at each of the heat injection depths (Fig. 5c–h). Reviewing the time series data in the 3-D plots (Supplement), the spreading of the heat pulse from the heat injection depth can be clearly detected, indicating how the heat pulse moves along the established flow line. The thermistor highlighted in blue in the 3-D plot was the sensor that showed the maximum temperature breakthrough curve and clearly shows a different orientation to the most likely flux direction (black arrow) as determined by the DREAM algorithm.

Some discrepancies in the direction of the modelled flow and differences in flux magnitude at each heat injection depth may be attributed to (1) placement orientation and the angle of the sensor positions of the Hot Rod relative to the flow conditions established in the tank, (2) boundary conditions in the tank to establish flow and (3) the fact that the optimisation routine determines the best fit of all the observed data in a 3-D volume around the heat injection depth rather than at a specific point.

The difference in flux magnitude at each heat injection depth may be attributed to the number of sensors that were used in the optimisation routine. For example, at the heat injection depth R2 (140 mm) there are three sensor arrays (an array being eight sensors positioned horizontally around the central carbon-fibre rod) vertically above R2 and four sensor arrays vertically below R2. In comparison, at heat injection depth R1 (65 mm) there is only one sensor array vertically above R1 and six sensor arrays below R2, which may limit

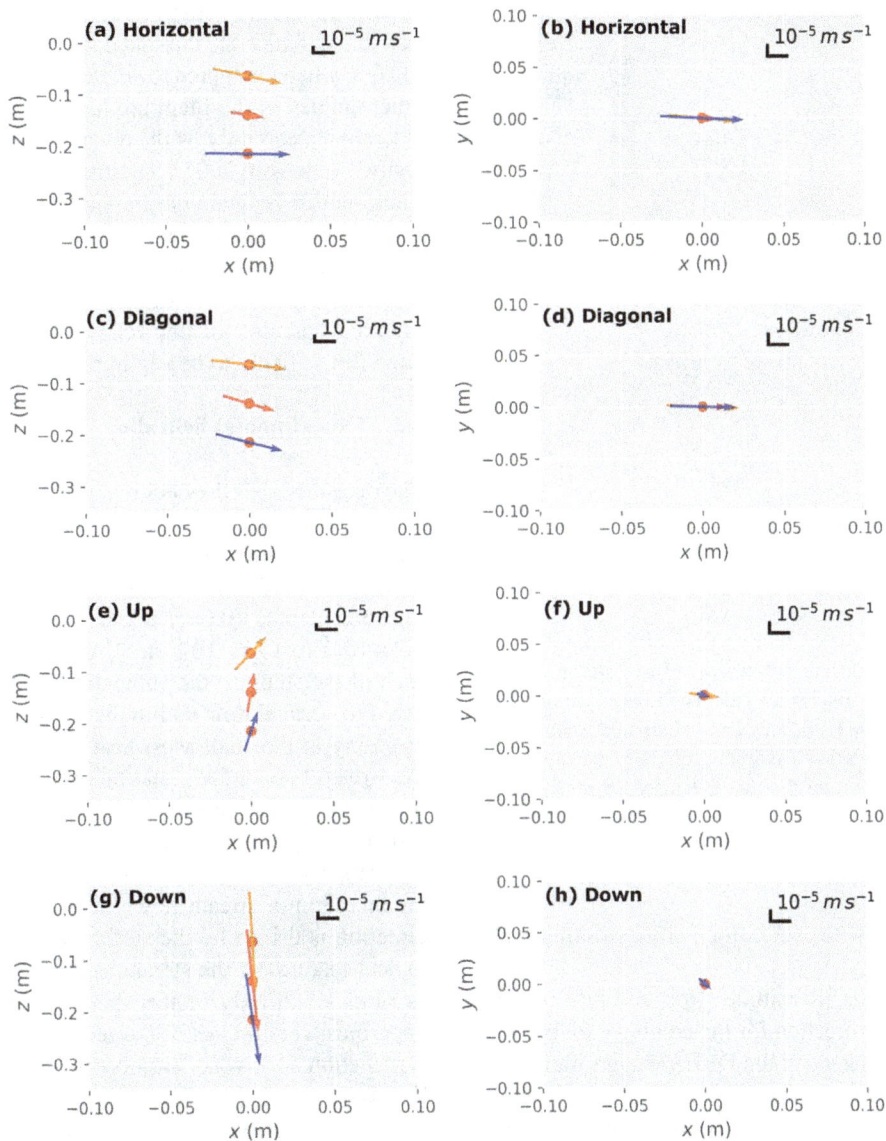

Figure 5. Calculated fluxes and directions for the four flow scenarios at each of the heat injection depths: (**a, b**) horizontal, (**c, d**) diagonal, (**e, f**) upward and (**g, h**) downward flow scenarios.

the optimisation routines for particular flow conditions i.e. strongly upwards flow.

In the case of no-flow conditions established in the sand tank (Fig. S2), the optimisation routine fitted the measured temperature breakthrough data; however, on closer inspection of the 3-D time series plot it was evident based on the uniform heat plume around the heat injection point during the injection period that heat transport was by conduction only. Absence of clear advective movement of the heat pulse and a calculated flux less than about $\sim 10^{-6}$ m s^{-1} indicated the lower limit of the active heat pulse sensor.

The flux magnitude ($\|\boldsymbol{q}\|$) of the different flow scenarios and different flow intensities calculated based on different heat injection depths of the HPS (grey bars) compared to the

fluxes determined based on Darcy's law (hydraulic gradient and hydraulic conductivity) and the measured discharge from the tank using an ultrasonic flowmeter is shown in Fig. 6 and Table S2.

The measured saturated hydraulic conductivity according to the KSAT meter for the sand tank sand was 4.95×10^{-4} m s^{-1}. The measured saturated thermal conductivity of the sand was 3 W m^{-1}°C^{-1} (using the Decagon KD2 Pro), whilst the average modelled κ_0 from all of the flow scenarios in the sand tank was 3.8 W m^{-1}°C^{-1}. The thermal conductivity is strongly influenced by the porosity, and therefore some differences can be expected due to changes to the particle density with the packing of the sediment, where by thermal

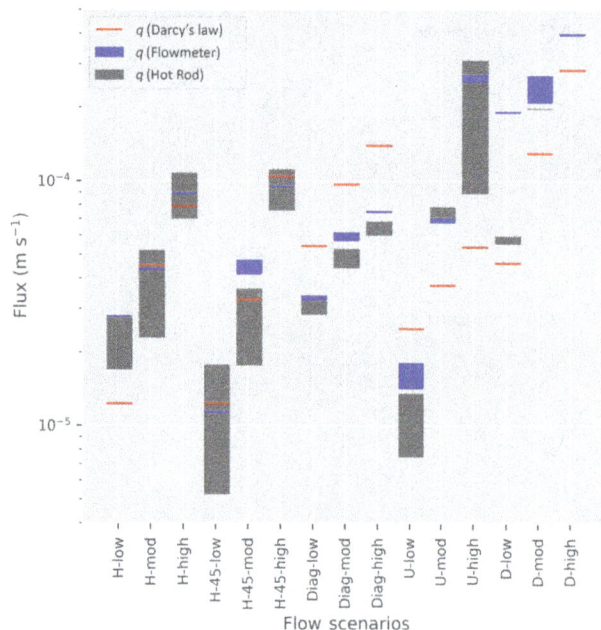

Figure 6. Fluxes of the different flow scenarios, listed along the x axis, and different flow intensities calculated based on different heat injection depths of the HPS (grey bars) compared to fluxes determined based on Darcy's law (hydraulic gradient and hydraulic conductivity – red dashes) and the flux calculated from the measured discharge from the tank using an ultrasonic flowmeter and the cross-sectional area of the tank (blue dashes).

conductivity increases with decreasing porosity (Smits et al., 2010).

Histograms of the flux magnitude ($\|q\|$) and flux components in the x, y and z direction for the combination of most likely parameter values used in the DREAM algorithm were generated for each measurement. The results from heat injection depth R2 for the horizontal flow scenario is shown in Fig. S3, which shows that the distribution is tightly constrained. This is also evident in cross correlation plots of the flux magnitude, thermal conductivity and the specific heat capacity (Fig. S4).

Constraining the range of the thermal conductivity values used in the optimisation routine to the known measured thermal conductivity of the sand from the KD2 Pro instrument showed little impact on the calculated flux magnitude. However, comparing the modelled breakthrough curves from two optimisations when the range in thermal conductivity was limited to the measured known thermal conductivity ($3\,\mathrm{W\,m^{-1}\,^\circ C^{-1}}$) in one model and in the other model it used a value of $3.52\,\mathrm{W\,m^{-1}\,^\circ C^{-1}}$ (for a best fit from a range of values from 2.5 to $4.5\,\mathrm{W\,m^{-1}\,^\circ C^{-1}}$) showed there were subtle differences between the modelled breakthrough curve peak and falling limbs (Fig. 7).

Our study found that the inclusion of longitudinal and transverse thermal dispersion had less than a 2 % difference on the mean calculated fluxes. The study by Rau et al. (2012)

determined Darcy velocities derived from heat experimentation that included the thermal dispersivity term differed by up to 20 % when compared to solute experimentation. However, other studies in the literature have shown that there is considerable uncertainty on the magnitude of the thermal dispersivity (Anderson, 2005). Thermal dispersivity has also been found not to be scale-dependent such as solute dispersivity because heat transport happens through the pore water and through the sediment matrix (Vandenbohede et al., 2009). Therefore, given the scale that we are working at (few centimetres) and also the low velocities, the effect on the calculated flux is likely to be negligible.

3.2 Experimental field site

The measured 3-D flow fields at the experimental site showed considerable variability in the direction of flow and flux magnitude over $\sim 0.20\,\mathrm{m}$ depth of the streambed. The flux magnitude at the six stations along the river at the three heat injection depths (0.065, 0.140 and 0.215 m) ranged from 4.2×10^{-6} to $4.26 \times 10^{-5}\,\mathrm{m\,s^{-1}}$ (mean: $1.6 \times 10^{-5}\,\mathrm{m\,s^{-1}}$). At each of the stations, the component of horizontal flow compared to vertical flow within the streambed was dominant. It was only at the shallowest heat injection depth, just below the stream bed surface, that there was a greater component of vertical flow. The results from two of the six stations are shown in Fig. 8, where the Hot Rod was positioned in the streambed such that the x axis of the figure was aligned to the direction of stream flow. In such environments the flow direction is driven by the surface flow and the geomorphological features of the streambed. Small bed structures such as ripples will only impact the hyporheic flow field in the uppermost centimetres of the sediment, which is a plausible explanation as to what was observed from the outputs of the Hot Rod.

Measured κ_0 values for the sediment collected at the six stations from 0 to 0.2 m depth ranged from 0.9 to $2.84\,\mathrm{W\,m^{-1}\,^\circ C^{-1}}$, (mean: $2.07\,\mathrm{W\,m^{-1}\,^\circ C^{-1}}$) compared to the values that were determined by the optimisation routine, which ranged from 1.0 to $3.74\,\mathrm{W\,m^{-1}\,^\circ C^{-1}}$ (mean: $2.91\,\mathrm{W\,m^{-1}\,^\circ C^{-1}}$). Physical observation of the sediments collected at the experimental site stations showed that the streambed material was quite heterogeneous and also contained varying proportions of organic matter. Higher clay content and organic matter in the sediment would cause a higher volumetric heat capacity compared to sandy sediments. The volumetric heat capacity also increases with increasing moisture content and particle density (Abu-Hamdeh, 2003; Barry-Macaulay et al., 2013; Jury and Horton, 2004). The assumption of uniform thermal properties of the 3-D volume around the heat injection point is likely to contribute to the uncertainty in the flow direction. For example, Su et al. (2006) used numerical simulations to show that differences in the thermal properties of the sediment around a flow sensor can lead to incorrect velocity estimates and

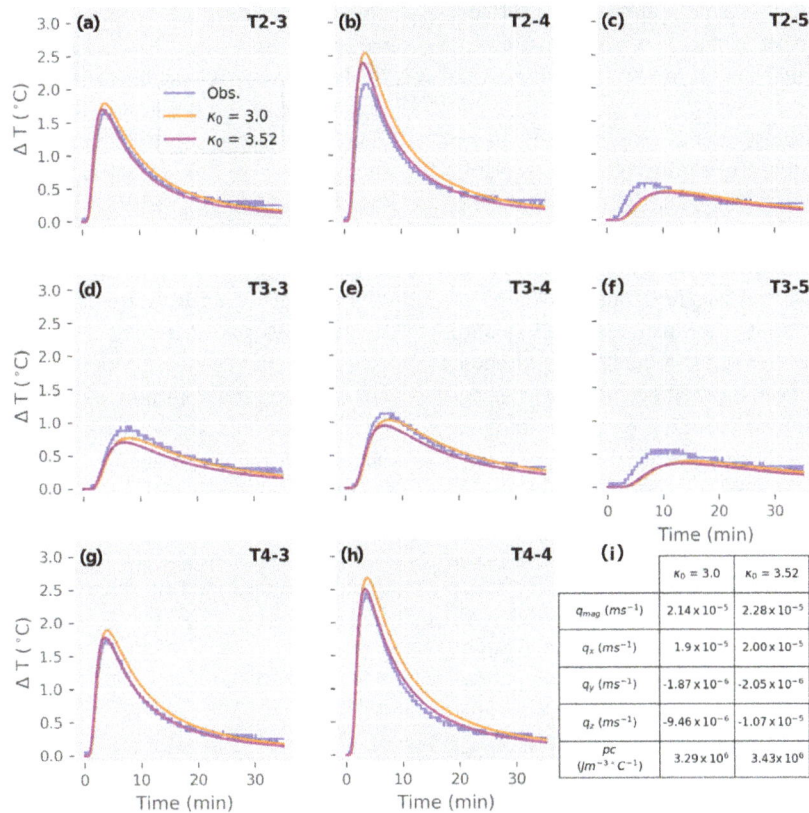

Figure 7. Comparison of the observed (blue line) and modelled breakthrough curves for selected temperature sensors ($Tx - x$) when the thermal conductivity, κ_0, of the sediment has a known value of $3\,\mathrm{W\,m^{-1}\,°C^{-1}}$ (orange graph) and when the model is provided with a range of plausible values from 2.5 to $4.5\,\mathrm{W\,m^{-1}\,°C^{-1}}$ resulting in $3.52\,\mathrm{W\,m^{-1}\,°C^{-1}}$ (red graph).

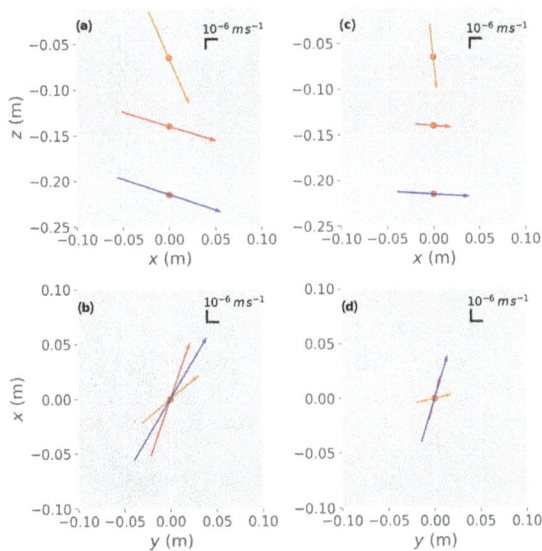

Figure 8. Calculated fluxes and flux directions at the three heat injection depths from two of the stations at the experimental field site. **(a, b)** Station 1 and **(c, d)** Station 2. The x axis in the figures is positive in the direction of streamflow.

in particular differences in the horizontal and vertical flux components. Additional measurements of the streambed sediments would provide a tighter constraint on the parameter set used in the optimisation; however, the parameter estimation and uncertainty analysis routine does successfully fit the measured data to provide a reliable estimate of the flux and its direction. Refer to Banks et al. (2017) for all of the temperature data files from the sand tank and experimental site.

4 Conclusions

Despite the early pioneering work of Lewandowski et al. (2011) and Angermann et al. (2012b) for the concept of a 3-D active heat pulse sensor to determine flux and direction in the shallow streambed, their studies experienced a number of shortcomings that are related to the design of the instrument and the analysis of the data. This included (1) a limited number of sensors and spatial positions around the heating element; (2) weakness with the sensor sticks wobbling and therefore poorly constrained sensor positions in relation to the heating element; (3) limited constraint on the input functions to the heat transport equation, i.e. not knowing the current input; and (4) lack of a suitable optimisation routine to determine the most likely set of parameters to constrain the data and an uncertainty analysis on the flux magnitude and its direction.

The rigidity and robustness of the Hot Rod and use of heating elements at three different vertical positions provided a method to examine how the flux and its direction varied vertically with depth beneath the streambed interface at individual locations in a range of different environmental settings and sediment types. The use of two horizontal spacings between the heating elements and thermistors as well as additional thermistors at multiple angles to the heating elements increased confidence in the measurement of heat transport processes and tightened the optimisation routine of the temperature data. The addition of the measured input of energy at the heating element in the heat transport equation as a series of discrete heat pulses over the injection period provided one less unknown variable to calibrate against. In many of the experiments conducted in the sand tank and at the experimental site, the optimisation routine using the DREAM algorithm showed that the most likely flux direction from the heat injection depth was not towards the sensor that showed the maximum temperature breakthrough because it uses all of the sensor temperature breakthrough curves in the analysis. The 3-D time series plot was a valuable tool in assessing this result and it also showed whether heat transport was dominated by diffusion and/or conduction with radial symmetry around the heating element or whether there was convective heat flow. This interrogation process was found to be critical in the data assessment to ensure that the model did not overfit the measured data with unrealistic physical values for the sediment and heat transport conditions.

The laboratory and experimental field site applications using the DREAM algorithm for parameter estimation and uncertainty analysis demonstrated the performance of the active heat-pulse-sensing instrument (the Hot Rod) to measure the multi-directional 3-D-flow fields and fluxes in the near-surface streambed. Active heat pulse sensing provides a number of advantages over other approaches that have investigated hyporheic exchange, including the low cost of data collection and the rapid assessment of small physical processes that can be undertaken on a reach scale. Marzadri et al. (2013) showed that the hyporheic residence time, which is influenced by the streambed physical morphology and in-stream flow discharge, ultimately determines the spatially complex patterns of the time-varying thermal regime within the hyporheic zone. The short-duration active heat pulse sensing helped overcome some of the challenges in measuring the water temperatures because of the stronger signal from the heat pulse.

Most other studies that use heat as a tracer assume 1-D flow only and the lateral or horizontal component is not considered. Studies that have identified the geometry of the subsurface flow field using a polynomial model fitted to the amplitude ratio of the vertical temperature profiles were only

able to determine the deviation from one-dimensional vertical flow (Munz et al., 2016). Errors in the vertical component of flow have been shown to progressively increase with the magnitude of the horizontal flow component (Lautz, 2010). The 3-D analysis routine and sensor arrangement applied in this study were able to capture all three components of the flow field around the point of observation. The importance of capturing the multi-directional flow field was clearly demonstrated in the sand tank under an extensive range of flow conditions that would be anticipated in a dynamic stream environment. Measurements of the hydraulic gradient and characterising the physical properties of the streambed sediment are also important in understanding the dynamic exchange processes within the hyporheic zone and the very transient nature of such environments.

The concept and design of the active heat-pulse-sensing instrument could also be adapted to other hydrological research areas, including the measurement of shallow interflow along hillslopes and discharge from groundwater seeps and springs.

Competing interests. The authors declare that they have no conflict of interest.

Acknowledgement. We are grateful to Flinders University South Australia for a small grant to develop the HPS Hot Rod and all Faculty of Science and Engineering technical workshop staff for their assistance in hardware development and construction. Funding support from the Australian Research Council (ARC) Linkage Project LP150100588 is acknowledged. Additional funding through the Australia–Germany Joint Research Cooperation Scheme of Universities Australia and the German Academic Exchange Service (DAAD, grant no. 57216806) provided support for fieldwork collaboration between the research institutes.

Edited by: Thom Bogaard

References

Abu-Hamdeh, N. H.: Thermal properties of soils as affected by density and water content, Biosyst. Eng., 86, 97–102, https://doi.org/10.1016/S1537-5110(03)00112-0, 2003.

Anderson, M. P.: Heat as a ground water tracer, Ground Water, 43, 951–968, 2005.

Angermann, L., Krause, S., and Lewandowski, J.: Application of heat pulse injections for investigating shallow hyporheic flow in a lowland river, Water Resour. Res., 48, W00P02, https://doi.org/10.1029/2012WR012564, 2012a.

Angermann, L., Lewandowski, J., Fleckenstein, J. H., and Nützmann, G.: A 3D analysis algorithm to improve interpretation of heat pulse sensor results for the determination of small-scale flow directions and velocities in the hyporheic zone, J. Hydrol., 475, 1–11, https://doi.org/10.1016/j.jhydrol.2012.06.050, 2012b.

Bakker, M., Caljé, R., Schaars, F., van der Made, K.-J., and de Haas, S.: An active heat tracer experiment to determine groundwater velocities using fiber optic cables installed with direct push equipment, Water Resour. Res., 51, 2760–2772, https://doi.org/10.1002/2014WR016632, 2015.

Ballard, S.: The in situ permeable flow sensor: a groundwater flow velocity meter, Ground Water, 34, 231–240, https://doi.org/10.1111/j.1745-6584.1996.tb01883.x, 1996.

Banks, E. W., Shanafield, M. A., and Cook, P. G.: Induced temperature gradients to examine groundwater flowpaths in open boreholes, Groundwater, 52, 943–951, https://doi.org/10.1111/gwat.12157, 2014.

Banks, E.: Temperature time series data for active HPS Hot Rod, Flinders University, https://doi.org/10.4226/86/5aab1b67337bb, 2017.

Barry-Macaulay, D., Bouazza, A., Singh, R. M., Wang, B., and Ranjith, P. G.: Thermal conductivity of soils and rocks from the Melbourne (Australia) region, Eng. Geol., 164, 131–138, https://doi.org/10.1016/j.enggeo.2013.06.014, 2013.

Boulton, A. J., Findlay, S., Marmonier, P., Stanley, E. H., and Valett, H. M.: The functional significance of the hyporheic zone in streams and rivers, Annu. Rev. Ecol. Syst., 29, 59–81, 1998.

Briggs, M. A., Lautz, L. K., McKenzie, J. M., Gordon, R. P., and Hare, D. K.: Using high-resolution distributed temperature sensing to quantify spatial and temporal variability in vertical hyporheic flux, Water Resour. Res., 48, W02527, https://doi.org/10.1029/2011WR011227, 2012.

Brunke, M. and Gonser, T.: The ecological significance of exchange processes between rivers and groundwater, Freshwater Biol., 37, 1–33, 1997.

Caissie, D.: The thermal regime of rivers: a review, Freshwater Biol., 51, 1389–1406, https://doi.org/10.1111/j.1365-2427.2006.01597.x, 2006.

Constantz, J., Stewart, A. E., Niswonger, R., and Sarma, L.: Analysis of temperature profiles for investigating stream losses beneath ephemeral channels, Water Resour. Res., 38, 1–13, https://doi.org/10.1029/2001WR001221, 2002.

Constantz, J., Eddy-Miller, C. A., Wheeler, J. D., and Essaid, H. I.: Streambed exchanges along tributary streams in humid watersheds, Water Resour. Res., 49, 2197–2204, https://doi.org/10.1002/wrcr.20194, 2013.

Constantz, J., Naranjo, R., Niswonger, R., Allander, K., Neilson, B., Rosenberry, D., Smith, D., Rosecrans, C., and Stonestrom, D.: Groundwater exchanges near a channelized versus unmodified stream mouth discharging to a subalpine lake, Water Resour. Res., 52, 2157–2177, https://doi.org/10.1002/2015WR017013, 2016.

de Marsily, G.: Quantitative Hydrogeology: Groundwater Hydrology for Engineers, Academic Press, Cambridge, Massachusetts, United States, 1986.

Domenico, P. A. and Schwartz, F. W.: Physical and Chemical Hydrogeology, vol. 1, Wiley, New Jersey, United States, 1998.

Engelhardt, I., Prommer, H., Moore, C., Schulz, M., Schüth, C., and Ternes, T. A.: Suitability of temperature, hydraulic heads, and acesulfame to quantify wastewater-related fluxes in the hyporheic and riparian zone, Water Resour. Res., 49, 426–440, https://doi.org/10.1029/2012WR012604, 2013.

Gandy, C. J., Smith, J. W. N., and Jarvis, A. P.: Attenuation of mining-derived pollutants in the hyporheic zone: a review, Sci. Total Environ., 373, 435–446, 2007.

Gomez-Velez, J. D., Harvey, J. W., Cardenas, M. B., and Kiel, B.: Denitrification in the Mississippi River network controlled by flow through river bedforms, Nat. Geosci., 8, 941–945, https://doi.org/10.1038/ngeo2567, 2015.

Greswell, R. B., Riley, M. S., Alves, P. F., and Tellam, J. H.: A heat perturbation flow meter for application in soft sediments, J. Hydrol., 370, 73–82, https://doi.org/10.1016/j.jhydrol.2009.02.054, 2009.

Harvey, J. W., Böhlke, J. K., Voytek, M. A., Scott, D., and Tobias, C. R.: Hyporheic zone denitrification: controls on effective reaction depth and contribution to whole-stream mass balance, Water Resour. Res., 49, 6298–6316, https://doi.org/10.1002/wrcr.20492, 2013.

Harvey, W. H. and Wagner, B. J.: Quantifying hydrologic interactions between streams and their subsurface hyporheic zones, in: Streams and Ground Waters, edited by: Jones, J. B. and Mulholland, P. J., Academic Press, San Diego, 3–44, 2000.

Jury, W. A. and Horton, R.: Soil Physics, 6th Edn., John Wiley, New Jersey, United States, 2004.

Kurth, A.-M., Weber, C., and Schirmer, M.: How effective is river restoration in re-establishing groundwater–surface water interactions? – A case study, Hydrol. Earth Syst. Sci., 19, 2663–2672, https://doi.org/10.5194/hess-19-2663-2015, 2015.

Lautz, L. K.: Impacts of nonideal field conditions on vertical water velocity estimates from streambed temperature time series, Water Resour. Res., 46, 1–14, https://doi.org/10.1029/2009WR007917, 2010.

Lewandowski, J., Angermann, L., Nützmann, G., and Fleckenstein, J. H.: A heat pulse technique for the determination of small-scale flow directions and flow velocities in the streambed of sand-bed streams, Hydrol. Process., 25, 3244–3255, https://doi.org/10.1002/hyp.8062, 2011.

Marzadri, A., Tonina, D., and Bellin, A.: Effects of stream morphodynamics on hyporheic zone thermal regime, Water Resour. Res., 49, 2287–2302, https://doi.org/10.1002/wrcr.20199, 2013.

Munz, M., Oswald, S. E., and Schmidt, C.: Analysis of riverbed temperatures to determine the geometry of subsurface water flow around in-stream geomorphological structures, J. Hydrol., 539, 74–87, https://doi.org/10.1016/j.jhydrol.2016.05.012, 2016.

Naranjo, R. C. and Turcotte, R.: A new temperature profiling probe for investigating groundwater-surface water interaction, Water Resour. Res., 51, 7790–7797, https://doi.org/10.1002/2015WR017574, 2015.

Rau, G. C., Andersen, M. S., McCallum, A. M., and Acworth, R.: Analytical methods that use natural heat as a tracer to quantify surface water–groundwater exchange, evaluated using field temperature records, Hydrogeol. J., 18, 1093–1110, https://doi.org/10.1007/s10040-010-0586-0, 2010.

Rau, G. C., Andersen, M. S., and Acworth, R. I.: Experimental investigation of the thermal dispersivity term and its significance in the heat transport equation for flow in sediments, Water Resour. Res., 48, 1–21, https://doi.org/10.1029/2011WR011038, 2012.

Rau, G. C., Andersen, M. S., McCallum, A. M., Roshan, H., and Acworth, R. I.: Heat as a tracer to quantify water flow in near-surface sediments, Earth-Sci. Rev., 129, 40–58, https://doi.org/10.1016/j.earscirev.2013.10.015, 2014.

Read, T., Bour, O., Selker, J. S., Bense, V. F., Le Borgne, T., Hochreutener, R., and Lavenant, N.: Active-distributed temperature sensing to continuously quantify vertical flow in boreholes, Water Resour. Res., 50, 3706–3713, https://doi.org/10.1002/2014wr015273, 2014.

Sellwood, S. M., Bahr, J. M., and Hart, D. J.: Evaluation of a discrete-depth heat dissipation test for thermal characterization of the subsurface, Geol. Soc. Am. Spec. Pap., 519, 67–79, https://doi.org/10.1130/2016.2519(05), 2016.

Shanafield, M., Pohll, G., and Susfalk, R.: Use of heat-based vertical fluxes to approximate total flux in simple channels, Water Resour. Res., 46, W03508, https://doi.org/10.1029/2009wr007956, 2010.

Shanafield, M., McCallum, J. L., Cook, P. G., and Noorduijn, S.: Variations on thermal transport modelling of subsurface temperatures using high resolution data, Adv. Water Resour., 89, 1–9, https://doi.org/10.1016/j.advwatres.2015.12.018, 2016.

Smits, K. M., Sakaki, T., Limsuwat, A., and Illangasekare, T. H.: Thermal conductivity of sands under varying moisture and porosity in drainage–wetting cycles, Vadose Zone J., 9, 172–180, https://doi.org/10.2136/vzj2009.0095, 2010.

Su, G., Freifeld, B., Oldenburg, C., Jordan, P., and Daley, P.: Interpreting Velocities from Heat-Based Flow Sensors by Numerical Simulation, Ground Water, 44, 386–393, https://doi.org/10.1111/j.1745-6584.2005.00147.x, 2006.

Vandenbohede, A., Louwyck, A., and Lebbe, L.: Conservative solute versus heat transport porous media during push-pull tests, Transport Porous Med., 76, 265–287, https://doi.org/10.1007/s11242-008-9246-4, 2009.

Vogt, T., Schneider, P., Hahn-Woernle, L., and Cirpka, O. A.: Estimation of seepage rates in a losing stream by means of fiber-optic high-resolution vertical temperature profiling, J. Hydrol., 380, 154–164, https://doi.org/10.1016/j.jhydrol.2009.10.033, 2010.

Vrugt, J. A., Robinson, B. A., and Hyman, J. M.: Self-adaptive multimethod search for global optimization in real-parameter spaces, IEEE T. Evolut. Comput., 13, 243–259, https://doi.org/10.1109/TEVC.2008.924428, 2009a.

Vrugt, J. A., ter Braak, C. J. F., Diks, C. G. H., Robinson, B. A., Hyman, J. M., and Higdon, D.: Accelerating Markov Chain Monte Carlo Simulation by differential evolution with self-adaptive randomized subspace sampling, Int. J. Nonlinear Sci., 3, 273–290, 2009b.

Vrugt, J. A., ter Braak, C. J. F., Gupta, H. V., and Robinson, B. A.: Equifinality of formal (DREAM) and informal (GLUE) Bayesian approaches in hydrologic modeling?, Stoch. Env. Res. Risk A., 23, 1011–1026, https://doi.org/10.1007/s00477-008-0274-y, 2009c.

Zarnetske, J. P., Haggerty, R., Wondzell, S. M., and Baker, M. A.: Dynamics of nitrate production and removal as a function of residence time in the hyporheic zone, J. Geophys. Res.-Biogeo., 116, 1–15, https://doi.org/10.1029/2010JG001356, 2011.

10

Event-based stochastic point rainfall resampling for statistical replication and climate projection of historical rainfall series

Søren Thorndahl[1]**, Aske Korup Andersen**[2]**, and Anders Badsberg Larsen**[2]

[1]Department of Civil Engineering, Aalborg University, Aalborg, 9220, Denmark
[2]Niras A/S, Aalborg, 9000, Denmark

Correspondence to: Søren Thorndahl (st@civil.aau.dk)

Abstract. Continuous and long rainfall series are a necessity in rural and urban hydrology for analysis and design purposes. Local historical point rainfall series often cover several decades, which makes it possible to estimate rainfall means at different timescales, and to assess return periods of extreme events. Due to climate change, however, these series are most likely not representative of future rainfall. There is therefore a demand for climate-projected long rainfall series, which can represent a specific region and rainfall pattern as well as fulfil requirements of long rainfall series which includes climate changes projected to a specific future period.

This paper presents a framework for resampling of historical point rainfall series in order to generate synthetic rainfall series, which has the same statistical properties as an original series. Using a number of key target predictions for the future climate, such as winter and summer precipitation, and representation of extreme events, the resampled historical series are projected to represent rainfall properties in a future climate. Climate-projected rainfall series are simulated by brute force randomization of model parameters, which leads to a large number of projected series. In order to evaluate and select the rainfall series with matching statistical properties as the key target projections, an extensive evaluation procedure is developed.

1 Introduction

In design of new and analysis of existing storm water drainage systems valid rainfall statistics are crucial. With climate changes anticipated to impact precipitation patterns, the historical rainfall statistics upon which the traditional design is based, is no longer valid for future design. There is therefore a need for climate projection of the rainfall statistics in order for these to represent the future loads on storm water drainage systems.

Traditionally many simple urban drainage systems are designed with intensity–duration–frequency (IDF) relationships, or types of design storms (e.g. Unit Hydrograph: Sherman, 1932; Chicago Design Storm, CDS: Keifer and Chu, 1957; SCS: NRCS, 1986) which represent statistics for rain with specific return periods. Climate projection of these types of design methods can be relatively simple, e.g. by multiplying the design rain by a bias climate factor (e.g. Semadeni-Davies et al., 2008; Olsson et al., 2009; Willems et al., 2012a; Willems, 2013b; Shahabul Alam and Elshorbagy, 2015), assuming that extreme rainfall events for a specific return period will be increased linearly with a given factor as a function of time. The most recognized approach for estimating climate factors is the downscaling of global circulation models (GCMs) and/or regional climate models (RCMs) (e.g. Wilby and Wigley, 1997; Fowler et al., 2007).

In general, statistical downscaling determines a statistical relationship between a large- and a local-scale climate variable based on historical records. The relationship can be used in a GCM/RCM to obtain local variables for a specific domain in a given time frame of climate projection (e.g. Wilby et al., 2002; Nguyen et al., 2007; Willems and Vrac, 2011; Willems et al., 2012b; Arnbjerg-Nielsen, 2012; Sunyer et al., 2015). The statistical downscaling approach requires long historical records of observations in order to establish the necessary statistical relationships. Based on various types of statistical downscaling assumptions and methods, climate factors for urban drainage design purposes (e.g for multiplication on IDF relationships) can be derived by statistically comparing contemporary climate conditions with projected future rainfall with regards to specific return periods, and aggregation levels (durations) or rainfall (e.g. Mailhot et al., 2007; Larsen et al., 2009; Madsen et al., 2009; Nguyen et al., 2009, 2010; Willems and Vrac, 2011; Olsson et al., 2012; Willems, 2013b).

Whereas a large proportion of the recent research described above has been conducted on estimating climate factors for design purposes, there is also a significant need, not only to describe future extremes (e.g. in the form of IDF relationships) but also to be able to project climate changes to continuous rainfall time series. Basically, simple design methods assume agreement between the return period of the rain intensity (for a given duration), and on the other hand the return period of the critical load in the drainage system (water level, flow, basin storage, etc.). Multiplication of climate factors to design storms, e.g. IDF relationships, is sufficient for many applications of urban drainage design; however, for more complex drainage systems with non-linear rainfall runoff response the simple design methods falls short. That is, for complex systems the return periods of the rainfall duration and intensity are not in agreement with the return periods of the corresponding drainage system state. Therefore, historical rainfall series (or climate-projected rainfall series) are required for complex systems in order to estimate maximum water levels in manholes, flooding, to estimate the return periods, and other loads on the drainage system such as outlet to recipient, inlet to wastewater treatment plants, combined sewer overflow, outlet flow, and pollutants loads in the future climate (e.g. Schaarup-Jensen et al., 2009; Thorndahl, 2009; Thorndahl et al., 2015).

According to Willems et al., (2012a, b) there are generally two methods that produce continuous climate-projected time series either by (1) stochastic rainfall generators which generate locally representative synthetic rainfall conditioned on climate variables in present and future climate or (2) statistical approaches to downscaling such as change factor, resampling or weather typing methods, in which future local rainfall is sought in historical rainfall records under equivalent historical climate conditions as projected in the future, or modified to represent future climate conditions.

In the literature, the most acknowledged methods for stochastically generating synthetic rainfall series are based on Poisson cluster processes and rectangular pulse models such as Bartlett–Lewis (Koutsoyiannis and Onof, 2001; Onof and Wheater, 1994, 1993; Segond et al., 2007; Onof and Arnbjerg-Nielsen, 2009; Paschalis et al., 2014; Kossieris et al., 2016) or Neyman–Scott (e.g. Entekhabi et al., 1989; Cowpertwait, 1991, 2010; Cowpertwait et al., 2002; Fowler et al., 2005; Burton et al., 2008; Paschalis et al., 2014; Sørup et al., 2016). Calibration of the generators is typically performed by comparing generated series to observed series and adjusting relevant parameters prior to climate projection. Methods for estimating point rainfall (e.g. Cowpertwait et al.,

1996; Marani and Zanetti, 2007; Onof and Arnbjerg-Nielsen, 2009) and spatially distributed rainfall or multi-site generators with spatial dependency (e.g. Kilsby et al., 2007; Burton et al., 2008; Sørup et al., 2016) have been applied. These methods have been shown to provide valid results for hourly or daily time steps but also have significant shortcomings in terms of modelling rainfall at a finer temporal resolution. For urban hydrological applications with fast rainfall response, a temporal resolution of input data down to 1–10 min is required (e.g. Schilling, 1991; Willems, 2000; Thorndahl et al., 2008, 2016, 2017). Because we are interested in maintaining the fine temporal resolution of observed rainfall series, generation of synthetic rainfall series using Poisson clusters is rejected here as an applicable method.

Change factor, resampling or weather typing methods (Willems et al., 2012a, b) of statistical downscaling outcomes of RCMs/GCMs can provide data in the required temporal resolution, since directly based upon historical records. Arnbjerg-Nielsen (2012) applied historical rain series originating from another geographical region, which had a climate analogue to the projected climate in order to obtain continuous representative rainfall series for future climate conditions. Zorita and Von Storch (1999), Olsson et al. (2009), Willems and Vrac (2011), and Ntegeka et al. (2014) used historical records of rain and modified these records to represent climate-representative continuous climate-projected rain series. Ntegeka et al. (2014) alternated the number of dry and wet days and used *quantile perturbation* (an advanced delta change method) to modify rainfall intensities. Olsson et al. (2009) applied the *delta change method* to multiply historical records with bias climate factors depending on rainfall intensity levels in order to fit projections of extreme, seasonal, and annual precipitation. This approach, however, was implemented without alternating the temporal variability and the seasonal distribution of events of the rain series and maintaining the chronology of the original series. This particular shortcoming might be problematic in order to project the frequency of extreme events sufficiently.

The approach presented in this paper is different from the methods presented above, although it can be considered as a variation of *resampling* combined with *stochastic generation*. Whereas other methods use other climate variables, e.g. pressure and temperature, as climate predictors, this approach aims at fitting statistical properties of climate-projected precipitation directly. In this case, these properties are derived from other studies of RCM projection (see Sect. 2 for details). The validity of the method therefore depends on whether the climate-projected target variables are comprehensive and detailed enough to project the future rainfall upon. The aim is to develop a generally adaptive method which can be applied to an arbitrary rainfall series and with different climate scenarios and projection period. in contrast to the studies described above, climate-projected time series are generated directly for urban drainage modelling purposes. The objective has been to develop a generally applica-

Table 1. The calculated Danish climate changes in annual and seasonal precipitation as well as extremes. The values are expressed as a multiplicative climate factor describing the difference between the reference period 1961–1990 and 2071–2100. The A1B scenario is presented in Olesen et al. (2014) and represents 14 regional climate model runs from the ENSEMBLES project. The climate factors from the two RCP scenarios are previously unpublished, but derived from the Euro-CORDEX-11 database (Jacob et al., 2014) and processed statistically for this paper. Standard deviation is listed in parentheses. The indices marked with bold are the ones used in this paper.

	Climate factors for the period 2071–2100		
Parameter	Scenario A1B (Olesen et al., 2014)	Scenario RCP4.5 (unpublished)	Scenario RCP8.5 (unpublished)
Annual precipitation	1.14 (±0.06)	1.08 (±0.06)	1.14 (±0.07)
Winter precipitation (DJF)	1.25 (±0.06)	1.12 (±0.06)	1.24 (±0.07)
Spring precipitation (MAM)	1.13 (±0.06)	1.13 (±0.08)	1.23 (±0.11)
Summer precipitation (JJA)	1.05 (±0.08)	1.06 (±0.18)	1.03 (±0.21)
Fall precipitation (SON)	1.13 (±0.06)	1.05 (±0.07)	1.09 (±0.13)
Events above 10 mm	1.37 (±0.12)	1.20 (±0.13)	1.35 (±0.14)
Events above 20 mm	2.50 (±0.14)	1.41 (±0.30)	1.80 (±0.40)
Max. daily precipitation	1.16 (±0.12)	1.12 (±0.09)	1.24 (±0.11)

ble method that can be used directly by practitioners and scientists within the field of urban drainage, who do not necessarily have detailed knowledge of climate projection, RCM's, downscaling, etc.

The procedure is divided into two major parts: (1) resampling of a historical point rainfall time series ("Method development": Sect. 3.1; "Results and evaluation": Sect. 4.1); and (2) climate projection of resampled time series ("Method development": Sect. 3.2; "Results and evaluation": Sect. 4.2).

The essential concept of the method is to stochastically generate a large number of either resampled historical series or climate-projected series, and to evaluate the statistical properties of the generated series against a number of key target variables. Rather than optimizing for the best parameter fit, the basic concept is to sample parameters from broad uniform distribution functions for each parameter and to either accept or reject each stochastically simulated series using a specified criterion. Repeating this procedure for a large number of realizations of rainfall series, it is possible to select a number of rainfall series which has a satisfying statistical representativeness in comparison with historical series or climate projection targets. The evaluation procedure is inspired by the generalized likelihood uncertainty estimation (GLUE) method (Beven and Binley, 1992; Thorndahl et al., 2008) and is presented in detail in Sect. 3.3.

The method assumptions and subjectivity are discussed in Sect. 5 and in Sect. 6 conclusions on this approach to climate projection of single-point historical rainfall series are provided.

2 Data

The development of the model is based on rain gauge data from Denmark and projection of Danish climate conditions,

but could easily be extended to other regions/countries of interest.

Specific statistical properties for the future precipitation in Denmark are necessary in order to climate project the resampled rainfall series. In Olesen et al. (2014) the Danish Meteorological Institute has collected and processed data from the ENSEMBLES project (http://www.ensembles-eu.org/, http://ensemblesrt3.dmi.dk/; Van der Linden and Mitchell, 2009; Boberg et al., 2010; Maule et al., 2013). The report includes projection of weather extremes (including precipitation) using the SRES A1B scenario (IPCC, 2007) and is produced from an ensemble of 14 regional climate models in the ENSEMBLES project. The RCMs are simulated for 1961–1990, 2021–2050, and 2071–2100, but in this case only the first and last time interval are applied. Table 1 presents annual and seasonal precipitation increment (expressed as a climate change factor) in 2071–2100 compared to the reference in 1961–1990. Furthermore, the report specifies changes in other climate indices. In the context of precipitation, the variables *number of events above 10 mm*, *number of events above 20 mm*, and *max. daily precipitation* are relevant (Table 1). In this paper these three variables are used to climate-project the resampled rainfall series, as they are considered important with regards to urban drainage modelling. Because the data from Olesen et al. (2014) represent the SRES scenarios (IPCC, 2007), new data representing the representative concentration pathway (RCP) scenarios (IPCC, 2013; Christensen et al., 2015) are developed for this paper. Daily RCM simulations from an ensemble of 14 models has been derived over Denmark from the Euro-CORDEX database (Casanueva et al., 2016; Jacob et al., 2014; Prein et al., 2016) and statistically processed by the same variables as in Olesen et al. (2014). Derived values are provided in Table 1. For the climate projections in this paper the RCP4.5 scenario is

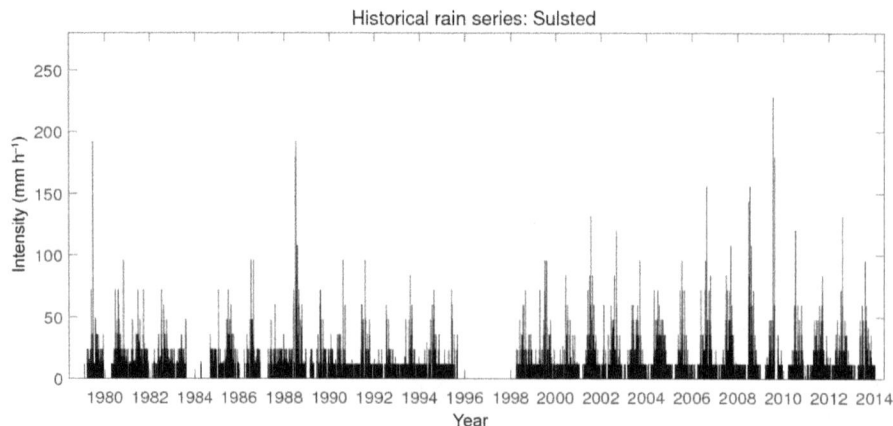

Figure 1. Measured time series of the Sulsted rain gauge. The temporal resolution of rainfall data is 1 min.

Table 2. Recommended climate factors for design of drainage systems in Denmark according to WPC (2008, 2014) and Gregersen et al. (2014b). The climate factors are valid for a duration of 1 h but also recommended for other durations up to 3 h. The indices marked with bold are the ones used in this paper. The standard deviations are not provided directly in the references, but estimated from tables and figures.

Return period (years)	Climate factors for the period 2071–2100		
	Scenario A2 (WPC, 2008)	Scenario RCP4.5 (WPC, 2014)	RCP8.5 (WPC, 2014)
2	1.20 (±0.1)	1.20 (±0.1)	1.45 (±0.1)
10	1.30 (±0.2)	1.30 (±0.2)	1.70 (±0.2)
100	1.40 (±0.3)	1.40 (±0.3)	2.00 (±0.3)

chosen throughout, but the paper could easily have been presented with other SRES or RCP scenarios.

The Water Pollution Committee of the Society of Danish Engineers has published reports (guidelines nos. 29 and 30) with recommendations for design of drainage systems considering climate change (WPC, 2008, 2014, background report: Gregersen et al., 2014b). Based also on the climate simulations of the ENSEMBLES project, the climate factors for drainage system design in Denmark are recommended (Table 2). Design rainfall, e.g. IDF relationships, with a specified return period is recommended to be multiplied by these climate factors. The values are derived for rainfall intensities over 1 h but also recommended for other durations (up to 3 h). In this paper these values are used to certify a correct representation of extreme events.

The rainfall series which are applied in this study has its origin in the rain gauge network of the Water Pollution Committee (WPC) of the Society of Danish Engineers. At present, the network consists of 145 tipping bucket rain gauges (DMI, 2014). The rain gauge no. 5047 located in Sulsted, North Jutland (lat 57.17, long 9.96), is applied since this is a station with a long recording time and few errors compared to

other gauge records. The gauge has been in operation over a period of 34 years from 1979 to 2014, but due to minor interruptions in the dataset, the effective length of the series is 32 full years. The interruptions do not affect the statistical calculations as these are excluded from the data before the calculations are performed. The time series of 1 min. values for the Sulsted rain gauge is shown in Fig. 1.

In the WPC rain gauge network the temporal resolution of data is 1 min. The start time of an event is determined at the minute of the first tip of 0.2 mm. All events therefore have initial values equivalent to a multiple of $0.2 \, \text{mm} \, \text{min}^{-1}$ ($12 \, \text{mm} \, \text{h}^{-1}$). These initial values are easily identifiable in Fig. 1. The end of an event is specified when there is no registered tip within 1 h. Using this definition of events, the minimum *inter-event time* (time between events) will be 1 h.

Using Danish rainfall data on a daily scale Gregersen et al. (2014a) have been able to identify multidecadal climate oscillations (Ntegeka and Willems, 2008; Willems, 2013a) as well as climate-related changes in precipitation patterns over the past 140 years. Nevertheless, since this paper is based on evidently shorter rainfall series, it is assumed that no significant trends or climate changes in this period are present. The historical records from the Sulsted series are therefore assumed to be stationary in terms of climate properties.

3　Method development

The procedure of the method is presented in two sections: the *resampling of historical rainfall series* (Sect. 3.1) and the *stochastic climate projection of resampled historical rainfall series* (Sect. 3.2). Since both methods involve random selection of events and brute force randomization of parameters there is a need for a unique method to evaluate the generated series against target values. This evaluation method is inspired by the GLUE methodology (Beven and Binley, 1992). The basic concept is to generate a large number of rainfall series and evaluate whether each generated series should be accepted or rejected based on an empirical likelihood (performance) measure based on individual criteria for each target value. For the accepted generated rainfall series a combined performance measure for each realization is calculated in order to find the rainfall series realization which in general fits the target values the best. This method is described in detail in Sect. 3.3.

3.1 Historical rainfall series resampling

The objective is to create synthetic rainfall series resampled stochastically from a historical series such that the synthetic and the historical series have the same statistical properties. The first step is to divide the historical rainfall series into smaller parts in order to describe variability of intensities, event duration, and time between events over the year. We chose to divide the series into four seasons (winter: DJF; spring: MAM; summer: JJA; autumn: SON), although a finer division (e.g. monthly) could have been implemented. Because the target projections (Table 1) are implemented in seasons, this is the one used. The summer precipitation in the synthetic rainfall series is thus generated based on statistics calculated for every summer period's precipitation in the historical rainfall series and correspondingly for the other seasons.

The stochastic generation (resampling) is based on the following:

1. Statistics of the inter-event time (also referred to as *rainfall intermittency*, e.g. by Molini et al., 2001, and Schleiss et al., 2011) using the definition of events presented in Sect. 2.

2. Sampling of rainfall events including original event durations and intensities randomly from the pool of historical rain events for each season.

The concept is outlined in Fig. 2.

The inter-event times(t_{ie}) for each season are approximated by a *two-component mixed exponential probability density function*:

$$\lambda f(t_{ie}) = p \left[\lambda_{a,ie} \exp\left(\lambda_{a,ie} t_{ie} \right) \right]$$
$$+ (1-p) \left[\lambda_{b,ie} \exp\left(\lambda_{b,ie} t_{ie} \right) \right], \tag{1}$$

where $\lambda_{a,ie}$ and $\lambda_{b,ie}$ are the rate parameters for two populations, "a" and "b", with different exponential distributions and p is the weight of population "a". This mixed distribution function was also applied by Rossi et al. (1984) and Willems (2000). Willems (2000) applied the distribution for fitting rainfall intensities arguing that the two distributions originated from two different types of storms (convective thunder storms and frontal storms respectively). The same rationale is applied here. The approximation to inter-event times for each season thus require approximation of three parameters, p, $\lambda_{a,ie}$, and $\lambda_{b,ie}$.

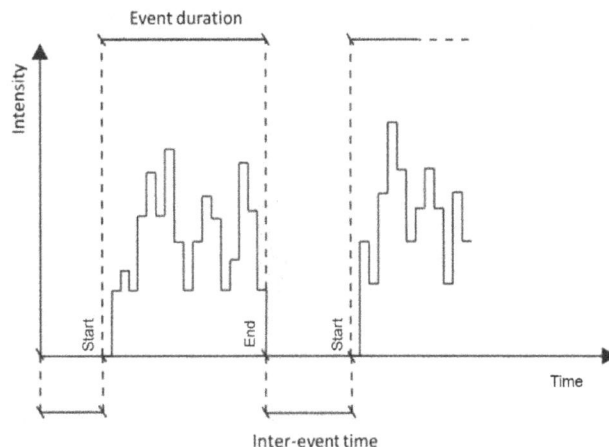

Figure 2. Diagram of the construction of the synthetic (resampled) rainfall series.

Molini et al. (2001) applied a Weibull distribution to describe the inter-event time of rainfall events. The Weibull distribution, along with exponential, gamma, and generalized Pareto distributions was also investigated for this paper, but was however outperformed by the mixed exponential distribution, especially in fitting both ends of the distribution.

As opposed to other rainfall generators which use a fixed timescale (e.g. Furrer and Katz, 2008), the time is sampled discontinuously in this case.

The sampling of the events is an automated process with random selection of events from the pool of historical rainfall events for each season. When sampling a specific event, the intensity sequence and consequently also the duration is maintained. Synthetic resampled time series are, therefore produced by random alternating sampling of the inter-event times and historical events from a specific season. It is possible to sample the same event more than once. The procedure is repeated until the length of the generated series corresponds to the length of the historical series or any other specified length shorter than the total length of the original series. The number and the chronology of events are therefore different from season to season and from year to year.

A vital assumption here is that events from the historical series can be sampled independently. Depending on the meteorological conditions at the time of a specific event there might potentially be some correlation to prior and posterior events due to short inter-event times. Extreme event statistics and development of IDF relationships from partial duration series in Denmark is also produced assuming independent events (Mikkelsen et al., 1998; Madsen et al., 2009), so in order to preserve this methodology, no inter-correlation between events has been implemented in the presented approach.

3.2 Climate projection and stochastic resampling of rainfall series

The climate-projected rainfall series is generated in three steps:

1. The inter-event time for each season is sampled using the same procedure as described in the previous section; however, the parameters of the mixed exponential distribution for each season are implemented as stochastic variables and thus sampled randomly from a uniform distribution with fixed upper and lower boundaries. This allows for different distributions of inter-event times than the ones used in the resampling of historical series. In the climate-projected series, it is thus possible to accommodate for climate changes in seasonal precipitation and the distribution between small and large events, by changing the number of events per season. As an example the method is able to accommodate a moderate increase of total summer precipitation, and at the same time a considerable increase in frequency and intensity of extreme events, with generally a lower number of total events in summer as a result.

2. Rainfall events are sampled from the pool of historical events for each season in the same way as described in Sect. 3.1. The duration of each event is not alternated under impact of climate change, since there is presently no evidence that single events will become shorter or longer in the future. This is obviously a crucial assumption, but nonetheless the best current estimate, which also has been applied by, for example, Olsson et al. (2009). The sampling of events is therefore done without alternating the events from the pool, other than multiplying by different change factors as presented below.

3. The climate projection of the generated time series is inspired by the delta change method. However unlike Olsson, the change factors are implemented as random variables. The change factor for a given rainfall intensity, i, is derived using the probability, $F(i)$, of that the intensity being less than or equal to i. For each season, the rainfall intensities from the original historical rain series are fitted to the same type of mixed exponential distribution (Willems, 2000a) as applied for fitting the inter-event times (Eq. 1):

$$F(i) = p\left[1 - \exp(\lambda_a i)\right] + (1 - p)\left[1 - \exp(\lambda_b i)\right], \quad (2)$$

where λ_a and λ_b are rate parameters for two populations "a" and "b", and p is the weight given to population "a". $F(i)$ has a range from 0 to 1.

For each season change factors are multiplied by intensities on the minute scale. The change factor as a function of intensity, $c(i)$, is thus calculated for each season

by a linear function:

$$c(i) = \alpha F(i) + \beta, \quad (3)$$

where α and β are random variables sampled from uniform distributions with fixed limits.

For each projected rainfall series there is a different value of α and β for each season. During the development of the procedure, the limits of the uniform distribution of α and β for each season were empirically selected starting with broad intervals which were reduced by discarding non-accepted runs (see below).

The total number of random variables for generating climate-projected stochastic rain series in the current setup with four yearly seasons is 20 (2×4 for the change factor plus 3×4 for the mixed exponential distributions).

3.3 Evaluation and optimization procedure

The governing assumption behind the resampling procedure is that the resampled rainfall series should have the equivalent statistical characteristics as the historical series on a number of key target variables. The climate-projected resampled series should therefore also have the equivalent statistical characteristics by means of a number of key target climate projections (as the ones presented in Tables 1–2). It is not a necessity that the same target variables are used to evaluate resampled historical rainfall series and the climate-projected series, but we chose to do so in this paper in order to keep the evaluation procedures the same regardless of generating series which should statistically represent historical series or climate-projected series. The key target variables are described in detail below:

1. Annual precipitation (ap). This target variable is included as it is a measure of the total "mass" balance. Since the individual years of the resampled and historical series are not directly comparable year by year, the mean of all years is applied as target variable.

2. Seasonal precipitation (sp). The mean seasonal precipitation is applied as a target variable in order to ensure same distribution between seasons in the resampled series. The four target parameters are labelled spwi, spsp, spsu, spau corresponding to winter, spring, summer, and autumn precipitation respectively.

3. Number of events above 10 mm per day (n10mm). This target variable provides a measure of the representation of extreme events.

4. Number of events above 20 mm per day (n20mm); same procedure as for no. 3.

5. Maximum daily precipitation (mdp, as a mean of the maximum day for all years). This target variable also certifies the representation of extreme events.

6. IDF relationships. The IDF relationships are traditionally applied in design of urban drainage systems and are therefore relevant to include as a target variable. In accordance with Table 2, it is chosen to use the mean rain intensity over a duration of 60 min for return periods of 2 and 10 years respectively as a target value. The two values are labelled d60T2 and d60T10 respectively.

The performance of each individual target variable is estimated using a simple ratio measure between the target value and the corresponding modelled value:

$$P_{i,j} = 1 - \frac{|T_i - M_{i,j}|}{T_i}. \tag{4}$$

Here $P_{i,j}$ is the individual performance parameter for target variable i (as presented above) corresponding to realization j, T_i is the target value, and $M_{i,j}$ is the modelled value of the target variable of the jth realization. For the evaluation of the resampled series against the historical series, $T_i = H_i$, where H_i is the value of the target variable of the historical series. With respect to the evaluation of the climate-projected rainfall series, where the target value is given by a climate factor (cf) multiplied by the target variable of the historical series,

$$T_i = \text{cf}_i \cdot H_i. \tag{5}$$

Thus the performance measure is

$$P_{i,j} = 1 - \frac{|\text{cf}_i \cdot H_i - M_{i,j}|}{\text{cf}_i \cdot H_i}. \tag{6}$$

Here P can vary between 0 and 1, where $P = 1$ corresponds to a perfect fit.

In order for a simulated rainfall series to be accepted $P_{i,j}$ has to be larger than a specified threshold. For the resampled historical series the acceptance criterion for the individual performance measures is fixed and has been chosen as $P_{\text{crit},i} = 0.90$, hence all 10 individual performance measures should exceed this value in order for the realization to be accepted (Table 4). This means that if a target value of just one of the 10 target values deviates more than 10 % from the value of the historical series, the realization is rejected.

For the climate-projected series, it is possible to estimate individual values of the performance using the standard deviations of the climate factors (cf) given in Tables 1 and 2:

$$P_{\text{crit},i} = 1 - \frac{2 \cdot \sigma_{\text{cf},i}}{\text{cf}_i}. \tag{7}$$

Assuming Gaussian distributed target variables, we will thus accept values which are within the 95 % confidence intervals of the distribution of each target variable. The acceptance criteria of the performance measure will thus be different for each target variable depending on the uncertainty (standard deviation) related to that specific climate projection (see Tables 1 and 2). The acceptance criteria for the performance

of each target variable are presented in Table 6 along with climate factors and standard deviation for each variable.

The combined performance measure P_j of each realization series (j) is estimated as

$$P_j = \sum_{i=1}^{I} w_i P_{i,j}, \tag{8}$$

where w_i is the weights of the individual performance measures, $\sum w_i = 1$, and I is the total number of individual performance parameters.

The individual weights are presented in Sect. 4.2 and Table 6. One could argue that each season should be given the same weight; however, because summer precipitation tends to be more important in terms of extreme events in Denmark this is given a higher weight. Moreover, because winter precipitation might be associated with larger measurement errors due to poor measurement of solid precipitation, this is given a smaller weight.

4 Results and evaluation

4.1 Historical rainfall series resampler

The synthetic resampled series are generated with the same total length as the original historical series – in this case 32 years.

The inter-event times for each season are sampled from the mixed exponential distribution as detailed in Sect. 3.1. The estimated parameters are presented in Table 3. By comparing the parameters, it is evident that there is a significant difference for each season. Therefore, it is important that the inter-event times are sampled individually for each season to ensure a representative number of events in the resampled rainfall series compared to the historical rainfall series. Figure 3 exemplifies empirical cumulative distribution functions for summer inter-events times for the historical series and for the fitted mixed exponential distribution of summer inter-event times. Furthermore, the empirical distribution from the resampled series with the best combined performance measure is presented ($P_j = 0.98$). Using the mixed exponential distribution, there is small underestimation of inter-event times between 1 and 6 h and an equivalent overestimation between 6 and 24 h. This is, however, insignificant in comparison to other fitted distribution functions and thus not considered a problem in random sampling of inter-event times from these distributions.

There is a stringent dependency between inter-event times and number of events in the rainfall series. In order to generate a valid and representational resampled rain series, the number of events series should correspond somewhat to the number of events in the historical rainfall series. Table 4 therefore includes the mean and standard deviation of the number of events per year even though the number of events

Table 3. The fitted rate and weight parameters for the mixed exponential distribution specified for each season.

Parameter	Winter	Spring	Summer	Autumn
Rate, population a, $\lambda_{a,ie}$, (days)	0.38	0.33	0.24	0.26
Rate, population b, $\lambda_{b,ie}$, (days)	4.87	4.46	3.00	2.90
Weight population a, p (–)	0.69	0.56	0.55	0.64

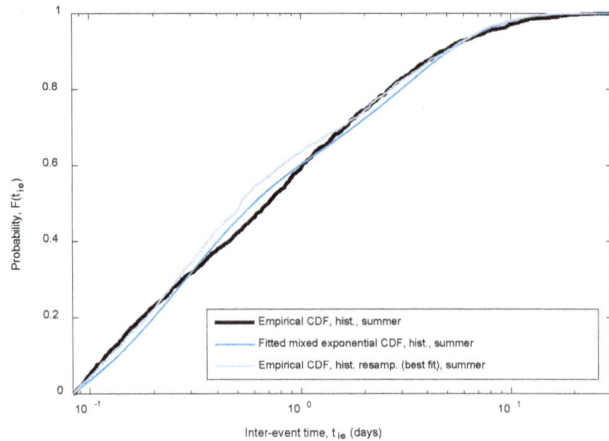

Figure 3. Example of cumulative distribution functions for summer inter-event times.

is not used as a target variable for estimating the individual performances.

The resampling of the observed rainfall series is performed generating 5000 different resampled rainfall series and assessing the performance of each generated series using the method described in Sect. 3.3. Out of the 5000 realizations of simulated series, 275 (5.5 %) are accepted using the criterion of a minimum individual performance measure ($P_{crit,i}$) of 0.90. The fact that all 10 individual performance measures have to be larger than the acceptance criteria has been shown to be a tough condition to fulfil. Often one or two of the 10 has a slightly lower value and the realization is thus rejected. On average the accepted realizations have a combined performance measure (P_j) of 0.95 (ranging between 0.92 and 0.98). Figure 4 presents a bar plot (blue shades) of each of the target variables for the historical series, the one resampled series with the highest combined performance measure, as well as the mean of the accepted resampled series (with uncertainty bounds indicating the minimum and maximum of the accepted series).

Generally there is a good agreement between the historical series and the accepted series on the target parameters with the highest weights, i.e. the seasonal precipitation. This is actually the case for the majority of the 5000 realizations; however, the performance measures becomes rather low if the extreme events are not represented correctly in the resampled series and they are in that case rejected. The variability between the resampled series is only due to the randomness

assembling events and inter-event times from the historical series because the mixed exponential parameters for each season are fixed corresponding to the fits (Table 3). The rejection of resampled series is therefore often due to either sampling of too few or too many "extreme" events within a season.

In many situations, only the one resampled series with the highest performance measure is of interest. Table 4, therefore, lists target values of the historical series and the resampled series with the highest performance measure (best fit). Besides the best combined performance measure of $P_j = 0.98$, the individual performance measures are given in the right column. In order not only to compare series on mean values, Table 4 also presents standard deviations describing the year-to-year variability over the total length of the series. Generally there is a satisfactory agreement (below 10 %) of both mean and standard deviations between the historical series and the "best" accepted resampled rainfall series.

To verify the representativeness of extreme rainfall, Fig. 5 (left) presents IDF relationships (from 10 to 360 min durations) for the historical and "best" resampled series for return periods of 2 and 10 years respectively. Grey areas represent the variability in all the accepted realizations. Generally, there is an acceptable agreement between the curves which verifies the resampling method. There is, however, a minor divergence for short durations of the 10-year return period. In general, the longer the return period the larger the divergence between the curves to be expected as a result of the random sampling of historical events in the generated series. Figure 6 shows the time series of the "best fit" resampled time series.

The overall assessment of the previous evaluation indicates that the rainfall resampler can represent the historical rainfall series well based on the selected performance parameters. Due to the stochasticity of the sampling of inter-event times and rainfall events, there is obviously some variability from year to year and from series to series, but because none of the target variables are significantly biased, the overall performance of the resampler is accepted. As it is possible to produce resampled rainfall series with the same statistics as the corresponding original historical series, the resampling algorithm will be applied to generate climate-projected rainfall series in the following section.

Table 4. Target variables (mean and standard deviation) and performance measures for the historical series and the one resampled series with the highest performance measure.

Target variable		Acceptance criteria and weights		Historical series (target)		"Best fit" resampled series		
		$P_{\text{crit},i}$	w_i	Mean	SD	Mean	SD	P_i
Annual no. of events				200.1	39.4	218.2	45.7	
Annual precipitation	ap (mm)	0.90		576.1	122.3	586.5	140.6	0.96
Seasonal precipitation, winter	spwi (mm)	0.90	0.05	90.9	36.5	101.6	26.7	0.99
Seasonal precipitation, spring	spsp (mm)	0.90	0.10	86.5	43.1	84.9	29.3	0.98
Seasonal precipitation, summer	spsu (mm)	0.90	0.25	213.0	57.0	209.1	75.3	0.98
Seasonal precipitation, autumn	spau (mm)	0.90	0.10	185.7	53.6	190.9	63.9	0.97
Annual number of events above 10 mm per day	n10mm (#)	0.90	0.17	16.0	4.5	15.8	5.4	0.99
Annual number of events above 20 mm per day	n20mm (#)	0.90	0.08	3.3	2.1	3.3	2.3	0.99
Annual maximum daily precipitation	mdp (mm)	0.90	0.08	35.2	12.7	32.0	12.0	0.99
Rain intensity for 60 min, $T = 2$ years	d60T2 (mm h^{-1})	0.90	0.08	15.7		16.4		0.95
Rain intensity for 60 min, $T = 10$ years	d60T10 (mm h^{-1})	0.90	0.08	28.4		28.7		0.99
Combined performance measure	P							0.98

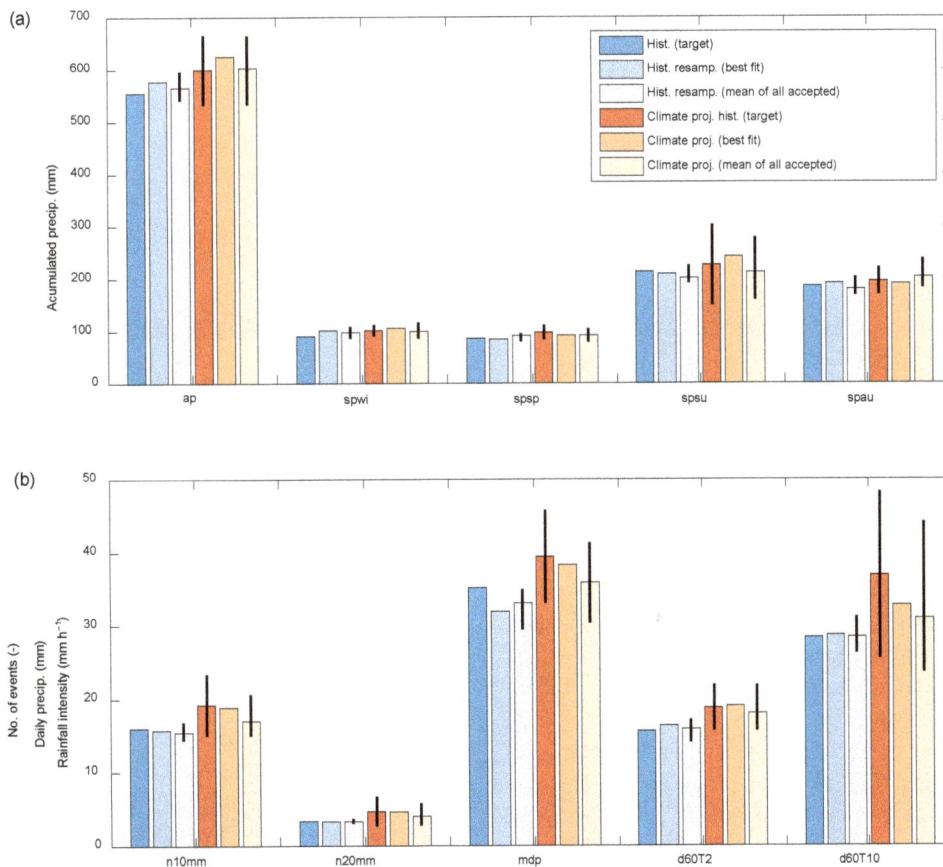

Figure 4. Target variables and their values for comparing historical series and resampled series (in blue shades) and climate-projected historical series and climate-projected and resampled series (in red shades). For the climate-projected target (deep red) the uncertainty bounds (black lines) represent 2 times the standard deviation of Tables 1 and 2. For the resampled series the uncertainty bounds represent the total range of the accepted realizations.

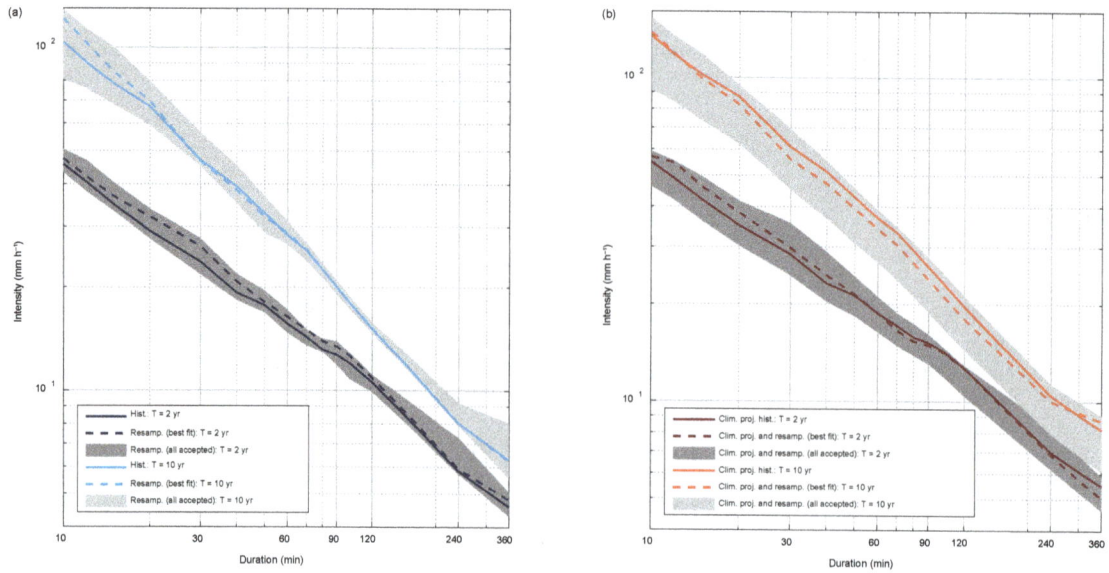

Figure 5. IDF curves for historical and resampled rainfall series **(a)** and climate-projected historical and resampled series **(b)**.

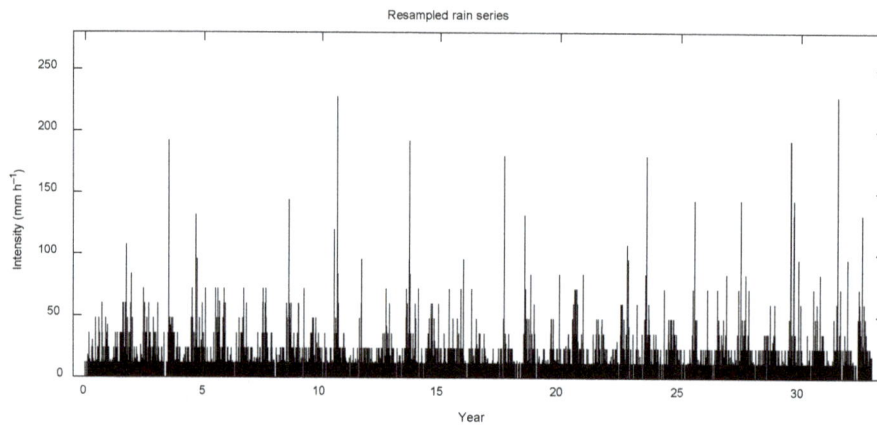

Figure 6. Time series example of resampled rainfall series. The temporal resolution of rainfall data is 1 min.

4.2   Climate-projected rainfall series

Figure 4 and Table 6 provides results for the climate-projected rainfall series. The target variables (climate-projected historical) are estimated using Eq. (5) and are thus the mean values of the historical series of Table 4 multiplied by the climate factors specified in Table 6. In addition Fig. 4 provides an uncertainty estimate on the target values obtained from the standard deviations of Tables 1 and 2.

Because the climate projection of rainfall series involves randomization of not only the event assembling but also randomization of *mixed exponential distribution* parameters and change factors as function of intensity for each season, the generation of rainfall series requires a larger quantity of realizations compared to the resampling of series described in the previous section. Therefore a total of 10 000 climate-projected rainfall series are generated. The acceptance cri-

teria implemented are, however, slightly different compared to the ones detailed in Sect. 4.1. In the evaluation of climate-projected rainfall series an individual acceptance criterion for each target variable is estimated using Eq. (7). For the 10 target variables the acceptance criteria range between 0.59 (n20mm) and 0.89 (spwi) as presented in Table 6. The total number of accepted realizations is 721 (7.2 %). The reason that a larger percentage is accepted here than in the previous section is that the acceptance criterion is somewhat softer encountering the uncertainty of climate factors. On average the accepted realizations have a combined performance measure (P_j) of 0.90 (ranging between 0.81 and 0.97).

Table 5 presents the range of mixed exponential distribution parameters as well as ranges of change factor parameters for the accepted climate-projected realizations for each season. Comparing with Table 3 (in which the parameter assessment is based on fitting the historical data) it is clear that

Table 5. Ranges of accepted parameter values for the mixed exponential distribution applied to sampling inter-event times and for the linear function applied to sample change factors for each season.

Parameter	Winter		Spring		Summer		Autumn	
	min	max	min	max	min	max	min	max
Rate, population a, $\lambda_{a,ie}$, (days)	0.32	0.44	0.27	0.39	0.20	0.28	0.23	0.29
Rate, population b, $\lambda_{b,ie}$, (days)	4.10	5.60	3.90	5.00	2.70	3.30	2.60	3.20
Weight population a, p (–)	0.63	0.74	0.50	0.61	0.51	0.60	0.60	0.68
Change factor slope, α (–)	0.000	0.050	0.000	0.050	0.000	0.025	0.000	0.049
Change factor intercept, β (–)	0.80	1.20	0.80	1.20	0.81	1.20	0.86	1.20

Table 6. Climate factors of target variables and minimum acceptance criteria of the individual performance parameters $P_{i,j}$, empirical combined performance measure weights (w_i), climate-projected target variables, and the corresponding values ($\pm$standard deviation) of the best-fit climate-projected series.

Target variable	Climate factors	Acceptance criteria and weights		Climate proj. hist. series (target)	"Best fit" climate proj. series		
	c_f	$P_{crit,i}$	w_i	Mean	Mean	SD	P_i
Annual no. of events					206.8	39.4	
Annual precipitation	ap (mm)	1.08 ($\pm$0.06)	0.89		599.6	629.8	147.3	0.96
Seasonal precipitation, winter	spwi (mm)	1.12 ($\pm$0.06)	0.89	0.05	101.8	105.8	28.7	0.93
Seasonal precipitation, spring	spsp (mm)	1.13 ($\pm$0.08)	0.86	0.10	97.7	92.2	39.8	0.99
Seasonal precipitation, summer	spsu (mm)	1.06 ($\pm$0.18)	0.66	0.25	225.8	242.1	89.3	0.99
Seasonal precipitation, autumn	spau (mm)	1.05 ($\pm$0.07)	0.87	0.10	195.0	189.8	65.9	0.90
Annual number of events above 10 mm per day	n10mm (#)	1.20 ($\pm$0.13)	0.78	0.17	19.2	18.9	5.5	0.98
Annual number of events above 20 mm per day	n20mm (#)	1.41 ($\pm$0.30)	0.57	0.08	4.7	4.6	2.7	0.99
Annual maximum daily precipitation	mdp (mm)	1.12 ($\pm$0.09)	0.84	0.08	39.4	38.3	14.3	0.92
Rain intensity for 60 min, $T = 2$ years	d60T2 (mm h^{-1})	1.20 ($\pm$0.10)	0.83	0.08	18.8	19.0		0.99
Rain intensity for 60 min, $T = 10$ years	d60T10 (mm h^{-1})	1.30 ($\pm$0.20)	0.69	0.08	36.9	32.8		0.96
Combined performance measure	P							0.97

the parameter values obtained by random sampling have a broader range, indicating that an accepted realization with a high performance value can be obtained from a broad range of parameter values. Scatter plotting the performance values as a function of parameter values (not shown) shows flat tops indicating that an equal performance can be obtained from low and high values within the range (uniform distribution). This means that there is a dependency between inter-event time parameters and chance factor parameters.

As seen in Table 5, the change factor is allowed to be both smaller and larger than 1. This allows for both decrease and increase in precipitation amounts in each seasons. The climate-projected precipitation can thus be obtained from an insignificant change in seasonal precipitation, but a rather large increase in extreme precipitation.

Generally there is an acceptable agreement of climate-projected target variables (*climate-projected historical*) and corresponding values for the climate-projected resampled series (red shades in Fig. 4). There is, however, slightly more deviation compared to the present-time simulations and larger ranges of target variable values. This is as expected since the climate projection includes more random parame-

ters and complexity as well as broader acceptance criteria. For the accepted realization with the highest performance measure, $P = 0.97$ (Fig. 4 and Table 6), there is a tendency for the target variables related to the extreme values to be marginally underestimated. This is inevitably a result of high weights given to the target values related to seasonal precipitation, especially summer precipitation. By changing the weights it would be possible to obtain more equal extreme values, however potentially at the expense of a poorer fit of the accumulated precipitation values.

In Fig. 5 (right) the IDF curves for the climate-projected series are shown. There is a slight underestimation of extremes for the 10-year return period, but an overestimation of the 2-year return periods on low durations. Since the total length of the series is 32 years, return periods larger than 10 years are not presented well, since they the associated with large uncertainties (see e.g. Thorndahl, 2009). The uncertainty bands (grey areas) however cover the climate-projected intensities. Figure 7 shows the time series of the "best fit" resampled time series.

The overall performance of the climate projection of resampled rainfall series is considered to be acceptable within

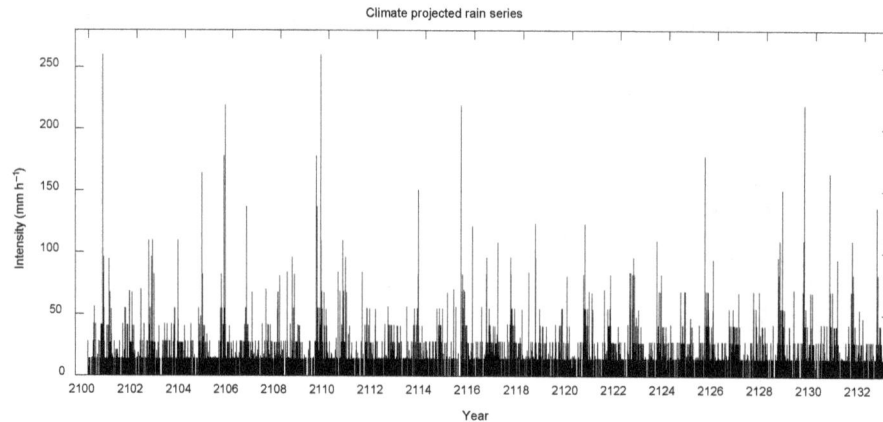

Figure 7. Time series example of climate-projected rainfall series. The temporal resolution of rainfall data is 1 min.

the range of uncertainties related to the climate projections. The introduction of 20 random variables and the random assembling of rain events obviously require many realizations in order to produce accepted rainfall series which have a satisfactory degree of agreement on all target parameters.

5   Discussion

The developed procedure obviously involves a large degree of subjectivity in the choice of processes and parameters to include. This section will discuss and argue for some of these choices.

The target variables have been chosen to represent both annual and seasonal precipitation as well as more extreme values. The choice of the 10 specific target variables is closely connected to the fact that this is what is currently available for Danish future climate conditions. However, when other, maybe more detailed, target variables becomes available, it would be possible to redo the generation of climate-projected rainfall series with other target variables. It was initially decided only to present values from the RCP4.5 climate scenario; however, the implementation of the method could just as well have been implemented with another RCP or SRES scenario. Another possibility could be to implement other durations and return periods than for 60 min durations for 2 and 10 years respectively in order to emphasize specific extremes further.

It is of utmost importance that the chosen target variables are representative of the future precipitation patterns and that they are comprehensive in the way that they cover both annual/seasonal variations and single events and the statistics related to these. In this paper, we chose only to include yearly mean values of target parameters (except for the target variables related to return periods), but it could also be relevant to apply the year-to-year variability as a target in itself in or-

der to certify the correct representativeness of the resampled series in comparison with the original historical series.

The weights applied in estimating the overall performance of resampled series are chosen in order to emphasize the accumulated precipitation values but, on the other hand, not neglect the extremes. Other weights could have been applied. One could imagine that the weights were chosen according to the purpose of use of the resampled and climate-projected series. If, for example, the series were to be used as an input to an urban drainage model simulating overflow from combined sewer systems to a recipient, it would probably be most important to have a good representation of the precipitation (event) totals. On the other hand if the purpose was simulating surcharge or flooding of a drainage system, the representation of extremes would be more important.

In the present approach a linear function and the probability of a given rainfall intensity for a given season is applied to derive the change factor as a function of intensity. The choice of parameters allows change factors to be both smaller and larger than 1. This might entail that the lowest fraction of intensities is allowed to be smaller in a future climate while the highest fraction of intensities will increase. Other continuous functions, rather than the applied linear function, might be an objective of future studies.

The proposed method applies two major assumptions which are relevant to discuss here. The first assumption is that events are sampled independently for each season. With inter-event times down to 1 h, this might constitute a problem in hydrological applications where the response time of the system in question is larger than 1 h. Hence, coupled events might impact the hydrological system response. The second assumption is that the duration of events does not change under changed climate signals. It has presently not been possible to find evidence for this contention in the scientific literature on climate change. Both of the assumptions are subject to further investigations.

6 Conclusions

This paper presented a procedure to generate both statistically representative resampled rainfall series from original historical rainfall series as well as climate-projected rainfall series, which includes the advantages in local historical rainfall series as well as projections on changes in rain patterns in the future climate.

The simulated rainfall series can represent the climate-projected target variables and it is shown possible to produce rainfall series which project not only accumulated seasonal precipitation but also extremes in correspondence with the climate projection of the RCP4.5 scenario. The procedure is generic, so if other climate scenarios and potentially other target variables for further precipitation patterns are available, the method will be able to adapt to these as well.

The procedure for generating resampled and climate-projected rainfall series fulfils a need for having local representative rainfall series which are valid both for the present and future climate. The series can be applied directly as inputs to urban drainage models in order to analyse the loads on a drainage system, e.g. combined sewer overflow, surcharge, storage filling, and flooding in the present and future climate.

Competing interests. The authors declare that they have no conflict of interest.

Acknowledgements. The authors would like to acknowledge Cathrine Fox Maule at the Danish Meteorological Institute for providing RCM data from EURO-CORDEX database.

Edited by: Nadia Ursino

References

Arnbjerg-Nielsen, K.: Quantification of climate change effects on extreme precipitation used for high resolution hydrologic design, Urban Water Journal, 9, 57–65, https://doi.org/10.1080/1573062X.2011.630091, 2012.

Beven, K. and Binley, A.: The future of distributed models: Model calibration and uncertainty prediction, Hydrol. Process., 6, 279–298, https://doi.org/10.1002/hyp.3360060305, 1992.

Boberg, F., Berg, P., Thejll, P., Gutowski, W. J., and Christensen, J. H.: Improved confidence in climate change projections of precipitation further evaluated using daily statistics from ENSEMBLES models, Clim. Dynam., 35, 1509–1520, https://doi.org/10.1007/s00382-009-0683-8, 2010.

Burton, A., Kilsby, C. G., Fowler, H. J., Cowpertwait, P. S. P., and O'Connell, P. E.: RainSim: A spatial-temporal stochastic rainfall modelling system, Environ. Model. Softw., 23, 1356–1369, https://doi.org/10.1016/j.envsoft.2008.04.003, 2008.

Casanueva, A., Kotlarski, S., Herrera, S., Fernández, J., Gutiérrez, J. M., Boberg, F., Colette, A., Christensen, O. B., Goergen, K., Jacob, D., Keuler, K., Nikulin, G., Teichmann, C., and Vautard, R.: Daily precipitation statistics in a EURO-CORDEX RCM ensemble: added value of raw and bias-corrected high-resolution simulations, Clim. Dynam., 47, 719–737, https://doi.org/10.1007/s00382-015-2865-x, 2016.

Christensen, O. B., Yang, S., Boberg, F., Maule, C. F., Thejll, P., Olesen, M., Drews, M., Sørup, H. J. D., and Christensen, J. H.: Scalability of regional climate change in Europe for high-end scenarios, Climate Res., 64, 25–38, https://doi.org/10.3354/cr01286, 2015.

Cowpertwait, P. S. P.: Further developments of the neyman-scott clustered point process for modeling rainfall, Water Resour. Res., 27, 1431–1438, https://doi.org/10.1029/91WR00479, 1991.

Cowpertwait, P. S. P.: A spatial-temporal point process model with a continuous distribution of storm types, Water Resour. Res., 46, 1–12, https://doi.org/10.1029/2010WR009728, 2010.

Cowpertwait, P. S. P., O'Connell, P. E., Metcalfe, A. V., and Mawdsley, J. A.: Stochastic point process modelling of rainfall. I. Single-site fitting and validation, J. Hydrol., 175, 17–46, https://doi.org/10.1016/S0022-1694(96)80004-7, 1996.

Cowpertwait, P. S. P., Kilsby, C. G., and O'Connell, P. E.: A space-time Neyman-Scott model of rainfall: Empirical analysis of extremes, Water Resour. Res., 38, 6-1–6-14, https://doi.org/10.1029/2001WR000709, 2002.

Danish Meteorological Institute (DMI): The rain gauge network of the Danish Water Pollution Committee of the Society of Danish Engineers, available at: https://www.dmi.dk/erhverv/anvendelse-af-vejrdata/spildevandskomiteens-regnmaalersystem/, last access: 31 August 2017.

DMI: Drift af Spildevandskomitéens Regnmålersystem, Technical report 15-03, Rikke Sjølin Thomsen (ed.), Danish Meteorological Institute, Copenhagen, Denmark, 2014.

Entekhabi, D., Rodriguez-Iturbe, I., and Eagleson, P. S.: Probabilistic representation of the temporal rainfall process by a modified Neyman-Scott Rectangular Pulses Model: Parameter estimation and validation, Water Resour. Res., 25, 295–302, https://doi.org/10.1029/WR025i002p00295, 1989.

Fowler, H. J., Kilsby, C. G., O'Connell, P. E., and Burton, A.: A weather-type conditioned multi-site stochastic rainfall model for the generation of scenarios of climatic variability and change, J. Hydrol., 308, 50–66, https://doi.org/10.1016/j.jhydrol.2004.10.021, 2005.

Fowler, H. J., Blenkinsop, S., and Tebaldi, C.: Linking climate change modelling to impacts studies: Recent advances in downscaling techniques for hydrological modelling, Int. J. Climatol., 27, 1547–1578, https://doi.org/10.1002/joc.1556, 2007.

Furrer, E. M. and Katz, R. W.: Improving the simulation of extreme precipitation events by stochastic weather generators, Water Resour. Res., 44, 1–13, https://doi.org/10.1029/2008WR007316, 2008.

Gregersen, I. B., Madsen, H., Rosbjerg, D., and Arnbjerg-Nielsen, K.: Long term variations of extreme rainfall in Denmark and southern Sweden, Clim. Dynam., 44, 3155–3169, https://doi.org/10.1007/s00382-014-2276-4, 2014a.

Gregersen, I. B., Sunyer, M. A. P., Madsen, H., Funder, S., Luchner, J., Rosbjerg, D., and Arnbjerg-Nielsen, K.: Past, present, and future variations of extreme precipitation in Denmark, DTU Environment, Kgs. Lyngby, Denmark, 2014b.

IPCC: Climate change 2007: the physical science basis summary for policymakers, Energ. Environ., 18, 433–440, https://doi.org/10.1260/095830507781076194, 2007.

IPCC: Climate Change 2013: The Physical Science Basis. Summary for Policymakers, IPCC, 1–29, https://doi.org/10.1017/CBO9781107415324, 2013.

Jacob, D., Petersen, J., Eggert, B., Alias, A., Christensen, O. B., Bouwer, L. M., Braun, A., Colette, A., Déqué, M., Georgievski, G., Georgopoulou, E., Gobiet, A., Menut, L., Nikulin, G., Haensler, A., Hempelmann, N., Jones, C., Keuler, K., Kovats, S., Kröner, N., Kotlarski, S., Kriegsmann, A., Martin, E., van Meijgaard, E., Moseley, C., Pfeifer, S., Preuschmann, S., Radermacher, C., Radtke, K., Rechid, D., Rounsevell, M., Samuelsson, P., Somot, S., Soussana, J. F., Teichmann, C., Valentini, R., Vautard, R., Weber, B., and Yiou, P.: EURO-CORDEX: New high-resolution climate change projections for European impact research, Regional Environmental Change, 14, 563–578, https://doi.org/10.1007/s10113-013-0499-2, 2014.

Keifer, C. J. and Chu, H. H.: Synthetic Storm Pattern for Drainage Design, Journal of the Hydraulics Division, 83, 1–25, 1957.

Kilsby, C. G., Jones, P. D., Burton, A., Ford, A. C., Fowler, H. J., Harpham, C., James, P., Smith, A., and Wilby, R. L.: A daily weather generator for use in climate change studies, Environ. Model. Softw., 22, 1705–1719, https://doi.org/10.1016/j.envsoft.2007.02.005, 2007.

Kossieris, P., Makropoulos, C., Onof, C., and Koutsoyiannis, D.: A rainfall disaggregation scheme for sub-hourly time scales: Coupling a Bartlett-Lewis based model with adjusting procedures, J. Hydrol., https://doi.org/10.1016/j.jhydrol.2016.07.015, in press, 2016.

Koutsoyiannis, D. and Onof, C.: Rainfall disaggregation using adjusting procedures on a Poisson cluster model, J. Hydrol., 246, 109–122, https://doi.org/10.1016/S0022-1694(01)00363-8, 2001.

Larsen, A. N., Gregersen, I. B., Christensen, O. B., Linde, J. J., and Mikkelsen, P. S.: Potential future increase in extreme one-hour precipitation events over Europe due to climate change, Water Sci. Technol., 60, 2205–2216, https://doi.org/10.2166/wst.2009.650, 2009.

Madsen, H., Arnbjerg-Nielsen, K., and Mikkelsen, P. S.: Update of regional intensity-duration-frequency curves in Denmark: Tendency towards increased storm intensities, Atmos. Res., 92, 343–349, https://doi.org/10.1016/j.atmosres.2009.01.013, 2009.

Mailhot, A., Duchesne, S., Caya, D., and Talbot, G.: Assessment of future change in intensity-duration-frequency (IDF) curves for Southern Quebec using the Canadian Regional Climate Model (CRCM), J. Hydrol., 347, 197–210, https://doi.org/10.1016/j.jhydrol.2007.09.019, 2007.

Marani, M. and Zanetti, S.: Downscaling rainfall temporal variability, Water Resour. Res., 43, 1–7, https://doi.org/10.1029/2006WR005505, 2007.

Maule, C. F., Thejll, P., Christensen, J. H., Svendsen, S. H., and Hannaford, J.: Improved confidence in regional climate model simulations of precipitation evaluated using drought statistics from the ENSEMBLES models, Clim. Dynam., 40, 155–173, https://doi.org/10.1007/s00382-012-1355-7, 2013.

Mikkelsen, P. S., Madsen, H., Arnbjerg-Nielsen, K., Jorgensen, H. K., Rosbjerg, D., and Harremoes, P.: A rationale for using local and regional point rainfall data for design and analysis of urban storm drainage systems, Water Sci. Technol., 37, 7–14, https://doi.org/10.1016/S0273-1223(98)00310-2, 1998.

Molini, A., La Barbera, P., Lanza, L. G., and Stagi, L.: Rainfall intermittency and the sampling error of tipping-bucket rain gauges, Physics and Chemistry of the Earth, Part C: Solar, Terrestrial & Planetary Science, 26, 737–742, https://doi.org/10.1016/S1464-1917(01)95018-4, 2001.

Nguyen, V. T. V., Nguyen, T. D., and Cung, A.: A statistical approach to downscaling of sub-daily extreme rainfall processes for climate-related impact studies in urban areas, in: Water Science and Technology: Water Supply, vol. 7, pp. 183–192, 2007.

Nguyen, V.-T.-V., Desramaut, N., and Nguyen, T.-D.: Estimation of urban design storms in consideration of GCM-based climate change scenarios, in Water and Urban Development Paradigms: Towards an Integration of Engineering, Design and Management Approaches – Proceedings of the International Urban Water Conference, 347–356, 2009.

Nguyen, V. T. V., Desramaut, N., and Nguyen, T. D.: Optimal rainfall temporal patterns for urban drainage design in the context of climate change, Water Sci. Technol., 62, 1170–1176, https://doi.org/10.2166/wst.2010.295, 2010.

NRCS: Urban Hydrology for Small Watersheds TR-55, USDA Natural Resource Conservation Service Conservation Engeneering Division Technical Release 55, 164, Technical Release 55, 1986.

Ntegeka, V. and Willems, P.: Trends and multidecadal oscillations in rainfall extremes, based on a more than 100-year time series of 10 min rainfall intensities at Uccle, Belgium, Water Resour. Res., 44, 1–15, https://doi.org/10.1029/2007WR006471, 2008.

Ntegeka, V., Baguis, P., Roulin, E., and Willems, P.: Developing tailored climate change scenarios for hydrological impact assessments, J. Hydrol., 508, 307–321, https://doi.org/10.1016/j.jhydrol.2013.11.001, 2014.

Olesen, M., Madsen, K. S., Ludwigsen, C. A., Boberg, F., Christensen, T., Cappelen, J., Christensen, O. B., Andersen, K. K., and Hesselbjerg Christensen, J.: Fremtidige klimaforandringer i Danmark, 2014.

Olsson, J., Berggren, K., Olofsson, M., and Viklander, M.: Applying climate model precipitation scenarios for urban hydrological assessment: A case study in Kalmar City, Sweden, Atmos. Res., 92, 364–375, https://doi.org/10.1016/j.atmosres.2009.01.015, 2009.

Olsson, J., Willen, U., and Kawamura, A.: Downscaling extreme short-term regional climate model precipitation for urban hydrological applications, Hydrol. Res., 43, 341–351, https://doi.org/10.2166/nh.2012.135, 2012.

Onof, C. and Arnbjerg-Nielsen, K.: Quantification of anticipated future changes in high resolution design rainfall for urban areas, Atmos. Res., 92, 350–363, https://doi.org/10.1016/j.atmosres.2009.01.014, 2009.

Onof, C. and Wheater, H. S.: Modelling of British rainfall using a random parameter Bartlett-Lewis Rectangular Pulse Model, J. Hydrol., 149, 67–95, https://doi.org/10.1016/0022-1694(93)90100-N, 1993.

Onof, C. and Wheater, H. S.: Improvements to the modelling of British rainfall using a modified Random Parameter Bartlett-Lewis Rectangular Pulse Model, J. Hydrol., 15, 177–195, https://doi.org/10.1016/0022-1694(94)90104-X, 1994.

Paschalis, A., Molnar, P., Fatichi, S., and Burlando, P.: On temporal stochastic modeling of precipitation, nesting models across scales, Adv. Water Resour., 63, 152–166, https://doi.org/10.1016/j.advwatres.2013.11.006, 2014.

Prein, A. F., Gobiet, A., Truhetz, H., Keuler, K., Goergen, K., Teichmann, C., Fox Maule, C., van Meijgaard, E., Déqué, M., Nikulin, G., Vautard, R., Colette, A., Kjellström, E., and Jacob, D.: Precipitation in the EURO-CORDEX 0.11 and 0.44 simulations: high resolution, high benefits?, Clim. Dynam., 46, https://doi.org/10.1007/s00382-015-2589-y, 2016.

Rossi, F., Fiorentino, M., and Versace, P.: Two-Component Extreme Value Distribution for Flood Frequency Analysis, Water Resour. Res., 20, 847–856, https://doi.org/10.1029/WR020i007p00847, 1984.

Schaarup-Jensen, K., Rasmussen, M. R., and Thorndahl, S.: To what extent does variability of historical rainfall series influence extreme event statistics of sewer system surcharge and overflows?, Water Sci. Technol., 60, 87–95, https://doi.org/10.2166/wst.2009.290, 2009.

Schilling, W.: Rainfall data for urban hydrology: what do we need?, Atmos. Res., 27, 5–21, https://doi.org/10.1016/0169-8095(91)90003-F, 1991.

Schleiss, M., Jaffrain, J., and Berne, A.: Statistical analysis of rainfall intermittency at small spatial and temporal scales, Geophys. Res. Lett., 38, L18403, https://doi.org/10.1029/2011GL049000, 2011.

Segond, M.-L., Neokleous, N., Makropoulos, C., Onof, C., and Maksimovic, C.: Simulation and spatio-temporal disaggregation of multi-site rainfall data for urban drainage applications, Hydrol. Sci. J., 52, 917–935, https://doi.org/10.1623/hysj.52.5.917, 2007.

Semadeni-Davies, A., Hernebring, C., Svensson, G., and Gustafsson, L. G.: The impacts of climate change and urbanisation on drainage in Helsingborg, Sweden: Combined sewer system, J. Hydrol., 350, 100–113, https://doi.org/10.1016/j.jhydrol.2007.05.028, 2008.

Shahabul Alam, M. and Elshorbagy, A.: Quantification of the climate change-induced variations in Intensity–Duration–Frequency curves in the Canadian Prairies, J. Hydrol., 527, 990–1005, https://doi.org/10.1016/j.jhydrol.2015.05.059, 2015.

Sherman: Streamflow from rainfall by the unit-graph method, Engineering News Record, 108, 1932.

Sørup, H. J. D., Christensen, O. B., Arnbjerg-Nielsen, K., and Mikkelsen, P. S.: Downscaling future precipitation extremes to urban hydrology scales using a spatio-temporal Neyman-Scott weather generator, Hydrol. Earth Syst. Sci., 20, 1387–1403, https://doi.org/10.5194/hess-20-1387-2016, 2016.

Sunyer, M. A., Hundecha, Y., Lawrence, D., Madsen, H., Willems, P., Martinkova, M., Vormoor, K., Bürger, G., Hanel, M., Kriauciuniene, J., Loukas, A., Osuch, M., and Yücel, I.: Intercomparison of statistical downscaling methods for projection of extreme precipitation in Europe, Hydrol. Earth Syst. Sci., 19, 1827–1847, https://doi.org/10.5194/hess-19-1827-2015, 2015.

Thorndahl, S.: Stochastic long term modelling of a drainage system with estimation of return period uncertainty, Water Sci. Technol., 59, 2331–2339, https://doi.org/10.2166/wst.2009.305, 2009.

Thorndahl, S., Beven, K. J., Jensen, J. B., and Schaarup-Jensen, K.: Event based uncertainty assessment in urban drainage modelling, applying the GLUE methodology, J. Hydrol., 357, 421–437, https://doi.org/10.1016/j.jhydrol.2008.05.027, 2008.

Thorndahl, S., Schaarup-Jensen, K., and Rasmussen, M. R.: On hydraulic and pollution effects of converting combined sewer catchments to separate sewer catchments, Urban Water Journal, 12, 120–130, https://doi.org/10.1080/1573062X.2013.831915, 2015.

Thorndahl, S., Balling, J. D., and Larsen, U. B. B.: Analysis and integrated modelling of groundwater infiltration to sewer networks, Hydrol. Process., 30, 3228–3238, https://doi.org/10.1002/hyp.10847, 2016.

Thorndahl, S., Einfalt, T., Willems, P., Nielsen, J. E., ten Veldhuis, M.-C., Arnbjerg-Nielsen, K., Rasmussen, M. R., and Molnar, P.: Weather radar rainfall data in urban hydrology, Hydrol. Earth Syst. Sci., 21, 1359–1380, https://doi.org/10.5194/hess-21-1359-2017, 2017.

Van Der Linden, P. and Mitchell, J. F. B.: ENSEMBLES: Climate Change and its Impacts: Summary of research and results from the ENSEMBLES project, Met Office Hadley Centre, Exeter, UK, 2009.

Wilby, R. L. and Wigley, T. M. L.: Downscaling general circulation model output: a review of methods and limitations, Prog. Phys. Geogr., 21, 530–548, https://doi.org/10.1177/030913339702100403, 1997.

Wilby, R. L., Dawson, C. W., and Barrow, E. M.: SDSM – a decision support tool for the assessment of regional climate change impacts, Environ. Model. Softw., 17, 145–157, https://doi.org/10.1016/S1364-8152(01)00060-3, 2002.

Willems, P.: Compound intensity/duration/frequency-relationships of extreme precipitation for two seasons and two storm types, J. Hydrol., 233, 189–205, https://doi.org/10.1016/S0022-1694(00)00233-X, 2000a.

Willems, P.: Compound intensity/duration/frequency-relationships of extreme precipitation for two seasons and two storm types, J. Hydrol., 233, 189–205, https://doi.org/10.1016/S0022-1694(00)00233-X, 2000b.

Willems, P.: Multidecadal oscillatory behaviour of rainfall extremes in Europe, Climatic Change, 120, 931–944, https://doi.org/10.1007/s10584-013-0837-x, 2013a.

Willems, P.: Revision of urban drainage design rules after assessment of climate change impacts on precipitation extremes at Uccle, Belgium, J. Hydrol., 496, 166–177, https://doi.org/10.1016/j.jhydrol.2013.05.037, 2013b.

Willems, P. and Vrac, M.: Statistical precipitation downscaling for small-scale hydrological impact investigations of climate change, J. Hydrol., 402, 193–205, https://doi.org/10.1016/j.jhydrol.2011.02.030, 2011.

Willems, P., Arnbjerg-Nielsen, K., Olsson, J., and Nguyen, V. T. V: Climate change impact assessment on urban rainfall extremes and urban drainage: Methods and shortcomings, Atmos. Res., 103, 106–118, https://doi.org/10.1016/j.atmosres.2011.04.003, 2012a.

Willems, P., Olsson, J., Arnbjerg-Nielsen, K., Beecham, S., Pathirana, A., Gregersen, I. B., Madsen, H., and Nguyen, V. T. V.: Impacts of Climate Change on Rainfall Extremes and Urban Drainage Systems, IWA publishing, London, 2012b.

WPC: Forventede ændringer i ekstremregn som følge af klimaændringer, Skrift nr. 29 (Anticipated changes in extrem precipitation as a result of climate change, Guideline no. 29), The Water Pollution Committee of the Society of Danish Engineers, Copenhagen, Denmark, 2008 (in Danish).

WPC: Opdaterede klimafaktorer og dimensionsgivende regnintensiteter, Skrift nr. 30 (Updated climate factors and rain intensities for design, Guideline no. 30), The Water Pollution Committee of the Society of Danish Engineers, Copenhagen, Denmark, 2014 (in Danish).

Zorita, E. and Von Storch, H.: The analog method as a simple statistical downscaling technique: Comparison with more complicated methods, J. Climate, 12, 2474–2489, https://doi.org/10.1175/1520-0442(1999)012<2474:TAMAAS>2.0.CO;2, 1999.

11

Climate response to Amazon forest replacement by heterogeneous crop cover

A. M. Badger[1] **and P. A. Dirmeyer**[1,2]

[1]George Mason University, Fairfax, Virginia, USA
[2]Center for Ocean–Land–Atmosphere Studies, Fairfax, Virginia, USA

Correspondence to: A. M. Badger (abadger@gmu.edu)

Abstract. Previous modeling studies with atmospheric general circulation models and basic land surface schemes to balance energy and water budgets have shown that by removing the natural vegetation over the Amazon, the region's climate becomes warmer and drier. In this study we use a fully coupled Earth system model and replace tropical forests by a distribution of six common tropical crops with variable planting dates, physiological parameters and irrigation. There is still general agreement with previous studies as areal averages show a warmer ($+1.4$ K) and drier (-0.35 mm day^{-1}) climate. Using an interactive crop model with a realistic crop distribution shows that regions of vegetation change experience different responses dependent upon the initial tree coverage and whether the replacement vegetation is irrigated, with seasonal changes synchronized to the cropping season. Areas with initial tree coverage greater than 80 % show an increase in coupling with the atmosphere after deforestation, suggesting land use change could heighten sensitivity to climate anomalies, while irrigation acts to dampen coupling with the atmosphere.

1 Introduction

1.1 Background information

The future of tropical forests is at risk in a warmer, more populous 21st-century world (Bonan, 2008a). Forests cover approximately 42 million km^2 in tropical, temperate and boreal regions, which is approximately 30 % of the Earth's land surface. Land use change (LUC) occurs on local scales, with real-world social and economic benefits, but can potentially cause ecological degradation across local, regional, and global scales (Foley et al., 2005). A large portion, almost 35 %, of the Earth's surface has already been modified for urban and industrial development, agriculture, and pasture land (Snyder, 2010). Worldwide changes to forests, woodlands, grasslands and wetlands are being driven by the need to provide food, fiber, water, and shelter (Foley et al., 2005). LUC has the potential to have a significant impact on land–atmosphere interactions and modify local climate conditions (e.g., Sun and Wang, 2011).

Loss of natural forests worldwide in the tropics during the 1990s was as high as 152 000 km^2 year^{-1}, and Amazonian forests were cleared at a rate of approximately 25 000 km^2 year^{-1} (Bonan, 2008a). By 1991, 426 000 km^2 of the Amazon forest had already been removed, approximately 10.5 % of the original forest area (Costa and Foley, 2000). More recent estimates suggest that by 2006, 663 177 km^2 of the Amazon forest had been removed (IBGE, 2006), with approximately an additional 60 000 km^2 deforested since 2006 (INPE, 2014). Nepstad et al. (2008) note that trends in Amazon economies, forests and climate could lead to the replacement or severe degradation of more than half of the closed-canopy forests of the Amazon Basin by the year 2030, even without including the impacts of fire or global warming. Snyder et al. (2004) acknowledge that wide-scale vegetation removal is unrealistic for most biomes, with the tropical forests being the lone exception.

It is clear that LUC in the Amazon region can have drastic consequences because of the role forests have in mediating the climate. Forests influence the climate through exchanges of energy, water, carbon dioxide, and other chemical species with the atmosphere (Bonan, 2008a). LUC has played a role in changing the global carbon cycle and, possibly, the global climate (Foley et al., 2005).

One of the most important roles that forests have in the climate system is their function in the carbon cycle. Forests sequester large amounts of carbon, storing approximately 45 % of all terrestrial carbon and contributing approximately 50 % of terrestrial net primary production (Bonan, 2008a). Bonan (2008a) also notes that carbon uptake by forests contributed to a residual 2.6 PgC year^{-1} terrestrial carbon sink during the 1990s, offsetting approximately 33 % of anthropogenic carbon emissions from fossil fuels and LUC, with deforestation releasing 1.6 PgC year^{-1} during the 1990s. The trees of the Amazon contain 90–140 billion tons of carbon, equivalent to approximately 9–14 decades of current global human-induced carbon emissions (Nepstad et al., 2008).

1.2 Design of previous modeling studies

Early total deforestation studies used coarse-resolution climate models that did not resolve the local features of deforestation, but may have given a reasonable representation of regional-scale changes. More recent experiments tend to have increased resolution and duration, a feature to be expected as computational resources have increased. The increased resolution and the associated ability to resolve small-scale features is desired to represent better the local dynamics involved with deforestation. With increased length of integration, the capability to reach a new equilibrium climate is greatly enhanced, and greater confidence in the significance of the results is obtained. Shorter simulations are likely missing some global features associated with Amazon deforestation that have not had a chance to develop in the model integration, particularly when ocean dynamics are not modeled.

A noticeable inconsistency among the simulations is the replacement vegetation used. The difference in using grassland, savanna, shrubs or bare soil as a substitute for tropical forests is not known, although some inherent differences may arise. Only one simulation, Costa et al. (2007), used a crop as replacement vegetation. Agricultural land cover should be the most realistic replacement vegetation from a socioeconomic standpoint, and may have different impacts than the aforementioned unmanaged replacement vegetation.

1.3 Results from previous modeling studies

Previous studies have reported a change in annual surface temperature from −1 to +3 °C. Several studies note that the change in temperature is statistically significant (Dickinson and Henderson-Sellers, 1988; Henderson-Sellers et al., 1993; McGuffie et al., 1995; Nobre et al., 1991; Shukla et al., 1990; Snyder et al., 2004; Lejeune et al., 2014). Dickinson and Henderson-Sellers (1988) add that while surface air temperature increases by 1–3 °C, the soil-surface temperature increased by 2–5 °C.

A common feature of previous studies is decreases in precipitation, although they are of varying intensity. Decreases in annual precipitation are typically found to be significant (Costa and Foley, 2000; Hasler et al., 2009; Henderson-Sellers et al., 1993; McGuffie et al., 1995; Nobre et al., 1991, 2009; Shukla et al., 1990; Lejeune et al., 2014). Nobre et al. (2009) points out a difference in precipitation change in simulations coupled with the ocean; the coupled model produced a rainfall reduction that is nearly 60 % larger than was obtained by use of an AGCM uncoupled from the ocean. As previously noted, the effect of different replacement vegetation may also play a role. Costa et al. (2007) found that changes in precipitation for 25, 50, and 75 % deforestation, respectively, were −6.2, −11.6, and −15.7 % for soybean land cover, which was significantly different than the +1.4, −0.8, and −3.9 % changes for pasture. Both Costa and Foley (2000) and Lejeune et al. (2014) note that the seasonality of the precipitation did not change significantly, with the rainy season and dry season remaining in the same periods.

Evapotranspiration decrease is a common finding of Amazon deforestation studies (Costa and Foley, 2000; Dickinson and Henderson-Sellers, 1988; Henderson-Sellers et al., 1993; McGuffie et al., 1995; Nobre et al., 1991; Shukla et al., 1990; Snyder et al., 2004). Costa and Foley (2000) found that the differences in evapotranspiration are statistically significant in all months. The decrease in transpiration of 53 % was much larger than the decrease in total evapotranspiration of 16 %; this indicates that evaporation from the surface can compensate for the drop in transpiration (Costa and Foley, 2000). Henderson-Sellers et al. (1993) noted that as the evaporation decreases, the near-surface specific humidity decreases. This result is of particular interest in the response of planetary boundary layer (PBL) growth.

Subsequent sections will describe the model of choice and associated simulations used in this study to analyze the local response to Amazon deforestation, along with a description of tropical crops incorporated into the model. Results detail the mean climate changes in temperature, precipitation, surface fluxes and modifications to the land–atmosphere coupling. The possible impacts and causes of these changes are discussed, as well as the role that irrigation plays in altering land–atmosphere coupling.

2 Methods

2.1 Model description

The model for this study is the Community Earth System Model (CESM) version 1.2.0 developed at the National Center for Atmospheric Research (NCAR). CESM is a coupled model system for simulating the Earth's climate and is composed of separate models simulating the Earth's atmosphere, ocean, land, land ice and sea ice (Vertenstein et al., 2013). Of the components available in CESM, the following were run in their default settings: the Community Atmosphere Model (CAM4), the Parallel Ocean Program (POP2), the Community Ice CodE (CICE4), and the River Transport Model (RTM) (see model documentation for full details).

The Community Land Model 4.5 (CLM4.5) incorporates recent scientific advances in the understanding and representation of land surface processes relevant to climate simulation (Oleson et al., 2013). CLM4.5 is a model developer's release that provides incremental improvements to CLM4.0 prior to the public release of CLM version 5. Land surface heterogeneity in CLM4.5 is accomplished with a nested sub-grid hierarchy in which grid cells are comprised of multiple land units, soil columns, snow columns, and plant functional types (PFTs) (Oleson et al., 2013). The PFT level, which also includes bare ground, is intended to capture the biogeophysical and biogeochemical differences between broad categories of plants in terms of their functional characteristics. Fluxes to and from the surface are defined at the PFT level, as well as the vegetation state variables, such as vegetation temperature and canopy water storage.

Each PFT is characterized by parameters that differ in leaf and stem optical properties to determine the reflection, transmittance and absorption of solar radiation (Oleson et al., 2013). Each PFT also has a specific root distribution to allow for root uptake of water from the soil. Different PFTs have aerodynamic parameters that determine heat, moisture and momentum transfers, and photosynthetic parameters that determine stomatal resistance, photosynthesis and transpiration. These parameterizations are used to represent optimally the behaviors of each PFT.

CLM4.5 includes a fully prognostic treatment of the terrestrial carbon and nitrogen cycles (Oleson et al., 2013). The model is fully prognostic for all carbon and nitrogen state variables in the vegetation, litter, and soil organic matter. The seasonal timing of new vegetation growth and litterfall for each PFT is also prognostic, responding to soil and air temperature, soil water availability, and day length. PFTs are classified into three distinct phenological types that are represented by independent algorithms: an evergreen type that has some fraction of annual leaf growth displayed for longer than 1 year; a seasonal-deciduous type with a single growing season per year controlled mainly by temperature and day length; and a stress-deciduous type with the potential for multiple growing seasons per year, controlled by temperature and soil moisture conditions.

CLM's default list of PFTs includes an unmanaged crop, essentially treated as a second C3 grass PFT (Levis et al., 2012; Oleson et al., 2013). In CLM4.5, a crop model based on the AgroIBIS (Kucharik et al., 2000) crop phenology algorithm has been added, consisting of three distinct phases. Phase 1 starts at planting and ends with leaf emergence; phase 2 continues from leaf emergence to the beginning of grain fill; and phase 3 starts from the beginning of grain fill and ends with physiological maturity and harvest.

CLM4.5 introduces three new agricultural PFTs: corn (CLM's only C4 crop), soybean, and temperate cereals, i.e., spring wheat and winter wheat (Levis et al., 2012). Temperate cereals represent wheat, barley, and rye, assuming that these three crops have similar characteristics and can be treated as one PFT. The changing of several PFT parameter values following AgroIBIS further distinguishes corn (a C4 crop), soybean, and temperate cereals from the existing unmanaged crop. The most notable difference in the model between C3 and C4 photosynthesis is that the C4 photosynthetic pathway allows for stomata to close more often, thus transpiring less, allowing for higher water-use efficiency in C4 plants. With the crop model active in CLM4.5, the vegetated land unit is split into unmanaged and managed parts. PFTs in the unmanaged land unit all share the same belowground properties per grid cell, including water and nutrients, while PFTs in the managed land unit occupy separate soil columns and do not interact with each other below the ground, and thus do not compete for water and nutrients. Having PFTs in separate managed land units allows for different management practices, such as irrigation and fertilization, for each crop PFT.

CLM4.5 simulates the application of irrigation as a dynamic response to simulated soil moisture conditions (Oleson et al., 2013). When irrigation is enabled, the crop area of each grid cell is divided into irrigated and rainfed fractions according to a gridded data set of areas equipped for irrigation. Irrigated and rainfed crops are placed on separate soil columns, so that irrigation is only applied to the soil beneath irrigated crops. In irrigated croplands, a check is made once per day to determine whether irrigation is required; this check is made in the first time step after 06:00 LT. Irrigation is required if crop leaf area is greater than zero, and water is the limiting factor for photosynthesis.

2.2 Tropical crops

In performing offline CLM4 simulations, the need to develop more realistic PFTs for the tropics became apparent. The tropical broadleaf evergreen tree PFT was initially replaced with the unmanaged crop PFT and C3 grass PFT. It was thought that there would be a reduction in leaf area index (LAI) when replacing the broadleaf evergreen trees; however, it was found that there was a drastic basin-wide increase in LAI. It was determined that the crop and C3 grass PFTs were parameterized solely for the mid-latitude conditions. The winter season temperature in the Amazon does not get cold enough to trigger senescence; the survival temperature for C3 grass is $-17\,°C$ and the establishment temperature for C3 grass is $15.5\,°C$, while the planting temperature for managed crops ranges from 7 to $13\,°C$. The Amazon has an annual average temperature of approximately $27\,°C$, meaning minimum temperature thresholds for each PFT are

Table 1. Key parameters used in developing CLM4.5 tropical crops. Planting dates are in the format of month-day (example: 4-15 is 15 April). "–" denotes a parameter that is not specified.

Parameters	C3 Crop	Corn	Spring Wheat	Winter Wheat	Soybean	Tropical Soybean	Tropical Corn	Tropical Corn (2)	Tropical Sugarcane	Tropical Rice	Tropical Cotton
Photosynthesis	C3	C4	C3	C3	C3	C3	C4	C4	C4	C3	C3
Max LAI	–	5	7	7	6	6	5	5	5	7	6
Max canopy top (m)	–	2.5	1.2	1.2	0.75	1	2.5	2.5	4	1.8	1.5
Last NH planting date	–	6-15	6-15	11-30	6-15	12-31	10-15	2-28	3-31	2-28	5-31
Last SH planting date	–	12-15	12-15	5-30	12-15	12-31	10-15	2-28	10-31	12-31	11-30
First NH planting date	–	4-01	4-01	9-01	5-01	10-15	9-20	2-01	1-01	1-01	4-01
First SH planting date	–	10-01	10-01	3-01	11-01	10-15	9-20	2-01	8-01	10-15	9-01
Min planting temp. (K)	–	279.15	272.15	278.15	279.15	283.15	283.15	283.15	283.15	283.15	283.15
Planting temp. (K)	–	283.15	280.15	–	286.15	294.15	294.15	294.15	294.15	294.15	294.15
GDD	–	1700	1700	1700	1900	2100	1800	1900	4300	2100	1700
Base temperature (°C)	0	8	0	0	10	10	10	10	10	10	10
Max day to maturity	–	165	150	265	150	150	160	180	300	150	160
Maximum fertilizer (kg N m^{-2})	0	0.015	0.008	0.008	0.0025	0.05	0.03	0.03	0.04	0.02	0.02
Leaf albedo – near IR	0.35	0.35	0.35	0.35	0.35	0.58	0.58	0.58	0.58	0.58	0.58
Leaf transmittance – near IR	0.34	0.34	0.34	0.34	0.34	0.25	0.25	0.25	0.25	0.25	0.25
Leaf transmittance – visible	0.05	0.05	0.05	0.05	0.05	0.07	0.07	0.07	0.07	0.07	0.07

always met. Another aspect is the greater moisture availability in most of the Amazon; plants are rarely stressed over most of the year by a lack of available moisture.

Using the Sacks et al. (2010) and Portmann et al. (2010) data sets of global crop distribution, it was determined that the most prevalent crops in and around the Amazon Basin are soybean, corn, cotton, rice and sugarcane. These crops were then selected as tropical crops to be added to CLM4.5. Two separate corn PFTs were added to simulate the two separate corn harvests that occur in the region. Given the long growing season in the tropics, after the first corn harvest of the year a second crop of corn is typically planted and harvested later in the year. For each crop added, a rainfed and an irrigated PFT were constructed based on irrigation data.

The new tropical crops are based on existing crops in CLM4.5, with adjustments to physiology parameters to get realistic behavior. Tropical soybean was based on the existing soybean PFT and tropical corn based on the existing corn PFT. Tropical sugarcane is derived from the existing corn PFT, tropical rice is a variation on the existing spring wheat PFT, and tropical cotton is similar to soybean. Sugarcane was based on corn because both are C4 plants and corn is the only C4 crop in CLM4.5. Rice was based on the existing spring wheat because they are both cereal grain crops. Cotton uses soybean as a basis because they are both bushy C3 crops, with neither being a cereal grain crop, as are the other C3 crops in CLM4.5. It is of note that sugarcane is a multi-year perennial crop, while all the other crops are annual; CLM4.5 does not currently have the capability to simulate perennial crops. Thus, sugarcane was modeled to have a planting date just after the previous harvest, with the intention of simulating perennial coverage with a decrease once a year when a portion of the sugarcane is typically harvested or replaced.

The Sacks et al. (2010) data were used to determine plant-

ing dates, growing degree days, maximum LAI and maximum number of days to plant maturity for the tropical crops being added; Table 1 shows original crop PFT and tropical crop PFT parameters that were modified. In addition to changing those physiology parameters, the albedo and radiative transmissivity of crop leaves were changed to match those of Bonan (2008b). The amount of fertilizer applied to each crop was modified to allow for a more realistic seasonal cycle. The goal of these new tropical crops is to provide a realistic physical seasonal cycle of planting, crop height, crop LAI and harvest time; compared to Sacks et al. (2010), the timing of planting and harvest are achieved, plant heights fall within the expected range (FAO, 2007) and LAI falls within the expected range of previous documentation.

The 5 min spatial resolution Portmann et al. (2010) data were regridded for use in CLM4.5. In the specified domain (85–35° W, 30° S–13° N), each CLM grid box having a total area of tree PFT (tropical broadleaf evergreen and tropical broadleaf deciduous) percentage (see Fig. 1 for the default PFT distribution) greater than zero was deforested; all existing PFTs in that grid box were removed. Each respective deforested grid box is checked for the presence of crops in the regridded Portmann data. If any crops are present in a deforested grid box, the acreage for each crop is used to determine the percent coverage, preserving the percentages in the deforested case. There is a maximum of five crops allowed in each CLM grid box. If all six crops are present, the lowest acreage crop is omitted. For deforested grid boxes with no crops present, a Cressman analysis is used to interpolate crop coverage from neighboring grid boxes. The calculated distribution of the tropical crops in the deforested case can be seen in Fig. 2, with 12.82 % of the area being soybean, 21.09 % for each corn crop, 14.77 % for sugarcane, 25.18 % for irrigated rice, and 5.04 % for cotton.

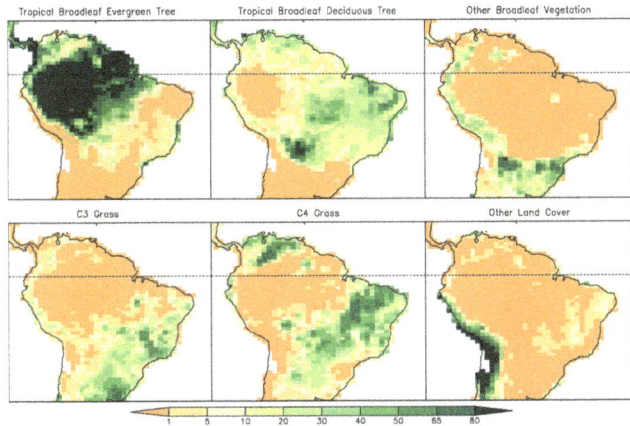

Figure 1. Distributions of the indicated plant functional types (PFTs) in the control simulation as a percentage of each grid box. "Other vegetation" includes C3 alpine grasses and bare soil.

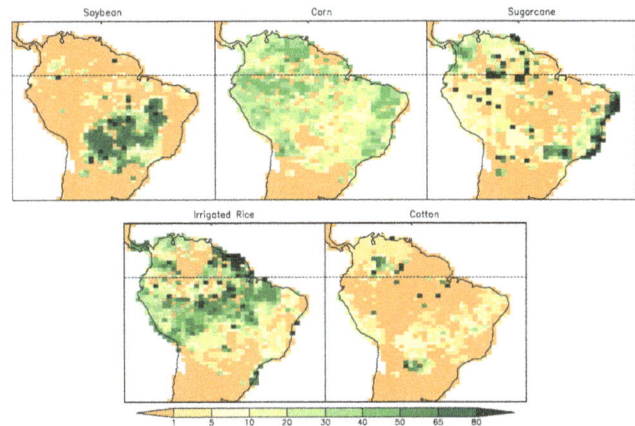

Figure 2. Distribution of each tropical crop as replacement vegetation in the Amazon region, with the color bar indicating the percentage of each grid box.

The initial seasonal mean changes to the land surface can be seen in Fig. 3. There is a basin-wide increase in surface albedo across the deforested region. In the closed canopy region where the highest percentages of broadleaf evergreen trees are located, there is a large reduction in both LAI and canopy height across all seasons. To the southeast of that region, an area where C4 grass was predominant, there is an increase in both LAI and canopy height in NDJFM, the main growing season of the dominant crops, soybean and rice, in that region. The other months show a general decrease in LAI and canopy height in that region.

The choice of these irregular seasons is based on the growing season of the tropical crops used as replacement vegetation. NDJFM largely coincides with crop growth in the region south of the Equator and planting north of the Equator. AMJ is the main growing season north of the Equator. JASO is predominantly a period after harvest has occurred and planting south of the Equator is taking place in the last month. Additionally, these seasons correspond to the seasons of peak precipitation, as NDJFM has precipitation predominantly south of the Equator, AMJ precipitation is centered on the Equator and extends into northern South America, and JASO is the driest period for the majority of the region, with precipitation centered over the northwestern portion of South America.

2.3 Model simulations

CESM with active components of CAM4, CLM4.5, POP2, CICE4 and RTM is used for the model simulations in this study. The simulations are run at an atmospheric model resolution of 0.9° × 1.25° and a nominal 1° ocean resolution grid with a displaced pole over Greenland for present-day (year 2000) initial conditions for greenhouse gas concentrations. Before starting the coupled runs, a spin-up simulation for the land surface was implemented to achieve a steady state

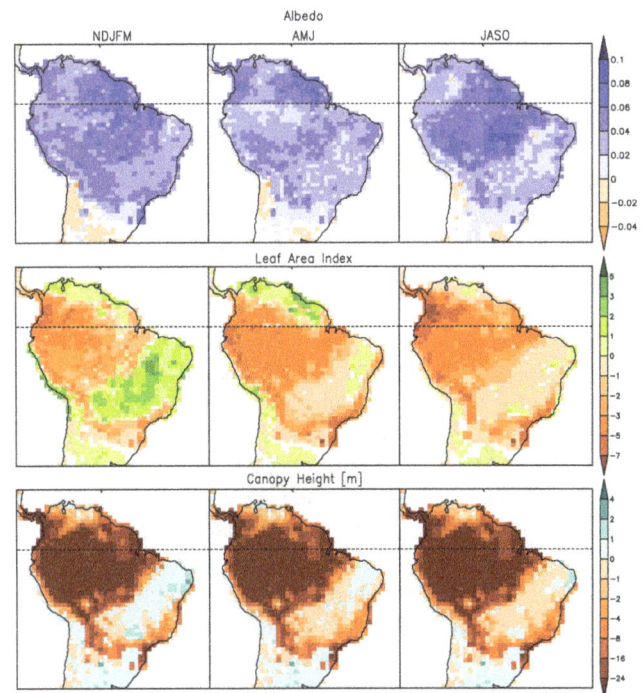

Figure 3. Changes to surface properties after deforestation in NDJFM, AMJ and JASO; albedo (top row), leaf area index (middle row) and canopy height (m) (bottom row). Shading indicates significance at the 95 % confidence level.

for the carbon and nitrogen processes of the interactive phenology. The CLM4.5 spin-up procedure consists of a 650-year offline simulation with present-day atmospheric forcing, achieved by repeatedly cycling through the Qian et al. (2006) input data set, years 1–600 are forced with years 1951–1990 and years 601–650 are forced with years 1951–2000; the last land state from the offline simulations is then used as the land initial condition in the coupled simulations. A separate spin-up simulation is done for each coupled experiment with matching PFT distributions. In the simulation utilizing tropical crops, the crop model and irrigation models are active. Each of the fully coupled simulations has a length of 250 years, in which only the last 125 years of monthly data are used for analysis. The control simulation uses the default PFT distribution (Fig. 1) and the deforested simulation used the crop PFT distribution in Fig. 2.

In all simulations, the fire module is turned off. When coupling CLM4.5 with CAM, specific humidity has been found to be too low over the Amazon region (W. Sacks and D. Lawrence, personal communications, 2013). Fires in CLM4.5 are invoked as a function of relative humidity, soil wetness, temperature and precipitation (Oleson et al., 2013). With low specific humidity, the relative humidity triggers the fire model in vast areas of the Amazon region, predominantly regions neighboring the closed canopy forests (grid boxes with greater than 60 % tree PFT). Along with a reduction in humidity, there is a decrease in precipitation that is enough to invoke fire in the closed canopy as well. From short coupled simulations, it was seen that fire occurs in year 1 along the edge of the closed canopy and LAI is reduced. LAI becomes significantly small in the northeast by year 4 and large reductions in LAI propagate westward into the closed canopy in subsequent years.

CLM4.5 was tested in short coupled simulations with the fire module both active and inactive. The results showed that canopy height was no longer decreasing with the fire module inactive, although the LAI was reduced by approximately 30 % from offline simulations. The LAI reduction is much more severe to both the canopy height and LAI with fire active. Reduced LAI in the coupled model presumably results from the low humidity and precipitation impacting the phenology algorithms previously discussed. Thus, it has been determined that the simulations used in this study should have the fire module turned off. The LAI impacts due to deforestation are still large and capable of producing a significant signal. In addition, the large changes exhibited in surface roughness also provide a boundary condition to the atmosphere capable of demonstrating the impacts of large-scale land use change.

3   Results

3.1   Temperature

As can be seen in Fig. 4, in the initially dense forest region there is an increase in surface temperature in all seasons; the majority of the region warms by 1–3 K, with the central region warming by more than 7 K. To the southeast, there is a region of temperature decrease, typically less than 3 K. This temperature decrease is largely over the region that was predominantly C4 grass. McGuffie et al. (1995) noted that changes in surface temperature over the deforested region are dipolar: an increase over the central and eastern Amazon and

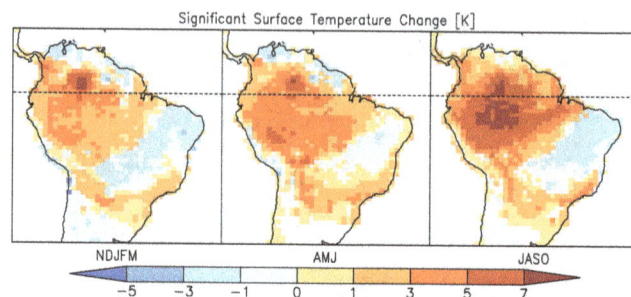

Figure 4. Change in surface temperature (K) for NDJFM, AMJ and JASO. Shading indicates significance at the 95 % confidence level.

a decrease to the southwest of the deforestation. The region of decrease is shifted eastward in these findings, but such a dipolar change has a precedent. Additionally, Lorenz and Pitman (2014) note a dipolar temperature change with a decrease in the east and an increase towards the west, which is noted as being directly related to the initial land–atmosphere coupling strength. Despite the region of cooling, the areal average for each season shows an increase: +0.8 K in NDJFM, +1.6 K in AMJ, and +2.1 K in JASO.

The contrast in temperature change between the densely forested and C4 grass areas becomes more apparent in the change in maximum monthly surface temperature. The forested region experiences an increase in all months, typically between 2 and 6 K. In the C4 grass area, the maximum monthly surface temperature decreases from August to January by 4–6 K, with the remaining months having a mixed change between −2 and 2 K. The same pattern tends to hold up for minimum monthly temperature, with the changes about half the magnitude. The overall range in extremes for the densely forested area increases by 2–4 K, while in the C4 grass area, the range of extremes is reduced by 2–4 K from August to January and increases by less than 2 K in the remaining months. It is worth noting that C4 grass in CLM can behave unrealistically at times by dying off and then regrowing a couple months later (Dirmeyer et al., 2013), which can affect surface temperature drastically.

The annual areal average increase in surface temperature of 1.4 K is consistent with previous modeling studies; Costa and Foley (2000) found a 1.4 K increase, Snyder et al. (2004) found a 1.5 K increase and Snyder (2010) found a 1.2 K increase. However, some studies found smaller or larger temperature increases: 0.6 K (Henderson-Sellers et al., 1993), 0.3 K (McGuffie et al., 1995), 0.3 K (Ramos da Silva et al., 2008), 2.5 K (Nobre et al., 1991) and 2.5 K (Shukla et al., 1990). The results in this study lie within the range of previous findings.

3.2   Precipitation

There is a significant decrease of at least 1 mm day^{-1} in precipitation over the originally densely forested region

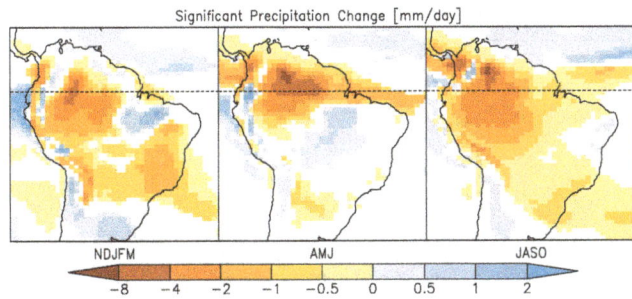

Figure 5. Change in precipitation (mm day^{-1}) for NDJFM, AMJ and JASO. Shading indicates significance at the 95 % confidence level.

throughout the year, with some areas experiencing decreases larger than 4 mm day^{-1}; see Fig. 5. The majority of this region sees decreases of more than 50 %. During NDJFM, when the majority of the Amazon region experiences at least 8 mm day^{-1} in precipitation in the control simulation, there is a largely statistically significant decrease in precipitation for the deforested region. An area of increase is present in a region that is mainly irrigated rice. During AMJ when precipitation is largely occurring within a few degrees of the Equator, there is a significant decrease across this region of the Equator, while a significant increase is present to the south. The driest season in the control simulation, JASO, has a significant decrease in precipitation over much of the deforested region. All seasons experience a decrease in the areal average: -0.27 mm day^{-1} in NDJFM, -0.37 mm day^{-1} in AMJ, and -0.44 mm day^{-1} in JASO.

Most of the precipitation changes can be explained by changes to convective precipitation, which decreases in all seasons (not shown), with the only exception being the region with irrigated rice. The reduction in convective precipitation suggests changes in flux partitioning at the surface may modify the properties and growth of the planetary boundary layer, as well as the land–atmosphere coupling in the region.

The decreases exhibited in this study are consistent with previous modeling studies; however, the magnitude of the decrease is smaller. This study found an annual areal average decrease of 0.35 mm day^{-1}, while previous studies found decreases of 0.7 mm day^{-1} (Costa and Foley, 2000), 0.4–0.7 mm day^{-1} (Hasler et al., 2009), 1.6 mm day^{-1} (Henderson-Sellers et al., 1993), 1.2 mm day^{-1} (McGuffie et al., 1995), 1.4 mm day^{-1} (Snyder et al., 2004; Snyder, 2010), and 0.8 mm day^{-1} (Werth and Avissar, 2002). The smaller decrease in precipitation may be due to previously mentioned model shortcomings with low humidity and less climatological precipitation in the region.

3.3 Radiation and fluxes

Net radiation is shown (Fig. 6) to be significantly reduced over the densely forest region in all seasons, typically by 30–

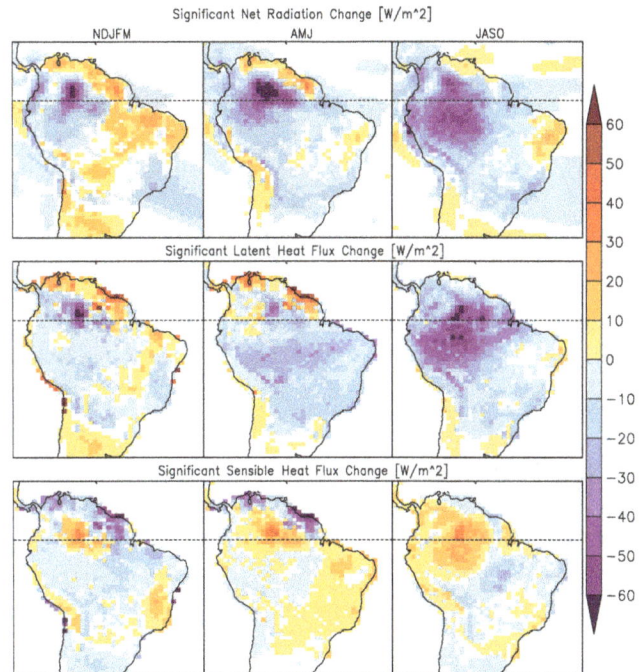

Figure 6. Changes in surface energy fluxes in NDJFM, AMJ and JASO; net radiation (W m^{-2}) (top row), latent heat flux (W m^{-2}) (middle row) and sensible heat flux (W m^{-2}) (bottom row). Shading indicates significance at the 95 % confidence level.

50 W m^{-2}. To the southeast over the C4 grass area, an increase is shown during NDJFM, changes between -10 and 10 W m^{-2} are present in AMJ, and decreases of 10 W m^{-2} exist in JASO. These changes are driven by changes to albedo (seen in Fig. 3) and impacts the partitioning of latent and sensible heat flux.

Latent heat flux is primarily reduced across the region in all seasons; the major exception is an increase during NDJFM in the former C4 grass area. Sensible heat flux increases in the formerly densely forested area in all seasons and is surrounded by a region of decrease in sensible heat flux. There is an increase in sensible heat flux in the southeast during both NDJFM and AMJ, while JASO has a mix of both increases and decreases, with most of the area not experiencing a significant change. The annual areal averages of latent heat flux and sensible heat flux both decrease, -8.1 and -1.7 W m^{-2}, respectively. This change in the fluxes has reduced the evaporative fraction in the region and indicates that the Amazon would shift to a drier climate.

Evaporative fraction (Fig. 7) is the ratio of latent heat flux to the sum of latent and sensible heat fluxes. After deforestation, nearly the entirety of the deforested region in AMJ and JASO have significant decreases in evaporative fraction, indicating a drier climate in the region. NDJFM experiences an increase in the evaporative fraction over a large portion of the area; this is due to it being the season of main crop growth over that area. The formerly densely forested region

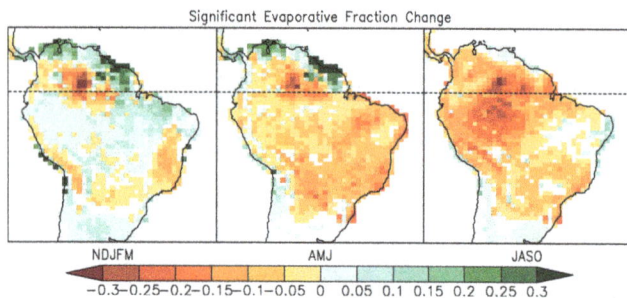

Figure 7. Changes to evaporative fraction in NDJFM, AMJ and JASO; shading indicates significance at the 95 % confidence level.

in NDJFM experiences a decrease in evaporative fraction; this is probably due to the deeper root profile of tree PFTs that would have access to a larger soil moisture reservoir.

3.4   Land–atmosphere coupling

A novel aspect of this study is an assessment of the impact on land–atmosphere coupling strength. A two-legged coupling metric (Guo et al., 2006; Dirmeyer, 2011) uses correlations between a land surface state variable (soil moisture) and surface flux (latent heat) as a means to assess terrestrial climate feedbacks, or a surface flux (sensible heat) and an atmospheric property (PBL height) for the atmospheric climate feedback. It is used here to describe the feedbacks present in the system and how they have changed after deforestation. Positive values in these two instances would imply that the land surface is controlling the feedback. We multiply these correlations by the standard deviation (SD) of the response variable (latent heat and PBL height, respectively) to determine the magnitude of the feedback (Guo et al., 2006). The significance of the control simulation coupling strength is based only on the correlation component, and the significance of the change in coupling strength is based only on the change in correlation (Dirmeyer et al., 2014).

In the terrestrial leg of the coupling (Fig. 8) for the control simulation, a large band of negative values during NDJFM corresponds to the heavy rains during that season when soil moisture is not a limiting factor for surface fluxes. As the rains shift throughout the year, this region shifts accordingly. During the drier seasons in the south, the sign switches to positive, an indication that soil moisture is controlling the latent heat flux (cf. Dirmeyer et al., 2013).

After deforestation, the previously densely forested areas become more strongly coupled throughout the year (Fig. 8). This is probably due to the shallower roots of crops, which have access to a smaller soil moisture reservoir. There are also large areas of decreased coupling, particularly over the southeast in JASO and south of the densely forested area in NDJFM. During AMJ, nearly the whole region sees an increase in coupling.

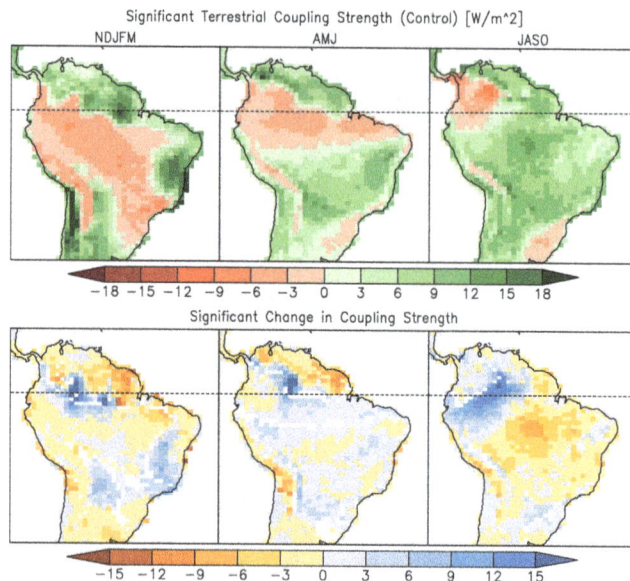

Figure 8. Terrestrial leg of coupling strength ($W\,m^{-2}$) between soil moisture and latent heat flux for the control simulation (top row) and change due to deforestation (bottom row) for NDJFM, AMJ and JASO. Shading indicates significance of the correlation component at the 95 % confidence level.

The changes in coupling can occur due to changes in the correlation, variability, or both. In NDJFM, the correlation increases in 54.8 % of the region and flux variability decreases in 56.4 % of the region. Neither component appears to be the leading agent of the changes; the changes in NDJFM (the rainy season) are largely atmospherically driven due to changes in precipitation. Areas with the largest reduction in precipitation have correlation increases; they also have increases in variability and are becoming more strongly coupled.

In AMJ and JASO, the changes in correlation are much larger: 69.5 and 76.6 % of the region have an increase in correlation, respectively. Increases in correlation alone do not necessarily imply increased coupling, as the combination of correlation and variance of the fluxes determines coupling strength (Dirmeyer, 2011; Dirmeyer et al., 2012). While the majority of the region in AMJ has stronger coupling, JASO has the majority of the region showing a decrease in coupling. JASO has a decrease in variability for 62.7 % of the region, with 46.2 % of the region having an increase in correlation and decrease in variability, largely taking place in the southeast, where there was lower initial tree cover. In contrast, the more densely forested regions largely experience an increase in correlation and an increase in variability.

For the atmospheric leg of the coupling, in the control run, the entire region is positively coupled based on the spatiotemporal correspondence between the two (Fig. 9). The areas of strongest coupling occur in locations that were initially

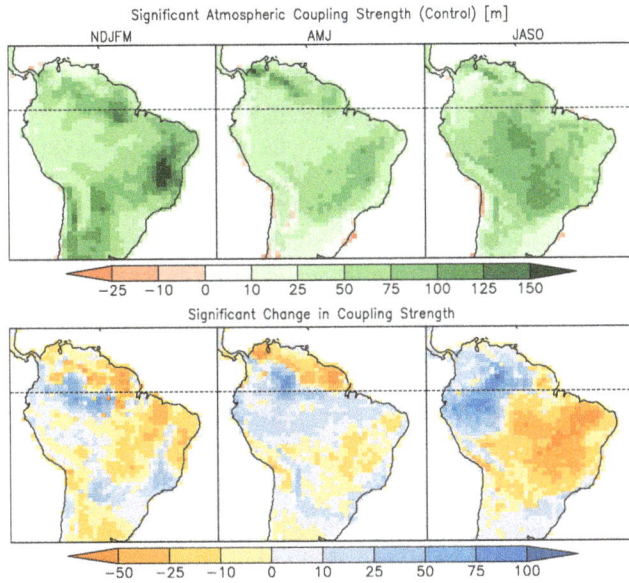

Figure 9. Atmospheric leg of coupling strength (m) between sensible heat flux and planetary boundary layer height for the control simulation (top row) and change due to deforestation (bottom row) for NDJFM, AMJ and JASO. Shading indicates significance of the correlation component at the 95 % confidence level.

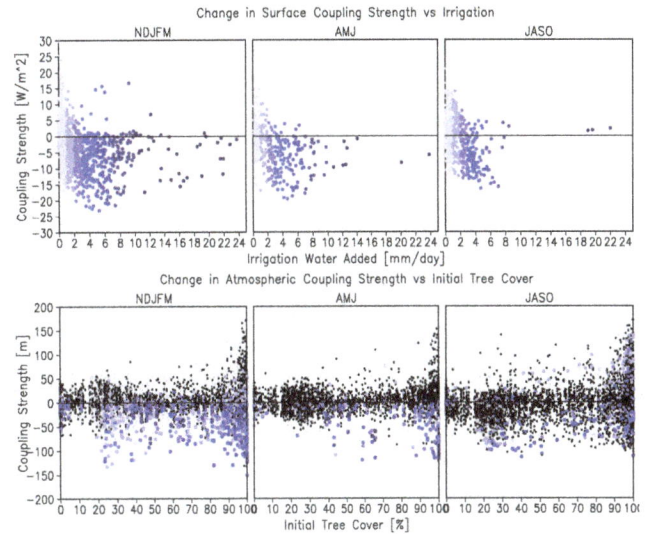

Figure 10. Top row: change in the terrestrial leg of coupling strength (Wm^{-2}) versus irrigation water added ($mm\,day^{-1}$) for irrigated grid boxes in NDJFM, AMJ and JASO. Bottom row: change in the atmospheric leg of coupling strength (m) versus initial tree cover percentage for NDJFM, AMJ and JASO. Shaded dots represent irrigated grid boxes, with the shading being equivalent to the shading for irrigation water added ($mm\,day^{-1}$) in the top row.

less tree-covered, as the dense canopy acts to dampen the coupling between surface sensible heat flux and PBL height.

In all seasons, the densely forested areas have an increase in coupling after deforestation (Fig. 9). The southeastern region largely experiences a decrease in coupling during all seasons. The largest contrast between the densely forested area and the southeast occurs in JASO, which is after most of the crops have been harvested and LAI is low.

For the atmospheric leg, the majority of the region either experiences an increase in both correlation and variability or a decrease in both. There are co-located correlation and variability increases over 31.0, 41.6, and 33.3 % of the region for NDJFM, AMJ and JASO, respectively. These regions are predominantly along the southeastern coast, where increased temperature and decreased precipitation occur, and in the previously forested areas. Regions experiencing decreases in both were 36.9, 29.7, and 40.2 % for those same seasons. These changes largely occurred in the southeastern area, where lower initial tree cover is located.

4 Discussion and conclusions

Replacement of natural vegetation with crops typical of tropical agriculture over the Amazon results in an albedo increase, lowering net radiation, which in turn modifies the surface fluxes. Latent heat flux is largely reduced across the domain, with the exception being the former C4 grass region in NDJFM; sensible heat flux has a more detailed spatial change with decreases in all seasons over the former densely forested

area and a seasonality to the changes in the surrounding regions. The areal averages for latent heat flux and sensible heat flux are reduced, but the evaporative fraction decreases, modifying the region toward a drier climate. Combining the surface temperature increase with the surface flux changes, a warmer, drier and deeper PBL results. There is a decrease in precipitation, largely due to decreased convection, which further alters flux partitioning due to reduced soil moisture. By modifying PBL properties and PBL growth, modified interaction between the PBL and the free atmosphere decreases vertical moisture transport and increases vertical heat transport. These changes in vertical transport provide a mechanism that can impact the circulation and may affect remote regions, with large-scale circulation changes enhancing the precipitation changes.

An added level of complexity that previous studies did not consider is irrigation. The irrigation impact is difficult to isolate, due to the grid boxes with irrigated rice also having other crops present. Irrigation adds water to the surface when water is a limiting factor for photosynthesis and can have an impact on land–atmosphere interactions. Irrigation does appear to have an impact on the coupling between land and atmosphere (Fig. 10). Irrigation is active in 8 months (OND-JFM in the Southern Hemisphere and JFMAM in the Northern Hemisphere) when rice is widely grown. In the months when irrigation is added, there is a negative correlation between irrigation water added and the change in the terrestrial leg of the coupling. The more irrigation water that is added, the less coupled the soil moisture becomes to the latent heat flux.

By affecting the surface coupling, irrigation can also impact the atmospheric leg of the coupling (Fig. 10). A negative relationship between irrigation water added and the change in SH-PBLH (atmospheric leg) coupling further shows that irrigation is modifying land–atmosphere interactions.

Although irrigation is shown to have an impact on the atmospheric leg of the coupling, the larger contributor appears to be the percentage of tree cover lost (Fig. 10). The coupling changes are largely the same for non-irrigated grid boxes with original tree percentage less than 80 %, typically between −50 and 50 m. JASO, the driest season, does have a larger spread, but comparable magnitudes of increases and decreases. When the initial tree cover is greater than 80 %, the coupling strength is predominantly increasing and has a greater magnitude of the change. This signal is also common in climate change scenarios driven by greenhouse gas increases (Dirmeyer et al., 2013), suggesting land use change could further amplify sensitivity to land surface anomalies in the tropics.

Irrigation largely decreases the coupling strength when the initial tree cover is less than 80 % and increases the magnitude of the change. When the initial tree cover is greater than 80 %, the grid boxes that experience a decrease in coupling are typically irrigated, with the more strongly irrigated grid boxes showing the largest decreases and less irrigated grid boxes showing an increase in coupling that is comparable to non-irrigated grid boxes. Just as with the terrestrial leg, more irrigation water added decreases the coupling strength of the atmospheric leg of the coupling.

Even using a realistic heterogeneous crop distribution in the Amazon region, there is still general agreement with previous modeling studies. The higher resolution and heterogeneity of the land cover show smaller-scale features and regions of opposite change, particularly in the southeastern Amazon, where the region has higher coverage of C4 grass. With crops being planted in different regions at different times of the year, a level of complexity not present in previous Amazon deforestation studies, and seasonality to land surface changes that were not previously modeled, are now seen.

A warming and drying of the region has impacted on how the land surface and atmosphere interact. By modifying the flux partitioning between latent and sensible heat fluxes, the region shifts to a drier climate with a warmer, drier and deeper PBL. By altering how the PBL grows, interaction with the free atmosphere is altered; this can lead to a warmer and drier atmospheric column above the region and may cause impacts to remote regions by modifying the general circulation and transports of moisture and heat. There is evidence that mesoscale responses of the atmosphere to land surface perturbations at low latitudes may not be well represented in climate models (e.g., Taylor et al., 2013); it would

be worthwhile to repeat tropical deforestation studies with cloud-resolving models in the future.

Remote impacts, such as modification to the African easterly waves and increased precipitation over the southwestern United States, have been found in these experiments, and will be discussed in a future paper (Badger and Dirmeyer, 2015). By employing a coupled ocean model, changes to sea surface temperature and the El Niño–Southern Oscillation have also been found and will be discussed in a later paper.

Acknowledgements. This research was supported by joint funding from the National Science Foundation (ATM-0830068), the National Oceanic and Atmospheric Administration (NA09OAR4310058), and the National Aeronautics and Space Administration (NNX09AN50G) of the Center for Ocean Land Atmosphere Studies (COLA). The lead author would like to acknowledge Sam Levis (NCAR) for help in getting CLM code modifications and the crop model to work in the coupled CESM, and to NCAR for supplying computing resources on the Yellowstone supercomputer.

Edited by: P. Gentine

References

Badger, A. M. and Dirmeyer, P. A.: Remote Tropical and Subtropical Responses to Amazon Deforestation, Clim. Dynam., 1–10, doi:10.1007/s00382-015-2752-5, online first, 2015.

Bonan, G. B.: Forests and Climate Change: Forcings, Feedbacks, and the Climate Benefits of Forests, Science, 320, 1444–1449, doi:10.1126/science.1155121, 2008a.

Bonan, G. B.: Ecological Climatology: Concepts and Applications, 2nd Edn., Cambridge University Press, New York, USA, 2008b.

Costa, M. H. and Foley, J. A.: Combined Effects of Deforestation and Doubled Atmospheric CO_2 Concentrations on the Climate of Amazonia, J. Climate, 13, 18–34, doi:10.1175/1520-0442(2000)013<0018:CEODAD>2.0.CO;2, 2000.

Costa, M. H., Yanagi, S. N. M., Souza, P. J. O. P., Ribeiro, A., and Rocha, E. J. P.: Climate change in Amazonia caused by soybean cropland expansion, as compared to caused by pastureland expansion, Geophys. Res. Lett., 34, L07706, doi:10.1029/2007GL029271, 2007.

Dickinson, R. E. and Henderson-Sellers, A.: Modelling tropical deforestation: A study of GCM land-surface parametrizations, Q. J. Roy. Meteor. Soc., 114, 439–462, doi:10.1002/qj.49711448009, 1988.

Dirmeyer, P. A.: The terrestrial segment of soil moisture–climate coupling, Geophys. Res. Lett., 38, L16702, doi:10.1029/2011GL048268, 2011.

Dirmeyer, P. A., Cash, B. A., Kinter, J. L., Stan, C., Jung, T., Marx, L., Towers, P., Wedi, N., Adams, J. M., Altshuler, E. L., Huang, B., Jin, E. K., and Manganello, J.: Evidence for Enhanced Land–Atmosphere Feedback in a Warming Climate, J. Hydrometeorol., 13, 981–995, doi:10.1175/JHM-D-11-0104.1, 2012.

Dirmeyer, P. A., Jin, Y., Singh, B., and Yan, X.: Trends in Land–Atmosphere Interactions from CMIP5 Simulations, J. Hydrometeorol., 14, 829–849, doi:10.1175/JHM-D-12-0107.1, 2013.

Dirmeyer, P. A., Wang, Z., Mbuh, M. J., and Norton, H. E.: Intensified land surface control on boundary layer growth in a changing climate, Geophys. Res. Lett., 41, 1290–1294, doi:10.1002/2013GL058826, 2014.

FAO: available at: http://ecocrop.fao.org, last access: May 2015, 2007.

Foley, J. A., DeFries, R., Asner, G. P., Barford, C., Bonan, G., Carpenter, S. R., Chapin, F. S., Coe, M. T., Daily, G. C., Gibbs, H. K., Helkowski, J. H., Holloway, T., Howard, E. A., Kucharik, C. J., Monfreda, C., Patz, J. A., Prentice, I. C., Ramankutty, N., and Snyder, P. K.: Global Consequences of Land Use, Science, 309, 570–574, doi:10.1126/science.1111772, 2005.

Guo, Z., Dirmeyer, P. A., Koster, R. D., Sud, Y. C., Bonan, G., Oleson, K. W., Chan, E., Verseghy, D., Cox, P., Gordon, C. T., McGregor, J. L., Kanae, S., Kowalczyk, E., Lawrence, D., Liu, P., Mocko, D., Lu, C.-H., Mitchell, K., Malyshev, S., McAvaney, B., Oki, T., Yamada, T., Pitman, A., Taylor, C. M., Vasic, R., and Xue, Y.: GLACE: The Global Land–Atmosphere Coupling Experiment. Part II: Analysis, J. Hydrometeor., 7, 611–625, doi:10.1175/JHM511.1, 2006.

Hasler, N., Werth, D., and Avissar, R.: Effects of Tropical Deforestation on Global Hydroclimate: A Multimodel Ensemble Analysis, J. Climate, 22, 1124–1141, doi:10.1175/2008JCLI2157.1, 2009.

Henderson-Sellers, A., Dickinson, R. E., Durbidge, T. B., Kennedy, P. J., McGuffie, K., and Pitman, A. J.: Tropical deforestation: Modeling local- to regional-scale climate change, J. Geophys. Res., 98, 7289–7315, doi:10.1029/92JD02830, 1993.

IBGE: Censo Agropecuário 2006, Insituto Brasileiro de Geografia e Estatística, Rio de Janeiro, Brazil, Tech. rep., 311.213.1: 63, 2006.

INPE: PRODES – Amazon deforestation database, INPE, São Jose dos Campos, Tech. rep., 2014.

Kucharik, C. J., Foley, J. A., Delire, C., Fisher, V. A., Coe, M. T., Lenters, J. D., Young-Molling, C., Ramankutty, N., Norman, J. M., and Gower, S. T.: Testing the performance of a dynamic global ecosystem model: water balance, carbon balance, and vegetation structure, Global Biogeochem. Cy., 14, 795–825, 2000.

Lejeune, Q., Davin, E. L., Guillod, B. P., and Seneviratne, S. I.: Influence of Amazonian deforestation on the future evolution of regional surface fluxes, circulation, surface temperature and precipitation, Clim. Dynam., 44, 2769–2786, 2014.

Levis, S., Bonan, G. B., Kluzek, E., Thornton, P. E., Jones, A., Sacks, W. J., and Kucharik, C. J.: Interactive Crop Management in the Community Earth System Model (CESM1): Seasonal Influences on Land–Atmosphere Fluxes, J. Climate, 25, 4839–4859, doi:10.1175/JCLI-D-11-00446.1, 2012.

Lorenz, R. and Pitman, A. J.: Effect of land–atmosphere coupling strength on impacts from Amazonian deforestation, Geophys. Res. Lett., 41, 5987–5995, doi:10.1002/2014GL061017, 2014.

McGuffie, K., Henderson-Sellers, A., Zhang, H., Durbidge, T., and Pitman, A.: Global climate sensitivity to tropical deforestation, Global Planet. Change, 10, 97–128, doi:10.1016/0921-8181(94)00022-6, 1995.

Nepstad, D. C., Stickler, C. M., Filho, B. S., and Merry, F.: Interactions among Amazon land use, forests and climate: prospects for a near-term forest tipping point, Philos. T. R. Soc. B, 363, 1737–1746, doi:10.1098/rstb.2007.0036, 2008.

Nobre, C. A., Sellers, P. J., and Shukla, J.: Amazonian Deforestation and Regional Climate Change, J. Climate, 4, 957–988, doi:10.1175/1520-0442(1991)004<0957:ADARCC>2.0.CO;2, 1991.

Nobre, P., Malagutti, M., Urbano, D. F., de Almeida, R. A. F., and Giarolla, E.: Amazon Deforestation and Climate Change in a Coupled Model Simulation, J. Climate, 22, 5686–5697, doi:10.1175/2009JCLI2757.1, 2009.

Oleson, K. W., Lawrence, D. M., Bonan, G. B., Drewniak, B., Huang, M., Koven, C. D., Levis, S., Li, F., Riley, W. J., Subin, Z. M., Swenson, S. C., Thornton, P. E., Bozbiyik, A., Fisher, R., Kluzek, E., Lamarque, J.-F., Lawrence, P. J., Leung, L. R., Lipscomb, W., Muszala, S., Ricciuto, D. M., Sacks, W., Sun, Y., Tang, J., and Yang, Z.-L.: Technical Description of version 4.5 of the Community Land Model (CLM), National Center for Atmospheric Research, Boulder, CO, USA, NCAR Technical Note, TN-503+STR, 2013.

Portmann, F. T., Siebert, S., and Döll, P.: MIRCA2000 – Global monthly irrigated and rainfed crop areas around the year 2000: A new high-resolution data set for agricultural and hydrological modeling, Global Biogeochem. Cy., 24, GB1011, doi:10.1029/2008GB003435, 2010.

Qian, T., Dai, A., Trenberth, K. E., and Oleson, K. W.: Simulation of Global Land Surface Conditions from 1948 to 2004. Part I: Forcing Data and Evaluations, J. Hydrometeor., 7, 953–975, doi:10.1175/JHM540.1, 2006.

Ramos da Silva, R., Werth, D., and Avissar, R.: Regional Impacts of Future Land-Cover Changes on the Amazon Basin Wet-Season Climate, J. Climate, 21, 1153–1170, doi:10.1175/2007JCLI1304.1, 2008.

Sacks, W. J., Deryng, D., Foley, J. A., and Ramankutty, N.: Crop planting dates: an analysis of global patterns, Global Ecol. Biogeogr., 19, 607–620, doi:10.1111/j.1466-8238.2010.00551.x, 2010.

Shukla, J., Nobre, C., and Sellers, P.: Amazon Deforestation and Climate Change, Science, 247, 1322–1325, doi:10.1126/science.247.4948.1322, 1990.

Snyder, P. K., Delire, C., and Foley, J.: Evaluating the influence of different vegetation biomes on the global climate, Clim. Dynam., 23, 279–302, doi:10.1007/s00382-004-0430-0, 2004.

Snyder, P. K.: The Influence of Tropical Deforestation on the Northern Hemisphere Climate by Atmospheric Teleconnections, Earth Interact., 14, 1–34, doi:10.1175/2010EI280.1, 2010.

Sun, S. and Wang, G.: Diagnosing the equilibrium state of a coupled global biosphere-atmosphere model, J. Geophys. Res., 116, D09108, doi:10.1029/2010JD015224, 2011.

Taylor, C. M., Birch, C. E., Parker, D. J., Dixon, N., Guichard, F., Nikulin, G., and Lister, G. M. S.: Modeling soil moisture-precipitation feedback in the Sahel: Importance of spatial scale versus convective parameterization, Geophys. Res. Lett., 40, 6213–6218, doi:10.1002/2013GL058511, 2013.

Vertenstein, M., Bertini, A., Craig, T., Edwards, J., Levy, M., Mai, A., and Schollenberger, J.: CESM User's Guide, CESM1.2 Release Series User's Guide, NCAR Technical Note, National Center for Atmospheric Research, Boulder, CO, USA, 2013.

Werth, D. and Avissar, R.: The local and global effects of Amazon deforestation, J. Geophys. Res., 107, 8087, doi:10.1029/2001JD000717, 2002.

12

Developing a drought-monitoring index for the contiguous US using SMAP

Sara Sadri, Eric F. Wood, and Ming Pan

Department of Civil and Environmental Engineering, Princeton University, 59 Olden St, Princeton, NJ 08540, USA

Correspondence: Sara Sadri (sadri@princeton.edu)

Abstract. Since April 2015, NASA's Soil Moisture Active Passive (SMAP) mission has monitored near-surface soil moisture, mapping the globe (between 85.044° N/S) using an L-band (1.4 GHz) microwave radiometer in 2–3 days depending on location. Of particular interest to SMAP-based agricultural applications is a monitoring product that assesses the SMAP near-surface soil moisture in terms of probability percentiles for dry and wet conditions. However, the short SMAP record length poses a statistical challenge for meaningful assessment of its indices. This study presents initial insights about using SMAP for monitoring drought and pluvial regions with a first application over the contiguous United States (CONUS). SMAP soil moisture data from April 2015 to December 2017 at both near-surface (5 cm) SPL3SMP, or Level 3, at $\sim$ 36 km resolution, and root-zone SPL4SMAU, or Level 4, at $\sim$ 9 km resolution, were fitted to beta distributions and were used to construct probability distributions for warm (May–October) and cold (November–April) seasons. To assess the data adequacy and have confidence in using short-term SMAP for a drought index estimate, we analyzed individual grids by defining two filters and a combination of them, which could separate the 5815 grids covering CONUS into passed and failed grids. The two filters were (1) the Kolmogorov–Smirnov (KS) test for beta-fitted long-term and the short-term variable infiltration capacity (VIC) land surface model (LSM) with 95 % confidence and (2) good correlation (≥ 0.4) between beta-fitted VIC and beta-fitted SPL3SMP. To evaluate which filter is the best, we defined a mean distance (MD) metric, assuming a VIC index at 36 km resolution as the ground truth. For both warm and cold seasons, the union of the filters – which also gives the best coverage of the grids throughout CONUS – was chosen to be the most reliable filter. We visually compared our SMAP-based drought index maps with metrics such as the U.S. Drought Monitor (from D0–D4), 1-month Standard Precipitation Index (SPI) and near-surface VIC from Princeton University. The root-zone drought index maps were shown to be similar to those produced by the root-zone VIC, 3-month SPI, and the Gravity Recovery and Climate Experiment (GRACE). This study is a step forward towards building a national and international soil moisture monitoring system without which quantitative measures of drought and pluvial conditions will remain difficult to judge.

1 Introduction

Drought is an extreme condition when water in one or a combination of water stores (e.g., river, lake, reservoir, snowpack, soil water or groundwater) or water fluxes (precipitation, evapotranspiration or runoff) drops below a defined condition for a prolonged period of time (Wilhite and Glantz, 1985; Wilhite, 2000; AMS, 2012). Such a water deficit evolves over weeks to months and can last for months and years. Drought's propagation is silent and often without warning until it impacts human lives and environmental activities (Tallaksen and Van Lanen, 2004). Drought conditions are related to water demand, so local water use plays a central role in defining conditions of scarcity and the resulting impacts. Wilhite and Glantz (1985) classified drought into meteorological, agricultural or hydrological, depending on whether the deficit is measured using precipitation, soil moisture or river discharge, respectively.

The reduced supply of precipitation (and subsequently soil moisture) for crops leads to an agricultural drought that impacts crop yield, inflicting enormous economic impacts on developed countries and the suffering of millions of people in less-developed regions of the world. In the US, since 1996, there has been at least one drought event per year except for the years 1997, 2001, 2004 and 2010, and each year drought cost between USD 1 billion and 14 billion in damages (in 2015 – adjusted dollars) (NOAA, 2018b). In California alone, the 2015 drought was estimated to cause USD 2–5 billion in damages to the agricultural sector (Howitt et al., 2015).

Although the impacts of drought are intimately linked to the vulnerability of a population to adverse conditions (UN/ISDR, 2007) and how society responds within the constraints of changing economies, the timely determination of the current level of agricultural drought aids the decision-making process in order to reduce its impacts. Scientifically based drought-monitoring tools and warning systems assist in the mitigation of the losses caused by droughts and the planing and management of water shortages that will accompany future droughts (Martinez-Fernandez et al., 2016). Such drought-monitoring tools are based on long-term observations of the hydrological variables such as precipitation, streamflow, soil moisture and groundwater.

Pluvial conditions are related to an abundance of precipitation and subsequently wet soil conditions that can adversely affect agriculture by waterlogging the fields or exacerbating flooding from additional rainfall. Thus, for monitoring extremes (either agricultural drought or pluvial conditions), realistic estimation of soil moisture at regional to continental scales is required. Soil moisture is the central source of information, since it reflects recent precipitation and antecedent soil conditions (Sheffield and Wood, 2011). In a sense, soil moisture captures the aggregate balance of all hydrological processes and represents available water, being a buffer between incoming precipitation and throughfall and evapotranspiration and drainage processes (Entekhabi et al., 1996). Unfortunately, soil moisture (and evapotranspiration) are among the least-observed components of the hydrological cycle, especially over large spatial and temporal scales (Reichle, 2017; Sheffield and Wood, 2011).

Many statistical measures or indices for extreme conditions have been developed in the US, particularly for drought conditions. This is due to the slow evolution of drought and its economic and social impact. Currently, no single drought index has been able to adequately capture the severity and intensity of drought and its impact on different groups of users (Heim, 2002). Heim (2002) gives an overview of the major 20th US drought indices. The most common ones are the standardized Precipitation Index (SPI), Palmer Drought Severity Index (PDSI), Standardized Runoff Index (SRI) and the U.S. Drought Monitor (DM or USDM).

The SPI is recognized by the World Meteorological Organization (WMO) as the standard index for quantifying and reporting meteorological drought. It is used to characterize drought on a range of timescales from 1 to 36 months. The raw precipitation is fit to an appropriate distribution function and is then transformed into a standardized normal distribution. The SPI index is expressed as the number of standard deviations by which the anomaly deviates from the long-term mean. On short timescales, the SPI is closely related to soil moisture, while at long timescales, it is related to groundwater. The advantages of the SPI include the following: it only relies on precipitation, it can characterize both drought and pluvial conditions, its computation over different timescales can be related to various water resource stores (such as soil moisture and groundwater), and it is more comparable across regions with different climates than the Palmer Severity Drought Index (PDSI). The key limitation of the SPI is the following: it is sensitive to the quantity of the data used. Usually, 30 years of monthly precipitation data are recommended for fitting the data. Additionally, the SPI is a meteorological tool that measures water supply but does not account for evapotranspiration. This limits its ability to capture the effect of increased temperatures (associated with climate change) on moisture demand and availability. Finally, the SPI does not consider the intensity of precipitation and how it impacts on runoff and streamflow. Overall, the SPI can provide information about anomalies in precipitation, so it needs to be used in combination with other information in order to be useful for agricultural drought assessment (NCAR, 2018).

The PDSI uses precipitation and an estimate of evaporation in conjunction with a water balance model to estimate relative soil dryness and potential evapotranspiration. The original formulation used only the temperature to estimate a potential evapotranspiration, but it is now recognized that an energy-based approach, such as the Penman–Monteith approach, is preferred (Sheffield et al., 2012; Mo and Chelliah, 2006). Since PDSI uses potential evapotranspiration and precedent (prior month) conditions, it takes into account the basic effect of global warming and is effective in determining long-term drought, especially over low and midlatitudes. Key limitations of the PDSI include that the PDSI is not as comparable across regions as the SPI and lacks the ability to handle winter-time conditions that include snowmelt and frozen precipitation, which makes its long-term monitoring problematic. Unlike SPI indices, the PDSI lacks multi-timescale features, making it difficult to correlate with specific water resources like runoff, snowpack and reservoir storage.

The SRI is based on the SPI and a model runoff. The strength of the SRI, as a runoff-based index, is that it can be used to forecast future runoff, and its predictability depends not only on climate outlooks, for which seasonal skill is generally low, but also on hydrologic initial conditions (e.g., spring snow state in the western US). The disadvantage of the SRI is similar to the disadvantage of using any modeled runoff; since modeled runoff cannot be verified everywhere, the runoff-based indices of the SRI reflect the customary uncertainties associated with model outputs (Shukla and Wood, 2008).

The USDM integrates several drought indices and professional input from all levels into a weekly operational drought-monitoring map product (Svoboda, 2000). The limitation of the USDM lies in its attempt to show drought at several temporal scales (from short-term drought to long-term drought) on one map product. Hence, the application of the DM is not for replacing any local or state information or subsequently declared drought emergencies or warnings but is rather for providing a general assessment of the current state of drought around the United States, its Pacific possessions, and Puerto Rico (Svoboda, 2000). Since the USDM relies on professional inputs from the field, it is difficult to have historical consistency (since the professionals change) or to provide forecasts.

Long-term and large-scale observations of soil moisture are scarce in the United States and elsewhere, so datasets produced by the North American Land Data Assimilation System (NLDAS) are valuable alternatives. Currently National Centers for Environmental Prediction (NCEP) offer an NLDAS drought monitor (NOAA, 2018a) based on four land surface models (LSMs): variable infiltration capacity (VIC), Noah, Mosaic and Sacramento. Sheffield et al. (2004) used simulations from the NLDAS VIC model forced with observed precipitation and near-surface meteorology to develop a drought index based on soil moisture. The approach Sheffield et al. (2004) took was to fit the VIC-simulated soil moisture to probability distributions, usually beta distributions, where the percentiles are translated to the index values that range from 0 to 1. Recent drought applications such as the VIC-based Princeton University drought and flood monitoring systems for Africa and Latin America (Sheffield et al., 2014) use the simulated soil moisture, which is mostly based on satellite precipitation (Princeton University Hydrology, 2013).

A major limitation of the indices discussed earlier, as well as of the LSM-based approaches, is a reliance on quality meteorological data. While precipitation is one of the best-observed variables, gauge observations are limited in many regions, especially in much of the developing world. Even when they are available, they are often not in near real time, preventing the computation of indices. This reveals one of the weaknesses of the above indices; their estimates rest on the availability and accuracy of the forcings, specifically precipitation (Reichle, 2017). In places such as the US, where the quality of the precipitation data is quite high, VIC quality is also relatively high (Pan et al., 2016). However, in regions with sparse networks or low accessibility, such as Africa, the VIC quality can be relatively low (Reichle, 2017). Additionally, intercomparison of the four NLDAS models showed that soil moisture differs considerably among models (Robock et al., 2000).

Heim (2002) summarizes four characteristics of a useful operational drought-monitoring system. These include the following: (1) the indices need to be available on a near-real-time basis, (2) the indices need to be monitored on a national scale, which will require the establishment of national networks for some variables, (3) complete and reliable historical

data are needed over a common reference period to allow the conversion of the observations into a meaningful form (such as a percentile ranking), and (4) the data need to be adjusted to remove non-climatic influences (such as those arising from water management practices; Friedman, 1957; Heim, 2002).

An alternative approach to using model-derived soil moisture for drought detection and prediction is satellite-derived soil moisture. There are currently four major satellite-based systems that provide soil moisture products at various spatial and temporal resolutions: MetOp with the advanced scatterometer (ASCAT; Brocca et al., 2010; Wagner et al., 2013), the Advanced Microwave Scanning Radiometer AMSR2 of the Japan Aerospace Exploration Agency (JAXA; Parinussa et al., 2015; Wu et al., 2015) with the C- and X-band passive radiometers on the GCOM-W1 satellite that is a follow-on to the AMSR-E sensor, which failed on 4 October 2011 and was part of NASA's Earth Observing System, ESA's Soil Moisture Ocean Salinity (SMOS) L-band radiometer (Pan et al., 2010; Kerr et al., 2012, 2016), and NASA's Soil Moisture Active Passive (SMAP) L-band radiometer (Entekhabi et al., 2010). The radar on SMAP failed after 3 months, but soil moisture estimates based on the radiometer continue to be produced.

Of particular interest, especially for applications in parts of the globe with sparse in situ data, is to have an SMAP-based monitoring product that expresses soil moisture in terms of probability percentiles for dry (drought) or wet (pluvial) conditions (Entekhabi et al., 2010). This study presents insights and the potential of using SMAP for monitoring drought and pluvial regions with a first application over the contiguous United States (CONUS). We fit the soil moisture data from SMAP at both the level 3 5 cm passive radiometer retrievals (SPL3SMP) and the level 4 root-zone product that assimilates the surface SPL3SMP into the Catchment LSM (SPL4SMAU) to beta distributions, construct probability distributions for warm and cold seasons, and measure the reliability of our estimates. Producing soil moisture drought indices at two different soil depths allow for the monitoring of agricultural drought in different stages of development (NDMC, 2018a). This is important, firstly because grid analysis showed that values of a full column soil moisture index can be less, similar, or more than near-surface soil moisture index values. Secondly, depending on the plant development stage, the surface soil moisture or root-zone soil moisture drought index can be more useful in agricultural management. For example, surface soil moisture is important in the germination stage but is less important for managing irrigation or in estimating yields. Deficient topsoil moisture at planting may hinder germination, leading to low plant populations per hectare and a reduction of final yield (NDMC, 2018a). At the same time root-zone moisture at this early stage may not affect final yield, but as the growing season progresses, it becomes more important for plant water needs.

The rest of this paper as follows: the SMAP data are discussed in Sect. 2.1, including a determination of whether

1006 days are sufficient for estimating a drought index. Section 2.2 develops the indices by fitting beta distributions, with upper and lower bounds, to the time series and using the percentiles as the index. Section 2.3 develops a numerical analysis of the adequacy of the SMAP data. In Sect. 3, results of adequacy tests are discussed, and comparisons are made to the currently available drought indices. To help relate the percentiles to the U.S. Drought Monitor, which uses levels D0–D4 to indicate severity, the percentiles are mapped. We also extended our indices to pluvial conditions similar to the maps from the Gravity Recovery and Climate Experiment (GRACE) and Princeton University. Conclusions are brought forth in Sect. 4.

2 Data and methods

2.1 SMAP data

Since April 2015, NASA's SMAP mission has been monitoring near-surface soil moisture, mapping the globe (between 85.044° N and S) using an L-band (1.4 GHz) microwave radiometer in 2–3 days depending on location. The SMAP mission provides a set of operational global data products that include the following:

- Level 3 (SPL3SMP) is a composite based on daily passive radiometer estimates of global land surface soil moisture (nominally 5 cm) that are resampled to a global, cylindrical 36 km Equal-Area Scalable Earth Grid, Version 2.0 (EASE-Grid 2.0; O'Neill et al., 2016). For this study, Version 4 of SPL3SMP is used, which is the release version from the very beginning of the launch of SMAP. The release number changes over time. The R16 version is the latest version released in June 2018. However, in all release versions of SMAP, including Version 4, regions with permanent snow and ice, frozen ground, excessive static or transient open water in the cell, excessive radio-frequency interference (RFI) in the sensor data, and heavy vegetation (vegetation water content $> 4.5\,\mathrm{kg\,m^{-2}}$) are masked out using a passive freeze–thaw retrieval based on the normalized polarization ratio (NPR). Given the 1000 km swath and 98.5 min orbit, the SPL3SMP retrievals are spatially and temporally discontinuous, with 2–3 day gaps depending on location.

- Level 4 (SPL4SMAU) provides estimates of global surface and root-zone soil moisture by assimilating the SMAP L-band brightness temperature data (for which SPL3SMP is the gridded version) from descending and ascending half-orbit satellite passes, every 3 h from approximately 06:00 to 18:00 LST (Local Solar Time), into NASA's Catchment LSM (Reichle, 2017; Reichle et al., 2015). The SPL4SMAU data product is gridded using an Earth-fixed, global, cylindrical 9 km EASE-

Grid 2.0 projection. The LSM component of the assimilation system is driven by a forcing data stream from the global atmospheric analysis system at the NASA Global Modeling and Assimilation Office (Rienecker et al., 2008). Additional corrections are applied using gauge- and satellite-based estimates of precipitation that are downscaled to the temporal and 9 km scale of the model forcing using the disaggregation methods described in Liu et al. (2011) and Reichle et al. (2011). The SPL4SMAU product provides global soil estimates for the surface (0–5 cm) and root zone (0–100 cm) and is an effort to provide continuous, daily information without discontinuous data restrictions due to gaps in the SPL3SMP soil moisture retrievals. Nonetheless, the only product that does not use ancillary meteorological data is the SPL3SMP soil moisture retrievals.

In this study, SPL3SMP products from the 06:00 LST retrievals and SPL4SMAU products from 06:00 LST retrievals are used in the analysis of the soil moisture drought index. Our SMAP data records are from 1 April 2015 to 31 December 2017, which is equivalent to 1006 days.

The approach selected here is somewhat similar to that of Sheffield et al. (2004), where the soil moisture time series are fit to a beta distribution (with upper and lower bounds), and the distribution percentiles are the index values. There are, however, differences in our approach to that of Sheffield et al. (2004). Firstly, the basis of the data used in Sheffield et al. (2004) was simulated soil moisture from VIC, while ours is remotely sensed data. Secondly, to calculate the bounds of beta distribution $[a, b]$, Sheffield et al. (2004) used the first (last) 10 % of the sorted soil moisture values linearly related to the empirical cumulative distribution function. In our study, this approach did not yield useful results with the estimated limits for a (b) for SMAP and often did not cover the full range of observed values, preventing interpretation of the historical data. Our methodology for obtaining beta distribution parameters a and b are discussed in this section.

As mentioned in the introduction by Heim (2002), one of the conditions for an index approach is complete and reliable historical data needed over a common reference period to allow the conversion of the observations into a meaningful form. The short SMAP record length of 1006 days, from 1 April 2015 to 31 December 2017, provides a statistical challenge in estimating the drought and pluvial indices, and thus the reliability assessments related to these extreme conditions are necessary. Therefore, to assess the data adequacy, we used a 1979–2017 VIC LSM simulation over CONUS. The VIC runs were carried out at a 4 km spatial resolution, and for the SPL3SMP comparisons, averaged up to 36 km. Here we refer to it as VIC near surface (VIC-ns). The SPL4SMAU is at 9 km spatial resolution, so VIC data were aggregated from 4 km computing grids and averaged over three soil layers with varying total soil thickness. We refer to it as VIC root zone (or VIC-rz). A statistical comparison is made between fitting a beta distribution to the VIC soil moisture values using only days when SPL3SMP soil moisture retrievals are available and fitting it to the complete 1979–2017 VIC data record. The Kolmogorov–Smirnov (KS) statistical test was used to evaluate the consistency of the beta fitted data. We made the assumption that if grids passed the consistency test using VIC data – i.e., the distributions from the SMAP-period record and the complete record were deemed the same statistically – then the SMAP time series over that grid was sufficient for providing an index. More discussion of these results is given in Sect. 3.

Furthermore, we looked at the frequency distribution of soil moisture data at each grid. The data seemed to be dominated by low soil moisture in the summertime and high soil moisture in the wintertime. Therefore, to capture this interseasonal behavior in soil moisture, we divided the record into a warm season (April–September) and a cold season (October–March). Dividing the year into warm and cold seasons enabled us to track the soil moisture dynamics, and thus the probability distribution and index, seasonally. Ideally, we would have divided it into monthly data but there are insufficient observations.

For our study period, each grid has between 144 and 329 SPL3SMP soil moisture retrievals during the warm season and from 16 to 272 retrievals during the cold season. Figure 1 shows that the number of overpasses per grid is related to the latitude, with higher latitudes having a higher number of overpasses, and to the season, with fewer values retrieved during winter, especially in the western US, due to snow cover and frozen ground. For the LSPL4SMAU root zone, there are 457 records for the cold season and 549 records for the warm season for each grid.

2.2 Fitting the beta distribution to the SMAP time series

The beta distribution is a family of continuous distributions with two shape parameters (p and q). It generalizes to a bounded distribution on the interval of $[a, b]$, where a and b usually take on the values of 0 and 1. The beta distribution is flexible enough to model a wide variety of shapes. In our study, we compared the beta distribution to several parametric distributions (including normal and Gumbel), but the beta distribution showed the best goodness of fit. Furthermore, given the bounded nature of the distribution, it is often used as the model of choice for modeling soil moisture time series (Sheffield et al., 2004). The general formula for the beta probability density function (pdf) is:

$$f(x) = \frac{(x-a)^{(p-1)}(b-x)^{(q-1)}}{B(p,q)(b-a)^{p+q-1}},$$
$$a \leq x \leq b, \quad p, q > 0, \tag{1}$$

where p and q are shape parameters, and a and b are lower and upper bounds, respectively, of the distribution. In the

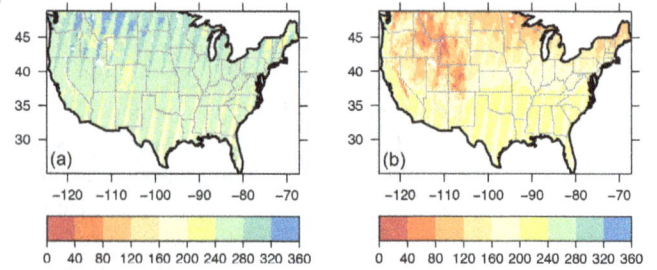

Figure 1. Number of retrievals for each season. **(a)** Warm season (1 April–30 September); **(b)** cold season (1 October–31 March).

case where $a = 0$ and $b = 1$, this becomes a standard beta distribution (NIST, 2013). $B(p, q)$ is a beta constant computed with the formula

$$B(p, q) = \int_0^1 t^{p-1}(1-t)^{q-1} dt. \tag{2}$$

A main challenge is to fit the four parameters of the beta distribution, given a set of empirical observations. Sheffield et al. (2004) used the method of moments to fit the beta distribution to historical soil moisture simulations from the VIC LSM. They computed the first three moments and minimized the difference between the distribution estimates and sample estimates, since they were over-constrained. We also used the standard method of moments to calculate the parameters p and q. But for each grid location, we fit the beta distribution to six sets of data related to the SPL3SMP product: (1) short warm season VIC, (2) short warm season SMAP (1 April–30 September for 2015, 2016 and 2017; 18 months), (3) long warm season VIC (1 April–30 September, 1979–2017; 129 months), (4) short cold season VIC, (5) short cold season SMAP (1 October–31 March, 2015–2016 and 1 October–31 December 2017; 15 months) and (6) long cold season VIC (1 October–31 March for 1979 and 2016 and 1 October–31 December for 2017; 126 months), using the first and second moments $\mu = p/(p+q)$ and $CV = \mu/\sigma$, where p and q are parameters and its standard deviation is defined as

$$\sigma = \sqrt{\frac{p \times q}{(p+q)^2 \times (p+q+1)}}. \tag{3}$$

For the SPL4SMAU root-zone soil moisture product, the beta distribution was fit to the warm season and cold season using all 457 and 549 records, respectively.

Figure 2 shows the 20th percentile, average and 80th percentile soil moisture data in the warm season and cold season for the SPL3SMP 5 cm soil moisture product, and this is shown similarly in Fig. 3 for the SPL4SMAU root-zone product after data were fit to the beta distribution.

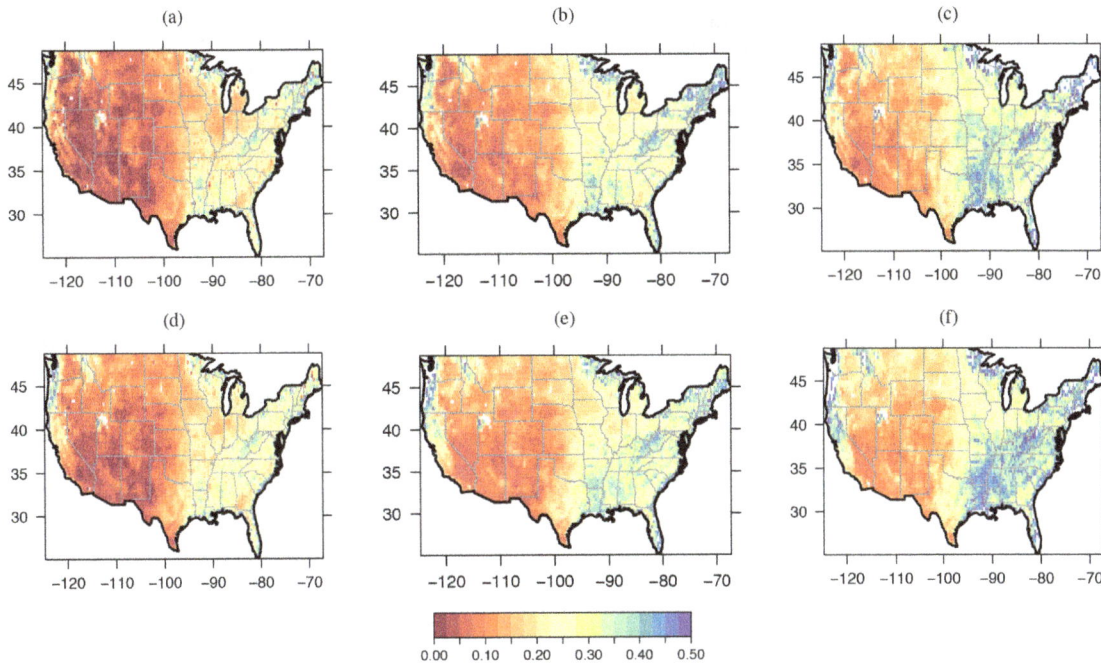

Figure 2. (a–c) SMAP soil moisture values for the warm season during summer for SPL3SMP top 5 cm soil moisture **(a)** at the 20th percentile, **(b)** at the average soil moisture, and **(c)** at the 80th percentile; **(d–f)** same as the top row, but for the cold season. Total period is from 1 April 2015 to 31 December 2017. The soil moisture unit is $m^3 m^{-3}$.

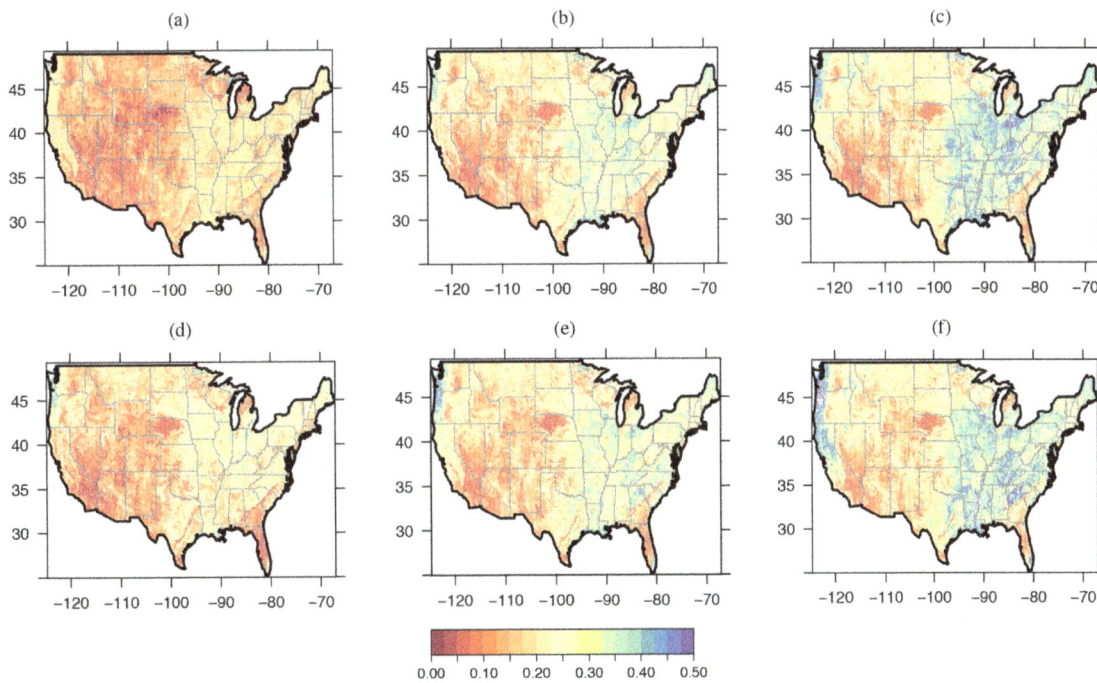

Figure 3. Same as shown in Fig. 2, but for SPL4SMAU (root-zone soil moisture).

2.3 Data adequacy filters

An insufficient SMAP record length may result in unreliable index values. To be meaningful in using short SPL3SMP data for making confident predictions, we will analyze which grids have the highest certainty in our SMAP drought index. That is, we perform adequacy analysis and define filters that separate grids with high reliability in drought monitoring and prediction from ones where we do not expect our predictions to be as accurate. We first define two filters which can separate the 5815 grids covering CONUS into grids that passed and failed quality control. The two filters are as follows:

1. the KS test for beta-fitted long-term and short-term VIC with 95 % confidence;

2. good correlation (≥ 0.4) between beta-fitted VIC and beta-fitted SPL3SMP.

Below we expand upon these two filters and then show how we used them to numerically find the best SPL3SMP filter. We also investigate if combinations of the filters are superior to the individual filters taken alone.

2.3.1   Kolmogorov–Smirnov (KS) filter

The KS test is a well-known nonparametric statistical test that compares whether two samples are coming from the same continuous distribution. We used the KS test for each grid, comparing the modeled beta distribution of the long-term VIC with the modeled beta distribution of the short-term VIC, in both warm and cold seasons. This shows if the long-term and short-term distributions are statistically indistinguishable. If this strong condition is satisfied for a grid, then it is reasonable to assume, for that grid, that the short SMAP time series would be consistent with a hypothetical long SMAP time series. The null hypothesis – that the underlying beta distribution of short-term soil moisture data is the same as the underlying beta distribution of long-term soil moisture data for VIC – is rejected for values of the KS statistic D that exceed a critical value at the 95 % significance level: $D_{\text{critical}} = \frac{1.36}{\sqrt{n}}$, where n is the number of observed variable (Lindgren, 1962).

2.3.2   Correlation filter

As mentioned earlier, one of the key assumptions of this paper is that if the beta distribution fit to the short-term VIC series is statistically consistent with beta fit to the long-term VIC time series, then we assume that the short-term beta-fitted SMAP series is consistent with the hypothetical long-term beta-fitted SMAP time series. This is possible because VIC modeled soil moisture is validated by ground measurements (Pan et al., 2016; Cai et al., 2017), and it is most plausible where the correlation between SPL3SMP and VIC is highest. Correlation maps are shown in Fig. 4 between SPL3SMP and the VIC-ns product for the warm season and cold season periods. This suggests another filter to use: require that the correlation of beta-fitted SPL3SMP and beta-fitted VIC soil moisture be relatively high. We examined the distribution of correlation values across all grids in order to pick the cutoff between high and low correlation. We chose the mean correlation, minus the standard deviation of correlation (across all grids), as a threshold. Thus grids whose correlation is close to average or better than the average pass the filter. For both the warm and cold seasons, this value was

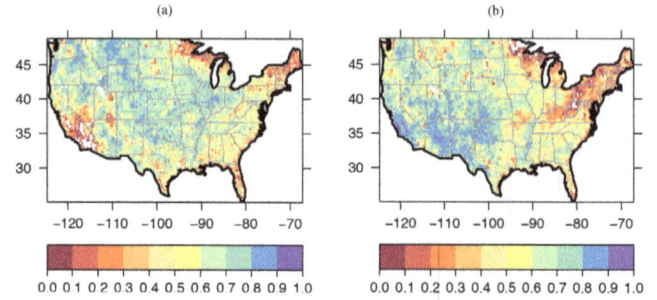

Figure 4. (a) Correlations (R) between VIC and SMAP beta models for the warm season (average $R = 0.57$) and **(b)** cold season (average $R = 0.56$). White regions signify a negative correlation.

very close to 0.4, and as a result, we picked this as the common threshold.

2.3.3   Mean distance (MD)

To evaluate whether the KS-based filter, the correlation filter or a combination of both is best, we define a simple mean distance (MD) metric. Assuming that a VIC index at 36 km resolution is the ground truth, we can calculate a distance between VIC and SMAP. For every day that SMAP provided a retrieval, if SMAP_i is the drought index percentile of grid i that passes the filter, and VIC_i is the VIC drought index percentile of the same grid, and in total n_g grids on day d passed the filter, then the MD_d is defined as the average of absolute distances between the SPL3SMP drought index percentiles and the VIC drought index percentiles. For the candidate date d and for a given filter,

$$\text{MD}_d = \frac{\sum_{i=1}^{n_g} |\text{VIC}_i - \text{SMAP}_i|}{n_g}. \tag{4}$$

In Eq. (4), VIC_i and SMAP_i are VIC and SMAP drought index values for grid i, n_g is the total number of grids that passed the filter, and MD_d is the mean distance for date d.

For each filter, the final pass and fail distance scores are calculated by averaging MD_d values over the number of days, especially for both dry or wet seasons:

$$\text{MD} = \frac{\sum_{i=1}^{n_d} |\text{MD}_d|}{n_d}, \tag{5}$$

where n_d is the total number of days for which the MD_d value is available. While n_g varies every day, since the number of overpasses varies every day, the value of n_d was constant (549 for warm season and 457 for cold season). The MD value obtained from grids that failed a filter is called MD_{fail}, and the MD value from grids that passed a filter is called MD_{pass}. For each filter a difference (Diff) was computed by reducing the MD_{pass} from the MD_{fail}: Diff $= \text{MD}_{\text{fail}} - \text{MD}_{\text{pass}} > 0$.

2.3.4 Combination filters

In addition to the KS filter and the correlation filter, we investigate two filters defined by the following combination rules:

- *Intersection filter.* A grid cell g passes the intersection filter if it passes both the KS filter *and* the correlation filter. Otherwise, it fails.

- *Union filter.* A grid cell g passes the union filter if it passes *either* filter or both filters. Note that using the union filter gives the best coverage of the grids throughout CONUS, while the intersection filter has the strongest requirements for passing.

3 Results and discussion

3.1 Data adequacy metrics

3.1.1 Correlation filter

Figure 4 shows that the average correlation for both warm and cold seasons is high and is around 0.6. During the warm season, the Central Valley and Southern California, Florida, the northeastern US, the north of Wisconsin, and Minnesota show poor correlation with VIC, at around 0.2. The extent of this poor correlation increases during the cold season for the northeastern US, Wisconsin and Minnesota. Snow season results in poor SMAP coverage during winter time in those areas. In addition, the low number of overpasses (presented in Fig. 1) during winter in the Northeast can play a role in the low amount of data and poor correlation during the cold season. Contrary to the warm season, southern California shows a high a correlation with VIC during the cold season, at around 0.9. We attribute this change from cold season to warm season in southern and southern-central California to the irrigation that SMAP picks up (Lawston et al., 2017) but VIC does not, since the version used here does not have water management effects. A land use and land cover map shows that about one-third of these areas are irrigated vegetation and another third are forests and woodlands (USGS, 2018). There are also as many as 2 million water wells in California that contribute to the irregularity of groundwater and affecting the soil moisture. They range from hand-dug, shallow wells to carefully designed large-production wells drilled to great depths (California Dept. of Water Resources, 2018). More data are needed before we can recognize further attributions to the low correlation between VIC and SMAP in that region. While systematic biases are not revealed in correlations, the temporal consistency among the time series is captured.

3.1.2 KS filter

Figure 5 shows which grids passed the 95 % KS test; there, we have confidence that the SMAP drought (pluvial) indices

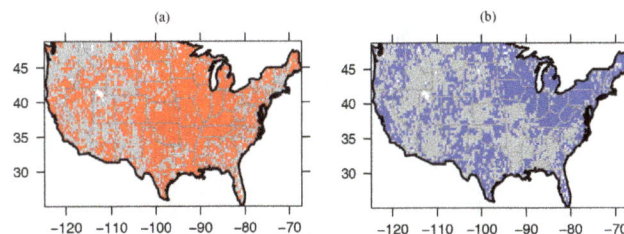

Figure 5. (a) Grids in red show areas whose short-term VIC in warm season data has the same underlying beta distribution as the long-term VIC in warm season data ($n = 3560$ or 68 % of grids are red); **(b)** the same as panel **(a)**, but for cold season period shown in blue ($n = 2927$ or 57 % of grids). Gray areas are grids where the short-term VIC does not have the same beta distribution as their long-term VIC.

provide reliable risk levels given the current period of record. The warm season shows 11 % more grids passing the adequacy test than the cold season. Note that as the record length gets extended, the above analysis needs to be repeated to see if the adequacy changes.

In the warm season, the majority of the grids whose underlying short-term and long-term beta distribution were different were in the western US. The low warm season correspondence in the Pacific Northwest (PNW) region is particularly apparent. The PNW region is covered by dense forests, mountain and heavily regulated agricultural lands by irrigation. This contributes to the fact that most grids in PNW do not pass the KS filter. A pattern of low correspondence over the major mountain areas (e.g., the Rocky Mountains, Sierra and Cascades) is also apparent, given the coarse SMAP brightness temperature (Tb) footprint and dense vegetation.

3.1.3 Combined filters

Figure 6 represents the results of correlation filter and KS filter together for both warm (panel a) and cold (panel b) seasons over all 5815 grids. We use these filters (passed and failed grids) on a daily basis for MD_d measures, though the value changes every day, depending on the number of overpasses for that date. Table 1 summarizes how many grids pass or fail each filter.

3.2 Evaluation of results under different filters

For each filter, the values of MD_d were averaged to calculate MD_{fail} and MD_{pass} for the whole CONUS over the 549 days of the warm season and 457 days of the cold season. The summary result of all four tests is shown in Tables 2 and 3. To test if having a filter is better than having no filter, for each season, we performed a two-sided null hypothesis. The tests used 95 % confidence limits between the MD of all grids – which was 22.7 in the warm season and 22.6 in the cold season – versus the MD of only passed grids. The results

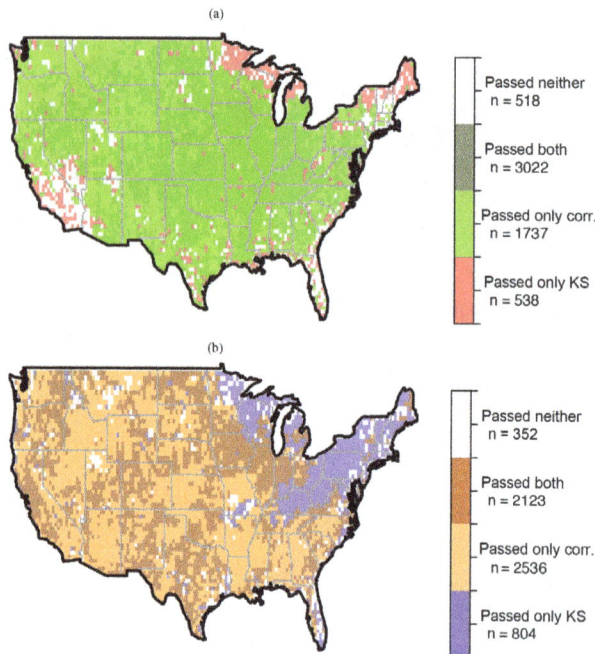

Figure 6. (a) Warm season grids that pass the correlation filter and/or the KS filter. Dark green grids include grids that pass intersection filters. **(b)** Cold season grids that pass the correlation filter and/or the KS filter. Dark orange grids include grids that pass intersection filters. In both figures, white grids show the grids that pass neither filter and will be cross-hatched in index maps.

Table 1. Number of grids, out of total 5815, that fail and pass the quality control for each filter.

n_g	KS filter	Correlation filter	Intersection filter	Union filter
Warm season fail	2255	1056	2793	518
Warm season pass	3560	4759	3022	5297
Cold season fail	2888	1156	3692	352
Cold season pass	2927	4656	2123	5463

Note: per day, the n_g numbers are less because of SMAP overpass missing grids.

Table 2. Mean distance (MD) of four tests averaged over 549 days of warm season. Diff is the difference between the first and second row.

	KS filter	Correlation filter	Intersection filter	Union filter
MD_{fail}	24.1	26.5	24.5	26.8
MD_{pass}	21.9	21.9	21.1	22.3
Diff	2.2	4.5	3.4	4.5

Table 3. Mean distance (MD) of four tests averaged over 457 days of cold season. Diff is the difference between the first and second row.

	KS filter	Correlation filter	Intersection filter	Union filter
MD_{fail}	22.8	29.0	24.1	29.2
MD_{pass}	22.4	21.2	20.1	22.1
Diff	0.4	7.8	4.0	7.1

showed that all four filters are significantly different than the MD of the whole CONUS. Thus, regardless of the type of the filter, having some sort of filter is better than having no filter.

In the warm season, the KS filter did better (i.e., larger Diff values or better skill in separating high- and low-performance grids) than the correlation filter for only 115 days out of 546 days, mostly in April. For almost half of the dates (260 days out of 546), the union filter did better than the correlation filter. This outperforming of the union filter occurs evenly throughout the warm season.

In the cold season, for only 48 days out of 457 days, the KS filter did better than the correlation filter, and for 198 days, the union filter did better than the correlation filter. These results suggest that for the cold season, the correlation filter is providing the most effective filter. However, if we only accept the grids that pass the correlation filter, we lose 804 grids. This area involved almost all of the northeastern coast and central East Coast as well as northern Wisconsin and northeastern Minnesota. However this is not a concerning problem for drought, since for most of the cold season these areas are covered by snow. We still decided to generate a cold season filter by including the KS filter with the correlation filter, thus we used the union filter for the cold season.

Three considerations for doing so are the following:

1. *The Diff values.* The correlation-filter Diff value and union-filter Diff value during the cold season are similar and close.

2. *The nature of our tests.* It is not that surprising that the correlation filter has a higher Diff than that of the union filter. The MD metric measures how the SMAP index resembles the VIC index. Thus, we find that the most important predictor is that the SMAP values should be correlated with the VIC values.

3. *Optimum coverage.* Although the cold season East Coast drought index is not a matter of concern for this study, cold season soil moisture variability can affect warm season soil moisture and consequently agricultural drought. The goal is to create a filter that does not lose important information while providing the best knowledge of soil moisture data.

During the warm season, most of the grids that failed the test were in southern California and southern Nevada, in the Northeast (New Hampshire, Massachusetts, and Connecticut), and in the Southeast along the eastern coast of Florida. These are attributed to both the lack of correlation between

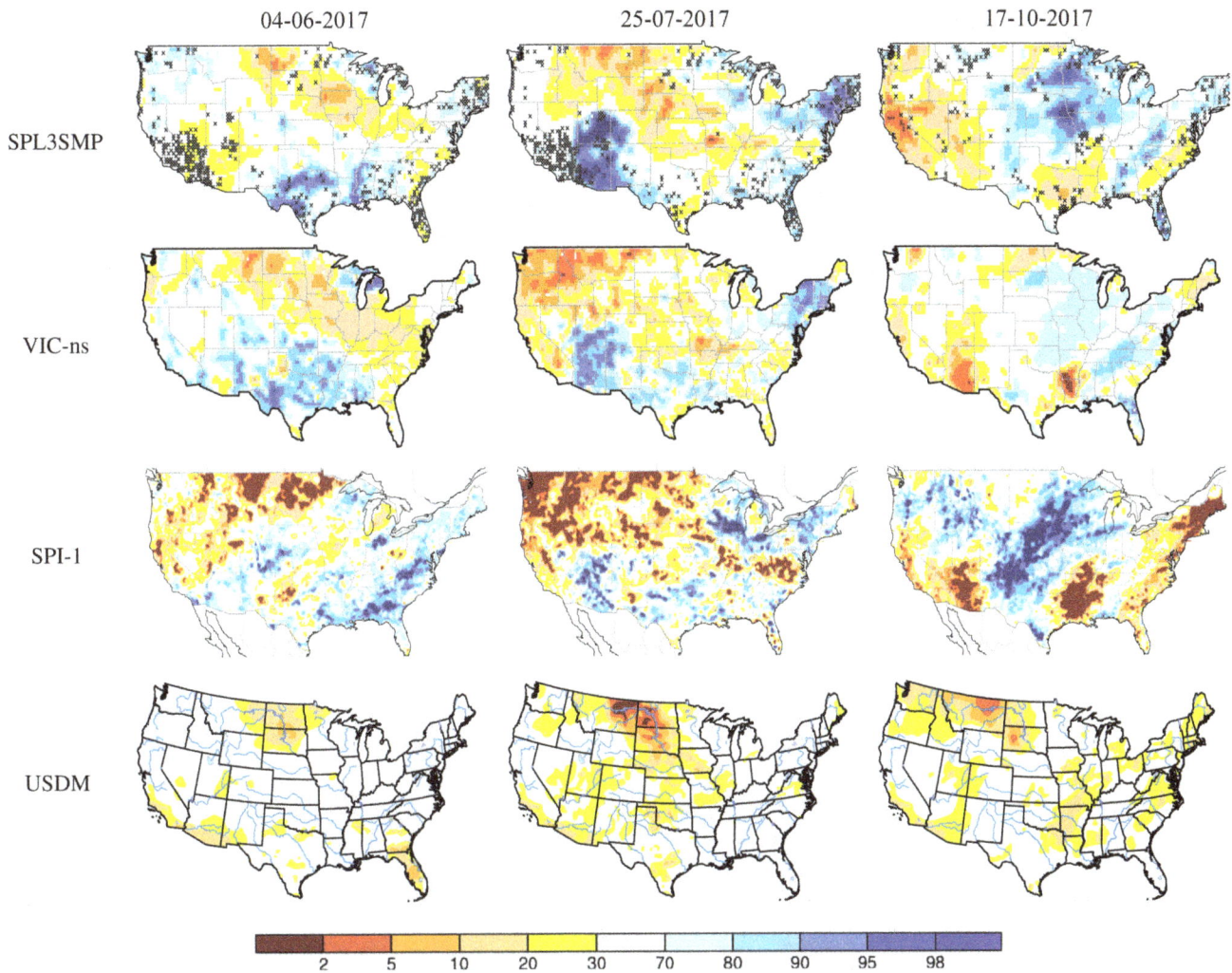

Figure 7. Comparison between SPL3SMP index map and VIC-ns, SPI-1 and USDM in 2017. The black x symbols in the SPL3SMP maps are the grids that passed neither filter and were shown as white grids in Fig. 6. For USDM, drought levels from 30 to 100 are shown in white.

SMAP and VIC and high variability between short-term and long-term soil moisture. These areas show non-stationarity in soil moisture, meaning that soil moisture distribution is subject to change over time, either due to climate or human interventions. During the cold season, most of the areas are covered using the union filter. However, as discussed, we use this filter with caution, knowing that at least according to our numerical analysis, the correlation filter did better than the union filter. The Great Lakes region, Minnesota, and the Mid-Atlantic region do not show a high correlation between VIC and SMAP in the cold season. Snow, heavy canopy and land development cause SMAP retrievals to have errors. In addition, this region does not have a good coverage of soil moisture and has a smaller number of retrievals per grid (Fig. 1). However, the KS filter complements the map by showing that the long-term and short-term VIC during cold season stays pretty stationary over time. This means that the soil mois-

ture in this area has been less subject to change during cold season at least for the past 40 years.

This information can be used to inform an interpretation of SMAP soil moisture percentiles maps based on < 10 years of data, as presented in Figs. 7 and 8 for a selection of soil moisture drought and flood indices. The grids that fail both KS and correlation tests (white grids in Fig. 6) will be flagged and are where we have the highest uncertainty of the quality of the data. This includes about 500 grids in the warm season and about 350 grids in the cold season over the CONUS.

3.3 Comparison of the drought indices

In Figs. 7 to 10, several indices are compared to the SMAP-based drought index. For the surface soil moisture index based on SPL3SMP, we provide a 3-day composite SMAP index to offer more continuous coverage. The union filter is applied to omit the grids that do not have reliable estimates.

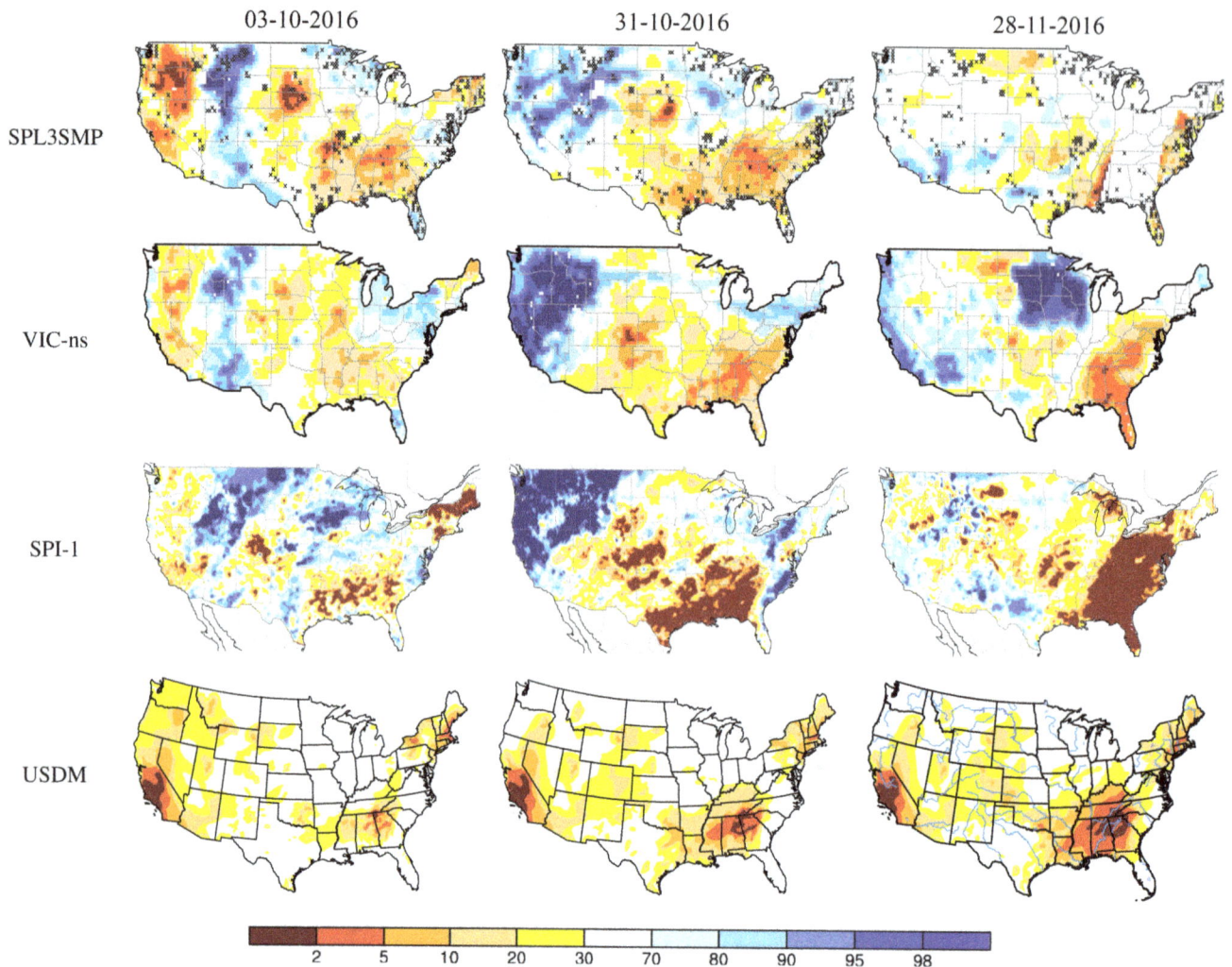

Figure 8. Comparison between SPL3SMP index map and VIC-ns, SPI-1 and USDM in 2016. The black x symbols in the SPL3SMP maps are the grids that passed neither filter and were shown as white grids in Fig. 6. For USDM, drought levels from 30 to 100 are shown in white.

Our index SPL3SMP index maps are compared with the 1-month SPI (SPI-1) index, a VIC-ns index and the USDM. For SMAP soil moisture index based on the SPL4SMAU, comparisons are made with a 3-month SPI (SPI-3) index and a GRACE satellite product. All the products except for GRACE were described in Sect. 1. GRACE is NASA's satellite system that detects small changes in the Earth's gravity field caused by the redistribution of water on and beneath the land surface. Combined with the Catchment LSM using an ensemble Kalman smoother for data assimilation (Zaitchik et al., 2008), GRACE maps root-zone soil moisture and groundwater transformed into percentiles (NDMC, 2018b).

Figures 7 and 9 show drought during the period from 4 June through 17 October 2017, for both the near surface and root zone. In this period, there was one agricultural drought event in Montana and North and South Dakota, with losses exceeding USD 1 billion across the United States

(NOAA, 2018b). The plains of eastern Montana experienced exceptional drought from July to October 2017, and in late October, drought started to end. The peak of the drought was in July 2017 when 20 % of Montana was in severe drought and 23 % of it was in moderate drought. Concurrently, 40 % of North Dakota was in extreme drought, while 70 % of the state was under some level of drought; similarly, 68 % of South Dakota was under severe drought (NOAA, 2018b). Both SPL3SMP and SPL4SMAU index maps seem to catch this drought event, although the event was more pronounced in the root zone than the surface. The maps of these two figures are also in general agreement. It is important to clarify that for 2017 period, the GRACE sensor was failing, and the resulting water storage observations were unreliable. Therefore, the last GRACE gravity field retrieval processing only goes through June 2017. Therefore, GRACE National Drought Mitigation Center (NDMC) results associated with

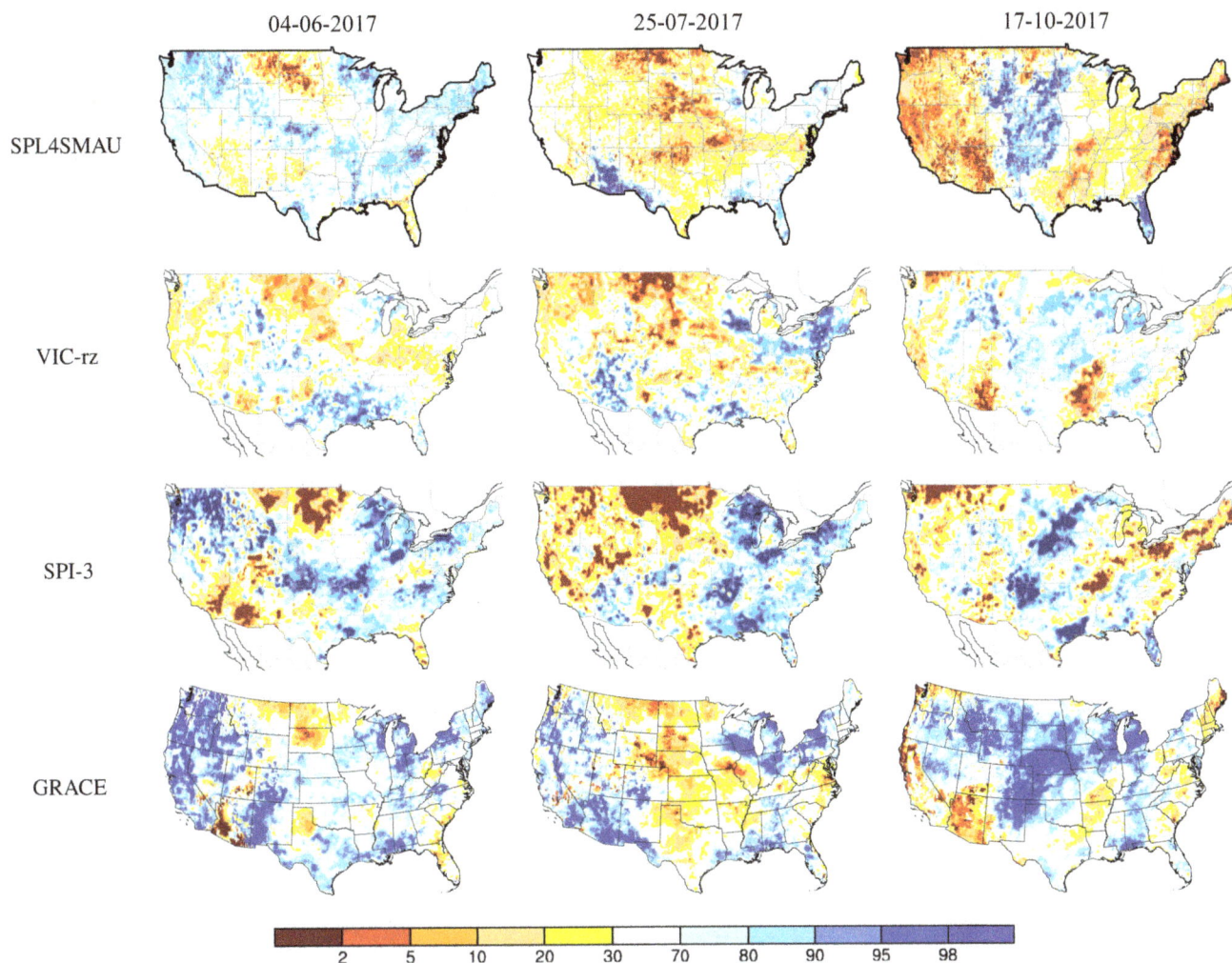

Figure 9. Comparison between SPL4SMAU index map and VIC-rz, SPI-3 and GRACE in 2017.

Fig. 9 are not consistent with other products and likely do not reflect actual GRACE observations for 2017.

In Figs. 8 and 10, drought during the period of 3 October to 8 November 2016 is shown for both near the surface and the root zone. In 2016, there were three drought events in the western, northeastern and southeastern parts of the US, which are captured by both SPL3SMP and SPL4SMAU index maps. The drought had mostly been alleviated in northern California by near-normal precipitation during the 2015–2016 winter and above normal precipitation in fall 2016. The extent that the drought persisted in Southern California after this period it is reflected in total column soil moisture rather than near-surface soil moisture (Fig. 9).

There is a high correspondence among the drought maps, particularly in the development of the drought in the southeastern US during October and November 2016. Due to heavy rainfall along the Mississippi River in November, the drought migrated eastwards. Also, by November 2016 the drought in Southern California was alleviated, which

is picked up by SPL3SMP, SPL4SMAU, VIC-ns and VIC-rz, SP-1 and 3, GRACE, and to a much lesser extent, by the USDM that showed an increasing area under drought on 28 November compared to SPL3SMP, SPL4SMAU, GRACE, or VIC-ns and VIC-rz. Additionally, for the maps that also include wetness (all except USDM), there is a high correspondence of pluvial regions (see Fig. 7).

Most of the grids where we do not have confidence in the accuracy of predictions are in Southern California and Nevada during the warm season (e.g., SPL3SMP index map on 4 June and 25 July 2017; Fig. 7). In fact, there is a visible discrepancy between SPL3SMP and VIC-ns index maps during that period in Southern California. We believe that this is due to a lack of correlation between SPL3SMP and VIC-ns in that area, since VIC does not model regulation. Human interference and the use of groundwater wells during the warm season can play a major part in what VIC models and SMAP see. For that reason, we think SMAP's metrics in the area are more accurate than those from VIC-ns.

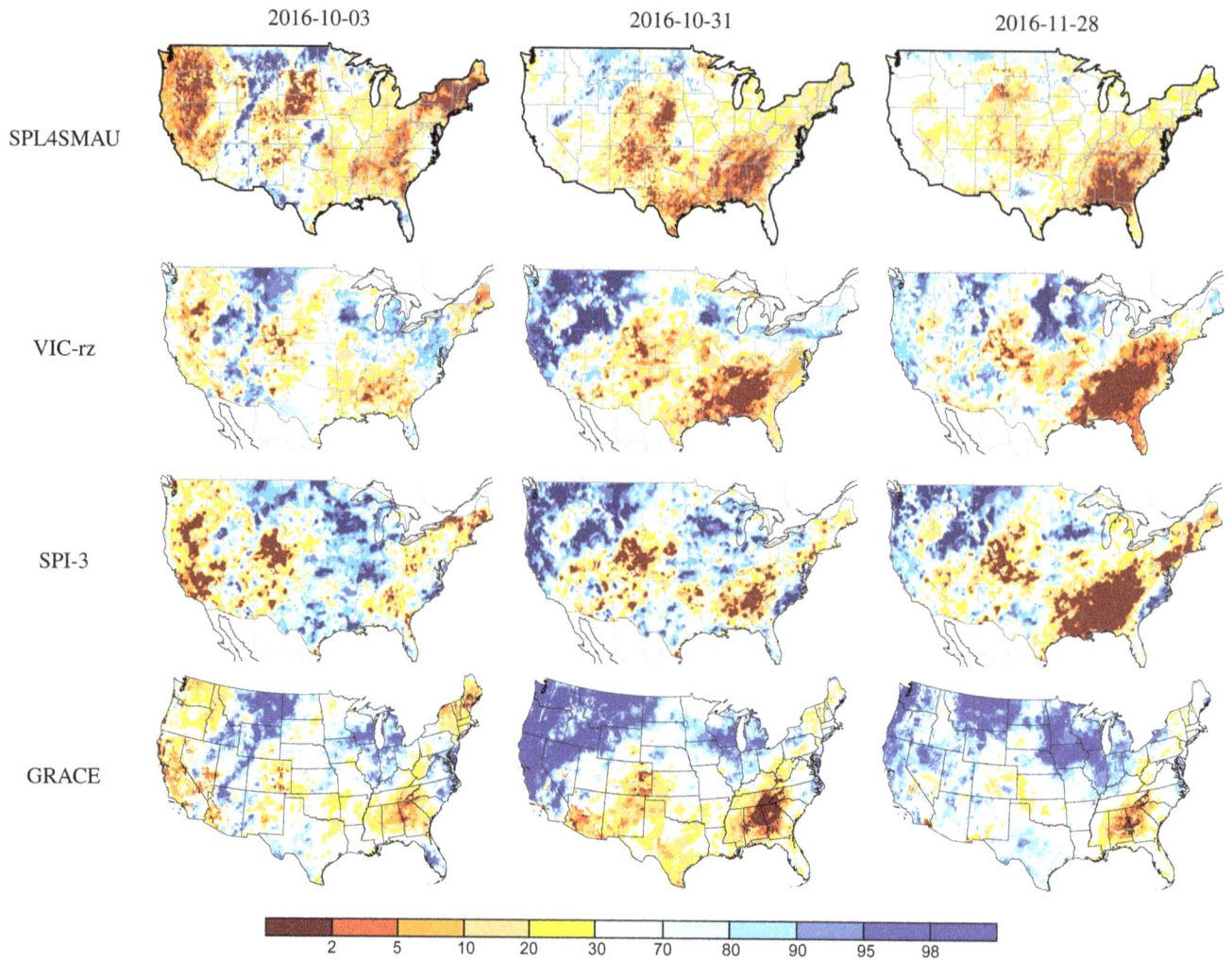

Figure 10. Comparison between SPL4SMAU index map and VIC-rz, SPI-3 and GRACE in 2016.

4   Conclusions

The drought index described in this study provides a reliable estimate of the state of drought on a daily basis for the CONUS, using SMAP. We fitted beta distributions to the SMAP data and used correlation, KS, and a combination of those two filters to numerically assess the adequacy of the short-term SMAP data for each grid cell. The areas that passed neither the KS nor correlation tests were flagged in the final SMAP drought index. These areas are grids where we have less confidence in reliable drought index estimates; they are non-stationary, and thus their soil moisture has been changing over the past 40 years. The flagged grids can be seen as an adjustment to the model to remove non-climatic influences or water management practices, although more in-depth research is needed to confirm such changes. Given the limited scope of the data, the results should be considered a demonstration of the reliability and usefulness of SMAP for

a drought-monitoring product and for implementation into an operational drought-monitoring tool.

Besides drought, SMAP can also identify regions of anomalously wet conditions that can be of great use to water and agricultural managers. Wet indices can indicate potential flood-prone conditions and regions can therefore be put on flood alerts if additional heavy rain occurs. Also, wet conditions can impact farm management, especially in the spring when sowing takes place or during the harvesting period.

Through comparing SMAP-based index maps for drought and wet conditions with other index products, we see a high similarity. Although there can be some errors at different levels, the overall evaluation reveals that SMAP-based drought products can be a viable alternative for drought monitoring in the US. This is advantageous, since SMAP is generated at a daily resolution with almost complete coverage every 3 days. This enables an observation of the effect of fluctuations in other hydrological variables, such as precipitation. In comparison, USDM, GRACE and the SPI have a low temporal resolution, which makes it difficult to study the shorter-term impacts from the other variables on soil moisture.

Both near-surface and root-zone soil moisture drought products can provide important information about the availability of soil moisture at the stage where plants develop in order to cultivate the optimum harvest. Future applications of this study can couple plant growth models with near-surface and root-zone soil moisture drought index products (NDMC, 2018a).

The soil moisture data are a culmination of all hydrological processes and represent available water from incoming precipitation and throughfall to evapotranspiration and drainage processes. The SMAP satellite provides global observations of soil moisture of unprecedented quality. Because SMAP monitors soil moisture directly and provides critical information for drought early warning, it is important that the future developments focus on drought assessment using SMAP in underrepresented parts of the world. Thus the results here provide significant support for a global SMAP drought and pluvial conditions monitoring system. Since SMAP data can be retrieved and maps can be generated in near-real time, it is very promising that a SMAP drought index product can be implemented operationally.

Author contributions. SS carried out the drought index development, developed the confidence analysis and online near-real-time website, and drafted the paper. MP prepared the VIC and SMAP soil moisture data as well as SPI drought index maps. EFW conceived of the study, supervised the project and helped to draft the paper. All authors read and approved the final paper.

Competing interests. The authors declare that they have no conflict of interest.

Acknowledgements. This work was supported by NASA grant CNV1003235. This paper benefited greatly from the reviewers' comments. We thank them for their time and support.

Edited by: Nunzio Romano

References

AMS: Drought, available at: http://glossary.ametsoc.org/wiki/Drought (last access: 30 April 2018), 2012.

Brocca, L., Hasenauer, S., Lacava, T., Melone, F., Moramarco, T., Wagner, W., Dorigo, W., Matgen, P., Martinez-Fernandez, J., Llorens, P., Latron, J., Martin, C., and Bittelli, M.: Soil moisture estimation through ASCAT and AMSR-E sensors: An intercomparison and validation study across Europe, Remote Sens. Environ., 115, 3390–3408, https://doi.org/10.1016/j.rse.2011.08.003, 2010.

Cai, X., Pan, M., Chaney, N. W., Colliander, A., Misra, S., Cosh, M. H., Crow, W. T., Jackson, T. J., and Wood, E. F.: Validation of SMAP soil moisture for the SMAPVEX15 field campaign using a hyper-resolution model, Water Resour. Res., 53, 3013–3028, 2017.

California Dept. of Water Resources: Wells, available at: https://water.ca.gov/Programs/Groundwater-Management/Wells, last access: 25 April 2018.

Entekhabi, D., Rodriguez-Iturbe, I., and Castelli, F.: Mutual interaction of soil moisture state and atmospheric processes, J. Hydrol., 184, 3–17, 1996.

Entekhabi, D., Njoku, E. G., O'Neill, P. E., Kellogg, K. H., Crow, W. T., Edelstein, W. N., Entin, J. K., Goodman, S. D., Jackson, T. J., Johnson, J., Kimball, J., Piepmeier, J. R., Koster, R. D., Martin, N., McDonald, K. C., Moghaddam, M., Moran, S., Reichle, R., Shi, J. C., Spencer, M. W., Thurman, S. W., Tsang, L., and Van Zyl, J.: The Soil Moisture Active Passive (SMAP) Mission Proc., IEEE, 98, 704–716, https://doi.org/10.1109/JPROC.2010.2043918, 2010.

Friedman, D. G.: The prediction of long-continuing drought in south and southwest Texas, Occasional Papers in Meteorology, p. 182, the Travelers Weather Research Center, Hartford, CT, USA, 1957.

Heim Jr., R. R.: A review of twentieth century drought indices used in the United States, B. Am. Meteorol. Soc., 83, 1149–1165, 2002.

Howitt, R. E., Medellin-Azuara, J., MacEwan, D., Lund, J. R., and Daniel, A.: Economic Impact of the 2015 Drought on Farm Revenue and Employment Sumner, Agricultural and Resource Update, Giannini Foundation of Agricultural Economics, University of California, Davis, USA, 2015.

Kerr, Y. H., Waldteufel, P., Richaume, P., Wigneron, J. P., Ferrazzoli, P., Mahmoodi, A., Al Bitar, A., Cabot, F., Gruhier, C., Juglea, S. E., Leroux, D., Mialon, A., and Delwart, S.: The SMOS Soil Moisture Retrieval Algorithm, IEEE T. Geosci. Remote, 50, 1384–1403, https://doi.org/10.1109/TGRS.2012.2184548, 2012.

Kerr, Y. H., Al-Yaari, A., Rodriguez-Fernandez, N., Parrens, M., Molero, B., Leroux, D., Bircher, S., Mahmoodi, A., Mialon, A., Richaume, P., Delwart, S., Pellarin, A. A. B. T., Bindlish, R., Rudiger, T. J. C., Waldteufel, P., Mecklenburg, S., and Wigneron, J.: Overview of SMOS performance in terms of global soil moisture monitoring after six years in operation, Remote Sens. Environ., 180, 40–63, https://doi.org/10.1016/j.rse.2016.02.042, 2016.

Lawston, P. M., Santanello Jr., J. A., and Kumar, S. V.: Irrigation Signals Detected From SMAP Soil Moisture Retrievals, Geophys. Res. Lett., 44, 11860–11867, 2017.

Lindgren, B.: Statistical Theory, Mac-millan, New York, USA, 1962.

Liu, Q., Reichle, R. H., Bindlish, R., Cosh, M. H., Crow, W. T., de Jeu, R., Lannoy, G. J. M. D., Huffman, G. J., and Jackson, T. J.: The contributions of precipitation and soil moisture observations to the skill of soil moisture estimates in a land data assimilation system, J. Hydrometeorol., 12, 750–765, 2011.

Martinez-Fernandez, J., Gonzalez-Zamora, A., Sanchez, N., and A Gumuzzio, .: Satellite soil moisture for agricultural drought monitoring: Assessment of the SMOS derived Soil Water Deficit Index (vol. 177, pg. 277, 2016), Remote Sens. Environ., 183, 368–368, 2016.

Mo, K. C. and Chelliah, M.: The modified Palmer Drought Severity Index based on the NCEP North American Regional Reanalysis, J. Appl. Meteor. Climatol., 45, 1362–1375, 2006.

NCAR: The Climate Data Guide: Standardized Precipitation Index (SPI), available at: https://climatedataguide.ucar.edu/climate-data/standardized-precipitation-index-spi, last access: 2 April 2018.

NDMC: Types of Drought, National Drought Monitoring Center, available at: https://drought.unl.edu/Education/DroughtIn-depth/TypesofDrought.aspx, last access: 17 December 2018a.

NDMC: Groundwater and Soil Moisture Conditions from GRACE Data Assimilation, the National Drought Mitigation Center University of Nebraska-Lincoln, available at: http://nasagrace.unl.edu/Archive.aspx, last access: 17 December 2018b.

NIST: Beta Distribution, available at: http://www.itl.nist.gov/div898/handbook/eda/section3/eda366h.htm (last access: 2 April 2018), 2013.

NOAA: NLDAS Drought Monitor Soil Moisture, available at: http://www.emc.ncep.noaa.gov/mmb/nldas/drought/, last access: 5 February 2018a.

NOAA: U.S. Billion-Dollar Weather and Climate Disasters, national Centers for Environmental Information (NCEI), available at: https://www.ncdc.noaa.gov/billions, last access: 4 October 2018b.

O'Neill, P., Chan, S., Njoku, E. G., Jackson, T., and Bindlish, R.: SMAP L2 Radiometer Half-Orbit 36 km EASE-Grid Soil Moisture, Distributed Active Archive Center Version 4, NASA National Snow and Ice Data Center, Boulder, Colo., USA, https://doi.org/10.5067/XPJTJT812XFY, 2016.

Pan, M., Li, H., and Wood, E.: Assessing the skill of satellite-based precipitation estimates in hydrologic applications, Water Resour. Res., 46, W09535, https://doi.org/10.1029/2009WR008290, 2010.

Pan, M., Cai, X., Chaney, N. W., Entekhabi, D., and Wood, E. F.: An initial assessment of SMAP soil moisture retrievals using high-resolution model simulations and in situ observations, Geophys. Res. Lett., 43, 9662–9668, 2016.

Parinussa, R. M., Holmes, T. R. H., Wanders, N., Dorigo, W. A., and de Jeu, R. A. M.: A Preliminary Study toward Consistent Soil Moisture from AMSR2, J. Hydromet., 16, 932–947, https://doi.org/10.1175/JHM-D-13-0200.1, 2015.

Princeton University Hydrology: http://stream.princeton.edu/CONUSFDM/WEBPAGE/interface.php?locale=en (last access: 15 December 2018), 2013.

Reichle, R. H.: Assessment of the SMAP Level-4 Surface and Root-Zone Soil Moisture Product Using In Situ Measurements, J. Hydrometeorol., 18, 2621–2645, 2017.

Reichle, R. H., Koster, R. D., Lannoy, G. J. M. D., Forman, B. A., Liu, Q., Mahanama, S. P. P., and Toure, A.: Assessment and enhancement of MERRA land surface hydrology estimates, J. Climate, 24, 6322–6338, 2011.

Reichle, R. H., Lucchesi, R., Ardizzone, J. V., Kim, G., Smith, E. B., and Weiss, B. H.: Soil Moisture Active Passive (SMAP) Mission Level 4 Surface and Root Zone Soil Moisture (L4SM) Product Specification Document, Tech. Rep. 10 (Version 1.4), NASA Goddard Space Flight Center, Greenbelt, MD, USA, 2015.

Rienecker, M., Suarez, M. J., Todling, R., Bacmeister, J., Takacs, L., Liu, H.-C., Gu, W., Sienkiewicz, M., Koster, R. D., Gelaro, R., Stajner, I., and Nielsen, J. E.: The GEOS-5 Data Assimilation System – Documentation of Versions 5.0.1, 5.1.0, and 5.2.0., NASA Technical Report Series on Global Modeling and Data Assimilation NASA/TM-2008-104606, NASA, vol. 28, 101 pp., 2008.

Robock, A., Vinnikov, K., Srinivasa, G., Entin, J., Hollinger, S., Speranskaya, N., Liu, S., and Namkhai, A.: The Global Soil Moisture Data Bank, B. Am. Meteorol. Soc., 81, 1281–1299, 2000.

Sheffield, J. and Wood, E. F.: Drought: Past Problems and Future Scenarios, 978-1-84971-082-4, EarthScan, London, UK, 2011.

Sheffield, J., Goteti, G., Wen, F., and Wood, E. F.: A simulated soil moisture based drought analysis for the United States, J. Geophys. Res., 109, D24108, https://doi.org/10.1029/2004JD005182, 2004.

Sheffield, J., Livneh, B., and Wood, E. F.: Representation of terrestrial hydrology and large scale drought of the Continental U.S. from the North American Regional Reanalysis, J. Hydrometeor., 13, 856–876, https://doi.org/10.1175/JHM-D-11-065.1, 2012.

Sheffield, J., Wood, E. F., Chaney, N., Guan, K., Sadri, S., Yuan, X., Olang, L., Amani, A., Ali, A., and Demuth, S.: A Drought Monitoring and Forecasting System for Sub-Sahara African Water Resources and Food Security, B. Am. Meteorol. Soc., 95, 861–882, 2014.

Shukla, S. and Wood, A. W.: Use of a standardized runoff index for characterizing hydrologic drought, Geophys. Res. Lett., 35, 1–7, 2008.

Svoboda, M.: An introduction to the Drought Monitor, Drought Network News, 12, 15–20, 2000.

Tallaksen, T. and Van Lanen, H. A.: Hydrological Drought, Processes and Estimation Methods for Streamflow and Groundwater, vol. 48, Elsevier Science, Amsterdam, the Netherlands, 2004.

UN/ISDR: Drought Risk Reduction Framework and Practices: Contributing to the Implementation of the Hyogo Framework for Action. United Nations Secretariat of the International Strategy for Disaster Reduction (UN/ISDR), Tech. Rep. 98+vi pp., United Nations Secretariat of the International Strategy for Disaster Reduction (UN/ISDR), Geneva, Switzerland, 2007.

USGS: National Gap Analysis Program, Land Cover Data Viewer,, available at: https://goo.gl/rntijg, last access: 30 April 2018.

Wagner, W., Hahn, S., Kidd, R., Melzer, T., Bartalis, Z., Hasenauer, S., Figa-Saldaña, J., de Rosnay, P., Jann, A., Schneider, S., Komma, J., Kubu, G., Brugger, K., Aubrecht, C., Züger, J., Gangkofner, U., Kienberger, S., Brocca, L., Wang, Y., Blöschl, G., Eitzinger, J., and Steinnocher, K.: The ASCAT Soil Moisture Product: A Review of its Specifications, Validation Results, and Emerging Applications, Meteorol. Z., 22, 5–33, https://doi.org/10.1127/0941-2948/2013/0399, 2013.

Wilhite, D. A.: Drought as a Natural Hazard: Concepts and Definitions, chap. 1, vol. I, National Drought Mitigation Center at

Digital Commons at University of Nebraska, Lincoln, Routledge, London, UK, 2000.

Wilhite, D. A. and Glantz, M. H.: Understanding the Drought Phenomenon: The Role of Definitions, Water Int., 10, 111–120, 1985.

Wu, Q., Liu, H., Wang, L., and Deng, C.: Evaluation of AMSR2 soil moisture products over the contiguous United States using in situ data from the International Soil Moisture Network, Int. J. Appl. Earth Obs., 45, 187–199, https://doi.org/10.1016/j.jag.2015.10.011, 2015.

Zaitchik, B. F., Rodell, M., and Reichle, R. H.: Assimilation of GRACE Terrestrial Water Storage Data into a Land Surface Model: Results for the Mississippi River Basin, Am. Meteorol. Soc., 9, 535–548, 2008.

13

Tributaries affect the thermal response of lakes to climate change

Love Råman Vinnå[1], **Alfred Wüest**[1,2], **Massimiliano Zappa**[3], **Gabriel Fink**[4], and **Damien Bouffard**[1,2]

[1]Physics of Aquatic Systems Laboratory – Margaretha Kamprad Chair, École Polytechnique Fédérale de Lausanne, Institute of Environmental Engineering, Lausanne, Switzerland
[2]Eawag, Swiss Federal Institute of Aquatic Science and Technology, Surface Waters – Research and Management, Kastanienbaum, Switzerland
[3]Swiss Federal Institute for Forest, Snow and Landscape Research, WSL, Birmensdorf, Switzerland
[4]Center for Environmental Systems Research, CESR, University of Kassel, Kassel, Germany

Correspondence: Love Råman Vinnå (love.ramanvinna@epfl.ch)

Abstract. Thermal responses of inland waters to climate change varies on global and regional scales. The extent of warming is determined by system-specific characteristics such as fluvial input. Here we examine the impact of on-going climate change on two alpine tributaries, the Aare River and the Rhône River, and their respective downstream peri-alpine lakes: Lake Biel and Lake Geneva. We propagate regional atmospheric temperature effects into river discharge projections. These, together with anthropogenic heat sources, are in turn incorporated into simple and efficient deterministic models that predict future water temperatures, river-borne suspended sediment concentration (SSC), lake stratification and river intrusion depth/volume in the lakes. Climate-induced shifts in river discharge regimes, including seasonal flow variations, act as positive and negative feedbacks in influencing river water temperature and SSC. Differences in temperature and heating regimes between rivers and lakes in turn result in large seasonal shifts in warming of downstream lakes. The extent of this repressive effect on warming is controlled by the lakes hydraulic residence time. Previous studies suggest that climate change will diminish deep-water oxygen renewal in lakes. We find that climate-related seasonal variations in river temperatures and SSC shift deep penetrating river intrusions from summer towards winter. Thus potentially counteracting the otherwise negative effects associated with climate change on deep-water oxygen content. Our findings provide a template for evaluating the response of similar hydrologic systems to on-going climate change.

1 Introduction

The thermal and hydrodynamic responses of lakes to climate change are considerably diverse. Observed responses vary on global, regional and even local scales (O'Reilly et al., 2015). Even neighboring freshwater bodies can react differently to a given increase in air temperature. This indicates that lake-specific characteristics will determine the response to climate change (for clarity and brevity, we refer to anthropogenic climate change simply as "climate change" or "climate" from now on). Local factors which affect climate warming of lakes include, among others, morphology (Toffolon et al., 2014), irradiance absorption (Kirillin, 2010; Williamson et al., 2015), local weather conditions (Zhong et al., 2016), stratification (Piccolroaz et al., 2015), atmospheric brightening (Fink et al., 2014a) and ice cover (Austin and Colman, 2007).

Throughflows affect epilimnion and hypolimnion temperatures of lakes. Studies of climate impact typically do not address these sorts of subtleties in lake dynamics due to a lack of data or difficulties in predicting future temperature and discharge conditions (Fenocchi et al., 2017). Several studies of large lakes suggest that major tributaries play only a minor role in climate-induced warming and deep-water oxygen renewal (Fink et al., 2014a; Schwefel et al., 2016). Medium- and smaller-scale lakes are, however, more abundant than large lakes (Verpoorter et al., 2014) and exhibit a greater degree of sensitivity to point sources of anthropogenic thermal input which can affect temperature and stratification (Kirillin et al., 2013; Råman Vinnå et al., 2017). Medium- and small-sized lakes also make a more significant contribution to the temperature-dependent global greenhouse gas budget (Holgerson and Raymond, 2016). Accurate prediction of climate change impacts therefore requires a more detailed understanding of small- to medium-scale lake and tributary systems.

Climate change exerts a dual influence on alpine rivers by introducing variation to both flow and temperatures. Discharge variation takes the form of less flow in summer and more flow in winter due to warmer high-altitude snow and ice melt/runoff regimes (Addor et al., 2014; Birsan et al., 2005), which also influence river temperature (Isaak et al., 2012; Van Vliet et al., 2013). Increased air temperature may also enhance erosion rates in river basins thereby supplementing river-borne suspended sediment loads (Bennett et al., 2013). River temperature and suspended sediment content determine water density and, by extension, the depth of river plume intrusions into downstream lakes or reservoirs. The depths and volumes of river intrusion plumes determine deep-water oxygen renewal, especially for deeper lakes where climate-related warming can reduce seasonal deep convective mixing and thereby deplete deep-water oxygen (Schwefel et al., 2016). Major (deep penetrating) river intrusion events typically occur due to flooding, which flush large sediment loads into the river (Fink et al., 2016). The frequency and volume of floods in the Alps are notoriously hard to predict, although a decrease in floods has occurred in association with recent warmer summers observed in the Alps (CH2011, 2011; Glur et al., 2013).

Recent model studies have identified inland waters as risk hotspots under expected climate change scenarios (IPCC, 2014). These systems require a more detailed analysis given their role in supporting fisheries, agriculture, drinking water supply, heat management and hydropower. This paper examines the complex interactions between tributaries and lakes in response to temperature increase and other modifications expected from climate change. Our objectives were to quantify the impact of specific climate change scenarios on (i) alpine tributaries and (ii) downstream peri-alpine lakes with a focus on river-borne suspended sediment concentration (SSC), water temperature, stratification and river intrusions.

We used coupled river–lake models to build on previous research by Fink et al. (2016). These authors investigated the effect of flood frequencies on deep-water renewal under established climate change scenarios. Their work did not generate tightly constrained estimates for flooding events. Our analysis therefore provisionally assumed that flood frequency does not change in the future. In addition to these sources of natural variation, our models addressed variation in river discharge regimes (i.e., daily mean level shift) under the specified A1B climate change scenario. These in turn affect SSC and thermal regimes for rivers and their associated downstream lakes. Furthermore, here we show that local point sources of anthropogenic thermal pollution can have

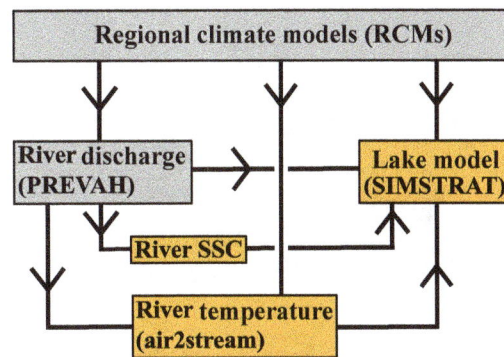

Figure 1. Schematic illustration of the one-way model chain of this study. Orange models represent modeling performed by this study, while grey models represent simulated data inputs obtained from external sources.

a major impact on the response of inland waters to climate change as previously suggested by Fink et al. (2014b).

2 Methods

2.1 Approach

The investigation of tributary influence on lake response to climate change followed these procedural steps:

i. Define river temperature and SSC models for two major alpine rivers and designate a one-dimensional lake model for a large- and medium-sized peri-alpine lake.

ii. Integrate model (i) with a river intrusion scheme: Fig. 1 shows the integration of the one-way component models.

iii. Obtain and apply estimates of future regional air temperature, tributary discharge and changes in local anthropogenic thermal emissions to both river and lake models.

iv. Identify patterns in model outputs of water temperature, SSC, lake stratification and river intrusion parameters (volume and depth).

2.2 Study area

This study examined two warm, monomictic, freshwater peri-alpine lakes in western Switzerland: Lake Biel (LB; 47°5′ N, 7°10′ E) and Lake Geneva (LG; 46°27′ N, 6°31′ E). Large tributaries originating in the Alps, the Aare River and the Rhône River, feed into LB and LG, respectively (Fig. 2).

LG is a large meso-eutrophic lake resting at 372 m elevation and covering an area of 580 km^2. It reaches a maximum depth of 309 m and holds a volume of 89 km^3 with an average hydraulic residence time of 11.5 years. Complete seasonal deep convective mixing only occurs on average every

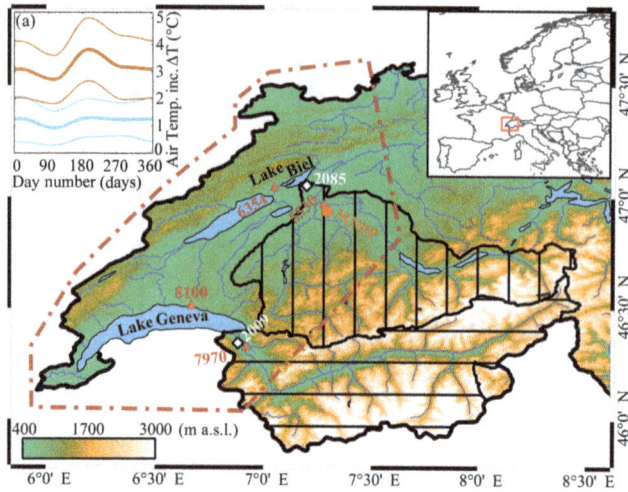

Figure 2. Study area and predicted regional air temperature increases. Elevation above sea level (green to white color ramp), locations and number of river stations (white diamonds) and atmospheric monitoring stations (red triangles), drainage area (Aare: vertical lines; Rhône: horizontal lines) and location of Mühleberg Nuclear Power Plant (MNPP, orange circle). Area covered by regional climate models (dark red dashed-dotted line) with **(a)** predicted air temperature increase ΔT in the near-future (blue, 2030–2049) and far-future (orange, 2080–2099) for medium (thick lines) and upper/lower estimates (thin lines) under the A1B emission scenario (CH2011, 2011).

fifth winter but is predicted to become less frequent with ongoing climate change (Perroud and Goyette, 2010; Schwefel et al., 2016). Whereas the global average lake surface temperature has increased by $\sim 0.34\,°C\ \mathrm{decade}^{-1}$ between 1985 and 2009 (O'Reilly et al., 2015), the Rhône River supplying $\sim 75\,\%$ of LG's inflow has experienced a temperature increase of $\sim 0.21\,°C\ \mathrm{decade}^{-1}$ from 1978 to 2002 (Hari et al., 2006).

LB is a 74 m deep, meso-eutrophic, medium-sized lake resting at an elevation of 429 m. It covers a surface area of $39.3\,\mathrm{km}^2$ and holds a volume of $1.18\,\mathrm{km}^3$ with a hydraulic residence time of 58 days. Complete deep convective mixing occurs every winter and effectively replenishes the oxygen-depleted deep-water. The Aare River provides $\sim 61\,\%$ of LB's inflow and experienced a $0.34\,°C\ \mathrm{decade}^{-1}$ increase in temperature from 1978 to 2002 (Hari et al., 2006). Several dams/lakes trap sediment along the upstream Aare course and increase sediment settling and water temperature prior to entering LB. The Mühleberg Nuclear Power Plant (MNPP), situated $\sim 19\,\mathrm{km}$ upstream from LB ($46°59'$ N, $7°16'$ E; Fig. 2) represents a point-source of thermal pollution. Planned for decommission in 2019, the plant emits $\sim 700\,\mathrm{MW}$ of heat into the Aare and substantially warms the river water (Råman Vinnå et al., 2017). The $\sim 8\,\mathrm{km}$ long Zihlkanal, LB's second largest tributary, supplies $\sim 32\,\%$ of the lake inflow and connects LB to Lake Neuchâtel (Fig. 2). This

tributary is neglected here since it mainly transports lake surface water, which has approximately the same temperature as LB surface water and thus without net heat effects.

2.3 River models

2.3.1 Temperature

Uncertainties concerning river morphology, heat fluxes, shadowing and atmospheric conditions such as wind speed and cloudiness (Caissie, 2006) pose a significant challenge to accurately model future river temperatures. Deterministic models typically require detailed knowledge unavailable for future climate scenarios. Regressions and stochastic models rely heavily on observed natural variability in a given time frame and typically do not include inputs representing additional or interacting physical processes. On their own, these sorts of "black box" models cannot balance trade-offs between constraints available from empirical data and the complexity offered by theoretical frameworks.

To overcome these limitations, we used the hybrid model air2stream (Toffolon and Piccolroaz, 2015). The model combines the simplicity of stochastic models with accurate representation of the relevant physical processes affecting temperature. Similar to the neural networks approach, the model calculates river water temperature (T_w) through a Monte-Carlo-like calibration process, which identifies optimal parameters for weighting physically dependent variables. We used the eight-parameter (a_1 to a_8) version of the model which incorporates air temperature (T_a) and river discharge (Q) as a function of time (t).

$$\frac{\Delta T_w}{\Delta t} = \frac{1}{\delta}\left\{ a_1 + a_2 T_a(t) - a_3 T_w(t) \right. \tag{1}$$

$$\left. + \theta\left[a_5 + a_6 \cos\left(2\pi\left(\frac{t}{t_y} - a_7\right)\right) - a_8 T_w(t)\right]\right\},$$

$$\delta = \theta^{a_4}, \quad \theta = Q(t)/\overline{Q}, \tag{2}$$

where t is expressed in years and t_y represents 1 year. Both the Aare and Rhône (stations nos. 2085 and 2009, respectively; Fig. 2) provided calibration (1990–1999) and validation data (2000–2009). Table 1 and Fig. 3 show best-fit parameters and model performance statistics. Model sensitivity to variation in T_w was assessed by removing MNPP thermal pollution as in Råman Vinnå et al. (2017) and repeating the calibration/validation for station no. 2085 (Table 1).

2.3.2 Suspended sediment concentration

Water density and intrusion depth of river water into downstream lakes is influenced by SSC. Intensive flow events create high levels of SSCs (Rimmer and Hartmann, 2014), as can exposure/erosion of sediment sources within the river basin through the so-called hysteresis effect, in which SSC varies for the same level of discharge (Tananaev, 2012). River

Table 1. air2stream river temperature model best-fit parameters and model performance statistics reported as coefficients of determination (R^2) and root mean square deviation (RMSD). Input parameters used in this study are shown in bold-faced type. The model was calibrated, validated and subjected to sensitivity tests using data from station no. 2085 (Aare River) representing past observed conditions and future predicted conditions assuming MNPP removal (No MNPP).

Parameter	Aare (no. 2085)		Rhône (no. 2009)
(unit)	Measurements	No MNPP	Measurements
a_1 (°C day^{-1})	**2.0316**	0.6434	**1.4927**
a_2 (day^{-1})	**0.2299**	0.3855	**0.2774**
a_3 (day^{-1})	**0.2267**	0.3177	**0.4133**
a_4 (−)	**0.0157**	0.5622	**0.6399**
a_5 (°C day^{-1})	**6.7022**	16.2387	**6.4792**
a_6 (°C day^{-1})	**4.4950**	9.9855	**2.3224**
a_7 (−)	**0.6066**	0.6066	**0.5244**
a_8 (day^{-1})	**0.7156**	1.5930	**1.0760**
	R^2 (−)		
Calibration[a]	0.97	0.96	0.94
Validation[b]	0.95	0.96	0.94
	RMSD (°C)		
Calibration[a]	0.83	0.95	0.52
Validation[b]	1.02	1.06	0.59

[a] 1990–1999; [b] 2000–2009

discharge regimes have been predicted to change in the future (Birsan et al., 2005), suggesting that SSCs will also change. To simulate future SSCs, we used the supply-based rating model described in Doomen et al. (2008), which Fink et al. (2016) adapted to the River Rhine.

The model consists of a base level SSC (g m^{-3}) function expanded to express erosion of sediment at high discharge and sediment accumulation at low discharge. The model is expressed as

$$\text{SSC}(t) = \tag{3}$$
$$m + b_1 Q(t)^{c_1} + d_1 d_2 b_2 (Q(t) - Q_{\text{th}})^{c_2} - b_3 (1 - d_2),$$

where b_x, c_x and m are adjustable parameters in combination with the threshold discharge (Q_{th}), which determines whether erosion or deposition occurs within the river. The parameters d_1 and d_2 control the deposition of (or erosion from) the river sediment storage (ψ (g)).

$$d_1 = \begin{cases} 0 : \psi = 0 \\ 1 : \psi > 0 \end{cases}, \tag{4}$$

$$d_2 = \begin{cases} 0 : Q \leq Q_{\text{th}} \\ 1 : Q > Q_{\text{th}} \end{cases}. \tag{5}$$

Erosion occurs if Q exceeds Q_{th} and the river basin contains erodible sediment ($\psi > 0$). Sedimentation occurs if Q

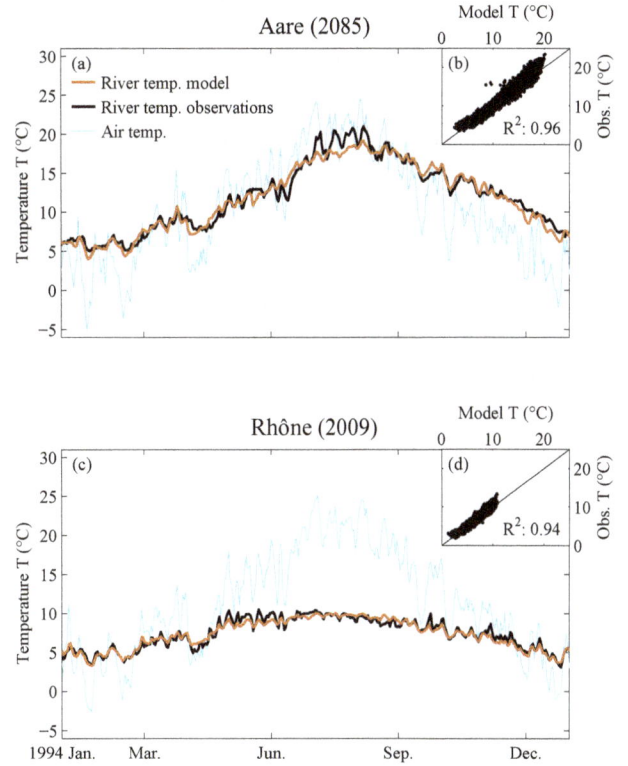

Figure 3. air2stream modeled (orange) and measured (black) temperature (T) compared to air temperature (blue) for **(a)** Aare River and **(c)** Rhône River in 1994. The insets **(b)** and **(d)** show modeled versus observed temperature from 1990 to 2009 with coefficient of determination (R^2).

is smaller than Q_{th}. The change in ψ over time can be formulated as

$$\frac{\Delta \psi}{\Delta t} = \left(b_3 (1 - d_2) - d_1 d_2 b_2 (Q(t) - Q_{\text{th}})^{c_2} \right) Q(t). \tag{6}$$

Parameters in Eqs. (3) to (6) were calibrated (2013) and validated (2014) through an evolutionary algorithm (Fink et al., 2016). Table 2 and Fig. 4 give model performance statistics and best-fit parameter values.

2.4 Lake model

We used the one-dimensional model SIMSTRAT (Goudsmit et al., 2002) to assess the impact of climate change on temperature and deep-water renewal in LB and LG. The model calculates heat fluxes and vertical mixing driven by wind and the internal wave field using a $k - \varepsilon$ turbulence closure scheme. It has been adapted to and validated for multiple lakes including Lake Zürich (Peeters et al., 2002), LG (Perroud and Goyette, 2010; Schwefel et al., 2016), Lake Neuchâtel (Gaudard et al., 2017), Lake Constance (Fink et al., 2014b; Wahl and Peeters, 2014) and LB (Råman Vinnå et al., 2017).

Table 2. River suspended sediment concentration (SSC) model best-fit parameters and model performance statistics reported as coefficients of determination (R^2) and root mean square deviation (RMSD).

Parameter (unit)	Aare (no. 2085)	Rhône (no. 2009)
m (g m^{-3})	8.8000	1.0000
b_1 (g day m^{-6})	0.2650	0.0006
c_1 (–)	0.6500	2.3200
b_2 (g day m^{-6})	0.0011	0.0010
c_2 (–)	2.3000	12.0000
b_3 (g m^{-3})	8.8000	2.0000
Q_{th} (m^3 day^{-1})	401	232
R^2 (–)		
Calibration[a]	0.20	0.74
Validation[b]	0.03	0.58
RMSD (g m^{-3})		
Calibration[a]	82	206
Validation[b]	217	222

[a] 2013; [b] 2014

Table 3. One-dimensional lake model SIMSTRAT best-fit parameters and model performance statistics reported as vertical volume-weighted averaged root mean square deviation (RMSD-V).

Parameter (unit)	Lake Biel	Lake Geneva
p_1 (–)	1.30	1.09
p_2 (–)	1.20	0.90
K (–)	0.70	1.40
q (–)	1.30	1.25
C_{Deff} (–)	0.0050	0.0020
C_{10} (–)	0.0016	0.0017
a_S (–)	0.0060	0.035
a_W (–)	0.0040	0.009
RMSD-V (°C)		
Calibration	0.73[a]	0.66[c]
Validation	0.68[b]	

[a] 1995–2004; [b] 2005–2015; [c] 1981–2012.

The model contains seven tunable parameters, including p_1 (irradiance absorption), p_2 (sensible heat flux) and K (vertical light absorption) for heat flux adjustments from the atmosphere to the lake. Momentum and kinetic energy transfer from the wind to internal waves is tunable by C_{10} (wind drag). The internal seiche energy balance can be adjusted through α (production), C_{Deff} (loss by bottom friction) and q (vertical distribution of turbulent kinetic energy). To include the effect of seasonally varying stratification strength, we followed Schwefel et al. (2016) and varied α: α_S for summer (April to September) and α_W for winter (October to March), where $\alpha_S > \alpha_W$. Here we used the best-fit parameter setup (Table 3) already established and validated for LG and LB by Schwefel et al. (2016) and Råman Vinnå et al. (2017).

Building upon the model developed by Råman Vinnå et al. (2017), we introduced an extended river intrusion scheme described in Appendix A1 (including sensitivity analysis). This scheme was chosen in order to include the effect of steep bathymetry on plume entrainment. Additionally, the robustness and simplicity of the intrusion scheme limits the uncertainty associated with more complex intrusion models including multiple parameters which can be hard to predict in the future. The entrainment of lake water into plunging underflows was modeled as proposed by Akiyama and Heinz (1984) with additional sedimentation of suspended load (Mulder et al., 1998; Syvitski and Lewis, 1992). The method addresses the transition of a homogenous open channel flow to a stratified underflow where entrainment and settling of sediment depend on bottom slope angle. The model scheme consists of (i) the homogenous region, where river water extends from the surface to the lake bed; (ii) the plunging region, where the plume separates from the lake surface and (iii) the underflow region, where the plume descends downslope while entraining surrounding water until it sepa-

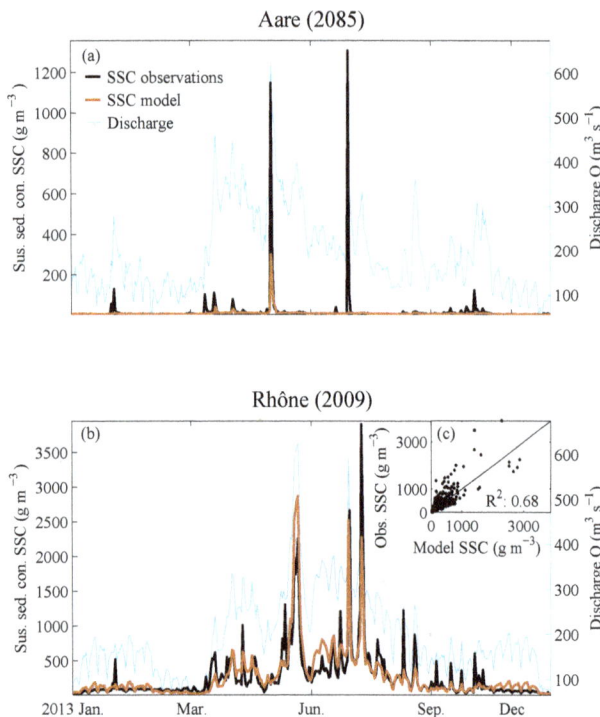

Figure 4. Modeled (orange) and measured (black) suspended sediment concentration (SSC) compared to river discharge Q (blue) for **(a)** Aare River and **(b)** Rhône River in 2013. The insert **(c)** shows modeled versus observed SSC for 2013 and 2014 in the Rhône River with coefficient of determination (R^2).

rates from the bottom and intrudes into the lake interior (Fink et al., 2016).

2.5 Data, hydrology and climate forcing

The models described above used hourly resolved data from 1989 to 2009 as inputs. For calibration/validation of river temperature, we used flow and temperature data from the Aare monitoring station no. 2085 (Fig. 2; 47°03′ N, 7°11′ E) and from the Rhône monitoring station no. 2009 (Fig. 2; 46°21′ N, 6°53′ E). The nearest meteorological stations, Mühleberg (no. 5530 in Fig. 2; 46°58′ N, 7°17′ E) for Aare and Aigle (no. 7970 in Fig. 2; 46°20′ N, 6°55′ E) for Rhône, provided air temperature data. Due to insufficient representation of high turbidity events, we calibrated/validated the SSC model with turbidity data converted to SSC with suspended sediment samples from 2013 and 2014.

The meteorological data used for SIMSTRAT included air temperature, vapor pressure, wind speed, solar radiation and cloud cover. These data were collected from the meteorological stations Cressier (no. 6354 in Fig. 2; 47°03′ N, 7°03′ E) for LB and Pully (no. 8100 in Fig. 2; 46°31′ N, 6°40′ E) for LG. Råman Vinnå et al. (2017) and Schwefel et al. (2016) provide additional information on climate data inputs to the one-dimensional model. The river intrusion scheme requires as input the slope angle traveled by the river underflow, which was obtained from a 25 m resolved digital height model (DHM25). Vertical temperature profiles, sampled at the deepest location of both lakes in January 1989, were used as initial conditions.

Van Vliet et al. (2013) suggested that river discharge and air temperature should be used while predicting future river temperatures. We incorporated recent findings of climate-induced changes in air temperature and river discharge regimes to model both future river temperature and SSC. Seasonal mean predictions for air temperature increase in western Switzerland (Fig. 2) were estimated from CH2011 (2011) for the A1B emission scenario (balanced use of renewable and fossil fuels) using results from 20 regional climate models. Flow projections were obtained from published results generated by the PREVAH (PREcipitation-Runoff-EVApotranspiration HRU Model) hydrological model (Viviroli et al., 2009) using a gridded configuration as described in Speich et al. (2015) and Kobierska et al. (2011). The model explicitly incorporates changes in glacial extent, snow melt, catchment runoff, floods and low water flows (FOEN, 2012; Bosshard et al., 2013; Speich et al., 2015). The PREVAH outcomes for the 1981–2009 period have been validated with data from 65 river gauges (Speich et al., 2015), including the two gauges upstream of LG (Rhône; no. 2009 in Fig. 2) and LB (Aare; no. 2085 in Fig. 2) used here.

2.6 Scenarios

Six different model scenarios were used to propagate climate change effects through the major tributaries and their associated downstream lakes. Model scenarios LG1 to LG3 represented LG while LB1 to LB3 represented LB (Table 4). Each scenario includes three time periods: a reference period (1990–2009), a near-future period (2030–2049) and a far-future period (2080–2099). The 20-year intervals allowed us to resolve natural variations on seasonal and shorter time scales. We initialized the models 1 year prior to the investigated period for each time frame (1989, 2029 and 2079) in order to remove effects of initial conditions.

Scenarios LG1 and LB1 excluded river inflow in order to isolate lake response to climate change from potential tributary influence. Scenarios LG2, LG3, LB2 and LB3 were used to differentiate between the effects of tributary temperature and SSC, and to provide model sensitivity estimates. The LB3 scenario excluded MNPP thermal pollution from near-future and far-future time periods but not from the reference period. The LB2 scenario included thermal pollution in modeling river water temperature. Scenarios LB2, LB3 and LG3 included SSC while LG2 did not. Low SSC values found in the Aare data resulted in negligible differences between models including and excluding SSC. Because they served primarily validation and sensitivity analysis purposes, the Aare/LB model results excluding particles and including/excluding MNPP thermal pollution (LB4 and LB5) are relegated to Appendix Fig. B1 and not discussed further. Scenarios LG3 and LB3 represent expected future developments.

The unmodified air temperature and modeled river discharge, temperature and SSC were used as inputs for the reference periods. Near-future and far-future models incorporated predicted changes in air temperature and river discharge, temperature and SSC with maximum, medium or mean, and minimum values serving as envelopes for each parameter (Figs. 2a and 5). This strategy gave nine simulations (three for scenarios LG1 and LB1, which exclude rivers; i.e., no variation in discharge nor river temperature) for each near-future and far-future time period. Predicted results included a total of 87 model runs. Upper, mean and lower impact estimates (described and interpreted below) were derived from the nine basic model runs.

3 Results

3.1 Rivers

The seasonality of predicted river discharge (Q) from FOEN (2012) varies with respect to the reference period 1990–2009 (Fig. 5a and b). The PREVAH model shows a future decrease in mean summer discharge (1 April to 30 September) for both the Aare ($-3.7 \, \text{m}^3 \, \text{s}^{-1} \, \text{decade}^{-1}$; no. 2085) and Rhône

Table 4. Model scenarios of climate change effects for near-future and far-future time periods, including (Inc.) and excluding (Exc.) the effects of rivers and suspended sediment. Thermal input from MNPP was also included/excluded. Most likely scenarios are shown in bold.

Lake	Exc. rivers	Inc. rivers	
		Exc. suspended sediment	Inc. suspended sediment
Geneva	LG1	LG2	**LG3**
		Inc. MNPP	Exc. MNPP
Biel	LB1	LB2	**LB3**

$(-3.8\,\mathrm{m^3\,s^{-1}\,decade^{-1}}$; no. 2009). The decrease in summer will be compensated by an observed increase in winter flow (1 October to 31 March) of the Aare $(+3.3\,\mathrm{m^3\,s^{-1}\,decade^{-1}})$ and Rhône $(+3.7\,\mathrm{m^3\,s^{-1}\,decade^{-1}})$. These results confirm previous findings presented by Addor et al. (2014) and Bosshard et al. (2013).

Regional air temperatures from the A1B emission scenario ($\sim +0.32\,°\mathrm{C\,decade^{-1}}$; CH2011, 2011; Fig. 2a) cause a predicted increase in mean annual water temperature (T) for both the Aare ($\sim +0.10\,°\mathrm{C\,decade^{-1}}$) and the Rhône ($\sim +0.08\,°\mathrm{C\,decade^{-1}}$). Both rivers experience seasonal variations in temperature increase similar to that predicted for air temperatures (Figs. 2a, 5e and f). The effect is strongest for the Aare during summer with warming of up to $+2.5\,°\mathrm{C}$ in water temperatures for the far-future time period relative to the reference period.

Thermal pollution from the MNPP in the Aare during the reference period (blue-green line in Fig. 5e; Råman Vinnå et al., 2017) causes approximately twice as much heating in winter relative to warming from climate change in the far-future. In summer, the relationship reverses with minor MNPP warming relative to that induced by climate change. The net effect of climate warming and MNPP decommission (i.e., removal of MNPP heat from near-future and far-future time periods) on the Aare is cooling in winter and warming in summer relative to the reference period (Fig. 5c).

Like river temperatures, SSCs depend on river discharge. Our model therefore show SSC increasing in winter and decreasing in summer due to shifts in the discharge regime (Fig. 5g and h). The model results for the Rhône exhibit a mean seasonal increase of $+14\,\mathrm{g\,m^{-3}\,decade^{-1}}$ in winter and a decrease of $-11\,\mathrm{g\,m^{-3}\,decade^{-1}}$ in summer. For reasons explained above (Sect. 2.2), results for the Aare show less variation, with a seasonal increase of $+0.3\,\mathrm{g\,m^{-3}\,decade^{-1}}$ in winter and a decrease of $-0.4\,\mathrm{g\,m^{-3}\,decade^{-1}}$ in summer. Altered temperature and SSC caused increases and decreases in water density for both rivers in winter and summer, respectively.

3.2 Lakes

Warmer air temperatures (Fig. 2a) predicted from climate change resulted in temperature increases in both LG and LB for all scenarios (Table 5). Models showed the highest warming rates in the epilimnion, intermediate values throughout

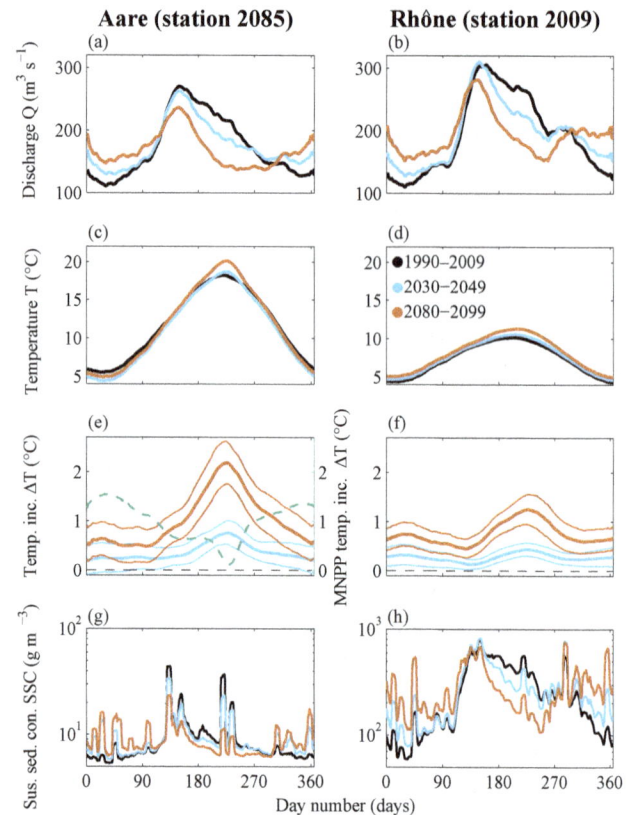

Figure 5. Modeled climate impact from scenarios LB3 (Aare River; **a, c, e, g**) and LG3 (Rhône River; **b, d, f, h**) displayed as daily average for reference (black, 1990–2009), near-future (blue, 2030–2049) and far-future (orange, 2080–2099) time periods. Discharge Q (**a** and **b**), net water temperature T (**c** and **d**) with anthropogenic heat from Mühleberg Nuclear Power Plant (MNPP) removed from near-future and far-future time periods, temperature increase ΔT (**e** and **f**) due to climate change (orange/blue), MNPP (blue-green) and modeled SSC (**g** and **h**). Maximum and minimum modeled values are marked by fine lines (**e** and **f**) and/or are omitted (**c, d, g,** and **h**) for clarity.

the metalimnion and the lowest rates in the hypolimnion (Table 5). We defined the epilimnion, metalimnion and hypolimnion using the water column stability method described in Råman Vinnå et al. (2017). The predicted warming of LG varied only slightly among the three different scenarios

Table 5. Change in temperature, length of the stratified period and depth of the thermocline (negative values correspond to a shallower thermocline) for each scenario listed in Table 4. Estimates given as mean of the daily difference between the reference period and the far-future time period. Temperature anomalies are volume-weighted and vertically averaged. Most likely scenarios are shown in bold.

Scenario	Temperature ($^\circ$C decade^{-1})			Stratification (days decade^{-1})	Thermocline (m decade^{-1})
	Epilimnion	Metalimnion	Hypolimnion		
Lake Biel					
LB1	0.19	0.16	0.13	1.5	−0.02
LB2	0.15	0.13	0.06	2.0	−0.07
LB3	**0.13**	**0.11**	**0.05**	**2.2**	**−0.13**
Lake Geneva					
LG1	0.17	0.13	0.07	2.9	−0.07
LG2	0.17	0.12	0.07	2.8	−0.06
LG3	**0.18**	**0.16**	**0.08**	**2.2**	**−0.04**

(Fig. 6a–c). Predicted warming of LB depends strongly on the scenario used (Fig. 6d–f).

Similar to the predicted warming patterns for rivers (Sect. 3.1), both lakes showed seasonally varying warming patterns. Reduced warming corresponds with periods of high river discharge (Fig. 5a and b). This cooling effect occurs primarily in winter and midsummer, and focussed in depth to the level of river intrusion (Figs. D1b, d and 7c–f). Model results showed a greater degree of fluctuations of the warming in LB than in LG. This probably results from the greater influence of the Aare on LB compared to that of the Rhône on LG, as LG has a longer hydraulic residence time. Scenario LB1, which excludes river intrusion, showed only limited seasonal variation in warming (Fig. C1c and e). According to these results, the closure of MNPP could offset climate-induced warming of LB by ~ 25 %.

Model results show that enhanced warming of the epilimnion relative to the hypolimnion strengthens stratification (Fig. 7g and h). This enhances the duration of stratification (for both lakes $\sim +2$ days decade^{-1}; Table 5) and slightly lifts the thermocline (in LB ~ -0.1 m decade^{-1} and in LG ~ -0.05 m decade^{-1}; Table 5). We used the Schmidt (1928) stability (S) to estimate the strength of stratification (J m^{-2})

$$S = \frac{g}{A_0} \sum_{z=0}^{z_{max}} (z - z_m)(\rho(z) - \rho_m) A(z) \Delta z. \qquad (7)$$

Equation (7) incorporates gravity ($g = 9.81$ m s^{-2}), depth (z), lake surface area (A_0), horizontal cross section area ($A(z)$), lake density ($\rho(z)$), maximum depth (z_{max}), mean lake density (ρ_m), lake volume (V) and volumetric mean depth (z_m) defined as

$$z_m = \frac{1}{V} \sum_{z=0}^{z_{max}} z A(z) \Delta z. \qquad (8)$$

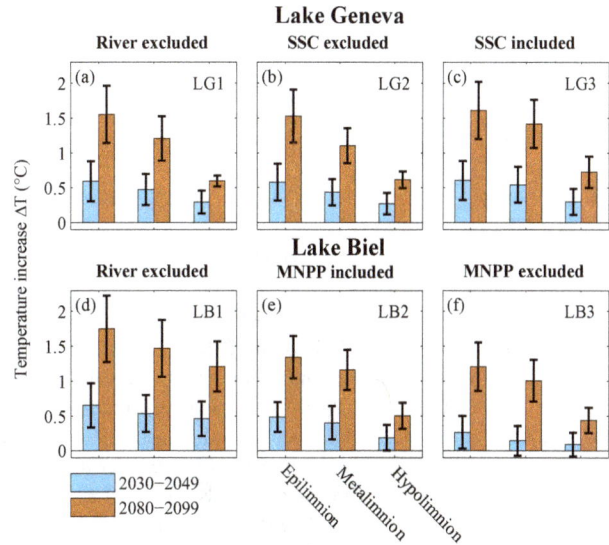

Figure 6. Temperature increase ΔT for near-future (blue) and far-future (orange) time periods relative to reference period temperatures, displayed as mean (columns) and standard deviation (black bars) calculated from the nine basic model runs in the near-future and far-future scenarios. Epilimnion (left pair of columns), metalimnion (middle pair) and hypolimnion (right pair) in LG (**a** to **c**) and LB (**d** to **f**). Graphs represent river intrusion excluded (**a** and **d**), river-borne SSC included (**c**, **e** and **f**) and excluded (**b**). Mühleberg Nuclear Power Plant (MNPP) heat release included in (**e**) and excluded in (**f**) from near-future and far-future time periods but retained for the reference period.

The duration of stratification was determined by counting the days when temperature differed by more than 1 °C between surface (2 m depth) and deep-water (280 m for LG; 50 m for LB) (Foley et al., 2012). The maximum water column stability expression $N^2 = -(g/\rho)\Delta\rho(z)/\Delta z$ (s^{-2}) was used to determine the thermocline depth.

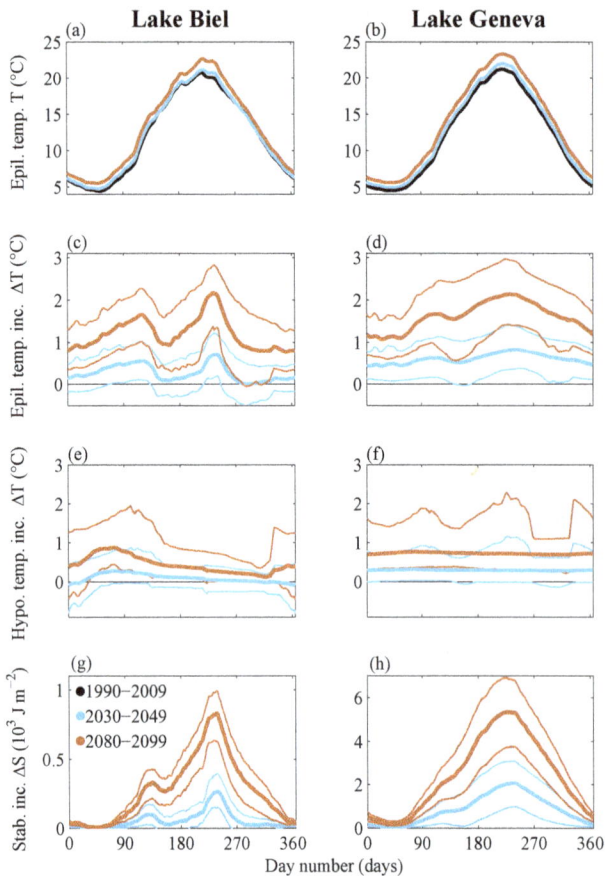

Figure 7. Modeled climate impact from scenarios LB3 (LB; **a**, **c**, **e**, **g**) and LG3 (LG; **b**, **d**, **f**, **h**) displayed as daily mean (thick lines) and maximum/minimum model values (thin lines) for near-future (blue, 2030–2049) and far-future (orange, 2080–2099) relative to the reference period (black, 1990–2009). Anthropogenic MNPP heat input entering LB has been excluded from near-future and far-future time periods but retained for the reference period. Temperature T (**a** and **b**), increase in temperature ΔT in epilimnion (**c** and **d**) and hypolimnion (**e** and **f**) as well as increase in stability ΔS (**g** and **h**).

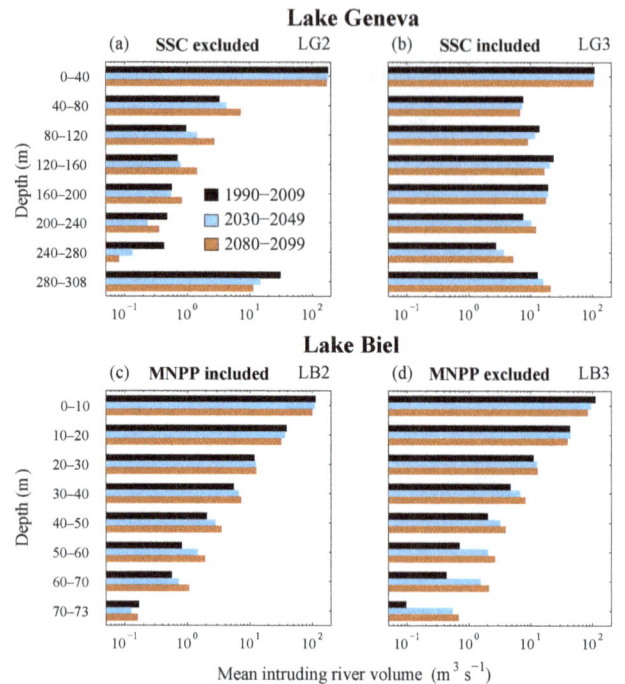

Figure 8. Modeled climate impact on intruding river volumes. Reference (black), near-future (blue) and far-future (orange) time periods for LG (**a**, **b**) and LB (**c**, **d**), including (**b–d**) and excluding (**a**) river-borne SSC and MNPP heat input included in (**c**) or excluded (**d**) from near- and far-future time periods but retained in the reference period.

The river intrusion depth is dependent on water density (temperature and SSC are dominant; dissolved solids are negligible). The Rhône is colder (Fig. 5c and d) and carries more suspended sediment (Fig. 5g and h) than the Aare. Reference period results showed that the Rhône intruded in LG at greater depths relative to depths of the Aare intrusion into LB (Figs. 8 and D1). Given the future change in river temperature and SSC, intrusion patterns will thus change as the densities of both the Aare and Rhône increase and decrease during respective winter and summer seasons (Sect. 3.1). This explains model results showing respective deeper and shallower intrusions during winter and summer for both rivers (Fig. D1).

Model results show that warming of the Rhône generally diminishes the amount of river water penetrating beyond

200 m depth in LG (Fig. 8a). Enlarged river flow in winter enhances SSCs and counteracts heating, thereby increasing the amount of river water intruding beyond 200 m depth (near-future ∼ 30 %; far-future ∼ 65 %; Fig. 8b). The difference in winter heating for the Aare and LB epilimnion (Figs. 5c, e and 7c) generally increased the amount of water penetrating into the hypolimnion (Fig. 8c). Decommission of the MNPP enhances temperature differentials between LB and the Aare, thereby increasing the amount of water reaching past 30 m in LB (near-future + ∼ 80 %; far-future + ∼ 120 %; Fig. 8d). In summary, the change in river discharge regime for the Aare and Rhône results in respective increase and decrease in winter and summer water density, resulting in a summer to winter shift of the amount of river water penetrating deeper than the metalimnion for both lakes.

4 Discussion

4.1 Rivers

Increases in air temperature expected from climate change modify tributary runoff. Less water is predicted to be bound in snow and ice at high elevation during winter and spring/summer floods will occur earlier (CH2011, 2011;

FOEN, 2012). The changed river discharge regime, appearing as increased flow in winter and decreased flow in summer (Fig. 5a and b), amplifies the increase and decrease in river temperature during respective summer and winter periods (Fig. 5e and f). Amplification results from (i) a smaller flow volume requiring less energy to heat and (ii) lower flow velocities which extend heat exposure. The PREVAH model predicts that the future discharge of the Aare in summer will be $\sim 20\%$ less than summer discharge in the Rhône. These results therefore suggest that the Aare summer conditions will be more impacted by climate change than the Rhône summer conditions (Fig. 5e and f).

Model results concerning discharge-dependent responses to climate-induced warming were consistent with previous findings reported by Isaak et al. (2012) and van Vliet et al. (2013). The river temperature increases predicted by this study ($+0.10\,°C\,decade^{-1}$ for the Aare and $+0.07\,°C\,decade^{-1}$ for the Rhône) were much smaller than past observed warming rates ($0.34\,°C\,decade^{-1}$ for the Aare and $0.21\,°C\,decade^{-1}$ for the Rhône; Hari et al., 2006). These differences may reflect contrasting reference periods with past observations conducted from 1971 to 2001 and modeled observations addressing 1990 to 2099. Past observations also incorporate effects of solar brightening during the 1980s (Fink et al., 2014a; Sanchez-Lorenzo and Wild, 2012; Wild et al., 2007), which led to additional warming of air and water.

Climate change effects aside, MNPP decommissioning in 2019 is predicted to decrease the temperature in the Aare by up to 4.5 °C at station no. 2085 (Råman Vinnå et al., 2017). The cooling effect of this plant closure primarily affects winter conditions when climate-change-induced warming is weaker and river flow is lower (Fig. 5e). The heating of the Aare and LB by MNPP heat emissions equates to approximately 1 decade of climate-induced warming of lake surface waters (O'Reilly et al., 2015; Råman Vinnå et al., 2017). This result highlights the role of point source thermal contributions in local climate impact assessments.

The amount of suspended sediment carried by rivers depends on both discharge and the amount of erodible sediment in the watershed (Fink et al., 2016). We used a supply-based sediment rating model subjected to a changing discharge regime to examine seasonal changes in suspended sediment for both the Aare and Rhône (Fig. 5g and h). Consistent with previous findings reported by Pralong et al. (2015), we predict an increase in SSC during winter and decrease in SSC in summer. This is caused by two phenomena associated with increased river discharge: (i) amplified river bed erosion linked to increased intensity of high discharge events carrying enhanced volumes of SSC and (ii) increase in the sediment available for erosion in the river catchment due to enhanced supply at low flow velocities.

Figure 4 and Table 2 show that the SSC model gives robust results for the Rhône (coefficient of determination $R^2 = 0.68$ from 2013 to 2014) but not for the Aare ($R^2 = 0.06$ from 2013 to 2014). The Aare includes several sediment-trapping reservoirs/lakes upstream of station no. 2085. Thus, peaks in SSC at station no. 2085 do not reflect watershed-scale discharge events (Fig. 4) but rather local precipitation and discharge events in the headwaters of a tributary (Saane River) to the Aare (Fig. 2). This tributary hosts few sediment traps and contributes $\sim 34\%$ of the downstream flow at station no. 2085. Given the limited impact of SSC on the Aare water density, models show only negligible impact on river intrusion depth and corresponding intruding volumes (Figs. 8c, B1c, e and D1c). The lower reaches of the Rhône are not dammed, thus adhering more directly to model assumptions and giving clearer results (Fig. 4).

High SSC events are usually associated with extreme floods (Fink et al., 2016), which are predicted to vary in alpine lake catchments with on-going climate change (Glur et al., 2013). The lack of constraints on extreme precipitation events introduces uncertainty into future flood frequency and magnitude predictions (CH2011, 2011). Shifts in river discharge regimes also depend on the amount of water bound in snow and ice as well as on the timing of spring/summer melt. Future climate scenarios predict that $\sim 30\%$ of the glacier mass will remain in the Aare and the Rhône catchments by the end of the 21st century (FOEN, 2012). Glacial meltwater is thus expected to continue to supply the Aare and Rhône throughout the time frames considered in this study. We thus assumed that the flood frequency remained unchanged, while the amplitude of the floods was adjusted in the future according to river discharge regime shifts predicted by FOEN (2012).

4.2 Lakes

All model scenarios showed that increased air temperature leads to warming of both lakes, especially of the epilimnion (Table 5, Fig. 6). Piccolroaz et al. (2015) showed that an increase in lake stability and earlier onset of stratification causes warming of surface waters due to the smaller volume undergoing warming and diminished heat transfer to the hypolimnion. The lake model used here showed an increase in stratification strength and a lengthening of the stratified period in both lakes (Table 5; Fig. 7g and h). Our results thus consistently support previous findings for LG reported by Foley et al. (2012), O'Reilly et al. (2015) and Schwefel et al. (2016).

Seasonal variations in warming of both epilimnion and hypolimnion (Fig. 7a–f) surpassed the seasonality of applied changes in air temperature (Fig. 2a). The model showed a decrease in warming during winter and midsummer, which corresponds to time periods of high river discharge from the main tributaries (Fig. 5a and b). This cooling effect was more effective for LB than for LG (Fig. 7) and appeared in all scenarios except for LB1 and LG1 (Fig. C1), both of which exclude coupled river effects. The extended seasonal variation in climate warming is thus driven by river discharge volume and temperature trends (Figs. 5 and 7). This response applies to aquatic systems in which a difference exists in temperature and heating regimes between rivers and lakes, but does not appear to affect water bodies with uniform temperature/heating regimes. Our results thus supports the hypothesis put forward by Zhang et al. (2014), stating that climate warming of lakes might be reduced and even reversed by addition of external water.

To investigate this effect, we varied the hydraulic residence time of LB and LG, while holding all other factors constant (Fig. 9). We implemented a stepwise reduction in LG size (to 1/80 of its original volume), simultaneously reducing hypsographic area but keeping maximum depth unchanged. Similar adjustments were made to LB to obtain corresponding hydraulic residence times. This stepwise approach required 972 additional model runs. These iterations showed that river water had to be cooler than lake water in order to generate a dampening effect for climate warming (Fig. 9a and d). Deep penetrations by large riverine volumes increase the cooling of the hypolimnion (Fig. 9b). The climate dampening effect is suppressed when river and epilimnion temperature are similar. MNPP thermal input creates such conditions in the Aare and therefore largely counteract the river cooling effect of the Aare on LB (Fig. 9c). For shorter residence times ($<\sim 1000$ days), rivers can exert influence if a significant temperature difference exists between river and lake waters. For longer residence times ($>\sim 1000$ days), tributaries cannot significantly offset climate effects in downstream water bodies.

Climate-induced warming of lakes (Schwefel et al., 2016), along with changing frequency or intensity of deep penetrating flood events (Fink et al., 2016) may curtail oxygen supply to deep lakes. Recent flood analysis has also indicated that input of river-borne organic matter increases respiration, causing a paradoxical net oxygen reduction within the intruding layer (Bouffard and Perga, 2016). Models showed respective winter increase and summer decrease in river water density relative to lake stability. This creates summer to winter seasonal shifts in deep intrusion dynamics for both lakes (Fig. D1), causing a net annual increase in the river water penetrating to deeper parts of both lakes (Fig. 8). An increase in Rhône SSC in winter represented the primary driver in LG (Figs. 8a, b and D1a, b), while the dominant factor in LB was Aare river temperature, which cooled in winter by increased discharge and removal of MNPP heat (Figs. 8c, d and D1c, d).

Fink et al. (2016) also found evidence that climate change will cause diminished deep river intrusion events in summer and enhanced intrusion in winter. They predicted an annual decrease in the amount of river water reaching the deepest parts of Lake Constance. The tributaries considered here differ from the Rhine River investigated by Fink et al. (2016) primarily in terms of temperature. The Rhône catchment for example rests at a mean elevation of 2127 m and includes greater glacial coverage (11 %), whereas the Rhine

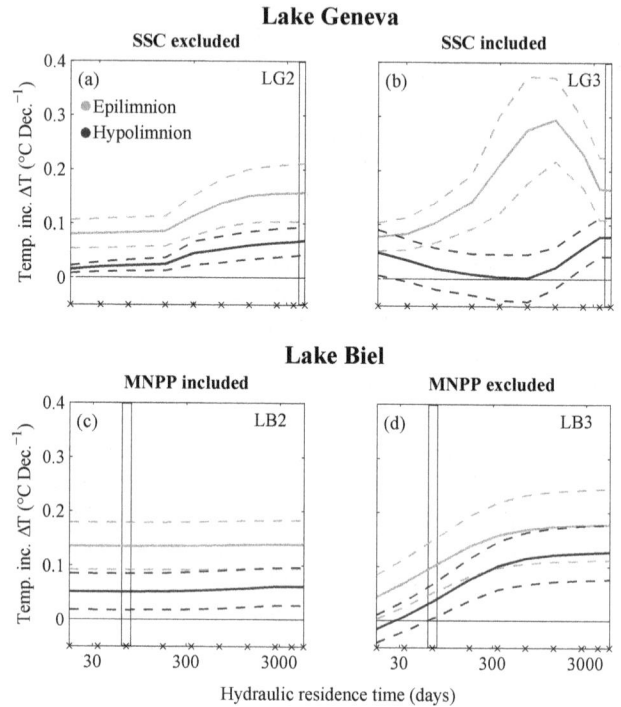

Figure 9. Variation in lake hydraulic residence times (changed lake volume) in response to modeled temperature increase (ΔT) in the epilimnion (grey) and hypolimnion (black) displayed as decadal mean (solid line) and standard deviation (dotted line) for LG (**a, b**) and LB (**c, d**). River-borne SSC included (**b**) and excluded (**a, c** and **d**), MNPP heat input included in (**c**) and excluded (**d**) from near-future and far-future time periods but retained for the reference period. Black x's mark modeled lake residence times, while full-height black rectangles mark the lakes' present-day residence times.

catchment has a mean elevation of 1771 m and only 1 % glacial coverage (www.hydrodaten.admin.ch). The closure of the MNPP and associated temperature decrease contribute to increase the volume/frequency of deep intrusions (Fig. 5). While Fink et al. (2016) focused primarily on flood frequencies, our models emphasized river discharge regimes and interacting river and lake temperature regimes. The annual increase in river penetration to depth predicted by our models suggests future increase in deep-water oxygen supply in similar tributary–lake systems. This prediction applies mainly to meromictic lakes such as LG. Analogous effects in holomictic lakes such as LB, which mix completely each winter, are less significant. Similar to findings of Fink et al. (2016), our models indicate that deep-water oxygen conditions will worsen during strongly stratified conditions due to seasonal shifts in deep river intrusions from summer to winter. Concluding, as river water density increases in winter, the volume of those intrusion events, which occurred in the reference period, will increase in the future. Likewise, high discharge events, which were previously unable to penetrate into the deep, are likely to do so in the future.

4.3 Model reliability

Predictions concerning the effect of climate change on rivers and lakes depend on (i) the choice of emission scenario, (ii) the accuracies of models linking climate to hydrology and climate to heat fluxes and (iii) natural variability of the system being investigated (Raymond Pralong et al., 2015). This section describes uncertainties and limitations of our approach.

Results of long-term forecasts (beyond 2050) depend strongly on representations of global greenhouse gas emission scenarios (FOEN, 2012). Given the uncertainties in future global climate policy, we chose a median scenario, which falls between the best (e.g., RCP3PD) and the worst case scenarios (e.g., A2) in terms of greenhouse gas emissions. A1B assumes a peak population at mid-century, balanced use of renewable and fossil fuels and rapid introduction of new technologies.

Estimates of future air temperatures and river discharge were obtained from a combination of regional climate models (RCMs; CH2011, 2011; FOEN, 2012). Uncertainties associated with individual RCMs were offset by combined forecasts from multiple-model chains. Numerous studies have performed detailed evaluations of uncertainty in air temperature and river discharge under established emission scenarios (RCP3PD, A1B, A2) and accounting for global–regional climate model interactions (Addor et al., 2014; Bosshard et al., 2011, 2013; CH2011, 2011).

The degree of accuracy with which model input parameters represent future conditions determines the accuracy of model predictions. We therefore ran the river temperature model with varying parameters to evaluate model sensitivity (Table 1) for different yet similar datasets. The air2stream parameter a_1 showed the greatest degree of sensitivity, varying within 3 orders of magnitude. The a_1 parameter, however, does not respond to variations in river discharge or air temperature (Eq. 1), which limits its sensitivity to climatic input data. The other parameters (a_2 to a_8) varied only within 1 order of magnitude (Table 1). The SSC model gives better results for the Rhône (coefficient of determination $R^2 = 0.68$ from 2013 to 2014) than for the Aare ($R^2 = 0.06$ from 2013 to 2014). Dams and reservoir infrastructure upstream of station no. 2085 along the Aare dampen sediment transport events and decouple them from regional discharge events (see above; Fig. 4). Given the relatively minor effect of SSC on the Aare water density, variation in the input parameter does not influence river intrusion depths (Figs. B1e–f and D1c–d). As with other vertical, one-dimensional models, SIMSTRAT cannot account for lateral heterogeneities in lakes. This inherent weakness in model design, however, does not significantly diminish the accuracy of model pre-

dictions concerning LB and LG (Råman Vinnå et al., 2017; Schwefel et al., 2016).

Of special importance for climate research in lakes is the sensitivity of models to shifts in the heat budget. Forcing parameters of importance, besides air temperature, include wind speed, solar irradiance, vapor pressure and light absorption. The sensitivity of SIMSTRAT to variable forcing has previously been established for lakes in Switzerland. Schmid and Köster (2016) demonstrated how solar brightening from 1981 to 2013 increased Lake Zürich surface warming comparable to heating by increased air temperature. Schwefel et al. (2016) revealed strengthening of the thermocline and decrease in the mean lake temperature by increased light absorption in LG, whereas a decrease in absorption had the reverse effect. As of yet, reliable predictions of wind speed, irradiance and vapor pressure under future climate conditions are not available for Switzerland (CH2011, 2011). Therefore, we use long-term (1981 to 2013) data from station no. 8100 (Fig. 2) as guidance for potential annual atmospheric forcing trends (Fig. A6; Table 6).

The sensitivity of SIMSTRAT was tested in LG by applying these trends, individually and combined, to the reference period. The increasing trend in air temperature was included for comparison, while no trend could be identified in cloud cover which was excluded. The decreasing trend in wind speed cooled the lake, while the increasing trend in irradiance and vapor pressure heated the lake comparable to air temperature (Table 6). By combining all trends, we obtained similar warming of the LG epilimnion ($\sim +0.38\,^\circ\mathrm{C\,decade^{-1}}$) as observed over land ($+0.38\,^\circ\mathrm{C\,decade^{-1}}$; 1985–2002; Wild et al., 2007) and globally in lakes ($\sim 0.34\,^\circ\mathrm{C\,decade^{-1}}$; 1985–2009; O'Reilly et al., 2015) as well as monitored in LG surface waters ($\sim +0.51\,^\circ\mathrm{C\,decade^{-1}}$, 1983–2000; Gillet and Quétin, 2006). The historical effect of increased air temperature caused ~ 40 to $\sim 70\,\%$ of the heating in the epilimnion/metalimnion and $\sim 240\,\%$ in the hypolimnion.

Here we include predictions of future temperature and precipitation. The extrapolation of observed atmospheric trends into the future is outside the scope of the present study. Yet, we expect our lake water temperature predictions for the near-future and far-future scenarios to underestimate the total heating in shallow water and overestimate warming of deep-water. Nonetheless, the solar brightening trend observed over Switzerland from 1980 to 2000, caused by a decrease in atmospheric aerosols, will not continue into the future (Sanchez-Lorenzo and Wild, 2012), thereby reducing the uncertainty of our predictions.

In this study we assumed that glacial meltwater feeding both the Aare and Rhône in summer will not disappear within the time frames considered. Loss of glacial sources would modify the discharge regime, especially in summer, which would affect accuracy of temperature, SSC and intrusion depth estimates. However, as stated in Sect. 4.1, FOEN (2012) predicts that the Aare and Rhône catchments will retain 30 % of their glacial masses by the year 2100.

Table 6. Observed trends in atmospheric forcing (1981 to 2013) at station no. 8100 (Fig. 2) per decade, and modeled temperature increase in Lake Geneva (LG) with forcing trends applied to the reference period (1990 to 2009).

Parameter	Observed atmospheric trend	Modeled LG temperature change ($°C$ decade^{-1})		
		Epilimnion	Metalimnion	Hypolimnion
Wind speed	-0.097 (m s^{-1} decade^{-1}); -5.7 (% decade^{-1})	-0.022	-0.149	-0.089
Shortwave irradiance	3.8 (W m^{-2} decade^{-1}); 2.6 (% decade^{-1})	0.134	0.131	0.027
Vapor pressure	0.26 (mbar decade^{-1}); 2.6 (% decade^{-1})	0.122	0.085	0.050
Air temperature	0.40 ($°C$ decade^{-1}); 3.7 (% decade^{-1})	0.149	0.101	0.017
All combined		0.379	0.147	0.007

These predictions support assumptions concerning the Aare and Rhône discharge regimes used here. Point sources/sinks of anthropogenic heat can affect inland water bodies response to climate change, as shown by the MNPP effects described here. Other changes in catchment management, such as hydropower damming would also alter river discharge regimes and by extension, temperatures, SSCs and deep-water renewal (Fink et al., 2016). Thus, the correctness of future climate change predictions depends on adequate accounting of regional anthropogenic factors affecting physical processes in the system under investigation.

5   Conclusion

Aquatic processes in lakes are the result of regional forcing and the upstream catchment environment. This study investigated the impact of climate change on inland waters by propagating climatic inputs through integrated fluvial-lacustrine systems. We fed predicted future climatic data into models for two connected river and lake systems in order to evaluate downstream thermal responses and how river discharge regime shifts might affect deep-water renewal in the lakes. Climate data propagated through discharge-dependent river temperature and SSC models, which were coupled to a one-dimensional lake model. We applied this approach for the two peri-alpine lakes Biel and Geneva.

The models showed that climate warming of rivers is enhanced in summer and diminished in winter due to future river discharge regimes with decreased flow in summer and increased flow in winter. This climate-caused alteration of the flow regime likewise increase and decreases the river-borne suspended sediment load in winter and summer, respectively.

Both lakes showed large seasonal temperature increases that could not be solely explained by climate-related (predicted) increases in air temperature. Instead, the lakes experienced a cooling effect associated with upstream tributaries, where responses to increasing future air temperatures differed from that of the lakes. The smaller Lake Biel showed

stronger response to this repressive effect of climate warming compared to the larger Lake Geneva. Predicted changes in Lake Biel strongly depend on the removal of upstream anthropogenic thermal emission into the Aare River. Local anthropogenic point sources of heat can thus rival climate change in their influence on lakes. This damping of climate warming depends on the lakes hydraulic residence times and requires adequate river/lake temperature differences. Our models indicate that tributaries can exert a system-wide influence on lakes with hydraulic residence times less than ~ 1000 days. Lake systems with longer residence times are resistant to tributary effects but may respond on a local level.

The combination of changes in river SSC and differential lake/river temperature/warming result in a seasonal shift of deep-water penetration (by rivers) into lakes. The volume of river water penetrating to deeper parts of lakes specifically decreases in summer and increases in winter. Higher rates of deep-water renewal can in turn enhance reoxygenation of the deepest reaches of lakes, which may otherwise experience lower oxygen concentrations under climate change.

Data availability. CTD profiles are available from the Office of Water Protection and Waste Management of the Canton of Bern (GBL/AWA) at: http://www.bve.be.ch/bve/de/index/wasser/wasser/messdaten.html (GBL/AWA, 2017). Meteorological data are available from the Swiss Federal Office of Meteorology and Climatology (MeteoSwiss) at: http://www.meteoswiss.admin.ch/home/services-and-publications/beratung-und-service/data-portal-for-teaching-and-research.html (MeteoSwiss, 2017). Tributary data are available from the Hydrology Department of the Swiss Federal Office for the Environment (FOEN) at: www.hydrodaten.admin.ch (FOEN, 2017a).

Future river discharge predictions are available from the Climate Change and Hydrology in Switzerland (CCHydro) project at: http://www.bafu.admin.ch/umwelt/index.html?lang=en (FOEN, 2017b). DHM25 model bathymetry data are available from the Swiss Federal Office of Topography (SwissTopo) at: https://shop.swisstopo.admin.ch/en/products/height_models/dhm25 (SwissTopo, 2017).

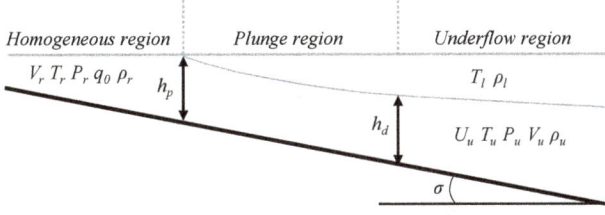

Figure A1. Illustration of river intrusion model.

Appendix A: River intrusion model

Figure A1 summarizes the river intrusion model. The depth where the river plume separates from the surface, the so-called plunge depth (h_p), depends on the slope angle (σ), gravity (g), coefficients ($S_1 = 0.25$, $S_2 = 0.75$; Ellison and Turner, 1959), bed friction ($f_t = 0.02$; Akiyama and Heinz, 1984), initial flow per unit width (q_0) $= V_r / W_r$ dependent on river discharge (V_r), river width ($W_r = 100$ m for the Aare and $W_r = 120$ m for the Rhône) and the relative density difference ($\rho' = (\rho_r - \rho_l(z_1))/\rho_l(z_1)$) between the homogenous river (ρ_r) and lake (ρ_l) with $z_1 =$ surface.

$$h_p = e_1 \left(\frac{f_t}{\sigma(z) S_2} \frac{q_0^2}{g \rho'} \right)^{1/3} + e_2 \left(\frac{q_0^2}{S_1 g \rho'} \right)^{1/3} \tag{A1}$$

The level of initial plume entrainment is treated differently on a gentle versus a steep slope. This is controlled by the two coefficients e_1 and e_2.

$$e_1 = \begin{cases} 1 : \sigma(z_1) < \sigma_c \\ 0 : \sigma(z_1) \geq \sigma_c \end{cases}, \tag{A2}$$

$$e_2 = \begin{cases} 0 : \sigma(z_1) < \sigma_c \\ 1 : \sigma(z_1) \geq \sigma_c \end{cases}, \tag{A3}$$

where the critical slope angle $\sigma_c = f_t S_1 / S_2$ distinguishes between gentle and steep slope designations. The initial height of the underflow (h_d) can then be written as

$$h_d(z_1) = h_p (1 + \gamma), \tag{A4}$$

where γ is the entrance mixing coefficient equal to ~ 0 for gentle slopes and increasing to larger values for steeper slopes. Here we find that a value of $\gamma = 0.1$ provides best results. The initial underflow temperature (T_u), velocity (U_u), particle content (P_u) and volume (V_u) is consequently expressed as a function of ambient lake water temperature (T_l), river temperature (T_r) and river particle content (P_r) in the homogenous region.

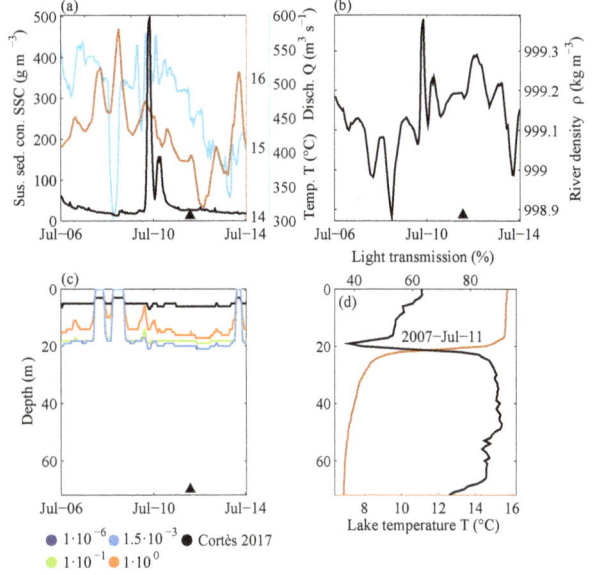

Figure A2. River intrusion entrainment sensitivity analysis for the LB/Aare system in July 2007. **(a)** SSC (black), temperature (T; orange) and river discharge (Q; blue) from Aare station no. 2085. **(b)** River density at station no. 2085 obtained from T and SSC in **(a)**. **(c)** River intrusion depth calculated from Appendix A with varying entrainment constant β (Eq. A13); light green denotes the value used in this study; black indicate intrusion depth modeled as in Cortés et al. (2014). **(d)** Vertical measurements of T and light transmission in LB for 11 July 2007. The triangles in **(a)–(c)** mark the time of the vertical profile in **(d)**.

$$T_u(z_1) = T_l(z_1) \frac{(h_d(z_1) - h_p)}{h_d(z_1)} + T_r \frac{h_p}{h_d(z_1)}, \tag{A5}$$

$$U_u(z_1) = (1 + \gamma) \frac{q_0}{h_d(z_1)}, \tag{A6}$$

$$P_u(z_1) = P_r \frac{h_p}{h_d(z_1)}, \tag{A7}$$

$$V_u(z_1) = V_r \frac{h_p}{h_d(z_1)}. \tag{A8}$$

Once the plume has passed through the plunge region into the underflow region, we express h_d, U_u, T_u and V_u as

$$h_d(z + 1) = h_d(z) + E(z) \Delta x, \tag{A9}$$

$$U_u(z + 1) = U_u(z) \frac{h_d(z)}{h_d(z + 1)}, \tag{A10}$$

$$T_u(z + 1) = T_l(z) \frac{(h_d(z + 1) - h_d(z))}{h_d(z + 1)} + T_u(z) \frac{h_d(z)}{h_d(z + 1)}, \tag{A11}$$

$$V_u(z + 1) = V_u(z) \frac{h_d(z)}{h_d(z + 1)}. \tag{A12}$$

where Δx is the horizontal distance between z and $z + 1$ and the entrainment factor (E) is expressed as a function of

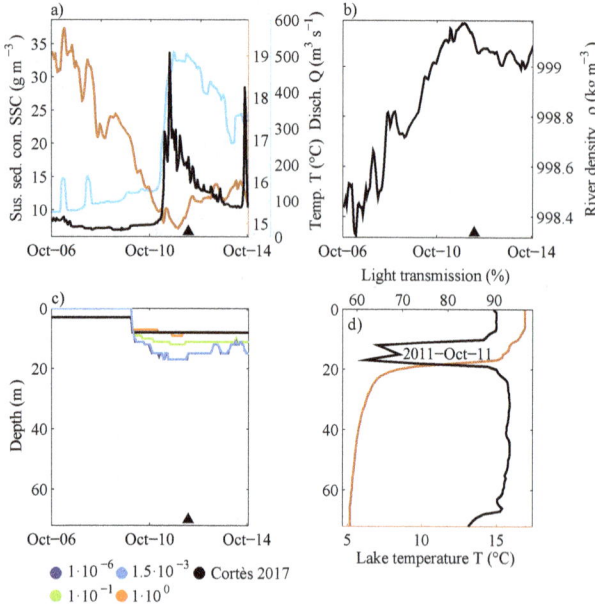

Figure A3. River intrusion entrainment sensitivity analysis for the LB/Aare system in October 2011. **(a)** SSC (black), temperature (T; orange) and river discharge (Q; blue) from Aare station no. 2085. **(b)** River density at station no. 2085 obtained from T and SSC in **(a)**. **(c)** River intrusion depth calculated from Appendix A with varying entrainment constant β (Eq. A13); Light green denotes the value used in this study; black indicate intrusion depth modeled as in Cortés et al. (2014). **(d)** Vertical measurements of T and light transmission in LB for 11 October 2011. The triangles in **(a)**–**(c)** mark the time of the vertical profile in **(d)**.

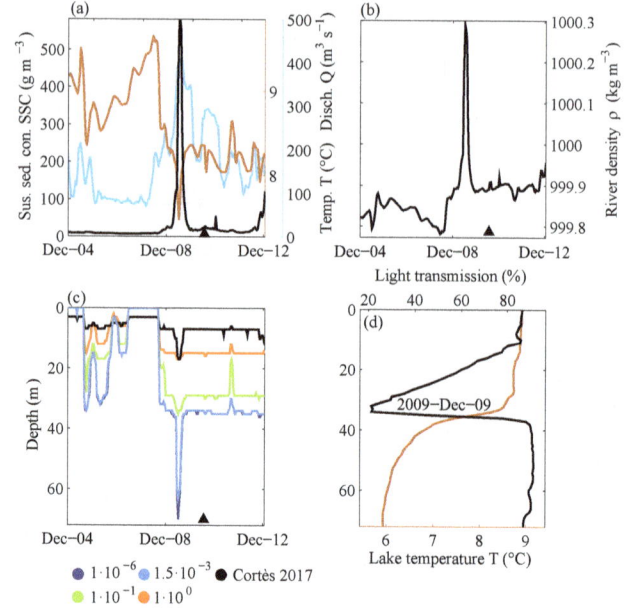

Figure A4. River intrusion entrainment sensitivity analysis for the LB/Aare system in December 2009. **(a)** SSC (black), temperature (T; orange) and river discharge (Q; blue) from Aare station no. 2085. **(b)** River density at station no. 2085 obtained from T and SSC in **(a)**. **(c)** River intrusion depth calculated from Appendix A with varying entrainment constant β (Eq. A13); light green denotes the value used in this study; black indicate intrusion depth modeled as in Cortés et al. (2014). **(d)** Vertical measurements of T and light transmission in LB for 9 December 2009. The triangles in **(a)**–**(c)** mark the time of the vertical profile in **(d)**.

the entrainment constant ($\beta = 0.0015$; Ashida and Egashira, 1975) and the Richardson number (R_i).

$$E(z) = \frac{\beta}{R_i(z)}, \tag{A13}$$

$$R_i(z) = \frac{f_t}{\sigma(z)S_2}. \tag{A14}$$

For P_u, we include a sedimentation term as proposed by Syvitski and Lewis (1992), which depends on the removal rate (r) and $\Delta t = \Delta x / U_u(z)$.

$$P_u(z+1) = \frac{h_d(z)}{h_d(z+1)}\left(P_u(z) - re_3 P_u(z)e^{-r\Delta t}\Delta t\right) \tag{A15}$$

Sedimentation occurs only if the plume velocity drops below a critical settling velocity (U_c) subject to the parameter e_3:

$$e_3 = \begin{cases} 1 : U_u(z) < U_c \\ 0 : U_u(z) \geq U_c \end{cases}. \tag{A16}$$

We set U_c equal to 0.46 m s^{-1} and r equal to 4.7 day^{-1} to represent medium-sized silt following Mulder et al. (1998). The plume travels downslope as long as the underflow plume

density (ρ_u) exceeds $\rho_l(z)$. Once $\rho_u \leq \rho_l(z)$, the plume raises from the slope and intrudes into the lake proper. The terms T_u and V_u were thus added to the lake model at this depth. Calculations excluded expressions for the settling of accumulated particles following plume intrusion, assuming that these exert only minor impacts on lake temperature and density.

The sensitivity of the river intrusion depth to entrainment of ambient water into the plume was tested by propagating a range of β (Eq. A13) values from 1 to 1×10^{-6} through model spaces composed of temperature, discharge and SSC data from the Aare (station no. 2085). Figures A2–A4 compare modeled intrusion depths to empirical estimates based on vertical temperature and light transmission data at the centre of LB (47°6′ N, 7°11′ E) collected shortly after major river intrusion events. Additionally, acoustic Doppler current profiler (ADCP) measurements of river plume intrusions in LB (47°5′ N, 7°12′ E; 2 km from the Aare inlet) were used for a temporal sensitivity analysis of the intrusion model (Fig. A5). Comparison of the modeled intrusion depth with light transmission depth (whose minimum value represents a proxy for actual river intrusion depth) suggests that $\beta = 0.0015$ offers an adequate representation of intru-

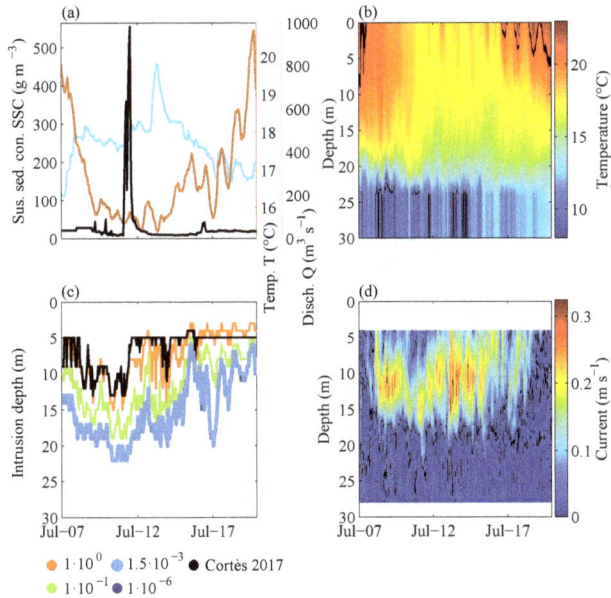

Figure A5. River intrusion entrainment sensitivity analysis for the LB/Aare system in July 2014. **(a)** SSC (black), temperature (T; orange) and river discharge (Q; blue) from Aare station no. 2085. **(b)** Lake temperature at M3 station. **(c)** River intrusion depth calculated as in Appendix A using river/lake density obtained from **(a)** and **(b)** with varying entrainment constant β (colored, Eq. 18); light blue denotes the value used in this study; intrusion depth (black) calculated with method described in Cortés et al. (2014). **(d)** Current speed obtained from ADCP at M3 station; velocities $> 0.15 \, \mathrm{m \, s^{-1}}$ are associated with the passing river plume.

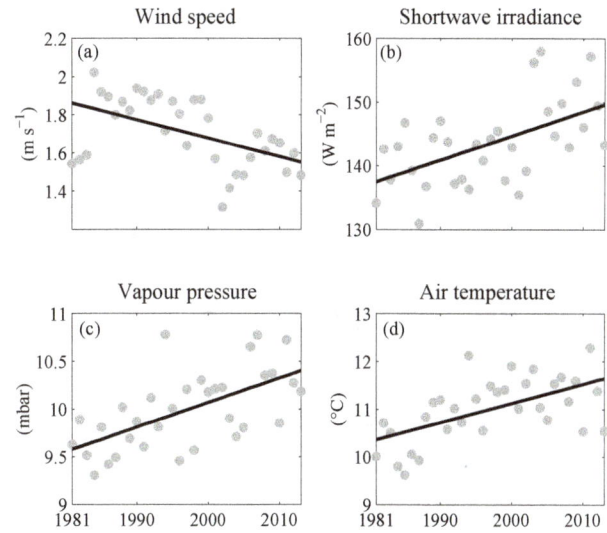

Figure A6. Annual atmospheric forcing (grey dots) of wind speed **(a)**, shortwave radiation **(b)**, vapor pressure **(c)** and Air temperature **(d)** at station no. 8100 (Fig. 2). The black line shows trends (Table 6).

sion depth. Larger β values generate intrusion depths shallower than the empirical reference points, whereas smaller β values exerted only a minor impact on deepening the intrusion depth. The intrusion model used here was compared (Figs. A2c–A5c) to the intrusion scheme proposed by Cortés et al. (2014), which produced an inferior result for LB.

Appendix B

Appendix C

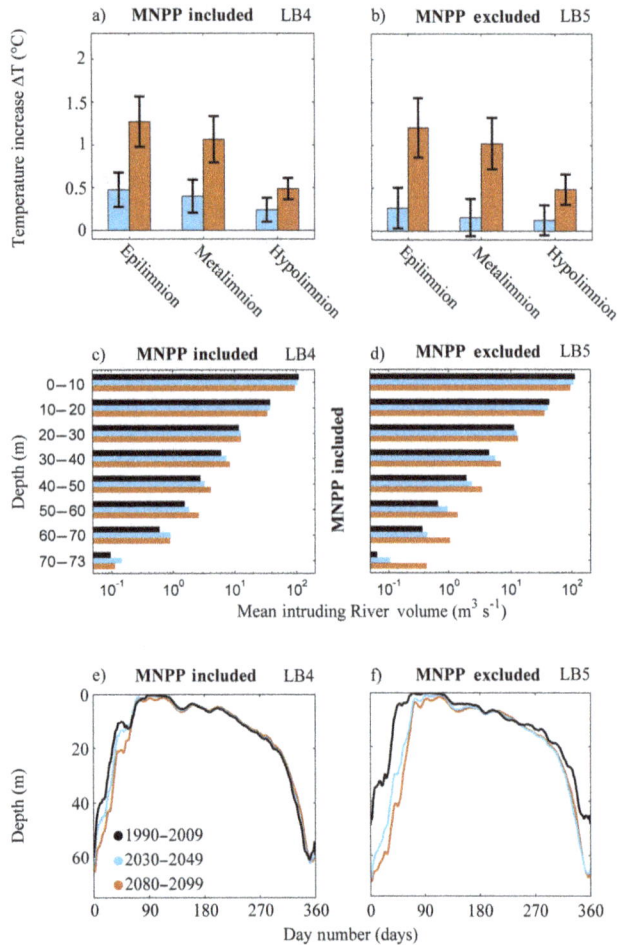

Figure B1. Modeled climate impact on LB excluding river-borne SSC. Temperature increase ΔT (**a** and **b**) displayed as means (bars) and standard deviations (black lines) in epilimnion (left bar group), metalimnion (middle bar group) and hypolimnion (right bar group); mean intruding river volume (**c** and **d**) and mean river intrusion depth (**e** and **f**). MNPP thermal input included (**a, c, e**) or excluded (**b, d, f**) in near-future (blue) and far-future (vermilion) time periods but retained in the reference period (black).

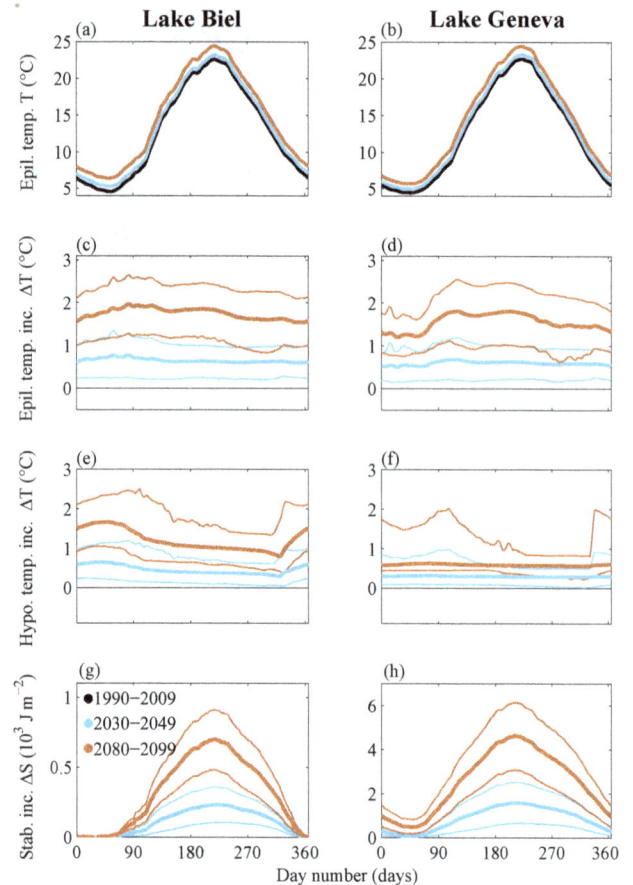

Figure C1. Modeled climate impact (river intrusion excluded) on LB (left column, scenario LB1) and LG (right column, scenario LG1) shown as daily mean (thick lines) and maximum/minimum model values (thin lines) for near-future (blue, 2030–2049) and far-future (orange, 2080–2099) time periods relative to the reference period (black, 1990–2009). Temperature T (**a** and **b**), temperature increase (ΔT) in the epilimnion (**c** and **d**) and hypolimnion (**e** and **f**) as well as increase in stability (ΔS; **g** and **h**).

Appendix D

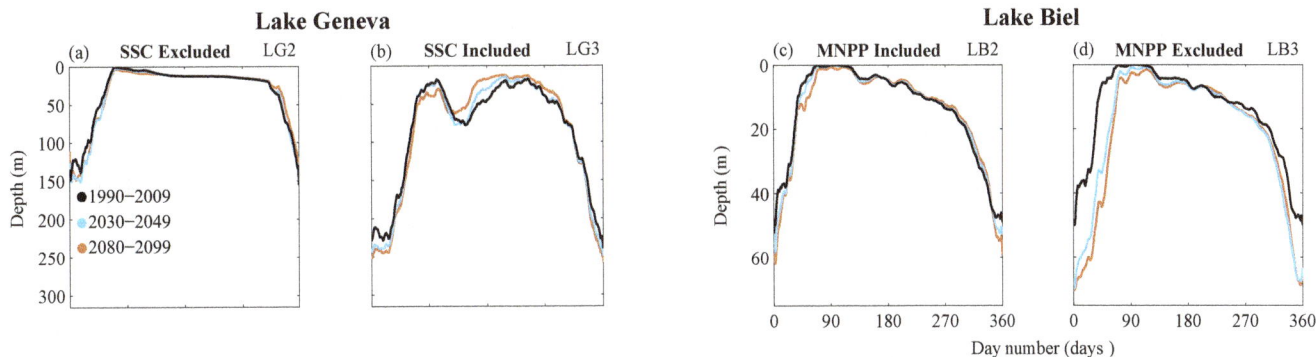

Figure D1. Modeled climate impact on mean river intrusion depth. Reference period (black), near-future (blue) and far-future (orange) time periods for LG (**a–b**) and LB (**c–d**) with (**b–d**) and without (**a**) river-borne SSC and MNPP thermal input included (**c**) or ex-cluded (**d**) from near-future and far-future time periods relative to the reference period.

Author contributions. LRV designed this study and preformed the modeling work; AW provided funding and supervision; DB contributed to the analysis of the result; GF adapted, calibrated and validated the SSC model; MZ provided river discharge predictions from the PREVAH model; all authors have contributed to the editing of this manuscript.

Competing interests. The authors declare that they have no conflict of interest.

Acknowledgements. This study is part of the *"Hydrodynamic Modelling of Lake Biel for Optimizing the Ipsach Drinking Water Intake"* project funded by Energy Service Biel (ESB). We are especially thankful to Andreas Hirt, Roland Kaeser and Markus Wyss for constructive collaboration. We thank the Office of Water Protection and Waste Management of the Canton of Bern (GBL/AWA) for providing their CTD profiles, the Swiss Federal Office of Meteorology and Climatology (MeteoSwiss) for providing meteorological data, the Hydrology Department of the Swiss Federal Office for the Environment (FOEN) for providing tributary data, the Climate Change and Hydrology in Switzerland (CCHydro) project for providing future river discharge predictions and the Swiss Federal Office of Topography (SwissTopo) for providing DHM25 model bathymetry data. We thank Bettina Schaefli at the University of Lausanne, Marco Toffolon and Elisa Calamita at the University of Trento, Nathalie Dubois at ETH Zurich, Robert Schwefel at EPFL Lausanne, Adrien Gaudard at Eawag and Stan Thorez at the University of Eindhoven for valuable insights. We furthermore thank Kei Ito (http://jfly.iam.u-tokyo.ac.jp/html/color_blind/) for valuable feedback on how to adapt our figures for color-blind readership.

Edited by: Anas Ghadouani

References

Addor, N., Rössler, O., Köplin, N., Huss, M., Weingartner, R., and Seibert, J.: Robust changes and sources of uncertainty in the projected hydrological regimes of Swiss catchments, Water Resour. Res., 50, 7541–7562, https://doi.org/10.1002/2014WR015549, 2014.

air2stream: air2stream model, available at: https://github.com/marcotoffolon/air2stream, last access: 2017.

Akiyama, J. and Heinz, S. G.: Plunging flow into a reservoir: theory, J. Hydraul. Eng., 10, 484–499, 1984.

Ashida, K. and Egashira, S.: Basic study on turbidity currents, P. Jpn. Soc. Civ. Eng., 1975, 37–50, 1975.

Austin, J. A. and Colman, S. M.: Lake Superior summer water temperatures are increasing more rapidly than regional air temperatures: A positive ice-albedo feedback, Geophys. Res. Lett., 34, L06604, https://doi.org/10.1029/2006GL029021, 2007.

Bennett, G. L., Molnar, P., McArdell, B. W., Schlunegger, F., and Burlando, P.: Patterns and controls of sediment production, transfer and yield in the Illgraben, Geomorphology, 188, 68–82, https://doi.org/10.1016/j.geomorph.2012.11.029, 2013.

Birsan, M.-V., Molnar, P., Burlando, P., and Pfaundler, M.: Streamflow trends in Switzerland, J. Hydrol., 314, 312–329, https://doi.org/10.1016/j.jhydrol.2005.06.008, 2005.

Bosshard, T., Kotlarski, S., Ewen, T., and Schär, C.: Spectral representation of the annual cycle in the climate change signal, Hydrol. Earth Syst. Sci., 15, 2777–2788, https://doi.org/10.5194/hess-15-2777-2011, 2011.

Bosshard, T., Carambia, M., Goergen, K., Kotlarski, S., Krahe, P., Zappa, M., and Schär, C.: Quantifying uncertainty sources in an ensemble of hydrological climate-impact projections, Water Resour. Res., 49, 1523–1536, https://doi.org/10.1029/2011WR011533, 2013.

Bouffard, D. and Perga, M.-E.: Are flood-driven turbidity currents hot spots for priming effect in lakes?, Biogeosciences, 13, 3573–3584, https://doi.org/10.5194/bg-13-3573-2016, 2016.

Caissie, D.: The thermal regime of rivers: a review, Freshw. Biol., 51, 1389–1406, https://doi.org/10.1111/j.1365-2427.2006.01597.x, 2006.

CH2011: Swiss climate change scenarios CH2011, C2SM, MeteoSwiss, ETH, NCCR Climate, and OcCC, Zurich, Switzerland, 88 pp., ISBN:978-3-033-03065-7, 2011.

Cortés, A., Fleenor, W. E., Wells, M. G., de Vicente, I., and Rueda, F. J.: Pathways of river water to the surface layers of stratified reservoirs, Limnol. Oceanogr., 59, 233–250, https://doi.org/10.4319/lo.2014.59.1.0233, 2014.

Doomen, A. M. C., Wijma, E., Zwolsman, J. J. G., and Middelkoop, H.: Predicting suspended sediment concentrations in the Meuse river using a supply-based rating curve, Hydrol. Process., 22, 1846–1856, https://doi.org/10.1002/hyp.6767, 2008.

Ellison, T. H. and Turner, J. S.: Turbulent entrainment in stratified flows, J. Fluid Mech., 6, 423–448, https://doi.org/10.1017/S0022112059000738, 1959.

Federal Office for the Environment (FOEN): Effects of climate change on water resources and waters, Synthesis report on "Climate Change and Hydrology in Switzerland" (CCHydro) project, Federal Office for the Environment, Bern, Umwelt-Wissen No. 1217, 74 pp., 2012.

Federal Office for the Environment (FOEN): Hydrological data and forecasts, Hydrology Department of the Swiss Federal Office for the Environment (FOEN), available at: www.hydrodaten.admin.ch, last access: December 2017a.

Federal Office for the Environment (FOEN): Climate Change and Hydrology in Switzerland (CCHydro) project model data, Hydrology Department of the Swiss Federal Office for the Environment (FOEN), available at: http://www.bafu.admin.ch/umwelt/index.html?lang=en, last access: December 2017b.

Fenocchi, A., Rogora, M., Sibilla, S., and Dresti, C.: Relevance of inflows on the thermodynamic structure and on the modeling of a deep subalpine lake (Lake Maggiore, Northern Italy/Southern Switzerland), Limnologica, 63, 42–56, https://doi.org/10.1016/j.limno.2017.01.006, 2017.

Fink, G., Schmid, M., Wahl, B., Wolf, T., and Wüest, A.: Heat flux modifications related to climate-induced warming of large European lakes, Water Resour. Res., 50, 2072–2085, https://doi.org/10.1002/2013WR014448, 2014a.

Fink, G., Schmid, M., and Wüest, A.: Large lakes as sources and sinks of anthropogenic heat: Capacities and limits, Water Resour. Res., 50, 7285–7301, https://doi.org/10.1002/2014WR015509, 2014b.

Fink, G., Wessels, M., and Wüest, A.: Flood frequency matters: Why climate change degrades deep-water quality of peri-alpine lakes, J. Hydrol., 540, 457–468, https://doi.org/10.1016/j.jhydrol.2016.06.023, 2016.

Foley, B., Jones, I. D., Maberly, S. C., and Rippey, B.: Long-term changes in oxygen depletion in a small temperate lake: Effects of climate change and eutrophication: Oxygen depletion in a small lake, Freshw. Biol., 57, 278–289, https://doi.org/10.1111/j.1365-2427.2011.02662.x, 2012.

Gaudard, A., Schwefel, R., Vinnå, L. R., Schmid, M., Wüest, A., and Bouffard, D.: Optimizing the parameterization of deep mixing and internal seiches in one-dimensional hydrodynamic models: a case study with Simstrat v1.3, Geosci. Model Dev., 10, 3411–3423, https://doi.org/10.5194/gmd-10-3411-2017, 2017.

GBL/AWA: Canton of Bern lake monitoring data, Office of Water Protection and Waste Management of the Canton of Bern (GBL/AWA), available at: http://www.bve.be.ch/bve/de/index/wasser/wasser/messdaten/Seen.html, last access: December 2017.

Gillet, C. and Quétin, P.: Effect of temperature changes on the reproductive cycle of roach in Lake Geneva from 1983 to 2001, J. Fish Biol., 69, 518–534, https://doi.org/10.1111/j.1095-8649.2006.01123.x, 2006.

Glur, L., Wirth, S. B., Büntgen, U., Gilli, A., Haug, G. H., Schär, C., Beer, J., and Anselmetti, F. S.: Frequent floods in the European Alps coincide with cooler periods of the past 2500 years, Sci. Rep.-UK, 3, 2770, https://doi.org/10.1038/srep02770, 2013.

Goudsmit, G.-H., Burchard, H., Peeters, F., and Wüest, A.: Application of $k - \varepsilon$ turbulence models to enclosed basins: The role of internal seiches, J. Geophys. Res., 107, 3230, https://doi.org/10.1029/2001JC000954, 2002.

Hari, R. E., Livingstone, D. M., Siber, R., Burkhardt-Holm, P., and Guettinger, H.: Consequences of climatic change for water temperature and brown trout populations in Alpine rivers and streams, Glob. Change Biol., 12, 10–26, https://doi.org/10.1111/j.1365-2486.2005.01051.x, 2006.

Holgerson, M. A. and Raymond, P. A.: Large contribution to inland water CO_2 and CH_4 emissions from very small ponds, Nat. Geosci., 9, 222–226, https://doi.org/10.1038/ngeo2654, 2016.

IPCC: Climate Change 2014: Synthesis Report, Contribution of Working Groups I, II and III to the Fifth Assessment Report of the Intergovernmental Panel on Climate Change, edited by: Core Writing Team, Pachauri, R. K., and Meyer, L. A., IPCC, Geneva, Switzerland, 151 pp., 2014.

Isaak, D. J., Wollrab, S., Horan, D., and Chandler, G.: Climate change effects on stream and river temperatures across the northwest U.S. from 1980–2009 and implications for salmonid fishes, Clim. Change, 113, 499–524, https://doi.org/10.1007/s10584-011-0326-z, 2012.

Kirillin, G.: Modeling the impact of global warming on water temperature and seasonal mixing regimes in small temperate lakes, Boreal Env. Res, 15, 279–293, 2010.

Kirillin, G., Shatwell, T., and Kasprzak, P.: Consequences of thermal pollution from a nuclear plant on lake temperature and mixing regime, J. Hydrol., 496, 47–56, https://doi.org/10.1016/j.jhydrol.2013.05.023, 2013.

Kobierska, F., Jonas, T., Magnusson, J., Zappa, M., Bavay, M., Bosshard, T., Paul, F., and Bernasconi, S. M.: Climate change effects on snow melt and discharge of a partly glacierized watershed in Central Switzerland (Soil-Trec Critical Zone Observatory), Appl. Geochem., 26, 60–62, https://doi.org/10.1016/j.apgeochem.2011.03.029, 2011.

MeteoSwiss: IDAWEB, Federal Office of Meteorology and Climatology (MeteoSwiss), available at: https://gate.meteoswiss.ch/idaweb/login.do, last access: December 2017.

Mulder, T., Syvitski, J. P. M., and Skene, K. I.: Modeling of erosion and deposition by turbidity currents generated at river mouths, J. Sediment. Res., 68, 124–137, https://doi.org/10.2110/jsr.68.124, 1998.

O'Reilly, C. M., Sharma, S., Gray, D. K., Hampton, S. E., Read, J. S., Rowley, R. J., Schneider, P., Lenters, J. D., McIntyre, P. B., Kraemer, B. M., Weyhenmeyer, G. A., Straile, D., Dong, B., Adrian, R., Allan, M. G., Anneville, O., Arvola, L., Austin, J., Bailey, J. L., Baron, J. S., Brookes, J. D., de Eyto, E., Dokulil, M. T., Hamilton, D. P., Havens, K., Hetherington, A. L., Higgins, S. N., Hook, S., Izmest'eva, L. R., Joehnk, K. D., Kangur, K., Kasprzak, P., Kumagai, M., Kuusisto, E., Leshkevich, G., Livingstone, D. M., MacIntyre, S., May, L., Melack, J. M., Mueller-Navarra, D. C., Naumenko, M., Noges, P., Noges, T., North, R. P., Plisnier, P.-D., Rigosi, A., Rimmer, A., Rogora, M., Rudstam, L. G., Rusak, J. A., Salmaso, N., Samal, N. R., Schindler, D. E., Schladow, S. G., Schmid, M., Schmidt, S. R., Silow, E., Soylu, M. E., Teubner, K., Verburg, P., Voutilainen, A., Watkinson, A., Williamson, C. E., and Zhang, G.: Rapid and highly variable warming of lake surface waters around the globe, Geophys Res Lett, 42, 10773–10781, https://doi.org/10.1002/2015GL066235, 2015.

Peeters, F., Livingstone, D. M., Goudsmit, G.-H., Kipfer, R., and Forster, R.: Modeling 50 years of historical temperature profiles in a large central European lake, Limnol. Oceanogr., 47, 186–197, https://doi.org/10.4319/lo.2002.47.1.0186, 2002.

Perroud, M. and Goyette, S.: Impact of warmer climate on Lake Geneva water-temperature profiles, Boreal Environ. Res., 15, 255–278, 2010.

Piccolroaz, S., Toffolon, M., and Majone, B.: The role of stratification on lakes' thermal response: The case of Lake Superior, Water Resour. Res., 51, 7878–7894, https://doi.org/10.1002/2014WR016555, 2015.

Råman Vinnå, L., Wüest, A., and Bouffard, D.: Physical effects of thermal pollution in lakes, Water Resour. Res., 53, 3968–3987, https://doi.org/10.1002/2016WR019686, 2017.

Raymond Pralong, M., Turowski, J. M., Rickenmann, D., and Zappa, M.: Climate change impacts on bedload transport in alpine drainage basins with hydropower exploitation, Earth Surf. Proc. Land., 40, 1587–1599, https://doi.org/10.1002/esp.3737, 2015.

Rimmer, A. and Hartmann, A.: Optimal hydrograph separation filter to evaluate transport routines of hydrological models, J. Hydrol., 514, 249–257, https://doi.org/10.1016/j.jhydrol.2014.04.033, 2014.

Sanchez-Lorenzo, A. and Wild, M.: Decadal variations in estimated surface solar radiation over Switzerland since the late 19th century, Atmos. Chem. Phys., 12, 8635–8644, https://doi.org/10.5194/acp-12-8635-2012, 2012.

Schmid, M. and Köster, O.: Excess warming of a Central European lake driven by solar brightening, Water Resour. Res., 52, 8103–8116, https://doi.org/10.1002/2016WR018651, 2016.

Schmidt, W.: Über die Temperatur- und Stabilitätsverhältnisse von Seen, Geogr. Ann., 10, 145–177, https://doi.org/10.2307/519789, 1928.

Schwefel, R., Gaudard, A., Wüest, A., and Bouffard, D.: Effects of climate change on deepwater oxygen and winter mixing in a deep lake (Lake Geneva): Comparing observational findings and modeling, Water Resour. Res., 52, 8811–8826, https://doi.org/10.1002/2016WR019194, 2016.

Simstrat: Simstrat source code, available at: https://github.com/Eawag-AppliedSystemAnalysis/Simstrat, last access: 2017.

Speich, M. J. R., Bernhard, L., Teuling, A. J., and Zappa, M.: Application of bivariate mapping for hydrological classification and analysis of temporal change and scale effects in Switzerland, J. Hydrol., 523, 804–821, https://doi.org/10.1016/j.jhydrol.2015.01.086, 2015.

Swiss Federal Office of Topography (SwissTopo): DHM25 model bathymetry data, available at: https://shop.swisstopo.admin.ch/en/products/height_models/dhm25, last access: December 2017.

Syvitski, J. P. M. and Lewis, A. G.: The seasonal distribution of suspended particles and their iron and manganese loading in a glacial runoff fiord, Geosci. Can., 19, 13–20, 1992.

Tananaev, N. I.: Hysteresis effect in the seasonal variations in the relationship between water discharge and suspended load in rivers of permafrost zone in Siberia and Far East, Water Resour., 39, 648–656, https://doi.org/10.1134/S0097807812060073, 2012.

Toffolon, M. and Piccolroaz, S.: A hybrid model for river water temperature as a function of air temperature and discharge, Environ. Res. Lett., 10, 114011, https://doi.org/10.1088/1748-9326/10/11/114011, 2015.

Toffolon, M., Piccolroaz, S., Majone, B., Soja, A.-M., Peeters, F., Schmid, M., and Wüest, A.: Prediction of surface temperature in lakes with different morphology using air temperature, Limnol. Oceanogr., 59, 2185–2202, https://doi.org/10.4319/lo.2014.59.6.2185, 2014.

Van Vliet, M. T. H., Franssen, W. H. P., Yearsley, J. R., Ludwig, F., Haddeland, I., Lettenmaier, D. P., and Kabat, P.: Global river discharge and water temperature under climate change, Glob. Environ. Change, 23, 450–464, https://doi.org/10.1016/j.gloenvcha.2012.11.002, 2013.

Verpoorter, C., Kutser, T., Seekell, D. A., and Tranvik, L. J.: A global inventory of lakes based on high-resolution satellite imagery, Geophys. Res. Lett., 41, 6396–6402, https://doi.org/10.1002/2014GL060641, 2014.

Viviroli, D., Zappa, M., Gurtz, J., and Weingartner, R.: An introduction to the hydrological modelling system PREVAH and its pre- and post-processing-tools, Environ. Model. Softw., 24, 1209–1222, https://doi.org/10.1016/j.envsoft.2009.04.001, 2009.

Wahl, B. and Peeters, F.: Effect of climatic changes on stratification and deep-water renewal in Lake Constance assessed by sensitivity studies with a 3D hydrodynamic model, Limnol. Oceanogr., 59, 1035–1052, https://doi.org/10.4319/lo.2014.59.3.1035, 2014.

Wild, M., Ohmura, A., and Makowski, K.: Impact of global dimming and brightening on global warming, Geophys. Res. Lett., 34, L04702, https://doi.org/10.1029/2006GL028031, 2007.

Williamson, C. E., Overholt, E. P., Pilla, R. M., Leach, T. H., Brentrup, J. A., Knoll, L. B., Mette, E. M., and Moeller, R. E.: Ecological consequences of long-term browning in lakes, Sci. Rep.-UK, 5, 18666, https://doi.org/10.1038/srep18666, 2015.

Zhang, G., Yao, T., Xie, H., Qin, J., Ye, Q., Dai, Y., and Guo, R.: Estimating surface temperature changes of lakes in the Tibetan Plateau using MODIS LST data, J. Geophys. Res.-Atmos., 119, 8552–8567, https://doi.org/10.1002/2014JD021615, 2014.

Zhong, Y., Notaro, M., Vavrus, S. J., and Foster, M. J.: Recent accelerated warming of the Laurentian Great Lakes: Physical drivers, Limnol. Oceanogr., 61, 1762–1786, https://doi.org/10.1002/lno.10331, 2016.

Permissions

The contributors of this book come from diverse backgrounds, making this book a truly international effort. This book will bring forth new frontiers with its revolutionizing research information and detailed analysis of the nascent developments around the world.

We would like to thank all the contributing authors for lending their expertise to make the book truly unique. They have played a crucial role in the development of this book. Without their invaluable contributions this book wouldn't have been possible. They have made vital efforts to compile up to date information on the varied aspects of this subject to make this book a valuable addition to the collection of many professionals and students.

This book was conceptualized with the vision of imparting up-to-date information and advanced data in this field. To ensure the same, a matchless editorial board was set up. Every individual on the board went through rigorous rounds of assessment to prove their worth. After which they invested a large part of their time researching and compiling the most relevant data for our readers.

The editorial board has been involved in producing this book since its inception. They have spent rigoroushours researching and exploring the diverse topics which have resulted in the successful publishing of this book. They have passed on their knowledge of decades through this book. To expedite this challenging task, the publisher supported the team at every step. A small team of assistant editors was also appointed to further simplify the editing procedure and attain best results for the readers.

Apart from the editorial board, the designing team has also invested a significant amount of their time in understanding the subject and creating the most relevant covers. They scrutinized every image to scout for the most suitable representation of the subject and create an appropriate cover for the book.

The publishing team has been an ardent support to the editorial, designing and production team. Their endless efforts to recruit the best for this project, has resulted in the accomplishment of this book. They are a veteran in the field of academics and their pool of knowledge is as vast as their experience in printing. Their expertise and guidance has proved useful at every step. Their uncompromising quality standards have made this book an exceptional effort. Their encouragement from time to time has been an inspiration for everyone.

The publisher and the editorial board hope that this book will prove to be a valuable piece of knowledge for researchers, students, practitioners and scholars across the globe.

List of Contributors

R. G. Knox
Massachusetts Institute of Technology, Cambridge, Massachusetts, USA
Lawrence Berkeley National Laboratory, Berkeley, California, USA

M. Longo
Harvard University, Cambridge, Massachusetts, USA
EMBRAPA Satellite Monitoring, Campinas, São Paulo, Brazil

K. Zhang
Harvard University, Cambridge, Massachusetts, USA
Cooperative Institute for Mesoscale Meteorological Studies, University of Oklahoma, Oklahoma, USA

N. M. Levine
Harvard University, Cambridge, Massachusetts, USA
University of Southern California, Los Angeles, California, USA

P. R. Moorcroft
Harvard University, Cambridge, Massachusetts, USA

A. L. S. Swann
University of Washington, Seattle, Washington, USA

R. L. Bras
Georgia Institute of Technology, Atlanta, Georgia, USA

Zhenchen Liu, Guihua Lu, Hai He and Zhiyong Wu
Institute of Water Problem, College of Hydrology and Water Resources, Hohai University, Nanjing, China

Jian He
Hydrology and Water Resources Investigation Bureau of Jiangsu Province, Nanjing, China

Hester Biemans
Alterra, Wageningen University and Research Centre, Wageningen, the Netherlands

Ashok Mishra
Agricultural and Food Engineering Department, IIT Kharagpur, Kharagpur, India

Bashir Ahmad
Pakistan Agricultural Research Council, Islamabad, Pakistan

Christian Siderius
Alterra, Wageningen University and Research Centre, Wageningen, the Netherlands

Environmental Economics and Natural Resources Group, Wageningen University, Wageningen, the Netherlands

Liang Chen, Paul A. Dirmeyer and Zhichang Guo
Center for Ocean-Land-Atmosphere Studies, George Mason University, Fairfax, Virginia, USA

Natalie M. Schultz
School of Forestry and Environmental Studies, Yale University, New Haven, Connecticut, USA

Anna Costa and Peter Molnar
Institute of Environmental Engineering, ETH Zurich, 8093 Zurich, Switzerland

Laura Stutenbecker
Institute of Applied Geosciences, Technische Universität Darmstadt, Darmstadt, Germany

Maarten Bakker and Stuart N. Lane
Institute of Earth Surface Dynamics, University of Lausanne, 1015 Lausanne, Switzerland

Tiago A. Silva and Jean-Luc Loizeau
Department F.-A. Forel for Environmental and Aquatic Sciences, University of Geneva, 1211 Geneva, Switzerland

Fritz Schlunegger
Institute of Geological Sciences, University of Bern, 3012 Bern, Switzerland

Stéphanie Girardclos
Department of Earth Sciences and Institute for Environmental Sciences, University of Geneva, 1205 Geneva, Switzerland

E. S. Garcia
Department of Atmospheric Sciences, University of Washington, Seattle, WA, USA

C. L. Tague
Bren School of Environmental Science and Management, University of California, Santa Barbara, CA, USA

A. B. M. Firoz, Alexandra Nauditt and Lars Ribbe
Institute for Technology and Resources Management in the Tropics and Subtropics (ITT), TH Köln, 50679 Cologne, Germany

Manfred Fink
Chair of Geographic Information Science, Department of Geography, Friedrich Schiller University Jena, 07743 Jena, Germany

Jérémy Chardon, Benoit Hingray and Anne-Catherine Favre
Univ. Grenoble Alpes, CNRS, IRD, Grenoble INP, IGE, 38000 Grenoble, France

Eddie W. Banks, Margaret A. Shanafield, Saskia Noorduijn, James McCallum and Okke Batelaan
National Centre for Groundwater Research and Training and the College of Science and Engineering, Flinders University, Adelaide, South Australia, Australia

Jörg Lewandowski
Department Ecohydrology, IGB, Leibniz-Institute of Freshwater Ecology and Inland Fisheries, Berlin, Germany
Geography Department, Humboldt University of Berlin, Berlin, Germany

Søren Thorndahl
Department of Civil Engineering, Aalborg University, Aalborg, 9220, Denmark

Aske Korup Andersen and Anders Badsberg Larsen
Niras A/S, Aalborg, 9000, Denmark

A. M. Badger
George Mason University, Fairfax, Virginia, USA

P. A. Dirmeyer
George Mason University, Fairfax, Virginia, USA
Center for Ocean–Land–Atmosphere Studies, Fairfax, Virginia, USA

Sara Sadri, Eric F. Wood and Ming Pan
Department of Civil and Environmental Engineering, Princeton University, 59 Olden St, Princeton, NJ 08540, USA

Love Råman Vinnå
Physics of Aquatic Systems Laboratory – Margaretha Kamprad Chair, École Polytechnique Fédérale de Lausanne, Institute of Environmental Engineering, Lausanne, Switzerland

Alfred Wüest and Damien Bouffard
Physics of Aquatic Systems Laboratory – Margaretha Kamprad Chair, École Polytechnique Fédérale de Lausanne, Institute of Environmental Engineering, Lausanne, Switzerland
Eawag, Swiss Federal Institute of Aquatic Science and Technology, Surface Waters – Research and Management, Kastanienbaum, Switzerland

Massimiliano Zappa
Swiss Federal Institute for Forest, Snow and Landscape Research, WSL, Birmensdorf, Switzerland

Gabriel Fink
Center for Environmental Systems Research, CESR, University of Kassel, Kassel, Germany

Index